T0337388

The Evolutionary Ecology of Plant Disease

Gregory S. Gilbert is Professor of Environmental Studies, University of California Santa Cruz, USA. He is a plant pathologist and forest ecologist, and his research interests include the dynamics of plant and fungal communities, as well as the application of evolutionary ecology to environmental problems. He directs the UCSC Forest Ecology Research Plot, and is a Research Associate at the Smithsonian Tropical Research Institute, Panama.

Ingrid M. Parker is Professor in the Department of Ecology and Evolutionary Biology, University of California Santa Cruz, USA. She is a plant evolutionary ecologist, and her research interests include plant disease ecology, the invasion of non-native species, the evolution of domestication, ecological restoration, and plant conservation. She is a Research Associate at the Smithsonian Tropical Research Institute, Panama.

The Evolutionary Ecology of Plant Disease

Gregory S. Gilbert
and
Ingrid M. Parker
University of California, Santa Cruz

Great Clarendon Street, Oxford, OX2 6DP,
United Kingdom

Oxford University Press is a department of the University of Oxford.
It furthers the University's objective of excellence in research, scholarship,
and education by publishing worldwide. Oxford is a registered trade mark of
Oxford University Press in the UK and in certain other countries

© Gregory S. Gilbert & Ingrid M. Parker 2023

The moral rights of the authors have been asserted

All rights reserved. No part of this publication may be reproduced, stored in
a retrieval system, or transmitted, in any form or by any means, without the
prior permission in writing of Oxford University Press, or as expressly permitted
by law, by licence or under terms agreed with the appropriate reprographics
rights organization. Enquiries concerning reproduction outside the scope of the
above should be sent to the Rights Department, Oxford University Press, at the
address above

You must not circulate this work in any other form
and you must impose this same condition on any acquirer

Published in the United States of America by Oxford University Press
198 Madison Avenue, New York, NY 10016, United States of America

British Library Cataloguing in Publication Data
Data available

Library of Congress Control Number: 2023930880

ISBN 978–0–19–879787–6
ISBN 978–0–19–879788–3 (pbk.)

DOI: 10.1093/oso/9780198797876.001.0001

Printed and bound by
CPI Group (UK) Ltd, Croydon, CR0 4YY

Links to third party websites are provided by Oxford in good faith and
for information only. Oxford disclaims any responsibility for the materials
contained in any third party website referenced in this work.

GSG
For my mentors Chun-Juan Wang, Jennifer Parke, and Jo Handelsman,
who shared with me their excitement of discovery and helped me find a less-traveled path through plant pathology.

IMP
For my mentors Douglas Schemske and Carla D'Antonio,
who shared with me their deep knowledge of evolution and ecology and their fascination for the natural world.

And for all our students and postdocs,
who have been with us on this journey and from whom we have learned so much.

Contents

Preface

Living through the COVID-19 pandemic forced us all to incorporate the principles of the evolutionary ecology of disease into our daily lives. The global disruption and personal devastation caused by a novel virus pushed us to learn about pathogen spillover, dispersal, survival, and reproductive rates, and about host responses, infectious periods, detection technologies, evolution of variants, and the science needed to figure it all out. Plant diseases may be a little less personal, but outbreaks disrupt food supply and safety, timber products, wildlands health, and international trade, and managing plant diseases comes at large economic and environmental costs. These truths make it easy to overlook the fact that plant pathogens are also normal, essential parts of all wild ecosystems, playing necessary roles in maintaining plant diversity, facilitating succession, and providing ecosystem services like nutrient cycling. Our world would be decidedly poorer without them.

Despite the importance of plant diseases, they remain invisible to most people. But once you start to notice, plant disease suddenly appears everywhere. This book is intended to get you, the reader, to notice what plant pathogens are doing all around you, wonder about them, and then think about plant disease through a framework of evolutionary ecology.

The two of us originally discovered the evolutionary ecology of plant disease through rather different scholarly traditions. Greg came into plant pathology through forestry and agriculture, where plant diseases were problems to be solved. Academic training in that tradition was rich in plant biology and microbiology. Ecology and evolution were included more as phenomena than as organizing frameworks, with genetics and epidemiology being necessary components in the pursuit of disease management. Ingrid came through plant population biology and evolutionary ecology, rich in theory and ideas and mathematical equations, with applications tied to biodiversity conservation. Academic training in that tradition looked at microbes only from a distance, and, in the rare instances when plant diseases were mentioned, they were usually treated as black-box drivers of ecological and evolutionary dynamics or as conservation threats. We independently developed deep interests in both traditions and pursued them through postdoctoral studies, expanding our languages and toolkits and frameworks in exciting ways. That eventually brought us together for more than two decades of fruitful collaboration.

We wrote this book in an effort to help people who, like us, come to the rich world of plant diseases with different motivations and from different scholarly traditions. It is based on decades of teaching about plant diseases to a heterogeneous group of students with backgrounds in agroecology, plant sciences, environmental policy, and ecology and evolutionary biology. We expect that most of the readers of this book will be upper-division college students, graduate students, and their instructors, and we are conscious that we are writing for multiple audiences from distinct traditions. We welcome those starting with applied perspectives in agriculture or forestry, with views shaped by ecological or evolutionary theory, and/or with a passion for conservation or environmental sustainability. We welcome all readers, whether concerned with intensive production agriculture, sustainable agroecosystems, urban forests, or wildlands, and we strive to highlight how an evolutionary ecology framework helps us take knowledge and experience from one kind of system and apply it to others. We draw broadly from an eclectic scientific literature and do our best to bring it together into a coherent framework.

The book is structured in two parts, with a lot of cross-talk among chapters. The first part opens with an introduction to fundamental concepts in plant-microbe symbiosis and evolutionary ecology (Chapter 1), and then continues with a series of six chapters that serve as entries into the basic biology of plants and pathogens (Chapters 2–7). The goal is to help the reader imagine what life is like from the perspective of a plant, a fungus, an oomycete, a virus, or a nematode living in antagonistic symbiosis, going beyond a bunch of life cycles to capture the "why" behind life histories. For some readers these organismal chapters will be familiar turf and can be skipped, while for others it may be their first time diving deeply into the microscopic realm of life. We do our best to make accessible the rather bewildering terminology of the microbial world in order to help open the doors to specialized scientific literature and to provide a foundation for chapters to come. Next comes a quick tour of the very many ways that diseases affect plants (Chapter 8), providing touchstones for the examples that litter the rest part of the book. The first part ends with an overview of the techniques and approaches that plant disease ecologists use in their work of discovery and disease monitoring (Chapter 9).

In the second part, we dive into the science and applications of the evolutionary ecology of plant–microbe interactions. We start with two chapters (10 and 11) that explore fundamentals of the temporal and spatial aspects of disease epidemiology, thinking about population dynamics of both pathogens and plants. Next we dive into the physiology and genetics of plant–pathogen interactions (Chapter 12), and investigate how those interactions are both drivers and consequences of evolution (Chapter 13). We then step out to look at community ecology from two viewpoints: first, how plant communities both respond to and influence the activities of pathogens (Chapter 14), and then how microbes interact with plants and with each other in the fascinating world of the plant microbiome (Chapter 15). The final chapters of the book aim to bring all previous components together to consider plant diseases in the context of global change (Chapter 16), and how applying an evolutionary ecology framework provides opportunities for creative and effective management of plant diseases (Chapter 17).

We have so many people to thank. Ian Sherman of Oxford University Press first encouraged us to undertake this project and has been an unwavering presence in the life of the book. We thank the anonymous reviewers of the original book proposal who shared both enthusiasm for the project and helpful critiques. Our project editors Charles Bath and Katie Lakina provided advice on many details along the way. Over 20 years of UC Santa Cruz students in ENVS 163 Plant Disease Ecology have helped shape the structure of the book and, since 2017, have provided specific suggestions and feedback on many aspects of the material as it was rolled out in draft form. Our understanding of plant–pathogen interactions has developed over many years, benefiting from collaborations and conversations with remarkable, thoughtful people across six continents—you know who you are! We apologize that the names are too many to print here, and we humbly ask that you accept our heartfelt gratitude.

We are indebted to the institutions that have supported our research, teaching, and intellectual growth. The departments of Environmental Studies and Ecology & Evolutionary Biology at the University of California Santa Cruz have been wonderful places to explore new ways of thinking and teaching. The living laboratories of the campus, including the UCSC Campus Natural Reserve and the Center for Agroecology Farm, have provided inspiration and resources for many of our ideas and projects. For decades, our second academic home has been the Smithsonian Tropical Research Institute in the Republic of Panamá, where we have explored plant diseases and evolutionary ecology in hyperdiverse ecosystems. We are also thankful to the CSIC *Estación Biológica de Doñana* in Sevilla, Spain, where we spent a wonderful sabbatical year of intellectually stimulating research and where this book project first began.

We are grateful for the generous funding that has supported our research and shaped our understanding of the world of plant disease. This includes grants from the National Science Foundation (DEB-1655896, DEB-1136626, DEB-0842059, DEB-0808337, DEB-0515520, DEB-9806517, DEB-9808501), the US Department of Agriculture (NIFA 2020-6713-31856, NIFA 2000-00891, and several cooperative agreements), and the California Department of Food and Agriculture (SCB16051), along with support from the Andrew W. Mellon Foundation, UCSC Committee on Research, UC-MEXUS, The Nature Conservancy, the Pacific Rim Research Program, the Jane Carver Fund, and the Jean H. Langenheim Endowed Chair, the Pepper–Giberson Endowed Chair, and the Robert Headley Presidential Chair for Integral Ecology and Environmental Justice.

In writing a book of this scope, in a field that is both deep-rooted and changing rapidly, there is great risk of oversimplifying or misrepresenting the details of any one of the many important topics we cover. Fortunately, a number of friends and colleagues were willing to lend a hand with invaluable editorial comments on individual chapters, including Emme Bruns, Akif Eskalen, Thomas Forge, Alejandra Huerta, Roger Innes, Timothy James, Joji Muramoto, Daniel Nickrent, Plex Sula, and Lincoln Taiz. These good souls share the credit for anything we got right, although we assure you that the errors are all our own.

We thank the many people who generously offered stunning photographs to bring plant diseases to life. We are grateful to the Kenneth G. Norris Center for Natural History at UCSC for supporting the artistic work of Josh Zupan to create the life cycle drawings in the book. We also recognize the hundreds of creative scientists around the world whose research has informed these pages, the ones highlighted explicitly as well as the massive quantity of important science that we would have liked to highlight, if only we could have done so within a volume of affordable size.

Our current and recent lab members provided critical feedback on earlier versions of nearly every chapter; we have benefited enormously from their intellectual engagement, courage in sharing their perspectives, and patience with their distracted advisors. We name them in alphabetical order, as they share equally in our gratitude: Erin Aiello, Asa Conover, Jon Detka, Taryn Farber, Sara Grove, Patrick Lee, Nicky Lustenhouwer, Shannon Lynch, Miranda Melen, Erica Mullins, Liz Rennie, Zackery Shearin, and Karen Tanner.

Finally, we are grateful to our friends and families for their support through the years and through all the ups and downs. Most importantly, we acknowledge our son Eli, who not only was a trusted field assistant for most of his childhood but also endured countless dinner-time discussions on the fine points of the plant immune system or the latest plant–soil feedback experiments. We thank him for keeping our lives filled with music and adventure, and we appreciate the grace with which he accepted this book into the family.

Greg Gilbert and Ingrid Parker,
Santa Cruz, California

Photo credit: Brent Dundor

Endorsements

"This outstanding book, by two long-standing leaders in the field, is both rich in content and a joy to read. It begins, 'This is a book about death, decay, and destruction. But in a good way.' Then it tells the story of plants and the pathogens that infect them by weaving together two perspectives: how to be a scientist, with chapters like 'How to do disease ecology' – and how to be a pathogen, with chapters like 'How to be a fungus' and 'How to be a virus.' In the end, most of the plants win, but so do most of the pathogens – and how those two outcomes are simultaneously possible is the true story of the book. Essential reading for students of all levels: from the scientifically curious layperson to the seasoned specialist."

— Charles Mitchell, PhD, University of North Carolina, Chapel Hill

"The clarity and ease with which the complexity of interactions among pathogen populations, plants, and biotic and abiotic factors are presented in this book could have only been achieved by the two phenomenal researchers, educators, and mentors who wrote it. Their passion for the advancement of science and society is reflected in each word, figure, and image that make up this book. This is a true learning tool, that will be used to introduce and attract young talent, scientists, and the public to learn more about the complexity of interactions in science and how multiple disciplines like ecology, evolution, plants, and plant disease are studied to provide solutions to complex environmental challenges. I can't wait to use it in introductory plant pathology courses and to share it with the public when they say, "You're a what? Plant pathologist? Now we just need it in Spanish too."

— Alejandra Huerta, PhD., North Carolina State University

"This treatise is impressive in its comprehensive and understandable introduction to the basic biology of diverse plant pathogens, and the complex interactions that they have with their plant hosts. It very successfully draws from a broad, yet eclectic, literature to illuminate the role of plant pathogens in many different contexts, ranging from production agriculture to natural ecosystems. The amazing wealth and diversity of information in this easy-to-read book will make a great introduction to plant pathology for those new to the topic and was a great refresher for someone such as myself who was rather familiar with the topic. It will be particularly valuable for plant pathologists wanting to learn more of the linkage of their work to mainstream evolutionary ecology and would be particularly valuable for those already well versed in ecological theory and practice to learn of the many exciting plant symbiont systems that might be fruitful to study. There is a lot to learn and think about in this book!"

— Steven E. Lindow, PhD., University of California, Berkeley

PART 1

Plant pathogens and disease

CHAPTER 1

Thinking like a plant disease ecologist

Gregory S. Gilbert and Ingrid M. Parker

1.1 Introduction

This is a book about death, decay, and destruction. But in a good way. Plant disease can be surprisingly fascinating, even beautiful. Which is why plant disease ecologists must often contain their excitement when a worried farmer, forester, or friend shows them a sick and dying plant. Why is this book "The Evolutionary Ecology of Plant Disease," rather than "Plant Pathology" or "Phytopathology"? We explore plant diseases through the theoretical framework of evolutionary ecology because we believe that both plant pathology and evolutionary ecology will be enriched by a deeper mutual engagement. Plant pathology offers a thorough understanding of mechanism, a wealth of case studies, and an experimental tradition that can help advance evolutionary ecology, while evolutionary ecology provides a robust theoretical framework that integrates genetics with conceptual and quantitative ecology. We offer this book in the tradition of previous authors who have also recognized the great benefits of bridging these two fields, such as Jeremy Burdon (1987) and Michael Milgroom (2015). We will not only look beyond agricultural crops to the role of plant disease in ecosystems all around us, but we will also use evolutionary ecology to unite different aspects of plant pathology in the service of scientific generalization and prediction.

Plant pathology is dedicated to solving problems of food security, forest health, and environmental protection. Because of its mission-driven nature, most of plant pathology has advanced through empirical, problem-driven science. The last century or so of plant pathology has taught us much about the "what" and "how" of plant–pathogen interactions. What organism causes sudden oak death? How does Tobacco mosaic virus infect tomato plants? What physiological and genetic mechanisms are involved in susceptibility to gray mold? How does the downy mildew pathogen spread among cucumber plants? What can we do to control it? How does drought moderate the impact of white pine blister rust on high elevation forests? How do populations of *Thanatephorus cucumeris* overcome resistance to root rot in host populations?

Evolutionary ecology, on the other hand, developed not as a science for solving problems but as a science driven by theoretical ideas, both conceptual and quantitative. This field takes as its primary motive an attempt to discover generalizable truths. For example, what determines the number of species in a community? Under what conditions will any two species coexist? Are more diverse communities more stable, or less stable? Under what conditions will one species have a large impact on the properties of a whole ecosystem? Why are some pathogens or herbivores highly specialized on one host species, while others use many different hosts? Evolutionary ecology includes many empirical studies on why species look and behave the way they do, how species interact, and the influence of the environment on species' traits, distribution, and abundance. However, it is rare for a study in evolutionary ecology to attain the high level of replication, of temporal and spatial scale, or degree of mechanistic understanding that has been possible through the intensive study of plant diseases of systems of great agricultural or economic importance.

The Evolutionary Ecology of Plant Disease. Gregory S. Gilbert and Ingrid M. Parker, Oxford University Press. © Gregory S. Gilbert & Ingrid M. Parker (2023).
DOI: 10.1093/oso/9780198797876.003.0001

By bringing these two traditions together, we have a lot to gain. In this introductory chapter, we provide a brief overview of what defines an evolutionary ecology perspective. We then review the foundational concepts of plant pathology, introducing some of the specialized terminology that will be used throughout the book, highlighted in **bold**. We also begin to explore how the features of a microbe's **life history**—the key aspects of its life cycle—affect its interactions with plants and are affected by its environment. We end this chapter with a road map for the rest of the book.

1.2 What is evolutionary ecology?

Ecology is the study of the distribution and abundance of organisms, along with the interactions among species and between organisms and their environment. Ecologists measure patterns of distribution of organisms, changes in numbers of individuals or biomass over time, exchanges of energy and nutrients through trophic interactions, outcomes of competitive or mutualistic interactions, and patterns of movement of individuals through space and time. Ecological interactions in part drive the distribution and abundance of organisms, and, therefore, disease. Ecologists often make the simplifying assumption that species have important traits that stay the same from the beginning to the end of a study and are consistent over space as well as time.

Evolutionary biologists focus mainly on understanding the processes that drive variation among organisms, to understand which traits are key to species' behavior and success, and how those traits came to be. Evolution takes place quickly, from generation to generation, as well as slowly, in lineages over millions of years. Evolutionary biology includes genetics and the study of heritable change we observe over the course of generations (**microevolution**), as well as patterns of change over the history of life (**macroevolution**).

Evolutionary ecologists combine contemporary and historical approaches by recognizing that how species interact with each other and with their environment is mediated by the traits of organisms, such as body plan, modes of dispersal, physiology and metabolism, chemical composition, and life-history strategies. Those phenotypic traits are often heritable and are the products of evolution. As the great biologist Theodosius Dobzhansky noted, "nothing in biology makes sense except in the light of evolution" (Dobzhansky 1950). *Evolutionary ecology recognizes that the outcome of ecological interactions is strongly shaped by evolutionary processes, and that evolutionary change takes place in an ecological arena.* For plant diseases, this means understanding that evolution in a pathogen to overcome the resistance of its host plant will immediately change the ecological impact of pathogen on plant, as well as the fitness of the pathogen itself, sometimes even leading to ecosystem-level changes. A simple evolutionary change can fundamentally change both plant and pathogen populations by changing the outcome of their interactions.

1.3 What is plant disease?

A plant is diseased when something (usually a parasite) disrupts the normal functions or form of the host plant, leading to impairment or death of the plant or parts of the plant. It is helpful to consider this definition in light of what disease is *not*. Injury is different from disease, because with disease there is a persistent association between the cause of disease and the host. Thus, a machete cut is an injury, not a disease. The word disease is sometimes used to describe impairment caused by environmental factors such as nutrient deficiencies, toxicity, or water stress. In this book we will focus exclusively on **infectious diseases**, meaning those caused by a biological agent. Is parasitism the same as disease? Parasites live in or on their hosts and gain nutrition from them. Disease is a special case of parasitism, where the effects on the host are more extensive or damaging than expected from the simple removal of nutrients from the host. When a **parasite** causes a disease on its host, it is called a **pathogen**.

1.4 How do we study plant disease?

Plant pathology has a long history and well-developed traditions of observational and experimental research. As is true for most ecological fields, the foundation of plant pathology is based in critical, systematic, detailed observations

Figure 1.1 Disease symptoms. (a) Necrosis and chlorosis associated with anthracnose of sycamore (*Platanus* hybrid) caused by the fungus *Apiognomonia veneta*. (b) Canker on *Ocotea whitei* (Lauraceae).
Photo by Gregory S. Gilbert.

of nature. We notice a disease through its **symptoms**—the observed effects on the plant that disrupt its normal function. Symptoms include discoloration from loss of chlorophyll (**chlorosis**) and the death of plant tissue (**necrosis**) (Figure 1.1a). Symptoms may include deformations like galls, cankers, or witches' broom, reductions in growth or yield, or changes in physiological functions (Figure 1.1b). Symptoms are what is expressed by the host plant—or what happens to the plant—as a result of the infection. Associated with symptoms are often **signs** of the pathogen, visible or otherwise detectable parts or products of the pathogen. Signs include the pathogen itself, reproductive structures, exudates produced by the pathogen, and, with recent technological advances, gene products of the pathogen that can be detected through transcriptomics.

The germ theory of disease is something we take for granted, but only 150 years ago not all scientists accepted that diseases were caused by pathogens and not by clouds of unhealthy air. Scientists also debated whether microbes and other organisms including flies could arise out of non-living materials in a process of "spontaneous generation." Several scientific advances of the 18th and 19th centuries led directly to the experimental approach that we use today to show that a microbe causes disease. In 1729, the Italian botanist/mycologist Pier Antonio Micheli showed that he could remove the spores from a species of fungus, place them on a slice of melon, and the same fungus would be produced. These experiments were a big step toward the final victory of the theory of biogenesis (*omne vivum ex vivo*, Latin for "all life [is] from life") over that of spontaneous generation. Louis Pasteur, the French "father of microbiology," put the nail in the coffin of the spontaneous generation idea with experiments comparing sterilized nutrient broth in open flasks to broth from which microbial spores were experimentally excluded using flasks with narrow, S-shaped necks. When microbes were excluded, the nutrient broth remained sterile. Importantly, Pasteur and others realized that biogenesis had implications for the cause of diseases in humans and other animals, which led to the development of germ theory. A German physician named Robert Koch (pronounced like Coke but with the "k" growled from the back of the throat) showed that the bacterium *Bacillus anthracis* was the cause of anthrax disease. This was the first demonstration that a specific microorganism caused a specific disease, and Koch went on to do experiments on guinea-pigs that showed conclusively that tuberculosis was not an inherited disease but rather was caused by infection with the bacterium *Mycobacterium tuberculosis*, a discovery for which he won the Nobel Prize in Physiology and Medicine. In 1882 he outlined the necessary elements to prove the pathogenicity of an organism, in what have become known as Koch's postulates (Box 1.1 and Figure 1.2).

Box 1.1 Koch's proof of pathogenicity

To show that a particular pathogen is the cause of a plant disease, it is not sufficient to show that it is present, because the microorganism may be only casually associated with the plant. Koch's four-step proof of pathogenicity provides a formal structure for establishing the cause of a disease (Figure 1.2).

1. There must be a consistent association between the microbe and the disease symptoms. This can be established through direct observation, if the microbe produces visible identifiable structures (signs) on the host, by culturing of microorganisms on culture media, or by using genetic probes.
2. The microbe must be isolated into pure culture, away from its host. This is usually accomplished by growing the microorganism on microbiological media, but in the case of obligate pathogens, purification of the putative pathogen (e.g., viruses) is considered sufficient.
3. Inoculation of the pure culture into a healthy host must produce the original disease symptoms. Care must be taken to provide the appropriate environmental conditions for disease development, or false-negative reactions are likely. It is critical to include control plants (mock inoculated in the same way as the target plants, but without the putative pathogen) to compare host responses to wounding with disease development.
4. The microbe must be re-isolated from the infected host, showing that it established infection and growth in the host, and that the symptoms were not due to a contaminating organism, or simply an effect of stress from wounding of the host. Control plants must not develop symptoms.

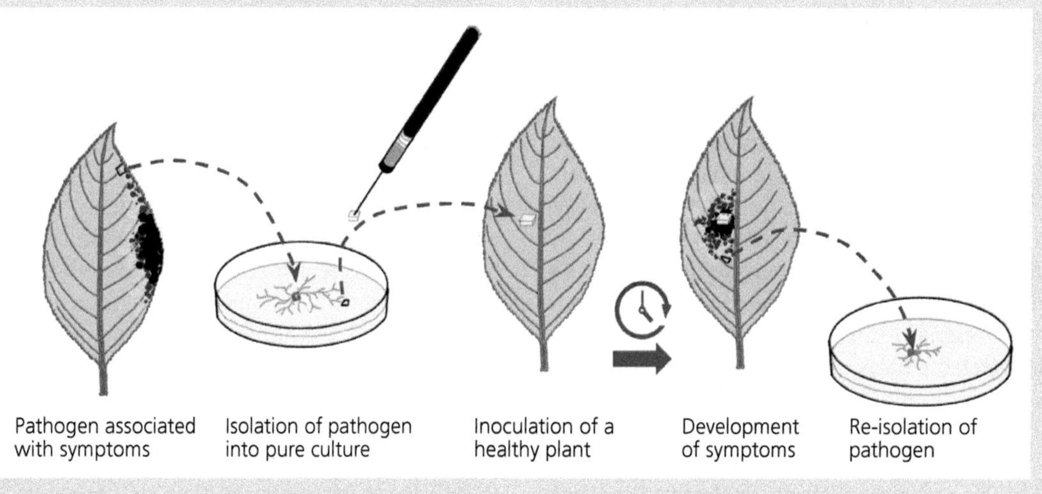

Figure 1.2 Koch's proof of pathogenicity is essential for showing what causes a disease. Constant association between a pathogen and symptoms, isolation of the pathogen into pure culture, reproduction of symptoms after inoculation of a healthy host, and re-isolation of the pathogen are the key steps.
Drawing by Ingrid M. Parker.

Koch's postulates for proof of pathogenicity are the gold standard for how to prove what organism causes a plant disease, although it needs some minor modifications for pathogens that cannot be grown in pure culture. The steps in proof of pathogenicity show that microbes found in association with a disease are the cause, rather than a consequence, of the disease. Pretty much all living plant and animal tissue is colonized by microorganisms, and diseased, weakened tissue is often susceptible to later colonization by opportunistic microbes. Koch's postulates allow us to be certain that a particular microorganism causes particular disease symptoms on its host and is not just a casual associate or a secondary invader of damaged tissues.

Plant disease ecologists have many scientific advantages over scientists who study diseases of humans or other animals. First, there is much greater social and ethical acceptance of doing experiments that cause harm to plants than to animals, so we can take plants, infect them with pathogens, and learn from them as we watch them get sick and die. Second, plants are much more amenable to large, replicated, controlled experiments. It is easy to grow hundreds or thousands of genetically similar plants for well replicated studies. Plants can be grown in greenhouses, growth chambers, pots, gardens, experimental plots, forest understories, and industrial-sized agricultural fields, or in tissue culture in a Petri dish. Plants stay where you put them, so you can experimentally control their density and spatial arrangements. Many plants grow quickly, so experiments can be repeated, and multiple generations of plants can be handled within the life cycle of a research grant. Plants can be bred through traditional crossing, or by using recombinant DNA technology to alter specific disease-related genes, allowing detailed study of the role of specific traits in disease interactions. Controlled inoculation studies can be done in elegant experiments that help elucidate the specific genetic and physiological mechanisms that drive interactions and ecological effects, or that measure evolutionary change across several generations. Simply put, plants are great for testing the really big ideas about the evolutionary ecology of diseases in both agriculture and wild ecosystems.

1.5 Epidemiology

Epidemiology, the study of the incidence and impact of disease across space and time, combines observational, experimental, and modeling approaches. Most studies on the distribution and spread of disease are observational; disease detection includes visual scouting for disease symptoms, molecular detection of pathogens, and remote sensing of physiological disturbances to crop plants and forests. Such observed patterns provide the raw data for sophisticated statistical analysis of patterns of change over time and space, shaping mathematical models that help us better predict how disease spreads.

Most of our understanding of plant epidemiology comes from studies in agriculture and forestry, with a goal of better preventing or controlling diseases. A few dramatic epidemics in wild systems (e.g., Chestnut blight and Jarrah dieback, Section 16.4) drew plant pathologists into research in wild ecosystems earlier, but plant disease ecology in wild systems really began to take form as a field in the 1980s and 1990s. This work developed in two areas—the ecological and evolutionary interactions of plant–pathogen systems at the population level, and the impact of disease on plant community structure, maintaining plant diversity, and driving succession of plant assemblages. Research in wild systems is closely linked to agroecological approaches to disease control in managed systems. Importantly, plant disease ecology has expanded our view of pathogens as parts of ecosystems; disease is bad for individual plants and for foresters, farmers, and consumers, but disease also plays a creative, natural, and crucial role in wild ecosystems and in the generation of ecosystem services such as decomposition and wildlife habitat.

1.6 Symbiosis: *It depends*

Plant–pathogen interactions are one example of a **symbiosis**: two different kinds of organisms living together in an intimate relationship (Figure 1.3). The outcome of any symbiotic relationship varies across a gradient in the degree of cost and benefit to each of the partners, ranging from parasitic to commensal to mutualistic. When a plant enters a symbiosis where it suffers a net cost and the partner derives a net benefit, the outcome is **parasitism**. Most plant disease is the result of a costly parasitic symbiosis between a plant and a much smaller partner that benefits from the symbiosis. Although the term "symbiont" should really apply to both the larger and smaller partners in a symbiosis, for simplicity we will adopt the common convention of referring to the plant partner in the symbiosis as the **host** and the smaller partner in the symbiosis as the **symbiont**. When the outcome of a parasitic symbiosis is plant disease, then the symbiont is a **pathogen**.

Symbiosis, as an idea and term, dates to the mid-19th century work on plant disease by two mycologists. In 1846 in England, Miles Berkeley, a prolific pioneer of fungal systematics (especially of lichens), drew on his skills as microscopist and mycologist to propose that the ravages of late blight of potato in Ireland was caused by the oomycete *Phytophthora*

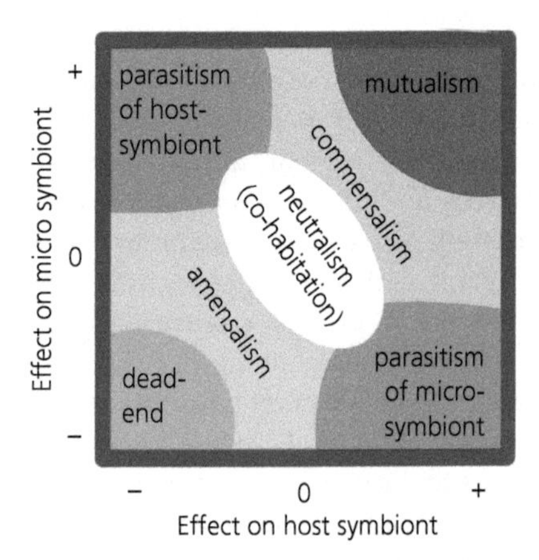

Figure 1.3 Schematic of gradient of outcomes of symbiotic species interactions. The outcome of a symbiosis falls along a separate gradient of negative to positive outcomes for each partner. The most common symbioses fall along the spectrum from parasitism (+/– or –/+) through commensalism (+/0 or 0/+) to mutualism (+/+). Amensal interactions (–/0 or 0/–), as well as negative-negative (–/–) symbioses, tend to be dead-end relationships. Coincidental, passive encounters without measurable consequences for either plant or microbe are probably common but not very interesting.
Drawing by Ingrid M. Parker.

infestans, rather than the prevailing view that the plant produced the microscopic balls and tubes as a result of physiological impairment (Box 4.1). His series of papers on Vegetable Pathology in *The Gardeners' Chronicle* through the 1850s and 60s outlined the life histories of a wide range of plant pathogens. Working in Germany and France, lichenologist Anton deBary coined the term symbiosis as a description of the biological relationship between *P. infestans* and the potato plant that leads to disease. This insight was an extension of his studies of lichens, which are intimate, mutualistic living arrangements between two very different organisms, algae and fungi, to create a third entity, a chimera of a photosymbiont and a heterotroph. The recognition that plant disease is also the product of a biological symbiosis between living plants and microbes, but with a different outcome than the stable lichen, opened the door for rapid advancement in the understanding of plant disease.

Many—maybe most—plant/microbe symbioses are **commensal**, meaning the microbe benefits but the plant does not really notice one way or the other. All plants and plant parts are covered and filled with microbial symbionts (you are too!). This is often called the plant **microbiome**, meaning the assemblage of microorganisms inhabiting the body of the host plant. The American Phytopathological Society has adopted *Phytobiome* as the name for their journal dedicated to research on the assemblage of microorganisms that live symbiotically with plants. Symbionts that live inside plant tissue are **endophytes**; symbionts that live on the surface of plant tissues are **epiphytes**. Endophytes can be either intercellular (living between the cells of the host plant) or intracellular (inside the plant cells) (Figure 1.4). Many endophytic and epiphytic symbiotic microbes depend entirely on the host plant to provide nutrition and habitat, but do not measurably affect the host plant. Such a positive/neutral symbiosis is commensal (Figure 1.3). Importantly, the physiological outcome for the host is often **context dependent**, and the same pair of microbe and plant may be mutualistic, commensal, or parasitic depending on the ontogenetic (developmental) stage of the host or the symbiont or on environmental conditions. For instance, some fungi form a mutualistic mycorrhizal association with plant roots, where the fungus receives carbon from the host plant and provides greater access to soil nutrients to the host. But under conditions of high soil nutrients the same association can be parasitic, with the fungus readily consuming host sugars but without providing a nutritional benefit in return, leading to loss of fitness in the plant.

What happens between a plant and a microbe is complicated, and "it depends." We have to keep in mind that to understand the ecological impacts or evolutionary dynamics of a symbiosis requires a disease ecologist to do several things: consider the life cycles of the symbiont and host, recognize that life histories and interactions fall along dynamic gradients and not into boxes, and appreciate that context matters.

1.7 The disease triangle

A fundamental framework in plant disease ecology is a conceptual model known as the **disease triangle** (Figure 1.5); disease development requires the combination of a **virulent pathogen**, a **susceptible host**, and an **environment favorable** to disease

Figure 1.4 Microbial endophytes in plants can be intercellular (in the apoplast between the plant cells) or intracellular (inside the plant cell). Some fungi penetrate the cell wall but not the plant cell membrane, instead invaginating it, so that the plant cell membrane closely surrounds the fungal hypha, facilitating nutrient exchange.
Drawing by Ingrid M. Parker.

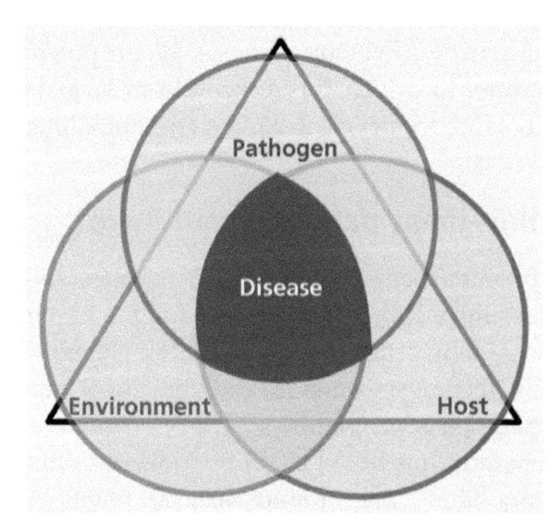

Figure 1.5 The disease triangle. Plant disease only develops when a virulent pathogen and a susceptible host come together under environmental conditions that are favorable for disease to develop.
Drawing by Ingrid M. Parker.

development. As we know from germ theory, infectious disease is caused by pathogens. Of course, no pathogenic organism can cause disease on all plant species; a **virulent** pathogen and a **susceptible** host must come together for disease to develop; virulence and susceptibility are determined by the genetic makeup of those organisms. Equally important is the environment. The host and the pathogen each have specific ranges of environmental conditions that allow them to thrive, and the symbiotic interaction between a pathogen and a host plant may only develop over some subset of the overlap between those environmental conditions. Because plants and their microbial pathogens are **poikilotherms** (unlike us; we are **homeothermic**), they are heavily influenced by external conditions. The role of the environment in plant disease development is even stronger than it is for diseases of people or other warm-blooded animals.

The disease triangle provides a framework for thinking about all the reasons that disease does *not* develop. Most pathogen species do not encounter all the possible host species on which they could cause disease simply because their geographic ranges do not overlap—ranges determined by evolution, history, and environmental conditions. When they do co-occur, and the spores of a virulent pathogenic fungus land on a susceptible host plant, environmental conditions that are too dry, too wet, too hot, or too cold for the spores to germinate would prevent **infection**, even though there is potential for disease development under the appropriate environmental conditions. Similarly, a fungus might infect but not cause disease in a host plant that is growing vigorously in moist, fertile soil in adequate sunlight, but the same fungus may kill a host plant should they meet under conditions that are more stressful to the plant. Changes in the environment over the duration of a symbiosis (e.g., summer drought following a wet spring) can change the nature of the interaction from benign to pathogenic.

Critically, "environment" does not mean only the abiotic factors like soil moisture and air temperature, and the "host" is never just a plant. *Pathogens interact with chimeras of the plant and its microbiome, including endophytic and epiphytic microbes*. Pathogens interact with the microbiome both directly and indirectly, through the effects of the microbiome on host physiology. Finally, although it is usually called the "disease" triangle, this model is applicable to all kinds of symbiosis—the outcome of any symbiotic interaction is dependent on environmental context.

1.8 Pathogen life-history strategies

Microorganisms that interact with plant material have three broad **life-history strategies**. If a microbe consumes plant material that is already dead, it is a **saprotroph**. This could be fungus decomposing a fallen leaf or the remains of last year's crop, or a polypore fungus hollowing out the center of a living tree, since the heartwood of a tree is just dead xylem! **Necrotrophs**, on the other hand, kill the plant tissue before or as they consume it. Some produce toxins that kill plant cells in advance of their growth, others may kill cells directly as they consume them through extracellular digestion. The necrotrophs then consume the dead material to grow and reproduce. Necrotrophs are reasonable analogs to herbivores. **Biotrophs** have a much more intimate interaction with host plants, requiring living host tissue to support their growth and reproduction. They obtain nutrients as parasites of living plant cells.

Many biotrophs are **obligate** biotrophs, meaning the only way they can complete their life cycle is on a living plant host. They may damage their host, causing disease by consuming energy that would otherwise go towards plant growth or seed production, or by physically disrupting plant structures when they make galls or pustules, but the host tissue must remain alive for the biotroph to reproduce. Necrotrophs, on the other hand, may be **facultative** pathogens—in the absence of a living susceptible host, they can complete their life cycles as saprotrophs, living in dead plant material. In many cases, facultative pathogens are only able to cause disease on stressed plants or those already damaged by pests or other pathogens. Some fungi may be biotrophs, necrotrophs, and saprotrophs at different parts of their life cycle. A fungus may live and grow for extended periods inside a host plant, obtaining nutrition but with minimal impact on host functions. At some point, the relationship with the host may change, leading it to kill host tissue, after which it can go on to live and reproduce in the dead host tissue. Such organisms are called **hemibiotrophs**.

Being an obligate biotroph is tricky business, but the rust and smut fungi, as well as many **fastidious** (i.e., picky eater) pathogens like some bacteria that live inside phloem or xylem cells, manage to pull it off. It requires getting nutrients from the host without triggering lethal host defenses and without killing the host tissue. Because this requires rather careful timing and placement of growth inside the plant, obligate biotrophs are generally able to attack only a narrow diversity of hosts. When potential hosts differ too much, it is difficult to strike just the right balance to survive in a plant without killing it.

1.9 How many pathogens are there?

The potential suite of pathogens for most plant species includes hundreds of fungi, bacteria, viruses, oomycetes, and nematodes. Fungi are usually the most common and the most important pathogens on plants, so we often (here and elsewhere in this book) begin by looking at fungi. Globally, there are around 400,000 plants and probably several million fungi (Blackwell 2011), although even defining what constitutes a fungal species is fraught (Steenkamp et al. 2018). Although only a subset of these millions of fungi are plant symbionts, and a smaller subset are pathogens, still the global diversity of pathogens probably far exceeds the diversity of their plant hosts. It is very hard to know how many kinds of pathogens cause disease on any given plant species.

The 10th edition of the *Dictionary of the Fungi* (Kirk et al. 2008) lists close to 100,000 species of fungi described by science, which is just a modest proportion of global fungal diversity. Mycologists have long tried to estimate how many species of fungi there are in the world by assuming the following relationship (e.g., Hawksworth 2001, Blackwell 2011):

$$\text{Total number of fungal species} = \text{Number of plant species} \times \text{Fungal species per plant species} \times \text{Adjustment for host sharing}$$

Unfortunately, the values used for the last two components of the calculation have been little more than informed guesses. Let's look at one approach to getting those data: compilations of published literature.

There is a massive scientific literature reporting on pathogens associated with plants. The USDA Fungus–Host Distributions database has gathered together much of that literature for fungi (and oomycetes) (https://nt.ars-grin.gov/fungaldatabases/fungushost/fungushost.cfm); as of April 2021, it includes 417,820 unique fungus–host combinations, including 99,062 fungus and 64,966 host names (L. Castlebury, pers. comm.). From this database you can readily look up which fungi are known from a particular host, or which hosts have been reported for a given fungus, and where. Calculating the number of fungal species per host is then a simple matter. For instance, the crop species maize, rice, soybean, lemons, and tomato have 988, 619, 597, 262, and 203 known fungal associates. There are parallel databases for other kinds of plant symbionts. From the Plant Virus Database (http://47.90.94.155/PlantVirusBase/#/), we can add some 57, 27, 95, 15, and 135 viruses to our crop species. Nemabase, compiled by UC Davis researchers (http://nemaplex.ucdavis.edu), adds another 149, 105, 80, 29, and 144 species of nematodes. However, there are significant limitations to such databases that are important to note if you are interested in using them for research.

First, lists like the USDA database usually include not only pathogens but all fungi associated with plants—harmful, benign, or even beneficial. Whether this is a problem or a strength depends on the question you want to ask. Second, fungal databases often do not adjust for synonymous fungal names, including taxonomic revisions or complexities like asexual anamorph (e.g., *Colletotrichum gloeosporioides*) and sexual teleomorph (e.g., *Glomerella cingulata*) names of the same fungus but published under its different names. This can produce a substantial overestimate of the number of fungal species on any given host.

Third, research effort varies considerably across plant hosts, and we know more about pathogens in regions with high concentrations of researchers. With an increase in the amount of research of any kind on a given plant, the number of fungal species reported from that plant also increases (Figure 1.6). Economically important crops and forest trees attract much more research than less important crops or wild plant species; crop species on average had 12-fold more citations than wild species. But important crops or forest trees planted widely around the world will also encounter broader sets of pathogens and interact with them under a greater range of environmental conditions. Indeed, broadly distributed host species can have up to an order of magnitude more known fungal pathogens than host plants with small geographic distributions (Miller 2012).

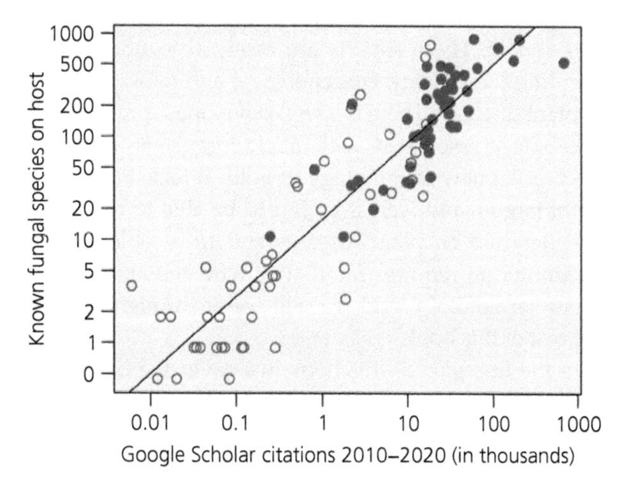

Figure 1.6 The number of fungi known from a plant species is a function of how intensely the species is studied. The number of fungal species reported for a host species in the USDA Fungus–Host Distributions database (accessed April 2021 and adjusted to account for double-listing of synonymous names) increases with the number of studies of any type on that host species as reported in Google Scholar from 2010 through 2020. Included in the analysis are 50 important crop plants in California (filled circles) and 50 wild species from the UCSC Santa Cruz campus (open circles). Three species with no reported fungi were assigned 0.5 fungal taxa for analysis. $\text{Log}_{10}(\text{\# fungi}) = -0.993 + 0.749 \log_{10}(\text{\# citations})$; $F_{1,98} = 484.2$, $P \leq 0.00001$, $R^2 = 0.83$. Figure by Gregory S. Gilbert.

1.10 Where we are heading: A roadmap to the book

This introductory chapter was a highly condensed overview of the basic ideas from phytopathology and evolutionary ecology that will pervade the rest of this book. We introduced some of the rich, specialized terminology to ensure we have a common language moving forward. Now it is time to jump in!

As plant disease ecologists, we are concerned with plants as hosts and their parasites, including fungi, oomycetes, viruses, bacteria, nematodes, and parasitic plants, roughly in descending order of importance as plant pathogens. The organisms we deal with include representatives of most of the major evolutionary branches of organisms, along with the tremendous diversity of ways those organisms make a living. In Part 1 of this book, we cover the basic organismal biology of different types of pathogens and their plant hosts. We explore the key morphological and physiological aspects of the hosts in "How to be a plant" (Chapter 2). Chapters 3–7 summarize the biology of the most important groups of plant pathogens. We recognize that many people coming into the study of plant disease ecology may have had little formal exposure to microbiology, and we believe that it is essential to understand pathogens as organisms to be able to think clearly about the evolutionary ecology of diseases. Our brief overviews focus on those aspects of basic biology of pathogens that are important to understand how they interact with host plants and how they grow, survive, reproduce, and spread. These aspects are essential to understanding the ecology, epidemiology, and evolution of plant diseases. There is a well-developed, precise vocabulary associated with microorganisms, and we present new terminology in bold. It feels like a lot of jargon, and, yes, it is! But to be able to read the literature on plant diseases and to be able to communicate requires familiarity with the appropriate terminology, and we will use it throughout the rest of this book.

In the first part of this book, we cover the basic organismal biology of different types of pathogens and their plant hosts. We explore the key morphological and physiological aspects of the hosts in "How to be a plant" (Chapter 2), and the basic biology of the most important groups of plant pathogens (fungi, oomycetes, bacteria, viruses, and nematodes) in Chapters 3–7. As a brief stylistic aside, when writing about formal taxonomic groups like the kingdoms or supergroups of Plants, Animals, Fungi, or Bacteria, the names are capitalized. But when we are referring to the organisms that belong to those groups, we will generally use the lower-case, colloquial terms fungi, bacteria, and plants, unless trying to make a specific taxonomic distinction.

After looking at each of the groups of organisms, we spend some time in Chapter 8 categorizing the different ways that plants and pathogens interact, exploring how to recognize different kinds of diseases, and what symptoms and signs tell us about these interactions in terms of life histories. In Chapter 9, we explore many key methods and approaches used to study plant diseases.

With those chapters as a foundation, Part 2 dives into thinking about the ecological aspects of plant diseases at the population level (Chapters 10 and 11), focusing on spatial and temporal patterns of disease, as well as mechanisms of spread of pathogens across space and time. These chapters introduce both the conceptual models and some basic quantitative aspects of plant disease epidemiology.

We then move on to look at plant–pathogen symbioses at a more intimate level, focusing on the genetics and physiology of interactions between plants and pathogens (Chapter 12). The focus here is to understand the physiological interactions and the underlying genetics in order to think clearly and critically about the evolution of the interactions and how the ecological outcomes of plant–microbe symbioses are shaped by environmental conditions. This leads directly into Chapter 13, which explores the evolution of plant–pathogen interactions in both wild and managed systems.

The last four chapters each provide a unique perspective on plant disease ecology. In Chapter 14 we look at plant pathogens as natural parts of wild plant communities—essential, creative components of ecosystem function and drivers of community structure, diversity, and dynamics. Pathogens are not villains, but key members of communities with outsized impacts. In Chapter 15, we look at how interactions between microbes shape disease outcomes. Multiple microbes can act synergistically to create disease complexes, and we can manipulate microbial interactions for biological control of diseases. Recent advances in understanding the plant microbiome are opening a new world of thinking about microbe–microbe interactions in the context of the evolutionary ecology of plant disease. Chapter 16 explores the implications of different aspects of global change on plant diseases. Plant diseases are ubiquitous in wild and managed ecosystems, and global change is transforming the landscape of plant disease, both by rapidly changing the geographic distribution of pathogens and plants and by changing environmental conditions in which interactions play out. Finally, in Chapter 17, we look at how to apply evolutionary ecology to the management of plant diseases. We explore how an integrated, systems-based approach creates opportunities for novel and sustainable disease management.

Further reading

Agrios, G. N. 2005. *Plant Pathology*, 5th ed. Elsevier, Burlington, MA.

Burdon, J. J. 1987. *Diseases and Plant Population Biology*. Cambridge University Press.

Milgroom, M. G. 2015. *Population Biology of Plant Pathogens. Genetics, Ecology, and Evolution*. APS Press, St. Paul, MN.

References

Berkeley, M. J. 1846. Observations, botanical and physiological, on the Potato Murrain. *Journal of the Horticultural Society of London* **1**:9–34 (reprinted as Phytopathological Classics number 38 by the American Phytopathological Society, 1948).

Blackwell, M. 2011. The Fungi: 1, 2, 3... 5.1 million species? *American Journal of Botany* **98**:426–438.

Burdon, J. J. 1987. *Diseases and Plant Population Biology*. Cambridge University Press.

Dobzhansky, T. 1950. Evolution in the tropics. *American Scientist* **38**:209–221.

Hawksworth, D. 2001. The magnitude of fungal diversity: the 1.5 million species estimate revisited. *Mycological Research* **105**:1422–1432.

Kirk, P. M., P. F. Cannon, D. W. Minter, and J. A. Stalpers. 2008. *Dictionary of the Fungi*, 10th ed. CABI, Wallingford, UK.

Milgroom, M. G. 2015. *Population Biology of Plant Pathogens. Genetics, Ecology, and Evolution*. APS Press, St. Paul, MN.

Miller, Z. J. 2012. Fungal pathogen species richness: why do some plant species have more pathogens than others? *American Naturalist* **179**:282–292.

Steenkamp, E. T., M. J. Wingfield, A. R. McTaggart, and B. D. Wingfield. 2018. Fungal species and their boundaries matter—Definitions, mechanisms and practical implications. *Fungal Biology Reviews* **32**: 104–116.

CHAPTER 2

How to be a plant

Gregory S. Gilbert and Ingrid M. Parker

In order to understand how plant pathogens use plants as a habitat and growth medium, as well as how pathogens affect plants, we need to understand some basic features of plant biology. These include how plant bodies are organized, how plants grow and reproduce, how they get their energy, how they transport water, nutrients, and sugars, and how plant hormones regulate physiology and development. In addition, we need to understand the basics of plant sensing and signaling and plant chemical defenses. These form the foundation for mechanisms of plant immune response and will be examined in detail in Chapter 12.

2.1 Plant cells and tissues

Unlike animals, plants have no specialized cells dedicated to immune response; every cell must defend itself against pathogens. The plant cell has all the same features as any eukaryotic cell—nucleus, mitochondria, ribosomes, etc. (Figure 2.1)—and it performs all the same functions, including the cellular respiration that uses oxygen and carbohydrates and produces energy. In addition to these universal cell functions, however, the plant cell has several unique features. First, plants photosynthesize. The cells of plants (and algae) contain chloroplasts, which are specialized organelles within which plant cells harvest light energy and convert it to chemical energy in the form of carbohydrates. Later in this chapter, we discuss photosynthesis in more detail.

Second, plant cells have cell walls. Unlike fungal or bacterial cell walls, those of plants are made of **cellulose**, **hemicellulose**, **pectin**, and frequently **lignin**, which is made up of chemically complex phenolic polymers. Lignins are an important component of wood. They are very resistant to degradation and can only be broken down by certain groups of bacteria and certain Basidiomycete fungi. The plant cell wall, in addition to playing central roles in plant structural support and water transport, is seen as a key feature in defense against pathogens, especially in woody plants. Through their thick cell walls, plant cells are linked via tiny passageways called **plasmodesmata**, which connect the cytoplasm and endoplasmic reticulum of adjacent cells. Some viruses are capable of moving from cell to cell via the plasmodesmata (Heinlein 2015).

A third special feature of plant cells is the prominence of the central **vacuole**, a liquid-filled organelle that is central to a number of plant functions including growth and development, structural integrity, and storage of all sorts of biomolecules including nutrients, enzymes, and waste products. Among the products stored in the vacuole are antimicrobial compounds and hydrolytic enzymes, suggesting that the vacuole may play an important role in defense against pathogens. One response to infection in some plants is the disruption of the vacuolar membrane, releasing antimicrobial compounds or mycotoxins and hydrolytic enzymes that result in **apoptosis** (programmed cell death), which isolates the spread of some pathogens such as viruses (Hatsugai and Hara-Nishimura 2010).

The plant body is made up of three tissue types—dermal tissue, ground tissue, and vascular tissue.

The Evolutionary Ecology of Plant Disease. Gregory S. Gilbert and Ingrid M. Parker, Oxford University Press. © Gregory S. Gilbert & Ingrid M. Parker (2023).
DOI: 10.1093/oso/9780198797876.003.0002

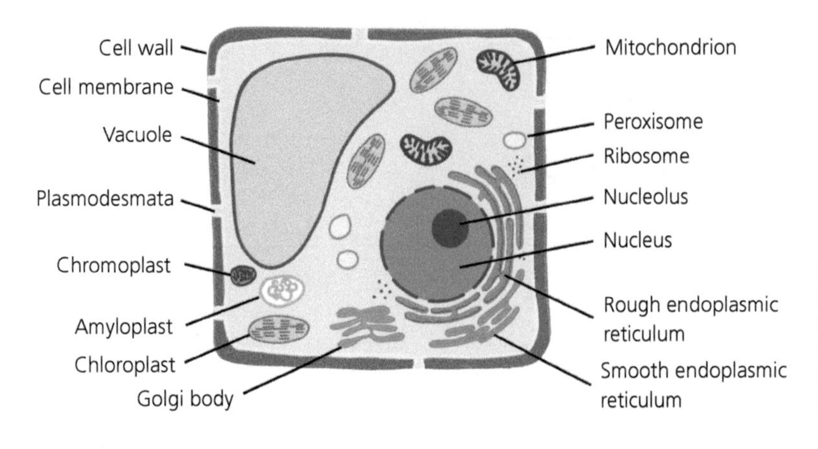

Figure 2.1 Structure of a plant cell. Features unique to plant cells include the chloroplast, the large central vacuole, and the cell wall. Adjacent cells are linked through their plasmodesmata.
Drawing by Ingrid M. Parker.

Figure 2.2 Trichomes of *Stachys bullata*.
Photo by Gregory S. Gilbert.

The **dermal tissue** is the outer layer of cells (epidermis) in leaves and stems. Specialized cells called **trichomes** protrude from the dermal tissue as tiny hairs or scales (Figure 2.2). The trichomes sometimes exude toxic chemicals to deter herbivores, as in the stinging nettle, *Urtica dioica*, and hairy trichomes can influence disease development by facilitating the attachment of spores to the leaf (Calo et al. 2006). The epidermis is covered by a waxy layer called the **cuticle,** which reduces water loss from the plant and protects the leaf from excess irradiation. The cuticle is also important in plant defense, because it forms a barrier to most pathogens. As we discuss in later chapters, there are different ways to get around the physical and chemical barrier of the cuticle. Some foliar fungal pathogens (e.g. *Fusarium*) penetrate the cuticle with cutinase, a special cutin-degrading enzyme that is upregulated in a fungal spore when it comes in contact with the epidermal cuticle (Serrano et al. 2015).

The **ground tissue** makes up most of the plant body; ground tissue functions in nutrient storage in vegetables such as potato, yam, and carrot, which are modified stems and roots. All photosynthesizing tissue is also made up of ground tissue, so most of the inside of a leaf is this tissue type. However, inside leaves, stems, and roots we also find **vascular tissue**, which is made up of xylem and phloem. **Xylem** cells, which form the conduits for water transport, are dead when they are mature, and the cell walls of the cylindrical cells line up to form a continuous tube. Water passes from roots to leaves

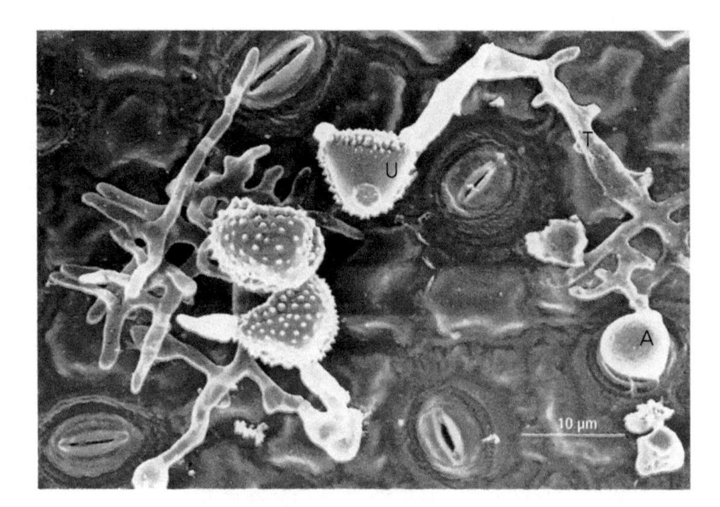

Figure 2.3 Coffee rust urediniospores (U) germinate, grow germ tubes (T), and then infect the leaf by producing plugs (appressoria, A) over stomata and penetrate them with hyphae.
Photo by M.C. Silva, licensed under a Creative Commons Attribution-NonCommercial 4.0 International License.

through this continuous tube. The morphology of xylem cells in angiosperms and gymnosperms differs in important ways; **tracheids** are found in all vascular plants, whereas angiosperms have also evolved wider **vessels** that provide more efficient water transport. Unlike xylem, **phloem** tissue is still alive when mature—with pairs of differentiated cells: a sieve tube element lacking a nucleus with a companion cell that controls the activities of the sieve tube. Sugars and other nutrients move from cell to cell of the sieve tube through the pores of the sieve plates that separate them.

2.2 Water and nutrient transport

Plants need to stay fully hydrated not only because all biochemical reactions take place in water, as in any organism, but also because water is what gives a plant its structural integrity. Each cell is filled with water that pushes against the structure of the cell wall, generating **turgor pressure.**

In a process called **transpiration**, water exits the plant through holes called **stomata** in the epidermis of the leaf. The closing of the stomata is regulated by pairs of specialized **guard cells**. Guard cells flatten together when they lose turgor pressure, closing the stomata. Plants can't avoid losing water through their stomata, because they need to let in CO_2 molecules for photosynthesis. Stomata are key openings in leaf surfaces that allow easy access for many plant pathogens to get inside the plant interior (Figure 2.3).

As described by the cohesion–tension theory, water is drawn up through the plant because as water molecules leave through the stomata other water molecules are pulled up to replace them. The cohesion of the water molecules binds them together in a single column, which is under negative pressure, or tension. Water moves up through a plant rapidly, on the order of centimeters per minute in the xylem of trees, and this movement doesn't require any energy investment on the part of the plant.

Xylem is vulnerable to a series of bacterial, fungal, and oomycete pathogens that cause **vascular wilt** diseases (Yadeta and Thomma 2013). The pathogens enter the xylem through the roots and travel up through the conduits (Figure 2.4). When they proliferate in those conduits, they block the flow of water. When the leaves can't replace the water lost to transpiration and lose turgor pressure, they wilt.

Sugars, along with amino acids, other nutrients, and certain hormones, are not transported in the xylem but instead in the phloem. It is a much slower process, and movement is from source (areas where sugars are loaded into the phloem) to sink (areas where sugars are being used). Stored carbohydrates in modified roots can be a source in the spring when new growing shoots represent a sink. Later in the summer, the photosynthesizing leaves are a

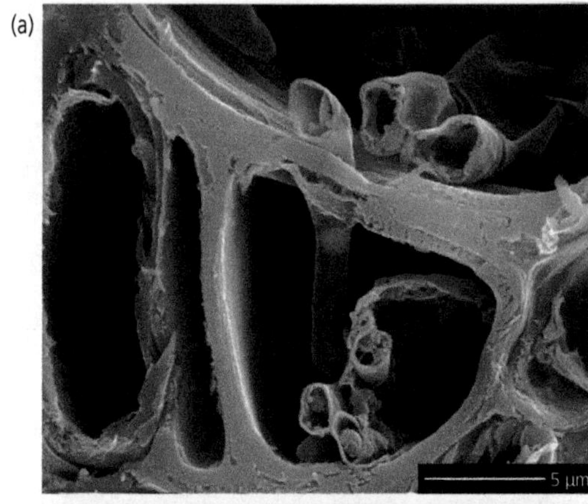

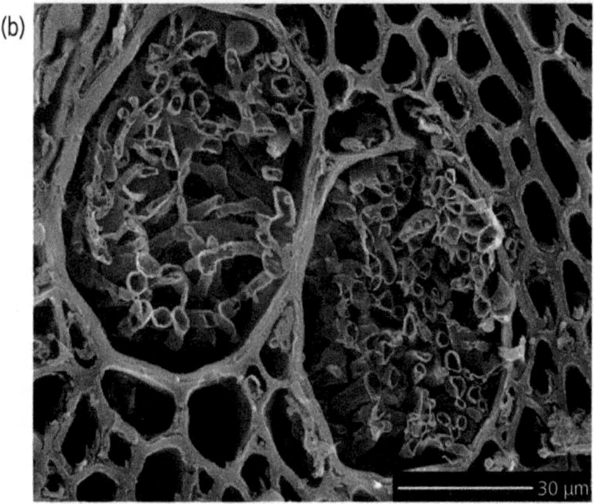

Figure 2.4 Fungal hyphae of *Fusarium oxysporum* growing inside the xylem of *Solanum pimpinellifolium*. (a) A fungal hypha moving through a vessel wall, (b) hyphae completely occluding two xylem vessels.

Photos courtesy D. Caldwell, © APS. Reproduced, by permission, from Caldwell, D., and Iyer-Pascuzzi, A. S. 2019. A scanning electron microscopy technique for viewing plant–microbe interactions at tissue and cell-type resolution. Phytopathology 109:1302–1311. doi: 10.1094/PHYTO-07-18-0216-R.

source, and sugars will move in the opposite direction, away from the leaves and toward a sink such as roots or flowers. In addition to growing parts of the plant, symbiotic fungi like **mycorrhizae** can also represent a sink for the plant, which provides carbohydrates to these fungi.

2.3 Photosynthesis

Photosynthesis is how plants harness the energy of the sun to turn CO_2 gas into sugar. This all goes on in the **chloroplast**, the organelle that gives plant cells

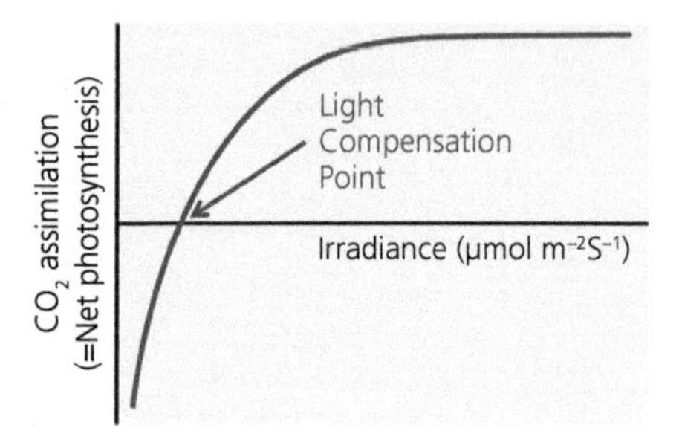

Figure 2.5 Light response curve. Net photosynthesis (measured as carbon assimilated) as a function of the amount of photosynthetically active radiation reaching a leaf. To survive and grow, a plant must be above the zero line.

Drawing by Ingrid M. Parker.

their green color. There are two distinct processes in photosynthesis. In the first (called the "light reactions"), photons of light are captured by systems of pigment molecules and used to excite electrons; the energy of these electrons is captured by an electron transport chain and stored as both ATP and NADPH. The complex molecules involved in the capture and transport of electrons are embedded in the **thylakoid membranes** within the chloroplast. In the second process of photosynthesis, called the "Calvin cycle," the ATP and NADPH produced by the light reactions are used in the fixation of CO_2, which is catalyzed by the enzyme RUBISCO. RUBISCO is the most common protein on earth, comprising up to over half of all the protein in a leaf (Ellis 1979). The biochemical reactions of the Calvin cycle take place in the liquid matrix of the chloroplast.

At a physiological level, you can think of a plant's ability to grow, reproduce, and respond to pathogen attack as limited by its photosynthetic rate. Photosynthetic rate increases with the amount of light that reaches the chloroplasts (Figure 2.5); a plant in the shade may not be able to tolerate pathogen infection as well as one in the sun because it lives close to its **light compensation point**—the point at which the rate of photosynthesis (CO_2 fixation) equals the rate of respiration (CO_2 release). Below this threshold, not enough carbon is fixed by photosynthesis to fuel the plant's own metabolic needs. Net photosynthetic rate is quantified using an instrument that measures gas exchange through the stomata; the light compensation point is the light level at

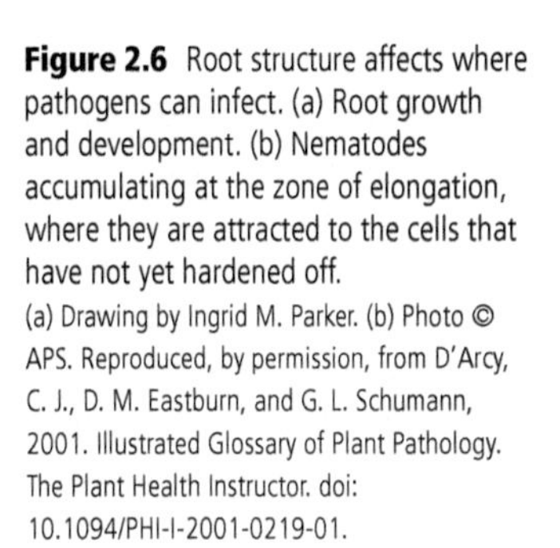

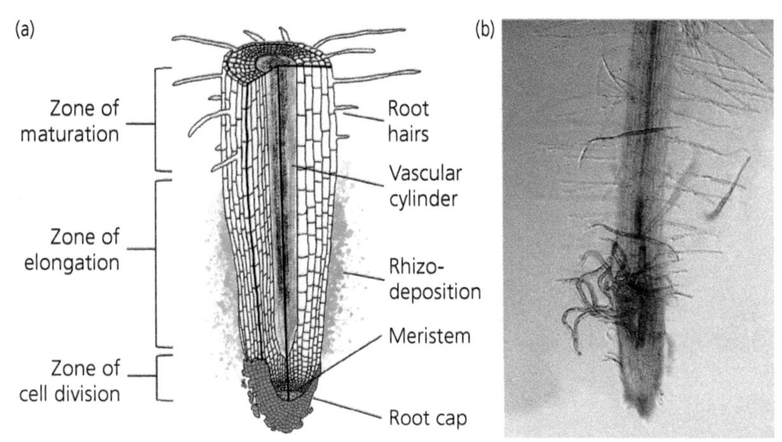

Figure 2.6 Root structure affects where pathogens can infect. (a) Root growth and development. (b) Nematodes accumulating at the zone of elongation, where they are attracted to the cells that have not yet hardened off.
(a) Drawing by Ingrid M. Parker. (b) Photo © APS. Reproduced, by permission, from D'Arcy, C. J., D. M. Eastburn, and G. L. Schumann, 2001. Illustrated Glossary of Plant Pathology. The Plant Health Instructor. doi: 10.1094/PHI-I-2001-0219-01.

which CO_2 is given off by the leaf, rather than being taken up. Photosynthetic rate may also be limited by the leaf's access to CO_2, which as explained above really means limited by water. Photosynthetic rate may also be limited by the nutrient status of the plant, particularly by access to nitrogen, which is used in large quantities in RUBISCO, or by access to phosphorus, which is needed for ATP.

2.4 Plant growth

In a plant, new cells are produced in **meristems,** where undifferentiated cells divide by mitosis. Unlike a frog or a human, plants are **modular**, meaning that they are made up of an indeterminate number of repeating components. Each separate branch on a plant and each root has at its tip a **apical meristem** producing new cells. These meristematic cells (which are equivalent to the stem cells of animals) grow and elongate, then differentiate into different cell types (epidermal cells, photosynthesizing cells, xylem cells, etc.). If the apical meristem at the tip of a branch is removed, for example by an herbivore or a gardener, buds lower down on the branch will begin to grow into new branches. This feature allows many plants to survive the die-back of individual branches or even the loss of whole segments to disease. In grasses, intercalary meristems are located at the stem internodes and the bases of leaves, where they are protected from fire and the sharp teeth of grazers.

In roots, it is easy to observe the structured way in which plant cell division and growth occurs (Figure 2.6A). Protecting the growing root tip is a **root cap** of dead cells, and just behind it is the apical meristem, where cell division occurs. In the **zone of elongation**, cells expand lengthwise in a process that is powered by the influx of water into the cell, similar to the growth of fungal hyphae. After cell growth is complete, the cells mature into vascular tissue, cambium, and epidermal cells including **root hairs**. About 40% of the carbon fixed by a plant through photosynthesis is sent to the roots, and about 27% of that is then lost from the root and deposited into the rhizosphere (Jones et al. 2009). That is about 11% of the total carbon from photosynthesis, and rhizodeposition is responsible for the loss of a similar percentage of total plant nitrogen. The zone of elongation, which has not yet toughened up because cells are still expanding, is rich in leaked carbon and nitrogen and is particularly vulnerable to infection by pathogens. The leaked nutrients can establish a gradient in soil that attracts motile pathogens like oomycetes and bacteria through chemotaxis, or fungi through chemotropism, making the zone of elongation a key site of disease initiation (Figure 2.6B).

Woody plants such as trees, in addition to primary (or "apical") meristems, grow in girth from a secondary (or "lateral") meristem, also called a **cambium** layer (Figure 2.7). In a tree, the vascular cambium produces new phloem cells to the outside

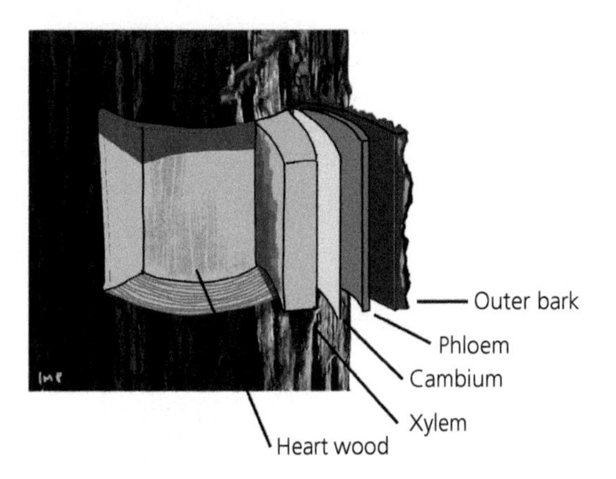

Figure 2.7 Tree growth and the structure of wood.
Drawing by Ingrid M. Parker.

and new xylem cells to the inside, creating new sapwood. The tree grows outward while the xylem at the center of the tree is filled with resin, becoming heartwood; remember that all xylem conducting cells (vessel members and tracheids) are already dead when mature. Some woody species in seasonal environments, where a wet period of the growing season is followed by a dry period, produce annual rings. Rings occur because wider xylem cells are produced under wet conditions and have a lighter color than the narrower xylem cells put down under dry conditions.

The most important functions of a tree are happening just under the surface. To the outside of the vascular cambium is another layer, the cork cambium, which produces the protective material of the bark of the tree. Because the phloem and the vascular cambium—both critical tissues—are located just under the bark of the tree, the tree is vulnerable to anything that damages its outer edge. Pathogens that cause canker diseases disrupt these tissues, killing a branch or even the whole tree by cutting off the flow of nutrients.

2.5 Plant reproduction

Plants have a wide range of **life histories** and reproductive strategies. The life cycle of flowering plants generally begins with a germinating seed, followed by vegetative growth of the plant until it matures to produce flowers for sexual reproduction, resulting in fruits that contain the next generation of seeds (Figure 2.8). Some plants reach sexual maturity and complete their life cycle in a single growing season (annuals), while others live many years (perennials) and reproduce multiple times (**iteroparous**). A few **semelparous** species like yucca (*Agave americana*), the suicide tree (*Tachigali versicolor*), and the talipot palm (*Corypha umbraculifera*) may live for decades before flowering once and then dying. In addition, many plant species are able to reproduce asexually. Like the nopal cactus (*Opuntia ficus-indica*), many succulents can sprout a new plant from fragments of branches or leaves. Some plants make asexual seeds (e.g., dandelions), while in perennial turf grasses, new shoots (**ramets**) are produced from underground runners. Many shrubs and trees are able to resprout from their root systems after disease, fire, timber harvest, or storms; redwood trees (*Sequoia sempervirens*) and poplar trees (*Populus* spp.) do this regularly.

Sexual reproduction in plants is remarkable and diverse. Plants demonstrate what we call "alternation of generations," the transition between a haploid phase (or generation) (called a **gametophyte**) and a diploid phase (**sporophyte**). Note that the production of gametes within the haploid generation is by mitosis, unlike in humans, where sperm and eggs are produced by meiosis within a diploid body. Fertilization of gametes then leads to the growth of a diploid stage. The haploid stage can be large and long-lived, as in some algae and non-vascular plants such as mosses. In contrast, in seed plants, which represent a more derived lineage, the haploid stage is greatly reduced, and the trees and herbs with which we are familiar are the diploid sporophyte form. Meiosis occurs inside the male and female cones (in a gymnosperm) or the male anthers and female ovaries (in an angiosperm, Figure 2.9). Pollen grains are the male haploid phase, while the female haploid phase, called the megagametophyte, remains inside the female cone (gymnosperm) or ovary (angiosperm).

The flowers of angiosperms are one of the great innovations in the history of plant evolution. With their bright colors, alluring scents, production of nectar, and intriguingly diverse shapes and patterns, many flowers show spectacular adaptations

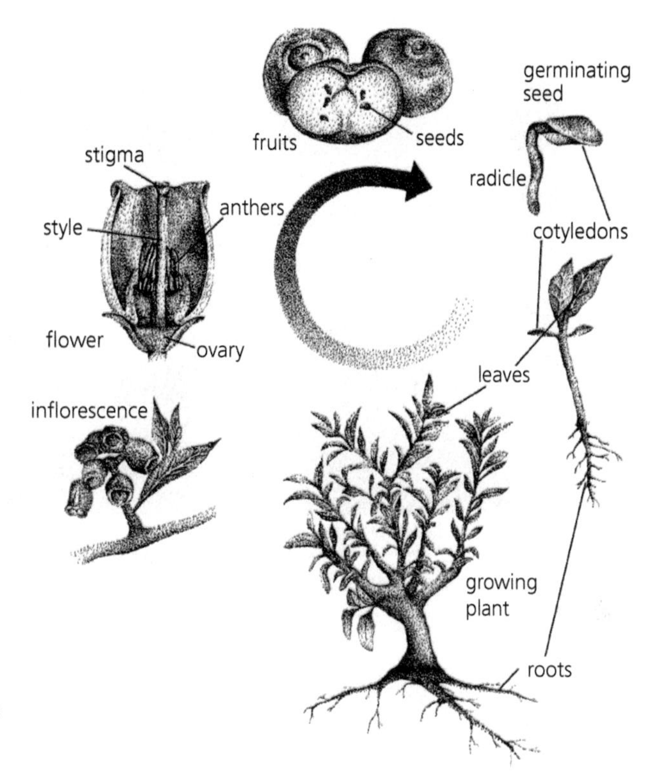

Figure 2.8 The life cycle of a flowering plant usually starts with the germination of a seed. The seedling draws energy from stored reserves in the cotyledons of the embryo or the endosperm of the seed, producing a radicle (root), and hypocotyl (shoot). The juvenile plant grows roots, stems, and leaves until it attains the size needed to reproduce sexually. The plant then reproduces sexually by producing flowers, which once pollinated, produce fruits. Seeds develop within the fertilized ovules inside the fruits. Once dispersed, seeds may germinate immediately, remain dormant for months until the start of a new growing season, or remain dormant and viable for many years, until appropriate conditions induce them to germinate.
Drawing by Josh Zupan.

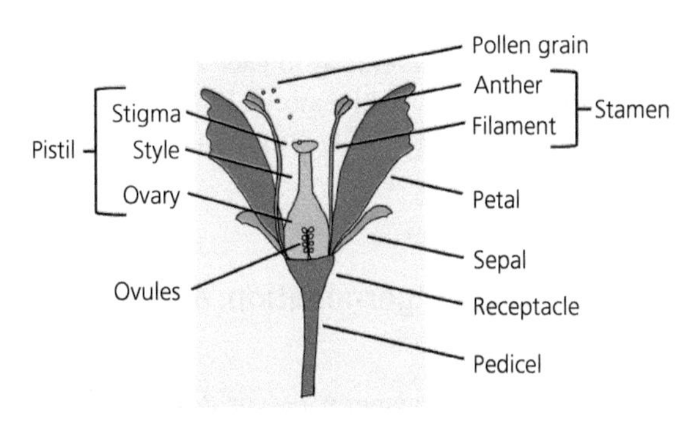

Figure 2.9 Parts of a flower.
Drawing by Ingrid M. Parker.

for attracting the insects, birds, bats, and other animals that pollinate them. Some pathogens have exploited the directed movement of pollinators from flower to flower for their own dispersal. In fact, several dozen viruses are exclusively transmitted via pollen and seeds (Card et al. 2007). Another type of pollinator-dispersed pathogen, the anther smut, infects flowers and converts the anthers into structures producing fungal spores (Figure 2.10). These spores hitch a ride with visiting pollinators to the next flower. The fact that anther smuts effectively castrate the flowers they infect has led researchers to call them "plant venereal diseases" (Antonovics 2005). Other types of pathogens take advantage of both the resources available in flowers as well as this facilitated dispersal, with the result that flowers show a range of defensive traits such as the production of antimicrobial compounds in nectar and volatile compounds that act as antibacterial fumigants (Stephenson 2012).

Note that not all plants are pollinated by animals. Plant species that are wind pollinated do not invest in attractive structures; instead, their flowers produce massive amounts of pollen to maximize the probability of reaching the ovule of another plant by riding a random wind current. In addition, plants show a remarkable diversity of **breeding systems**. While some species have separate male and female individuals, most plants have

Figure 2.10 *Microbotryum silenes-inflatae*, the silene anther smut shown here on *Silene acaulis*, converts anthers to spore-producing structures. Anthers in the healthy flower on the left are covered in pollen; on the right, they bear fungal spores.
Photo by Michael Hood.

male and female flowers on the same plant or male and female parts in the same flower (as in Figure 2.9). Some species are completely unable to fertilize ovules with their own pollen because they have a genetic self-incompatibility system. In contrast, other species are nearly always self-fertilized because they have tiny flowers with anthers that shed pollen directly onto their stigmas. Flower traits such as the proximity of male and female parts in the flower, or the relative timing of pollen maturity and stigma receptivity, can favor either selfing or outcrossing. The degree of selfing and outcrossing is called the **mating system**.

One reason plant mating systems are important is that the pattern of genetic diversity in a species is heavily influenced by its mating system. A species that is highly outcrossing is expected to show lots of genetic diversity both within and between individuals, especially in a large population. A species that is highly self-fertilizing will have almost no genetic diversity within an individual—that is, its genome will be almost entirely homozygous—but separate populations or separate lineages will be genetically differentiated from each other. This is what happens with inbred lines produced in the breeding of agricultural varieties. A lack of genetic diversity within an individual can make it vulnerable to pathogen attack, and a lack of genetic diversity among individuals in a population makes that population vulnerable to disease epidemics, as we will see in later chapters. Notice that the genetic consequences of self-fertilization are not the same as the genetic consequences of asexual reproduction, like the clonal propagation of the nopal cactus. Asexual offspring are all identical to each other and to the parent plant, but, unlike inbred lines, they may show lots of genetic diversity within the genome of an individual.

2.6 Fruits, seeds, germination, and development

The ovary of an angiosperm flower develops into a **fruit**. Fruits can take many forms, from the dry capsule of an iris, to the nut (acorn) of an oak, to the winged samara of a maple. However, the fruits that are eaten by animals (**frugivores**, including us) are fleshy, full of nutrients, and often sweet. The structure of an edible fruit may be derived from the ovary alone, like a tomato, or may also involve supporting tissues, like an apple, a strawberry, or a pineapple. Pathogens that compete with frugivores for the flesh of fruits must be able to live in a substrate with relatively high sugar content (Section 8.6).

Ovules develop into seeds; in an angiosperm this occurs within the fruit, and in a gymnosperm it occurs within female cones. Some pathogens, like the pollen-transmitted viruses, infect seeds while they are still actively developing. Once a seed is fully mature, it is covered by a hard seed coat, or **testa**, which protects the seed from desiccation, and the seeds of many species can stay alive for years, even decades. The seed coat also plays an important role in resistance to pathogens, with both physical and chemical traits that help protect the seed from infection. Many plant species exhibit **seed dormancy**, periods when seeds are incapable of germinating. Dormancy can be either innate (internal to the organism) or enforced (caused by external conditions). Seeds that are buried too deep or in waterlogged soil experience enforced dormancy because they lack conditions suitable to begin germination. In contrast, innate dormancy can be caused by impermeable seed coats or by molecular germination inhibitors; innate dormancy can be broken through scarification in the gut of an animal, freeze-thaw cycles, or by leaching by snow melt or rain.

When seeds begin to germinate, they suddenly become very vulnerable to soilborne pathogens. Young seedlings have soft tissues that are not yet well defended chemically or physically. Perhaps for this reason, pathogens that kill young seedlings often are capable of infecting a very wide range of host species. Damping-off pathogens including oomycetes (e.g., *Pythium* spp.) and true fungi (e.g., *Fusarium* spp.) can cause the loss of entire fields of crops and devastating disease impacts at the population level (Section 8.4).

2.7 Plant hormones

Growth processes are regulated by plant hormones, which, like animal hormones, are signal molecules produced in one location and which cause a response at the cellular level at a separate, target location (Table 2.1). Innate dormancy, described above, is an example of a plant physiological process that is regulated by a hormone; the hormone **abscisic acid** acts as a growth inhibitor in the seed, although it has other functions as well, including the regulation of the closing of stomata in response to drought. The hormone **ethylene** regulates ripening of fruit. The hormone **auxin** regulates many critical processes in plant growth, including the elongation of individual cells. Auxin is produced by the apical meristem of the shoot and moves down from cell to cell through special auxin transporters on the plasma membranes. Auxin from the tip is responsible for the suppression of buds farther down the stem, resulting in apical dominance. In contrast, **cytokinins** are produced at the root tip and travel up in the xylem. Together, auxin and cytokinins determine organ development, and the biotechnology process of plant tissue culture relies on the delivery of precise ratios of these two hormones at different stages to make a plant callus differentiate into roots and shoots (Section 17.3).

Table 2.1 Plant hormones, what they do, and their chemical structures

Abscisic Acid	Auxin	Cytokinins	Gibberellin	Ethylene
Seed growth inhibitor; stomate closing; bud dormancy; stress responses 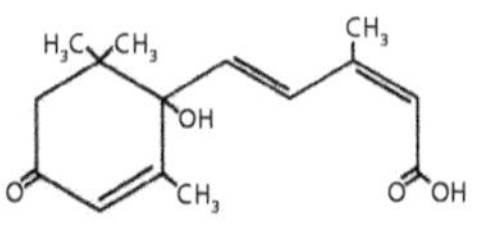	Cell elongation, division, and differentiation; apical dominance; tropism 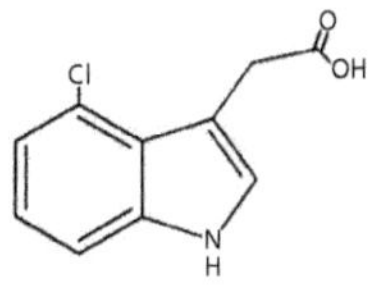	Cell division and differentiation; apical dominance; leaf senescence 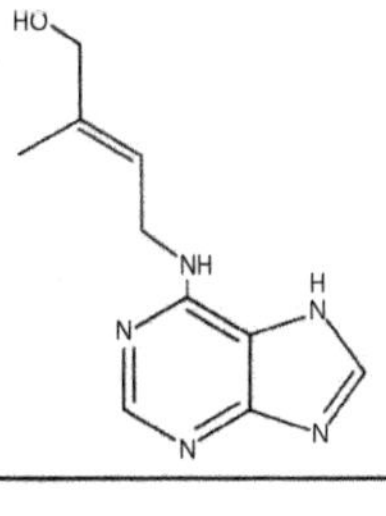	Seed germination; stem elongation; dormancy; flowering; leaf and fruit senescence 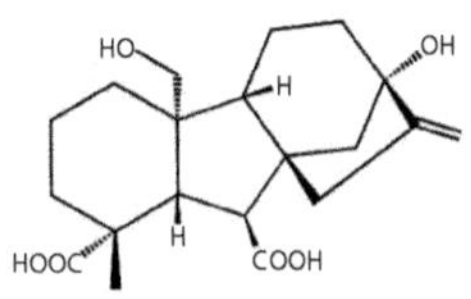	Fruit ripening leaf abscission; flower opening 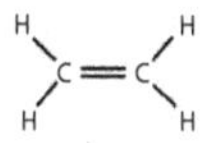

Pathogens have evolved the ability to manipulate plants by mimicking their hormones. For example, in crown gall disease, bacterial genes inserted into the plant genome result in the production of auxins and cytokinins that cause the plant to build an abnormal outgrowth like a tumor. The hormone **gibberellin**, which regulates seed germination and the etiolation response, was first discovered by Japanese scientist Eiichi Kurosawa in seedlings infected by the fungal pathogen *Gibberella fujikuroi*. Kurosawa called the disease caused by *G. fujikuroi* "foolish seedling disease," because the infected seedlings grew tall and gangly. The pathogen produces gibberellins, which mimic plant hormones and distort plant growth. It was Kurosawa's fungal strain of *G. fujikuroi* that was used to chemically isolate the chemical compound that we now call gibberellin.

2.8 Major groups of plants

Early land plants lacked organized vascular systems, relying on direct absorption of water into the plant tissue; today we still have many nonvascular plants including the mosses, liverworts, and hornworts. The evolution of vascular systems that include xylem and phloem allowed plants to become larger and more complex; **vascular plants**, which include ferns (which reproduce by making spores) together with seed plants, now make up most of plant life on earth. There are two major lineages within the seed plants, the Gymnosperms and the Angiosperms (Figure 2.11). **Gymnosperms** (which includes conifers and the adorable ginkgo) produce "naked" seeds that are not enclosed in a protective structure, although they may be borne in a cone. In contrast, **Angiosperms** (flowering

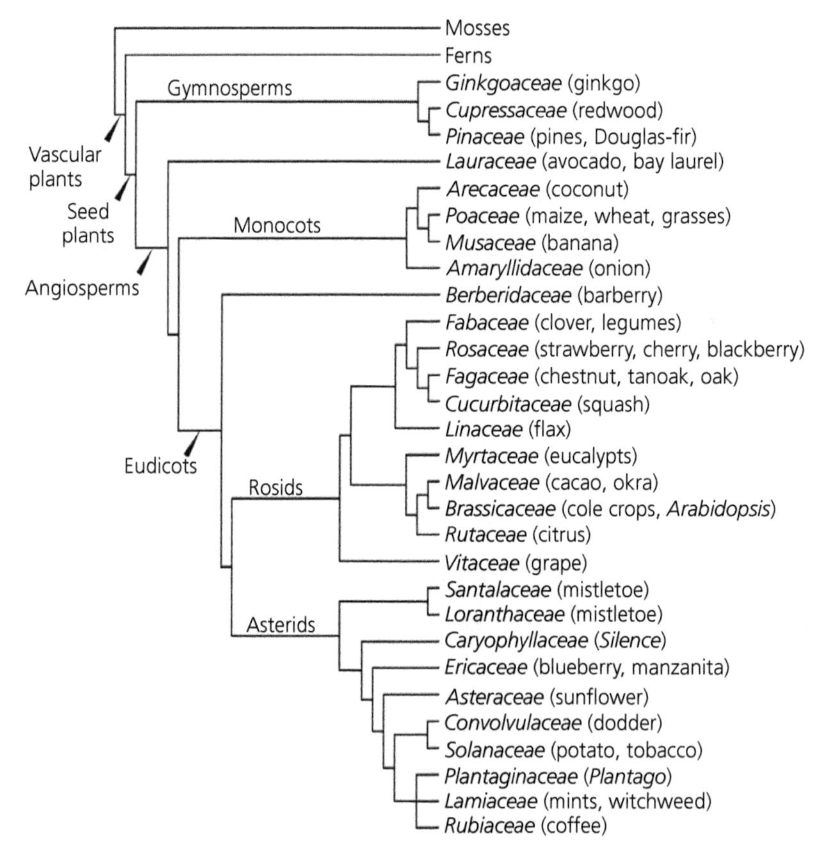

Figure 2.11 Phylogenetic relationships among the major families of plants discussed in this book. For some guidance on reading phylogenetic trees, see Section 13.7.
Graphic by Gregory S. Gilbert.

plants) produce their seeds encased in ovules. While Gymnosperms are important components of some forest ecosystems, nearly all crops and most wild species are Angiosperms. There are an estimated 435,000 species of land plants (Enquist et al. 2019), which include about 20,000 non-vascular plant species and 1000 Gymnosperms, leaving over 400,000 Angiosperms; this reflects a tremendous diversifying radiation of Angiosperms.

Plants were traditionally placed into a systematic classification based on similarities in their reproductive structures. In recent decades, analysis of evolutionary relationships based on molecular data—mostly through comparison of DNA sequences—has confirmed much of the traditional, morphologically based classification system. However, this work has also shown how some traditional groups were **polyphyletic**, with similar structures that were the products of convergent evolution rather than common descent. Some of the polyphyletic groups actually had hidden subgroups with deeply divergent evolutionary histories that were not apparent just from their flowers. For example, the Monocots (Figure 2.11), were long thought to be a sister clade to the Dicots, and many generations of students learned about the monocots and dicots as a duality; however, we now understand the Monocot clade to be nested within multiple separate lineages that used to be grouped together within the Dicots. In other words, the traditional "dicots" are a polyphyletic grouping.

A large consortium of plant systematists called the Angiosperm Phylogeny Group was created to produce a consensus, higher-level classification of flowering plants, using mostly DNA sequence data. In 2016 they published the most recent framework, known as APG IV (Angiosperm Phylogeny Group et al. 2016). APG IV classifies Angiosperms into 64 orders that include 416 families. Having a classification system that reflects evolutionary relationships is valuable for many reasons. As we will see, close relatives tend to have similar traits—chemical composition, growth habits, environmental requirements, physiological mechanisms—that are important to ecological performance and to plant–pathogen interactions (Section 13.7).

As new molecular tools and more extensive sampling help systematists learn more about evolutionary relationships among taxa, the taxonomic names are revised. Sometimes a single genus or species is split into two, and sometimes multiple specimens that were described as different species turn out to be the same. Keeping track of the **synonyms** can be daunting, since there are currently over 350,000 scientifically accepted names for plants. Established in 2020, the go-to source for correct classification and taxonomic names of plants is the World Flora Online (http://www.worldfloraonline.org/). They keep track of all the name changes, so we don't have to!

This book often talks about **plant families**. Because so many important plant traits are held in common by many members of a family, botanists (along with ecologists, agronomists, gardeners, foragers, and all other types of plant enthusiasts) use plant families as a short-hand for the suite of traits that characterize plants of interest. For example, the Brassicaceae include the mustards, cole crops such as cabbage and broccoli, plus the model plant *Arabidopsis thaliana*, and they all share a pungent flavor that derives from isothiocyanates (Section 12.2). The Poaceae is the grass family and includes grains like wheat and barley, as well as many ecologically important species such as *Ammophila arenaria* (Section 14.7). The Fabaceae is the legume family, which includes beans, lentils, and clovers and is known for its nitrogen-fixing symbiosis with rhizobia (Section 15.3). The Solanaceae contains many species with toxic alkaloids, like deadly nightshade, locoweed, and tobacco, as well as potato and tomato (now doesn't that make you curious?).

Whereas all plants are susceptible to disease, some families of plants feature more prominently in this book than others. Many of the plant families you will encounter are presented in Figure 2.11, showing their evolutionary relationships following the APG IV framework. Some are of particular agronomic significance, like the Rosaceae, or of ecological significance, like the Pinaceae. Some genera and species have been developed as model systems due to particular traits that make them amenable to study, like *Arabidopsis*. Others have become important model systems for plant disease ecology because of the idiosyncratic histories of the individual scientists who studied them first, such as flax (*Linum* and allies), *Festuca*, *Silene*, *Plantago*,

Ammophila, Avena, Microstegium, Notholithocarpos, and *Trifolium.*

Well, that was a very brief review indeed, only scratching the surface of the many fascinating topics in plant physiology and development that form the basis of a mechanistic understanding of plant responses to pathogens. Later chapters return again to all of the material covered here.

Further reading

Levitin, E. and K. McMahon. 2020. *Plants and Society*, 8th ed. McGraw-Hill, New York.

Taiz, L., I. M. Møller, A. Murphy, and E. Zeiger. 2022. *Plant Physiology and Development*, 7th ed. Oxford University Press.

References

Angiosperm Phylogeny Group, M. W. Chase, M. Christenhusz, M. Fay, J. Byng, W. S. Judd, D. Soltis, D. Mabberley, A. Sennikov, P. S. Soltis, and P. F. Stevens. 2016. An update of the Angiosperm Phylogeny Group classification for the orders and families of flowering plants: APG IV. *Botanical Journal of the Linnean Society* **181**: 1–20.

Antonovics, J. 2005. Plant venereal diseases: insights from a messy metaphor. *New Phytologist* **165**:71–80.

Calo, L., I. García, C. Gotor, and L. C. Romero. 2006. Leaf hairs influence phytopathogenic fungus infection and confer an increased resistance when expressing a *Trichoderma* α-1, 3-glucanase. *Journal of Experimental Botany* **57**:3911–3920.

Card, S. D., M. N. Pearson, and G. R. G. Clover. 2007. Plant pathogens transmitted by pollen. *Australasian Plant Pathology* **36**:455–461.

Ellis, R. J. 1979. The most abundant protein in the world. *Trends in Biochemical Sciences* **4**:241–244.

Enquist, B. J., X. Feng, B. Boyle, B. Maitner, E. A. Newman, P. M. Jørgensen, P. R. Roehrdanz, B. M. Thiers, J. R. Burger, and R. T. Corlett. 2019. The commonness of rarity: Global and future distribution of rarity across land plants. *Science Advances* **5**:eaaz0414.

Hatsugai, N. and I. Hara-Nishimura. 2010. Two vacuole-mediated defense strategies in plants. *Plant Signaling & Behavior* **5**:1568–1570.

Heinlein, M. 2015. Plasmodesmata: Channels for viruses on the move. In: M. Heinlein, editor. Plasmodesmata: Methods and Protocols, pp. 25–52. Springer, New York.

Jones, D. L., C. Nguyen, and R. D. Finlay. 2009. Carbon flow in the rhizosphere: carbon trading at the soil–root interface. *Plant and Soil* **321**:5–33.

Serrano, M., F. Coluccia, T. Martha, F. L'Hardion, and J. Métraux. 2015. The cuticle and plant defense to pathogens. *Frontiers in Plant Science* **5**:6–13.

Stephenson, A. G. 2012. Safe sex in plants. *New Phytologist* **193**:827–829.

Yadeta, K. and B. Thomma. 2013. The xylem as battleground for plant hosts and vascular wilt pathogens. *Frontiers in Plant Science* **4**:97.

CHAPTER 3

How to be a fungus

Gregory S. Gilbert and Ingrid M. Parker

Fungi rule as the most important plant pathogens. The complexity of fungal life histories, as well as the sheer diversity of fungal species, is remarkable when you consider the simplicity of their basic approach to life. Fungi are **eukaryotes** that obtain their water and nutrition through direct absorption from the **substrate** in which they are growing; that means fungi live in and eat their way through their food. The basic body units of fungi are tubes called **hyphae** (singular, hypha) that grow, absorb nutrients, and form reproductive structures. Fungi reproduce by making **spores** that can disperse to fresh substrate or wait out periods of adverse conditions. At the most basic level you can think of fungi as little more than balls (spores) and tubes (hyphae). Very creative, resilient, beautiful, and diverse balls and tubes that are also the most important and common plant pathogens.

If we want to understand how fungi cause disease, we must first be comfortable thinking about how fungi make a living. This chapter provides a brief overview of how fungi grow, how they reproduce both asexually and sexually, and how they are affected by their environment. We also briefly survey the major taxonomic groups of importance in plant–fungal symbioses.

3.1 What are fungi made of?

Those balls and tubes are living, growing cells with complex structures that affect how they interact with their environment and with other organisms. Fungal hyphae have rigid cell walls like plants, but unlike in plants these cell walls are made of **chitin**, **β-glucans**, and **glycoproteins**, surrounding a semipermeable **plasma membrane** stabilized by **ergosterol**. Chitin is a long-chain polymer of *N*-acetylglucosamine that forms rigid, high-tensile crystalline nanofibrils; it is also found in insect exoskeletons, octopus beaks, and fish scales. Glycoproteins are threaded throughout the wall structure and anchor it to the plasma membrane. β-Glucans are also found in some bacteria (but without the 1,6 branches) and in cereals (alternating as β-1,3 and β-1,4 chains), but the forms found in fungi are unique to them. What is most important here is to note that the cell walls of fungi are made of very different materials than the cellulose-rich cell walls of the plants they attack.

3.2 How do fungi grow?

Spores germinate to produce a **germ tube**, which then grows to become a hypha. As the hyphae grow and fill with new cytoplasm, nuclei divide and a microtubule motor controls nuclear migration along the hypha to maintain a relatively constant density of nuclei.

Hyphae grow from their tips at rates as fast as several centimeters per day under optimal nutrient, temperature, and moisture conditions (Lew 2011). The leading hyphae explore the substrate in search of nutrients; older hyphae form an interconnected network. The sum of all the hyphae is called **mycelium** (Figure 3.1); we often use the term **thallus** to refer to the entire fungal body, and sometimes the term mycelium is a synonym for the fungal thallus. The high surface area of the mycelium is ideal for extracting nutrients from its substrate, which

The Evolutionary Ecology of Plant Disease. Gregory S. Gilbert and Ingrid M. Parker, Oxford University Press. © Gregory S. Gilbert & Ingrid M. Parker (2023).
DOI: 10.1093/oso/9780198797876.003.0003

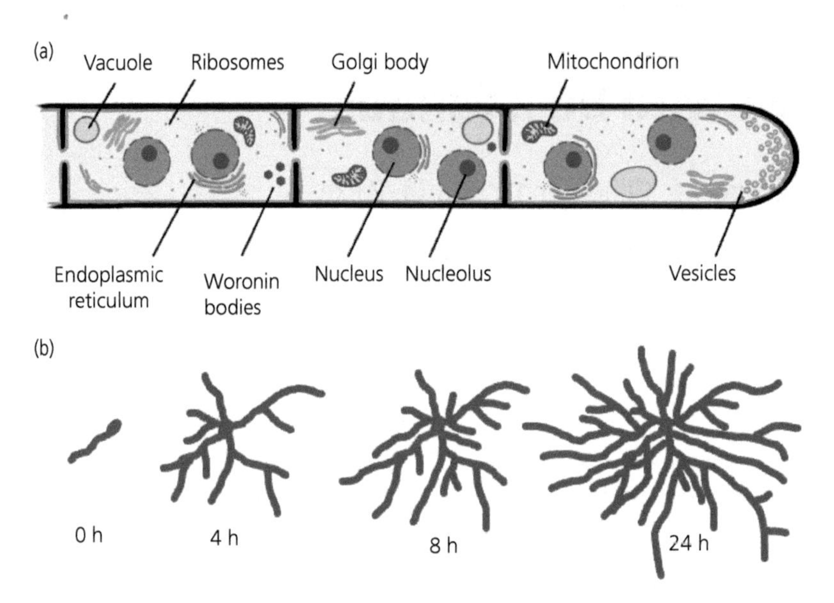

Figure 3.1 Structure and growth of fungal hyphae. (a) A fungal hypha, showing septa with pores, cell wall, and organelles. (b) Tracing of growing fungal mycelium over 24 h after a single spore germinated.
Drawings by Ingrid M. Parker and Gregory S. Gilbert.

fuels growth at the hyphal tips. Hyphae branch as they go, creating new hyphal tips that help optimize how well the fungus can explore the available substrate. A hypha does not stretch to grow, but rather wall vesicles deposit cell wall components at the hyphal tip; the cytoplasm, vesicles, and organelles move toward the growing tip through a combination of active transport along microtubules (and their associated molecular motors) and mass flow. Mass flow functions through an osmotic pressure gradient inside the hypha which pushes water into the hypha and toward the growing tip; this means that hyphae need moist conditions to grow. Nearly all plant pathogenic fungi are filamentous and follow this general growth pattern, although there are a few groups of odd-ball fungi that instead grow by budding (**yeasts**) or as single cells that alternate between motile and sessile phases (chytrids). When mycelium is growing in a laboratory on solid growth media it is often called a **colony**.

Hyphae of most fungi are not endless, open tubes, but instead are regularly divided by cross-walls called **septa** (singular is septum) (Figure 3.1a). However, the septa are incomplete—open pores in the septa allow movement of cytoplasm, nuclei, and most organelles. That makes fungi **functionally coenocytic**, meaning that sections between septa are not separated from adjacent sections by a plasma membrane. However, the pores between sections can be closed very quickly with floating plugs called Woronin bodies, effectively creating separate compartments in the hypha on either side of the septum. This provides a mechanism to prevent loss of cytoplasm if the hypha is damaged, and it allows the separate compartments to develop differentially and function independently when that is to the benefit of the fungus; this is key, for instance, in creating reproductive structures. Importantly, the organelles are not simply sloshing around randomly in a cytoplasmic soup, but their position and distribution is controlled through several mechanisms, including a microtubular motor system. For instance, the density of nuclei is highly regulated, often, but not always, with one nucleus per section.

3.3 How do fungi feed?

Remember, fungi live inside their food. Fungi are **osmotrophic** heterotrophs—they feed by secreting a diversity of extracellular **enzymes** into their substrate, which break down complex polymers like starches, cellulose, lignin, and proteins into simple sugars, amino acids, and fatty acids. Those simpler compounds are then brought inside the hyphae by membrane-bound transporter proteins (Talbot 2010) where they can be metabolized to produce energy or can be converted into structures and useful chemical products. Fungi absorb essential minerals (like N, P, K, Ca, Mg) in dissolved ionic forms from their substrate. These ions, as well as small organic molecules, can pass down a simple concentration gradient from outside the hypha (higher concentration) to inside (lower concentration) across the cell membrane by facilitated diffusion, moving through gated, water-filled pores created by transmembrane proteins. Some fungi produce **mycotoxins** that kill or weaken the host in which they live, making the host an easier environment to colonize and consume. Many fungi excrete organic acids that make minerals more soluble and more readily absorbed; this is part of how **mycorrhizal fungi** help provide their host plants with greater access to soil nutrients (Landeweert et al. 2001).

The key to the efficiency of fungal foraging for nutrients is the very high surface area of the hyphal structure. Fungal hyphae have an average diameter of about 5 μm, much smaller than even a plant root hair, which may measure 15 μm across and grow to 1500 μm long. A single hypha can grow 1500 μm in an hour, and nine hyphae can fit in the same cross-section area as a single root hair. The greater surface-to-volume ratio of fungal hyphae than plant root hairs gives a three-fold greater absorptive surface area in fungi than in plants in the same volume of substrate.

Hyphae grow so the fungus can either obtain more nutrients or reproduce. As a fungus consumes nutrients, the concentration of nutrients in the substrate around it will become depleted. This will create a nutrient gradient, with greater concentration of nutrients at a distance than close by. Fungal foraging efficiency is boosted because fungi can detect chemical gradients and grow towards greater concentrations of nutrients. Such directional growth along a chemical gradient is called **chemotropism**. Growth toward a chemical stimulus is positive chemotropism, such as growth toward a leaking wound on a plant root; growth away is negative chemotropism, such as growth away from toxic chemicals.

Because hyphae branch as they grow, each branch can explore the substrate separately and contribute to the nutrition of the mycelium as a whole. If different hyphae end up in different environmental conditions that demand different functions, **gene expression** can vary in different parts of the same mycelium. For instance, the "Honey fungus" *Armillaria mellea*, an important pathogen of trees, will produce (1) mycelium that causes white rot inside the woody trunk, (2) a cream-colored mycelial fan scavenging on the surface of the wood under the trunk, (3) thick black rhizomorphs that extend through the soil to infect neighboring trees, and (4) a cluster of mushrooms. All four of these components are made of hyphae that comprise one mycelium, just expressing genes differently in different places. Each piece of hypha in this mycelium can independently change gene expression and grow to create any other component.

Cut a worm in half and only the head part will continue to grow and thrive, because the tail can't regenerate what is needed in the head. But all parts of a fungus are **totipotent**, like a stem cell; even the smallest piece of a fungus can grow to complete a fungal life cycle. This means that all parts of a fungus can serve as **propagules**, defined as organisms or parts of an organism that can give rise to a new individual in a new place. Mycelium embedded in plant tissue, or hyphal fragments growing in soil particles, can help disperse fungi or allow them to survive a cold winter or a drought. Even just a hyphal tip—the single section at the growing point of a hypha—can grow to form mycelium if it finds itself with a nutritious substrate and moisture. Once growing, hyphae generally have **indeterminate growth,** so a single clonal mycelium has the potential to grow for a long time and reach huge proportions. At the extreme, a single clonal individual of *Armillaria gallica* covered 15 ha, weighed an estimated 10,000 kg, and remained genetically stable for at least 1,500 years (Smith et al. 1992). Such **clonal** reproduction with indeterminate growth is not only important in the ecology of plant pathogens but is useful to plant pathologists (Box 3.1).

Box 3.1 Growing fungi in the lab

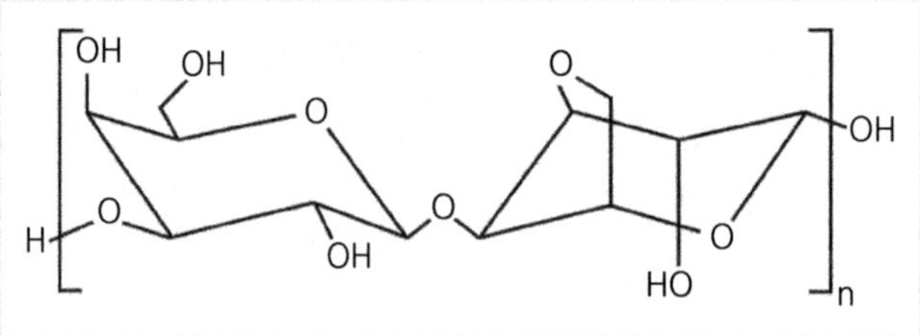

Figure 3.2 One unit in the repeating chain of an agarose molecule.
Figure by Gregory S. Gilbert; created using BioRender.com.

Many kinds of fungi are easy to grow in a laboratory, which is useful when you want to work with them in experiments. At a minimum, fungi need access to a carbon source, a nitrogen source, and water. Mycologists often use extracts of natural substrates to provide the necessary nutrition; three of the most common growth media for fungi include potato-dextrose agar, made from the broth of boiled potatoes plus some glucose; malt extract agar, based in an extract of germinated grains; and corn meal agar, made from boiled corn meal. The extracts provide all the nutrients many fungi need to live; agar is added as a gelling agent to make it easier to handle the fungi.

Agar is a natural vegetable gelatin that is extracted from the cell walls of some Rhodophyta (red algae); it is a mixture of long-chain polysaccharide agarose (Figure 3.2) and short-chain agaropectin. Agar has two properties that make it useful for preparing growth medium for fungal cultures. First, whereas agar dissolves in water when heated above about 85° C, it does not solidify until it cools to about 50° C, when it traps water within the coiled agar molecules to form a gel. This asymmetry in melting and solidifying temperatures is called **hysteresis**, and it makes it particularly useful for creating solid media with high water content at normal growth temperatures (from freezing to 40° C). Second, very few fungi or bacteria can break down agar, so it retains its structure even while microbes are consuming the other nutrients mixed into the agar growth medium. Gelatin, made from animal products, can be degraded by many fungi and bacteria as they grow, turning the growth medium into a soupy mess.

A small bit of fungal hyphae or spores (**inoculum**) can be transferred to agar growth media, allowing the fungus to grow in pure culture, away from the host plant. The hyphae grow into the medium, secreting enzymes and absorbing nutrients. As nutrients are consumed, the mycelium expands radially as hyphae grow toward areas with higher nutrients (Figures 3.1b and 3.3). Many fungi will produce asexual or sexual spores when growing in the agar growth medium, facilitating identification or the production of inoculum for experiments.

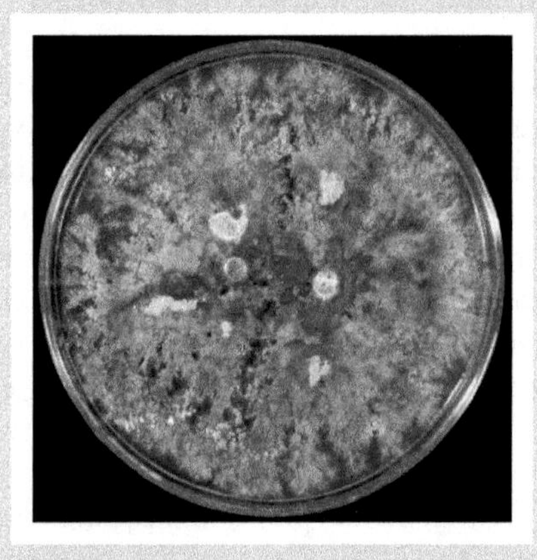

Figure 3.3 Colony of the fungus *Entoleuca mammatum* growing in a Petri dish filled with malt extract agar growth medium.
Photo by Gregory S. Gilbert.

Not all kinds of fungi can be grown in culture; for instance, obligate parasites including the rusts, powdery mildews, and arbuscular mycorrhizae cannot easily be grown away from their host plants. In some cases, such fastidious fungi can be grown by embedding pieces of sterilized plant tissue in agar or even by placing inoculated pieces of sterilized, excised plant tissue in a moist chamber, such as a plastic bag or box.

3.4 Spores for fungal dispersal and survival

Most fungi produce one or more kinds of spores (Figure 3.4). Spores are propagules that contain everything a fungus needs to produce a new hypha: a nucleus, other essential organelles, nutrient reserves, and cytoplasm, surrounded by a plasma membrane and cell wall. Some spores are unicellular with a single nucleus, some unicellular with multiple nuclei, and others are septate with multiple nuclei. Fungal spores are often uninucleate or, if septate with multiple nuclei, are usually **homokaryotic**, meaning they contain nuclei of only one **haplotype**, and multiple nuclei in a spore are mitotic clones, as in a human embryo. Note that in a few fungi, notably the mycorrhizal Glomeromycota, spores can be multinucleate and **heterokaryotic** (a collection of nuclei of different haplotypes). Spores serve for both dispersal and survival, and may be the product of mitosis or meiosis (Section 3.6). Many fungi can produce multiple kinds of spores with different traits that determine their ecological functions.

Spores usually have low water content and a very low rate of metabolic activity, which allows them to remain dormant until they encounter appropriate moisture and nutrient conditions. They often have large concentrations of nutrient reserves stored as lipids, glycogen, and trehalose. Trehalose, which is a disaccharide formed from two glucose units, is not only a nutrient reserve, but it also protects organelles from freezing or desiccation by retaining water to form a protective gel phase inside the spore. When the spore is rehydrated or thawed, the trehalose can release the water, organelles can resume their functions, and the trehalose can be consumed as an energy source. Spores with thick walls, dark pigmentation that protects against UV damage, and chemical traits that permit extended dormancy are associated with long-term survival, with some spores remaining viable and dormant for decades. Some fungi produce **chlamydospores**, which are large, thick-walled resting spores, often produced asexually by modifying part of a hypha. Some rust and smut fungi (which are basidiomycetes) produce a specialized chlamydospore called a **teliospore**, which serves both for survival and for the initiation of sexual reproduction. The impressive lifespan of these teliospores was demonstrated by Garber et al. (1978), who were able to grow a viable fungal culture from an herbarium specimen infected with anther smut (see Figure 2.9) collected in 1893!

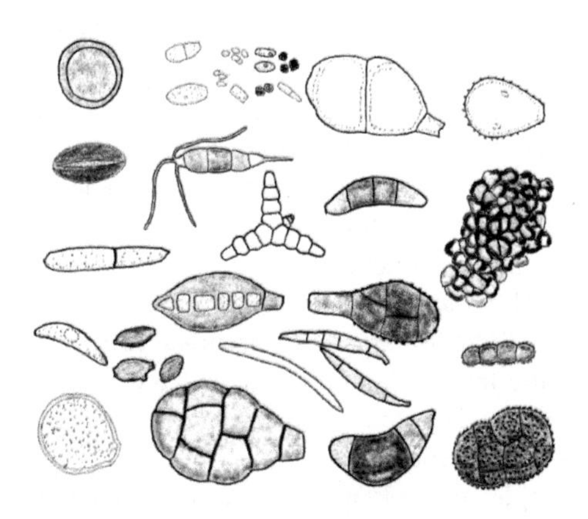

Figure 3.4 Fungal spores come in a tremendous diversity of sizes and shapes.
Drawing by Gregory S. Gilbert.

Other spores may be short-lived and function mainly for short-distance dissemination and infection of new hosts. Spores can have a wide variety of traits that facilitate dispersal (Figure 3.4). Small, dry spores, sometimes with rough or ornamented surfaces, are readily taken up in air currents and dispersed passively. Spores with appendages are particularly suited for dispersal in moving water. How and where spores are produced also affects how they are dispersed, with some spores ejected forcibly with hydraulic pressure for short-distance dispersal, others released in clouds of dry masses for wind dispersal, and others splashed out by rain drops. Sometimes spores are produced in sticky masses that are readily picked up by insect or other animal vectors. How and where spores are produced (e.g., embedded in plant tissue or on leaf surface), as well as the traits of the spores themselves, give clues to the ecology of that fungal species.

3.5 Hyphal fusion and genetic diversity

Spores of fungi are usually **unicellular** and **uninucleate** (single cell with a single nucleus), or if

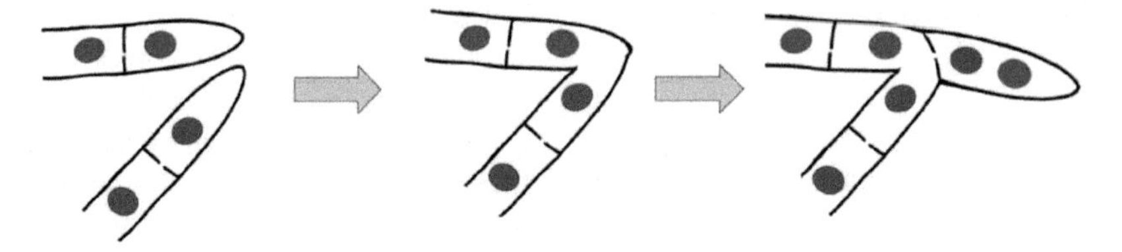

Figure 3.5 Formation of a dikaryotic hypha from the anastomosis of two homokaryons. From left, hyphae show chemoattraction of two homokaryotic hyphae, followed by initial contact, and anastomosis with plasmogamy. The hypha then continues to grow as a dikaryon.
Drawing by Ingrid M. Parker.

multicellular, they are **homokaryotic** (nuclei are mitotic copies). The hypha produced from this spore is then homokaryotic, and each nuclear mitotic division produces genetic clones as the hypha grows. So how do fungi swap genes with other individuals to generate and maintain genetic diversity? Hyphae can sometimes fuse with genetically different hyphae of the same fungal species. As mycelium grows through the substrate, the hyphae are likely to encounter other hyphae that grew from different spores of the same species. If the hyphae are of compatible types (**vegetative compatibility** is determined by having matching alleles at several genetic loci), they may undergo **anastomosis**, or hyphal fusion. In this process, cell walls break down, plasma membranes fuse, and the cytoplasm from the two hyphae merge, a process called **plasmogamy** (Figure 3.5).

Anastomosis means that not only do the two cytoplasms merge, but the nuclei of two different genetic individuals (different **haplotypes**) then occupy the same hypha. In this way, homokaryotic mycelia have now fused to create a **heterokaryon.** When the resulting heterokaryon has exactly two nuclei per cell, it is called a **dikaryon**. The nuclei remain separate and haploid, replicate through mitosis, and can each express the genes they carry. Fungi in the phyla Basidiomycota and Ascomycota, which include most of the fungal diversity and the great majority of fungal plant pathogens, comprise the subkingdom Dikarya (Figure 3.6). The subkingdom takes this name because the Dikarya share an important **dikaryotic** state (heterokaryosis with one haplotype from each of two parent hyphae) in their life cycles. In humans, it would be as if the sperm and egg came together to produce a structure in which their two haploid nuclei coexisted but did not fuse to make a diploid zygote. Many ascomycetes can grow well as homokaryons for most of their life cycle, and most form dikaryons only when producing sexual reproductive structures. Basidiomycetes, on the other hand, grow poorly as homokaryons, and spend most of their life cycle as stable dikaryons. Both phyla have specialized structures to maintain the dikaryons (**croziers** in Ascomycota and **clamp connections** in Basidiomycota). Clamp connections associated with septa in dikaryotic hyphae in many Basidiomycetes provide exquisite control of nuclear distribution, ensuring that, as dikaryotic hyphae grow, each new section contains exactly one of each of the different types of nuclei. A few Basidiomycetes, including the humungous fungus *Armillaria gallica*, have fused nuclei and spend most of their life cycle as diploids, instead.

Genetic diversity can also accumulate through mutation, such that a homokaryotic mycelium derived from a single spore can still give rise to heterokaryons. The hyphae of fungi are filled with haploid nuclei that are individually replicating and expressing their genes; the genetic diversity among the nuclei can provide flexibility and redundancy. Even a lethal **recessive mutation** in a gene could persist and spread through a growing mycelium if it is masked by functioning genes in the **wild-type** nuclei. The mutant nuclei can continue to replicate until they are isolated into homokaryotic spores, when the lethal mutation is uncovered; therefore, production of haploid spores is important in weeding out deleterious mutations.

A single, heterokaryotic mycelium can benefit from expression of genes from different haplotypes in different hyphae. The mycelium may be growing simultaneously under a range of different environmental conditions, and a particular genetic variant may be beneficial in one part of the mycelium, while a different variant is beneficial under different conditions. This provides the opportunity for **sectoring**, which results from the appearance of genetically distinct morphs, separation of homokaryons from heterokaryons, or different ratios of component nuclei in different parts of the same mycelium (Maheshwari 2005).

Compare this heterokaryotic existence, where homologous chromosomes are separated into separate haploid nuclei that co-habit the same hypha, to that of diploid organisms including most animals and plants, where two homologous chromosomes are held within a single nucleus. In diploids, the dominance relationship between alleles at the same locus on the homologous chromosomes determines what gene products are made, and thus the phenotype of the organism. It is fascinating to think about how gene regulation in fungi is different; how physiology, growth, and the basic functions of the cell are controlled by the genes of both nuclei, and how they interact.

3.6 Sexual and asexual reproduction

Recombination, which only occurs during sexual reproduction, both drives and constrains many evolutionary processes (Chapter 13). Therefore, understanding whether, where, how, and how often a fungal species reproduces sexually is important. For sexual reproduction to occur, first the two haploid genomes need to be in the same hyphal segment (accomplished, as we've seen, through plasmogamy), and then the two haploid genomes need to fuse into a single **diploid** nucleus, which is **karyogamy**. In basidiomycetes, karyogamy is long delayed after plasmogamy (Figure 3.7), and the somatic mycelium is nearly always dikaryotic (each section with two nuclei). Ascomycetes, on the other hand, often have extensive, long-term mycelial growth as homokaryons (each section with one nucleus) (Figure 3.8). Plasmogamy to create dinucleate hyphal segments is then needed to initiate sexual reproduction in ascomycetes, and is followed soon after by karyogamy.

Fungi do not have males and females, but rather two or more **mating types** that are determined by alternate alleles for a mating-type gene (or genes). To accomplish karyogamy, two fungal individuals must be sexually compatible, which means the two (haploid) nuclei must have different mating-type alleles. Most Basidiomycetes have two mating types (**bipolar**). Some mushrooms are **tetrapolar**, with mating-type alleles distributed across two genes. Ascomycetes can have anywhere between one and many mating-type genes. Having more mating-type genes provides more possible sexually compatible combinations—multiple (sometimes even many) sexes! Some **homothallic** fungi are able to undergo self-fertilization, although this is not as common as it is with plants (Section 2.5).

Fungal spores can be either sexual (**meiospores**) or asexual (**mitospores**). Meiospores contain nuclei that are the products of meiosis from a **diploid** nucleus. Meiospores are the products of sexual recombination between the two parent strains (**haplotypes**), and so hyphae that grow from meiospores are genetically different from their parents and from each other. Meiosis occurs within different structures and at different points in the life cycle in different groups of fungi, and we will go over some of the details for each group below (Section 3.7).

Unlike meiospores, mitospores are clones of the parent haploid nuclei. Mitospores (also called **conidia**) are produced when haploid nuclei divide mitotically to make identical copies, which are then packaged up inside a plasma membrane and cell wall together with cytoplasm, organelles, and nutrients. They are generally simple to make quickly and in large quantities, and they can be produced in a wide variety of ways. The fungal spores that cause allergies are mostly airborne mitospores produced in great quantities. Most ascomycetes make mitospores, but only a few basidiomycetes do.

For many fungi, the asexual form is much more commonly encountered than the sexual form. Often fungi were even given different taxonomic names for the two forms, not always knowing they were the same organism! This two-name tradition can cause confusion when reading the literature or conversing about pathogens (Box 3.2). Previously, fungi that were only known to make mitospores (and never meiospores) were called **deuteromycetes** or **imperfect fungi** (imperfect because they lacked a sexual stage). The terms "deuteromycetes" and

Box 3.2 Why do fungi often have two names?

Like plants, Fungi have traditionally been classified and named according to the reproductive structures they make. From a systematics perspective, this makes sense because sexual reproductive structures are often broadly evolutionarily conserved traits (flower traits are the basis of much classification in plants). But many fungi are much more commonly seen reproducing asexually than sexually, and it is not always clear that a particular sexual form is produced by the same fungal entity as a particular asexual form. As a result, many fungi have historically been given two names, one applied to the asexual form and one applied to the sexual form. For instance, an important pathogen of clovers produces abundant mitospores on a type of conidiophore that has traditionally been associated with the name *Stemphylium botryosum*. The same organism will occasionally reproduce sexually, producing an ascoma that has been associated with the name *Pleospora herbarum*. There are long histories in the literature for each of these names and they have been useful for communications. Although we know that they are just two different reproductive strategies in the life cycle of the same fungus, mycologists adopted a two-name system, where the mitosporic state was called the **anamorph** and the meiosporic state was called the **teleomorph**. To refer to the whole fungus (both states together), they used the term **holomorph**, and the holomorph always took the name of the teleomorph. Thus, in our example, the correct holomorph name of the fungus would be *Pleospora herbarum*. One would then talk about the *Stemphylium* state of *Pleospora herbarum*.

This double-naming solution had its problems. Most importantly, although the sexual reproductive structures were generally pretty good indicators of evolutionary relationship, the asexual structures were not—the same type of conidiophore and conidia might show up associated with a number of very different types of ascomata; that is, there was a lot of convergent evolution in asexual reproductive structures. That means that while asexual taxonomy was useful for communication, it was terrible as a proxy for systematics based on evolutionary relationships. So finally, in the 2011 Melbourne Code, a revision of the *International Code of the Nomenclature for Algae, Fungi and Plants* adopted the radical "One fungus, one name" rule, eliminating the labels of anamorph and teleomorph. This is a tremendous step forward, but the transition is far from smooth or complete, with a great deal of committee work still to come. Sorting out the correct "one name" is a major undertaking and includes not only logistical issues (there are nearly 100,000 named fungi!) but also a number of tricky questions such as exactly how to assign a name when only DNA evidence, and no physical sexual state, suggests placement of a fungal taxon into a genus that is otherwise delimited by a particular sexual reproductive structure. So, since the existing literature uses a mix of names, and the new nomenclature (naming rules) is likely to continue settling out for a number of years, we need to be able to work with the different combinations of names that come our way.

In the meantime, there are some great resources to help sort out fungal names, with three official repositories of fungal names:

Index Fungorum (http://www.indexfungorum.org/, Royal Botanic Gardens, Kew, UK)

MycoBank (http://www.mycobank.org, International Mycological Association, Centraalbureau voor Schimmelcultures in The Netherlands)

Fungal Names (https://nmdc.cn/fungalnames/, Institute of Microbiology, Chinese Academy of Sciences)

"imperfect fungi" are no longer used, but we mention them because you encounter them in older scientific literature.

It is important to remember that many plant pathogenic fungi have both asexual and sexual reproduction happening regularly—sometimes simultaneous, sometimes sequentially. Both are intrinsic parts of complex life cycles, playing important roles in pathogen survival, dispersal, and infection of the host.

3.7 Major groups of Fungi

The kingdom Fungi is highly diverse, and it contains the most pathogens, the most well-studied pathogens, and arguably the most economically and ecologically important pathogens of plants. About 100,000 species of Fungi have been named by **mycologists**, but based on the rate of discovery of previously unknown species, estimates suggest there are probably a few million species altogether.

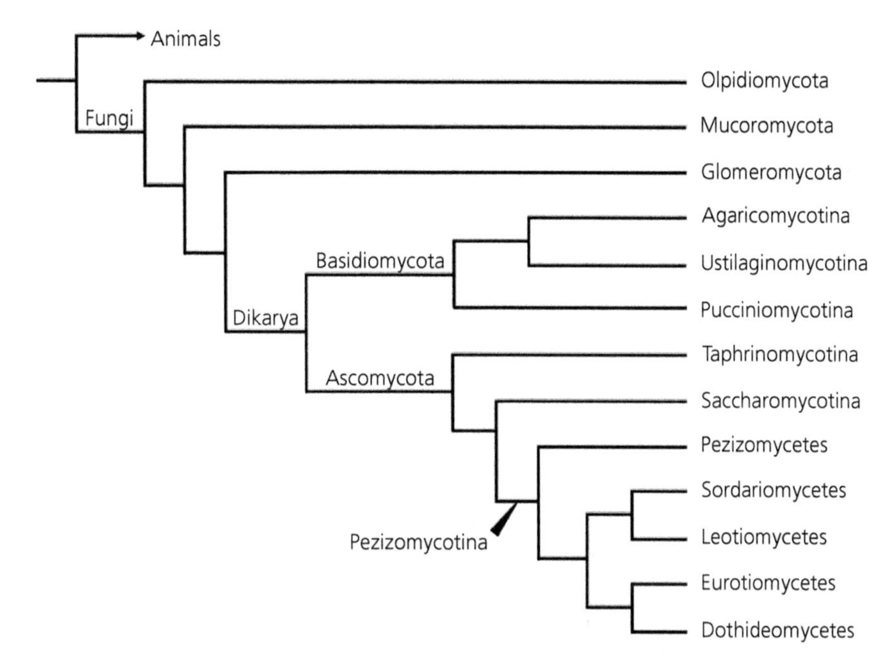

Figure 3.6 Higher-level phylogenetic arrangement of the major groups of Fungi engaged in plant–fungal symbioses. The primitive Olpidiomycota are the only Fungi that retain motile zoospores. Mucoromycota and Glomeromycota (which are sometimes considered part of Mucoromycota) lack septa and so have coenocytic hyphae. The Dikarya have septate hyphae that are dikaryotic in critical phases of the life cycle. The subkingdom Dikarya includes two divisions, the Basidiomycota and the Ascomycota. The Basidiomycota include the mushrooms and polypore fungi (Agaricomycotina), as well as two major groups of obligate biotrophs, the rusts (Pucciniomycotina) and the smuts (Ustilaginomycotina). The Ascomycota include the evolutionarily basal Taphrinomycotina, the Saccharomycotina (yeasts), and the tremendously diverse Pezizomycotina, which includes the vast majority of all plant pathogens. Branch lengths on the tree are not to evolutionary scale.
Figure by Gregory S. Gilbert.

Five divisions (equivalent to phyla, ending in -mycota) are of particular interest in the ecology of plant–fungal interactions. The Basidiomycota and Ascomycota (which together form the Dikarya) include most plant pathogenic Fungi, while three groups of more primitive Fungi (the Olpidiomycota, Mucoromycota, and Glomeromycota) also include important plant symbionts (Figure 3.6). Without attempting to survey the entire fungal kingdom, let's take a look at the major traits of these five divisions.

3.7.1 Basidiomycota

Quick, draw a picture of a fungus. Chances are you drew a mushroom, the sexual reproductive structure of a fungus in the subdivision of **Basidiomycota** called the **Agaricomycotina**. From a dikaryotic mycelium with sexually compatible nuclei, Basidiomycota produce sexual spores on specialized cells called **basidia** (Figure 3.7), within which karyogamy takes place. The resulting diploid nucleus then undergoes meiosis, producing four haploid nuclei. These four nuclei each migrate to four horns on the basidium (called **sterigmata**, singular sterigma), where they are packaged into uninucleate **basidiospores** and released, ejected with hydrostatic pressure, and then dispersed passively through the air.

In the Agaricomycotina the basidia are produced in one of three types of macroscopic **basidiomata**: in mushrooms (**agarics**), basidiospores are produced by a **hymenium** (single layer) of basidia on the surface of gills or pores on the underside of an open cap; in puffballs or **gasteroid** (stomach-like) fungi, basidiospores are produced inside a closed spherical body; and in **polypores** (which are often tough or woody, variously called bracket fungi, shelf

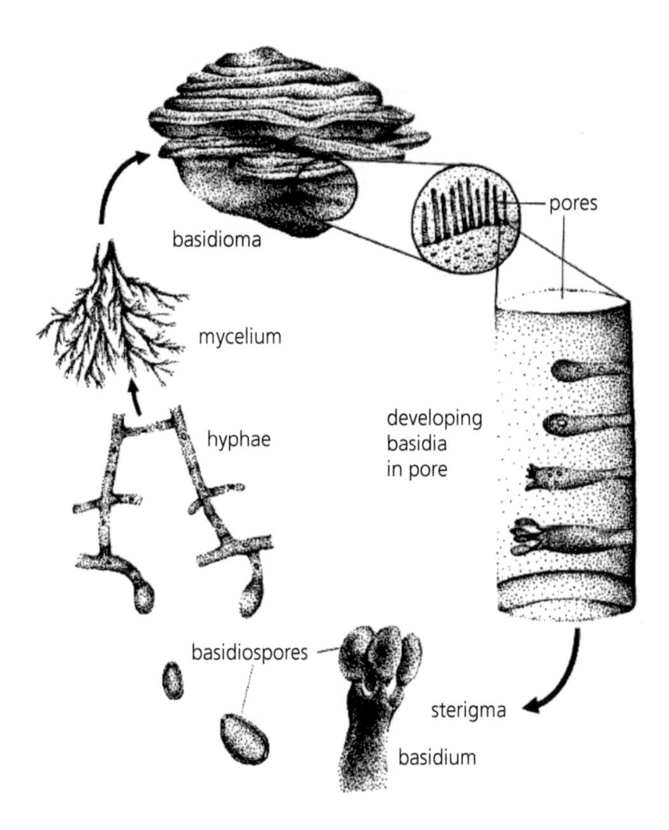

Figure 3.7 Generic life cycle and sexual reproduction of a Basidiomycete Fungus.
Drawing by Josh Zupan.

fungi, conks, or in Spanish, *orejas de palo*—tree ears), basidiospores are produced in pores or tubes on the underside. Whereas most Agaricomycotina are saprotrophs, some produce **ectomycorrhizal symbioses** with tree roots, and a few are important plant pathogens, like *Moniliophthora perniciosa* on cacao. Polypore fungi are important pathogens and decomposers of trees, in part because they are the only organisms able to degrade lignin, an important component of wood. The life cycle of a typical fungus in the Agaricomycotina is detailed in Figure 3.7.

The **smuts** (**Ustilaginomycotina**) are Basidiomycete fungi that specialize in growing on plant reproductive parts. Smuts grow as single, budding cells called **sporidia** that can fuse to form dikaryotic hyphae. The hyphae grow through the plant host as obligate biotrophs, eventually producing a gall that fills with asexual **teliospores**, a kind of thick-walled resting chlamydospore. The teliospores are dispersed to new plants through the wind, or sometimes by insect vectors. Under some conditions the teliospores germinate to produce a basidium, within which karyogamy and then meiosis occurs; the resulting haploid nuclei are packaged into uninucleate sporidia, which are functionally basidiospores.

The last group of Basidiomycetes are the **rust fungi** (**Pucciniomycotina**), the Rube Goldberg machines of the fungal world. They are all obligate biotrophs, and they may require two different, distantly related hosts and produce up to five different kinds of spores to complete their life cycle! Such a rust is **heteroecious** (two obligate plant host species) and **macrocyclic** (all five spore types) (Box 3.3). Some macrocyclic rusts are **autoecious**, producing all five spore types on a single host species. Other rusts are **microcyclic**, in that they do not produce pycniospores or aeciospores; because they lack an alternate host, these species are always autoecious. Other rusts lack the repeating stage of urediniospores, and are termed **demicyclic**.

Box 3.3 Life cycle of wheat rust

The life cycle of the wheat rust *Puccinia graminis* f. sp. *tritici* is the classic representative of a **macrocyclic** (all five spore types) **heteroecious** (two obligate plant host species) rust. **Teliospores** are the resting spores where the fungus can survive away from a living host, as well as the platform for sexual reproduction. Teliospores start off dikaryotic, but then karyogamy takes place within the spore to create diploid nuclei. The teliospore germinates to produce a basidium where meiosis takes place, resulting in four haploid basidiospores. Basidiospores infect leaves of the **alternate host** (or **aecial host**), producing monokaryotic hyphae that organize to produce **pycnia** that emerge from the leaf tissue. Haploid **pycniospores** are produced by the pycnia, which disperse; when they land on receptive hyphae on a pycnium of a different mating type, the pycniospore and receptive hyphae fuse to create a dikaryotic mycelium that extends through the leaf tissue. This mycelium then produces **aecia**, structures open to the air in which are produced dikaryotic **aeciospores**, which then disperse to the primary host. On the **primary host**, the dikaryotic mycelium produces **uredinia**, which generate dikaryotic **urediniospores**, which reinfect the primary host (the same individual or other individuals of the same species). This is often called the "repeating" stage. At the end of the host growing season, the rust produces **telia**, which produce dikaryotic teliospores as resting structures, setting the stage for sexual reproduction. The host on which the telia are produced is designated the primary host.

Trying to remember the spore types of heteroecious rusts can be bewildering, but mnemonics come to the rescue. The **A**lternate host is the one on which **A**ecia are produced. Aecia produce **A**eciospores which are **A**way spores, dispersing to infect the primary host. **U**redinia produce **U**rediniospores on the primary host, which then go back to reinfect the same primary host species—the **U**rediniospores reinfect yo**U**. The **T**elia produce **T**eliospores, which are **T**ime-out or resting spores, away from the host. Inside the **T**eliospores the haploid nuclei get **T**ogether through karyogamy to have sex. Sexual reproduction leads to **B**asidiospores, which then **B**egin the cycle again by infecting the alternate host (which conveniently, in the case of wheat rust, is Barberry). The monokaryotic mycelium produced from basidiospores produce **P**ycnia, which swap **P**ycniospores in order to **P**ick their mates by forming dikaryons. Got it?

3.7.2 Ascomycota

Ascomycetes are called sac fungi because they produce their meiospores (the sexual spores), called **ascospores**, in sacs called **asci** (Figure 3.8). Remember that ascomycetes spend much of their time as homokaryon mycelium, with a single haploid nucleus per hyphal segment. Plasmogamy leads to dinucleate hyphae, which are **ascogenous,** meaning they are capable of producing sexual ascospores; these fertile cells are intermixed and surrounded by non-fertile homokaryotic hyphae, and together they create a fruiting body called an **ascoma** (plural **ascomata**). Ascomata are generally small—barely visible to the naked eye—but can contain thousands of ascogenous hyphae arranged in a hymenium (a layer across the bottom of the ascoma). Karyogamy leads to diploid zygotes within individual cells of the ascogenous hyphae. Each zygote elongates into a sac-shaped **ascus**, and the nucleus undergoes meiosis to produce four haploid nuclei. Each haploid nucleus then undergoes mitosis, and the resulting haploid nuclei are packaged as individual **ascospores**—usually eight to an ascus. The ascospores are released (or ejected) from the ascomata and disperse.

The mitospores (asexual spores) are called conidia. Ascomycetes often produce conidia directly on hyphae (called **conidiophores**) that are not much different from everyday hyphae. Fungi that produce mitospores directly from mycelium in that way are given the informal names of **molds** or **hyphomycetes**. Those airborne mitospores are the spores responsible for "mold allergies."

Mitospores are produced in two basic ways, thallic or blastic, with lots of variations. **Thallic** mitospores are produced when complete septa divide up a hypha into small sections that then break apart as spores. ("Thallic" because the whole fungal body converts into a bunch of spores.) **Blastic** mitospores are extruded or ejected (blasted!) out of the conidiophore hyphae through small pores, through small side-tubes called **phialides**, or

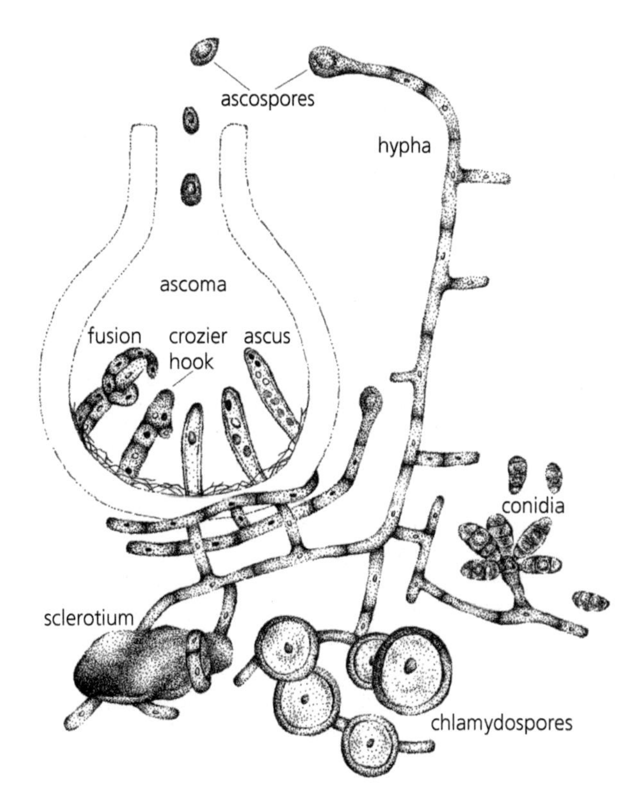

Figure 3.8 Generic life cycle and sexual reproduction of an Ascomycete Fungus.
Drawing by Josh Zupan.

sometimes on slightly more specialized, branching hyphae (Figure 3.9a, b, and c). Some fungi produce small, flask-shaped structures called **pycnidia** (barely visible to the naked eye and made of pigmented fungal hyphae, Figure 3.9d), or **acervuli**, which are structures made of a mixture of fungal hyphae and plant parts (Figure 3.9e); mitosis then happens inside the pycnidium or acervulus, and the resulting nuclei are packaged into mitospores. Mitospores can take a wide variety of forms from single-celled, colorless, smooth conidia to large, multicellular structures with dark pigments. Mitospores are usually homokaryotic, packaging individual nuclei into separate spores; one result is to break up heterokaryons as the mitospores disperse.

The division Ascomycota is divided into three subdivisions: the Taphrinomycotina, the Saccharomycotina, and the Pezizomycotina. Each of the three subdivisions contain plant pathogens, although by far most plant pathogens are in the Pezizomycotina. Here we very briefly review some of the main characteristics of the three subdivisions and a few of the major classes of fungi within the Pezizomycotina (Figure 3.6).

The **Taphrinomycotina** are called the leaf-curl fungi and are evolutionarily basal to the rest of the Ascomycota. These include a small number of plant pathogens (e.g., *Taphrina*) that manipulate the growth of their plant hosts to cause galls and deformed leaves, producing primitive asci naked on the leaf surface.

True yeasts (**Saccharomycotina**) grow mostly asexually by budding a new cell out from an older cell, rather than by making hyphae. Some yeasts are dimorphic and can grow either as yeasts or as filamentous fungi. This subphylum includes familiar yeasts like baker's yeast (*Saccharomyces cerevisiae*) and *Candida albicans* (which infects humans), many saprotrophs, and a few plant pathogens, such as *Dipodascus geotrichum*, which causes sour rot of post-harvest tomatoes.

The **Pezizomycotina** is the largest subphylum of fungi and includes all the fungi that produce

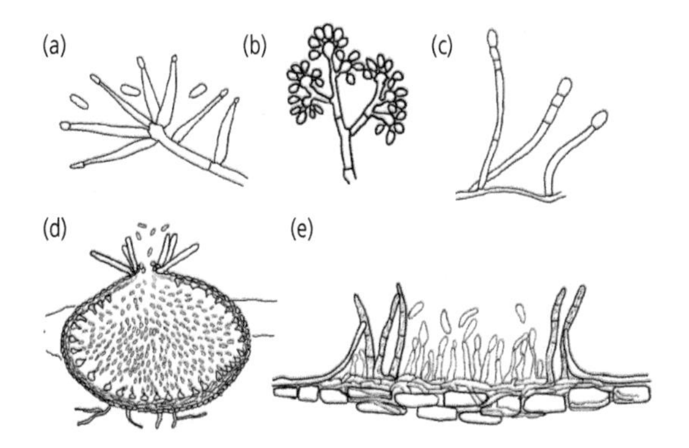

Figure 3.9 Some examples of common types of mitospore (conidium) production. Mitospores are sometimes produced from conidiophores that look only a little different from growing hyphae. (a) *Verticillium* produces conidia from whorls of elongated phialides. (b) *Botrytis* (anamorph of *Botryotinia*) produces branching conidiophores with swollen apical ampullae from which conidia are produced. (c) The powdery mildew *Oidium* (anamorph of *Erysiphe*) makes chains of conidia with the oldest conidium at the end of the chain. In other mitosporic fungi, the conidiophores are produced inside discrete structures. (d) *Phoma* (anamorph of *Didymella*) produces conidia from flask-shaped phialides that line the walls of a pycnidium; the pycnidium walls are made of fungal hyphae and are often embedded in plant tissue with the opening exposed to allow for mitospores release. (e) *Colletotrichum* (anamorph of *Glomerella*) produces masses of conidia in an acervulus, which erupts through and is delimited by the plant epidermis. Unlike pycnidia, acervuli do not have a defined wall of fungal hyphae, but do often have long, dark sterile hyphae sticking up near the edges.
Drawings by Gregory S. Gilbert.

ascomata for sexual reproduction. Because of their overwhelming importance as plant pathogens, we will look at five classes (ending with -mycetes) within the subphylum.

Fungi in the **Pezizomycetes** usually make fleshy ascomata that are disk-shaped **apothecia**, with asci in a hymenium on the upper surface; ascospores are discharged forcibly by rain drops. This group also includes some mycorrhizal fungi with modified ascomata that form a closed structure, including the hypogeous (below ground) structures known as truffles. This class includes many saprotrophs, but also pathogens like *Rhizina*, which causes root rot of many conifer seedlings.

Like the Pezizomycetes, almost all the **Leotiomycetes** produce apothecia. They include many important plant pathogens including *Botryotinia*, *Monilinia*, and *Sclerotinia* in the Letoiales, and *Rhytisma* in the Rhytismatales. Fungi in one order of Leotiomycetes, the Erysiphales (the **powdery mildews**), produce **chasmothecia**, closed spherical ascomata that produce asci inside.

The **Sordariomycetes**, which are sometimes called Pyrenomycetes, are an incredibly diverse group of fungi, with some 1300 genera! Fungi in this group have every possible fungal life history, but they are united by producing flask-shaped **perithecia** (or rarely spherical chasmothecia) as their ascomata. Single-walled (unitunicate) asci are arranged in a hymenium along the inner surface of the perithecium. The class includes a tremendous number of plant pathogens; some of the more notorious include *Nectria*, *Magnaporthe*, *Gibberella*, and *Claviceps*.

Like the Erysiphales (powdery mildews), the **Eurotiomycetes** usually produce chasmothecia as their ascomata. The most well-known and important fungi in this class belong to the order Eurotiales, the green, blue, and black molds known by their asexual reproductive states as *Penicillium* and *Aspergillus*. *Penicillium* is known for causing food spoilage and for producing antibiotics; *Aspergillus* is known for the production of carcinogenic aflatoxins in many crops, for causing disease in humans and other animals, and for its diverse industrial uses including the production of sake, a Japanese alcoholic beverage made by fermenting rice. Some Eurotiales have known sexual states, but many are thought to be entirely asexual.

Last are the **Dothideomycetes**, a tremendously diverse class of Ascomycetes, rivaling the 1300 genera in the Sordariomycetes. They differ from the Sordariomycetes by producing **pseudothecia** rather than perithecia. Pseudothecia are still round to flask-shaped, but the asci are not arranged in an organized hymenium and they are bitunicate, or double-walled. Bitunicate asci use water pressure to shoot out ascospores explosively. They include important plant pathogens such as *Venturia* (apple scab) and *Botryosphaeria*, as well as the common fungi known primarily by their asexual state *Alternaria*. This class also includes the epiphytic fungi known as sooty molds, in the Capnodiales.

3.7.3 Primitive Fungi

While Ascomycetes and Basidiomycetes occupy most of our attention as disease ecologists, three primitive divisions of Fungi deserve note here. Some primitive Fungi retain **unikont zoospores** (motile spores with a single posterior flagellum). Among the motile Fungi, only the **Olpidiomycota** are plant pathogens. Zoospores of genus *Olpidium* swim through water-filled pores in the soil to infect plant roots, similar to the heterokont zoospores of Oomycetes (Section 4.1). After growth in the plant root, the entire fungal thallus (which in this case is a mass of individual cells rather than hyphae) is converted to produce a resting spore. Resting spores can survive for years in the soil or germinate to release zoospores and cause additional infections.

Most Fungi in the **Mucoromycota** are saprotrophs, but a few, also called **zygomycetes**, cause plant diseases. Zygomycetes produce sexual spores called **zygospores**, formed following the fusion of two hyphae (called **gametangia**), after which karyogamy produces a diploid zygote cell (the zygospore). Zygospores are thick-walled resting spores, resistant to harsh conditions. When they germinate, the nuclei undergo meiosis to return to a haploid state. The hypha then produces an asexual **sporangium** at the end of a long stalk called a **sporangiophore**. Inside the sporangium, the nuclei are packaged into numerous single-celled haploid mitospores. When those spores are released, they germinate to produce haploid hyphae, which have almost no septa and so are coenocytic (like oomycetes, Section 4.1). The hyphae can then produce additional asexual sporangia, or, if they encounter a compatible strain, sexual zygospores. The most important plant pathogenic genus of Mucoromycota is *Rhizopus*, which causes soft rot of fruits and vegetables. For those familiar with tempeh, *Rhizopus* is also what ferments and holds the soybeans together in that delicious traditional Indonesian food.

Finally, the **Glomeromycota** form close symbiotic relationships known as **arbuscular mycorrhizae** with plant roots (Section 15.4). Like the Mucoromycota (within which Glomeromycota are sometimes classified), the hyphae of Glomeromycota are essentially aseptate and coenocytic. They reproduce by producing large, multinucleate, heterokaryotic mitospores called **glomerospores** (or commonly, just spores). As they form, the glomerospores acquire a stream of many nuclei from the hyphae (very different from spores in other Fungi). They have no known sexual reproduction. The Glomeromycota are **obligate biotrophs**, growing only in association with living host plants, and are not readily grown in laboratory culture.

3.8 Environmental envelope for growth

Fungal growth is strongly affected by the temperature, moisture, and pH of the environment in which they grow, so environmental conditions often determine where a particular fungus will thrive. With few exceptions, fungi are **obligately aerobic**. However, fungi are extremely diverse and physiologically versatile, so from the perspective of a plant pathologist, if the environment is suitable for a plant to survive, fungal pathogens will be there, too.

Fungi are able to grow over a broad range of temperature from near freezing to above human body temperature, but most fungi grow best at moderate temperatures (Figure 3.10a). The optimum, minimum, and maximum temperatures for growth vary tremendously among fungal species and strains (Pietikåinen et al. 2005).

Fungi require moisture and generally grow better under moist conditions. But fungi are diverse in their physiological requirements, and some species are able to grow under quite dry conditions (down to a water activity (a_W) of 0.8; Box 3.4) as found in stored grains, whereas others require free water (a_W = 1) for spores to germinate and grow (Figure 3.10b). Because oxygen moves through water much more slowly than through air, water films can generate anaerobic conditions that can impede fungal growth under very wet conditions (Figure 3.10b).

As osmotrophic heterotrophs, fungi require moisture to feed. Dry conditions reduce intracellular water content, which impedes enzyme activity, and

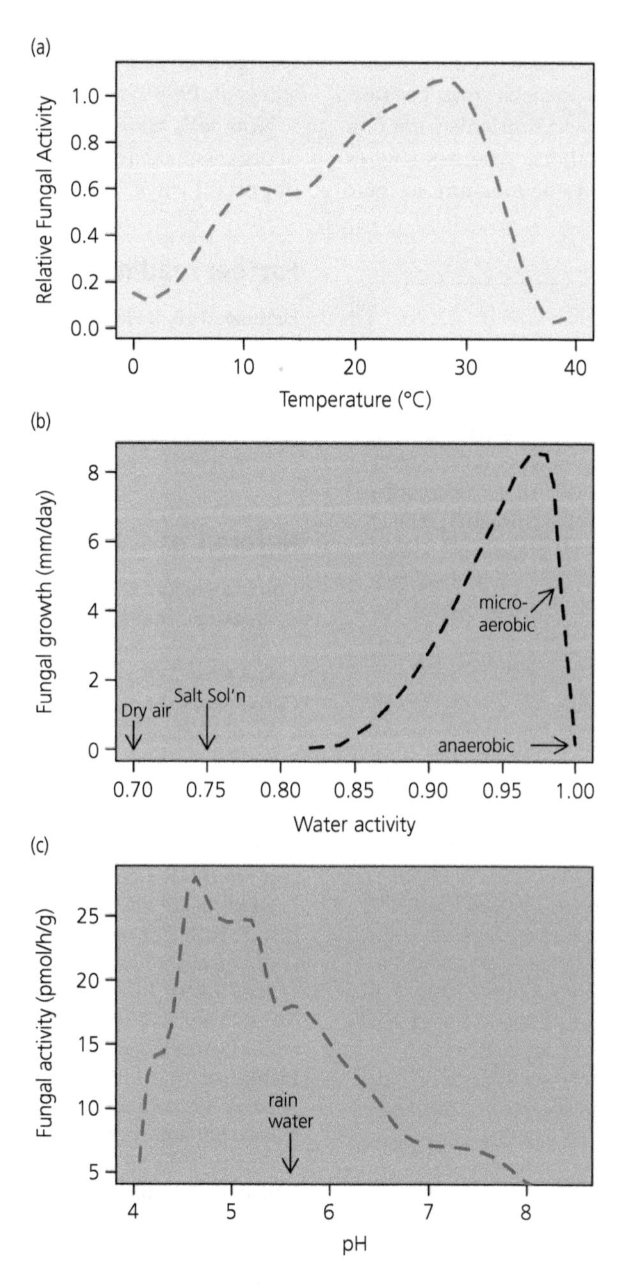

Figure 3.10 (a) Fungi grow best at moderate temperatures between 20 and 30° C; species vary greatly in their optima and limits. (b) Fungi generally grow better with more moisture, but under a water film (water activity of distilled water = 1.0) fungal growth is slowed because of lack of oxygen. (c) Fungi tend to prefer moderately acidic environments, but tolerate a broad range of pH conditions.

Figures by Gregory S. Gilbert, redrawn from (a) Pietikåinen et al. 2005, (b) Sautour et al. 2001, and (c) Rousk et al. 2009).

dry conditions can limit access to nutrients that diffuse through the substrate. Fungi may be more able to tolerate drier conditions than bacteria because fungi can translocate water and nutrients from one part of the mycelium to others, whereas single-celled bacteria are limited by the moisture in their immediate surroundings.

Box 3.4 Water activity

Water activity (a_W) is widely used in the food industry as a standard measure of the water available to microorganisms. Water activity is related to the water content of the substrate, the interaction of water with structure surfaces, and the solute content of the water. Water activity decreases roughly linearly as the solute (salt) content increases. Water activity is defined as the ratio of the partial pressure of the water to the partial pressure of pure water at that temperature and is measured using a dew point hygrometer or freezing points. As a ratio, a_W is unitless, but multiplying a_W by 100 gives you the equilibrium relative humidity as a percentage. This is useful experimentally, since you can maintain a desired specific relative humidity in a closed air space by allowing it to reach equilibrium with a salt solution of specific concentration (saturated salt has $a_W = 0.75$; the relative humidity above that solution would have an equilibrium relative humidity of 75%).

Water moves from moist areas with high a_W toward dry areas with low a_W. When all the water is available (i.e., free, pure water), $a_W = 1$. Dry air has a_W around 0.5 to 0.7. Most bacteria require $a_W > 0.9$, whereas most fungi can grow at $a_W > 0.8$. This is why honey and dried fruit ($a_W = 0.6$) do not get moldy but fresh fruits ($a_W > 0.97$) do. The record holder for cell division at the lowest a_W goes to the xerophytic, sugar-tolerant fungus *Xeromyces bisporus*, which can grow at $a_W = 0.61$ (Stevenson et al. 2015).

Fungi tend to prefer moderately acidic environments, but they are generally able to tolerate a wide range of pH conditions (Figure 3.10c). Because fungi tolerate low pH better than do bacteria, acidic soils have a 30-fold increase in fungal importance over bacteria compared to alkaline soils (Rousk et al. 2009) and plant pathologists often acidify culture media to selectively grow fungi while excluding contaminating bacteria. The pH of plant fluids ranges widely, varying among species and with soil pH (Cornelissen et al. 2010), and xylem pH can change from acidic to basic during drought (Wilkinson et al. 1998).

Now let's take a look at the Oomycetes, a group of organisms that seem a lot like Fungi, but differ in important ways.

Further reading

Heitman, J., B. J. Howlett, P. W. Crous, E. H. Stukenbrock, T. Y. James, and N. A. R. Gow, editors. 2018. *The Fungal Kingdom*. ASM Press, New York.

Moore, D., G. D. Robson, and A. P. J. Trinci. 2020. *21st Century Guidebook to Fungi*, 2nd edition. Cambridge University Press.

References

Cornelissen, J. H. C., F. Sibma, R. S. P. Van Logtestijn, R. A. Broekman, and K. Thompson. 2010. Leaf pH as a plant trait: species-driven rather than soil-driven variation. *Functional Ecology* **25**:449–455.

Garber, E. D., M. L. Baird, and L. M. Weiss. 1978. Genetics of *Ustilago violaceae*. II. Polymorphism of color and nutritional requirements of sporidia from natural populations. *Botanical Gazette* **139**:261–265.

Landeweert, R., E. Hoffland, R. D. Finlay, T. W. Kuyper, and N. van Breemen. 2001. Linking plants to rocks: ectomycorrhizal fungi mobilize nutrients from minerals. *Trends in Ecology & Evolution* **16**:248–254.

Lew, R. R. 2011. How does a hypha grow? The biophysics of pressurized growth in fungi. *Nature Reviews Microbiology* **9**:509–518.

Maheshwari, R. 2005. Nuclear behavior in fungal hyphae. *FEMS Microbiology Letters* **249**:7–14.

Pietikåinen, J., M. Pettersson, and E. Bååth. 2005. Comparison of temperature effects on soil respiration and bacterial and fungal growth rates. *FEMS Microbiology Ecology* **52**:49–58.

Rousk, J., P. C. Brookes, and E. Bååth. 2009. Contrasting soil pH effects on fungal and bacterial growth suggest functional redundancy in carbon mineralization. *Applied and Environmental Microbiology* **75**:1589–1596.

Sautour, M., P. Dantigny, C. Divies, and M. Bensoussan. 2001. A temperature-type model for describing the relationship between fungal growth and water activity. *International Journal of Food Microbiology* **67**:63–69.

Smith, M. L., J. N. Bruhn, and J. B. Anderson. 1992. The fungus *Armillaria bulbosa* is among the largest and oldest living organisms. *Nature* **356**:428–431.

Stevenson, A., J. Burkhardt, C. S. Cockell, J. A. Cray, J. Dijksterhuis, M. Fox-Powell, T. P. Kee, G. Kminek, T.

J. McGenity, and K. N. Timmis. 2015. Multiplication of microbes below 0.690 water activity: implications for terrestrial and extraterrestrial life. *Environmental Microbiology* **17**:257–277.

Talbot, N. J. 2010. Living the sweet life: How does a plant pathogenic fungus acquire sugar from plants? *PLoS Biology* **8**:e1000308.

Wilkinson, S., J. E. Corlett, L. Oger, and W. J. Davies. 1998. Effects of xylem pH on transpiration from wild-type and flacca tomato leaves: A vital role for abscisic acid in preventing excessive water loss even from well-watered plants. *Plant Physiology* **17**:703–709.

CHAPTER 4

How to be an oomycete

Gregory S. Gilbert and Ingrid M. Parker

This chapter introduces some really important plant pathogens that just don't quite fit in. These are the Oomycetes, often called "water molds." Oomycetes include the most socially and politically influential plant pathogen in history, *Phytophthora infestans*, which was the driver of the Irish Potato Famine (Box 4.1). They are also the cause of extensive death of seedlings and mature plants in both agriculture and forests.

So, what are Oomycetes? Well, they are *not* Fungi, although like Fungi they are Eukaryotes that grow as hyphae and make spores, and they have traditionally been studied by mycologists. But from an evolutionary perspective, they fall into a somewhat heterogeneous group of organisms currently referred to as **SAR**, which includes the three groups whose first letters provide the name for the group: Stramenopila, Alveolata, and Rhizaria. These three groups have some similarities, but may actually be **polyphyletic** (that is, they do not come from a common ancestor) (Box 13.3). Collectively these are often colloquially called protists. Oomycetes are the main plant pathogens in the SAR, and they fall into the group (some say kingdom) Stramenopila, which also includes diatoms and brown algae like giant kelp! We won't talk about the Alveolata, but the R part of the SAR triumvirate, the Rhizaria, include plant pathogens called Phytomyxea or Plasmodiophorids, and we'll check in briefly about them here too. The systematics of SAR is complicated and there is no current consensus about how they are related to each other. So, for now we will focus more on how they make a living than on how to classify them. Just please don't call them fungi—you are much more closely related to Fungi than are the Oomycetes. Should we call you a fungus?

Box 4.1 Late blight and the Irish Potato Famine

In the 1840s in Ireland, poor tenant farmers raised grain to pay rent to their English landlords, while they lived almost exclusively on potatoes as their staple food source. In 1845, potatoes began to die of a mysterious disease, which soon was destroying potato crops at a startling rate. This was late blight disease of potato, caused by the oomycete *Phytophthora infestans*, the most famous plant pathogen of all time. This pathogen was introduced into Ireland, probably from Mexico via infected potatoes from the United States (Goss et al. 2014). Under cool, moist conditions, the *P. infestans* life cycle can be completed in only five days, from zoospore germination, hyphal growth, sporangia production, to additional zoospore formation (Figure 4.2). The sporangia are spread by wind and water, and the chemotactic zoospores move through water films on leaves or in soil. Rain washes the zoospores from leaves into the soil, where they infect the developing tubers. Once infected, the tubers are quickly colonized by bacterial pathogens that convert them to stinking, rotting mush. In Ireland, the unusually cool, wet summers of 1845–47 propelled rapid population growth of the pathogen and a devastating epidemic of late blight that destroyed one-third to three-quarters of all potato production. In Ireland, this period is called *"An Gorta Mor"* (the Great Hunger) or *"An Drochshaol"* (the Hard Times). An estimated million people died of starvation, and an even larger number emigrated to the United States (Boyle and Gráda 1986). The human tragedy and social upheaval

The Evolutionary Ecology of Plant Disease. Gregory S. Gilbert and Ingrid M. Parker, Oxford University Press. © Gregory S. Gilbert & Ingrid M. Parker (2023).
DOI: 10.1093/oso/9780198797876.003.0004

caused by this introduced plant pathogen were transformative for both Irish and US history in ways not since seen.

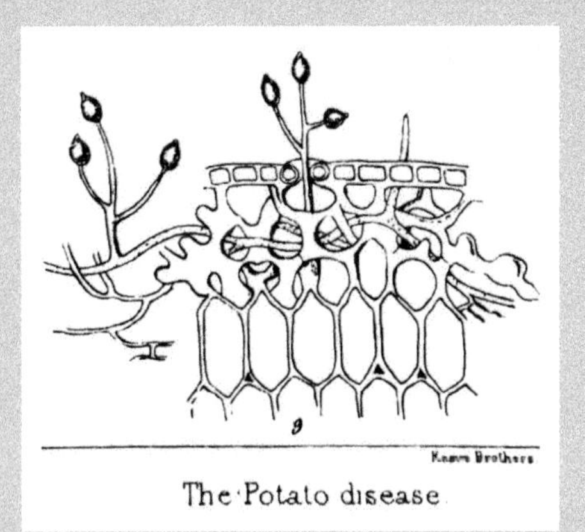

Figure 4.1 Drawing of *Phytophthora infestans* by the Rev. Miles Berkeley (1846). Coenocytic hyphae are growing through the plant leaf tissue, and emerging to produce sporangia, which disperse easily to new plants, creating new infections.

The late blight epidemic in Ireland was not only socially transformative but scientifically transformative as well. British clergyman and mycologist Miles Berkeley was convinced that the disease was caused by a pathogen (then called *Botrytis infestans* and thought to be a Fungus), and he published detailed sketches of it (Figure 4.1), some 15 years before Pasteur's experiments on germ theory. Subsequently, Anton de Bary's work placed it into the new genus *Phytophthora* and led him to coin the term "symbiosis," describing the intimate interaction between the *Phytophthora* and the potato plant (Goff 1982, Large 2003). The late blight epidemic in Ireland can arguably be called the start of plant pathology as a science.

4.1 Oomycete life cycle and structure

Oomycetes produce mycelium comprised of hyphae that are superficially similar to fungal hyphae—long tubes with high surface area that grow through their substrate, excrete enzymes that break the substrate down into smaller components, and then absorb the resulting sugars and amino acids. But that is where the similarities end. There are several really important differences from the perspective of a pathologist. First, their cell walls are made of cellulose, β-glucans, and the amino acid hydroxyproline; they do *not* contain chitin, which is the main component of cell walls in Fungi. Who else has cellulose cell walls? Plants do. So, the exposed part of an oomycete pathogen is chemically much more similar to the plant host than is a fungal pathogen. Second, hyphae of oomycetes are truly **coenocytic** because they generally lack septa (Figure 4.2). Cross-walls are limited to producing spores and walling off sections of damaged mycelium. Having coenocytic hyphae facilitates rapid movement of cellular components throughout the oomycete mycelium. Fungi are functionally coenocytic, because their septa are incomplete and allow movement of cellular components between cells (Section 3.2); however, only oomycetes are truly coenocytic.

Like all Stramenopila, Oomycetes are **heterokonts**, which indicates they have a motile stage in their lifecycle (called a **zoospore**) that has two different kinds of **flagella** (Figures 4.2 and 4.3). Oomycetes have a whiplash flagellum used to steer and a "tinsel" flagellum used to pull the zoospores through the water (*hetero* = different *kontos* = punting pole). The whiplash flagellum is smooth, and the tinsel flagellum is covered with short **mastigonemes** that stand perpendicular to the main flagellum and act as oars as the flagellum undulates. (A few Fungi (Chytridiomycota) also have zoospores, but they are unikont spores with only a whiplash flagellum.)

Zoospores are **mitospores** (asexual spores) capable of moving under their own power through water to disperse to a new host (Figure 4.3). They lack cell walls—they are just little packets of organelles within a cell membrane that live off their own energy reserves. Because they don't have a cell wall they need to be in water, so you would normally find zoospores in the water-filled pore spaces of saturated soils, or in a stream, or in free water sitting on a plant surface after a rain. Hence the name "water molds." Despite their simplicity, zoospores can have pretty complex behavior. Zoospores show **chemotaxis**, the ability to move directionally in response to a chemical gradient. For instance, zoospores travel through soil toward plant roots, moving up a concentration gradient of carbon exudates from the root. Once a zoospore arrives at

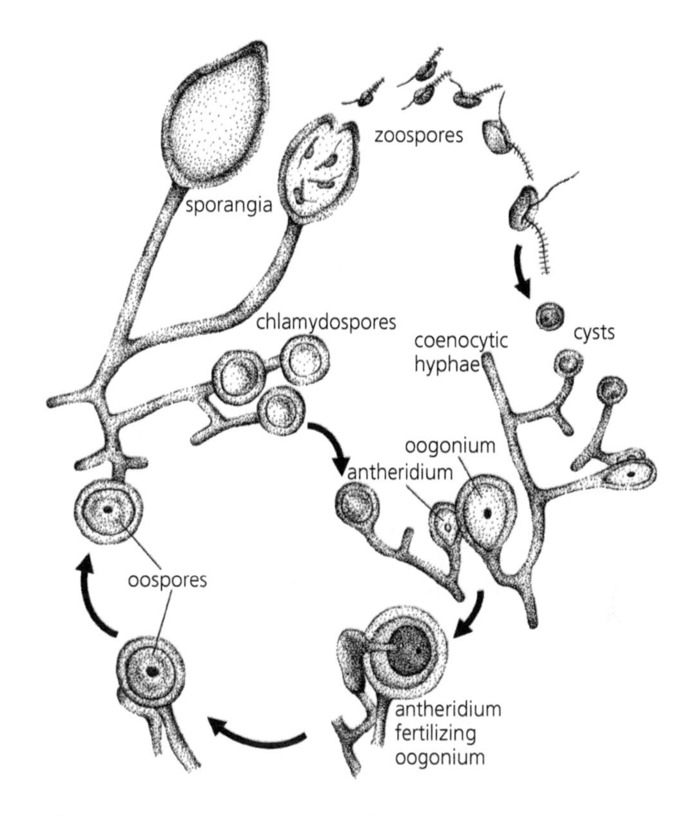

Figure 4.2 Generic life cycle of an oomycete (here based on the important pathogen genus *Phytophthora*).
Drawing by Josh Zupan.

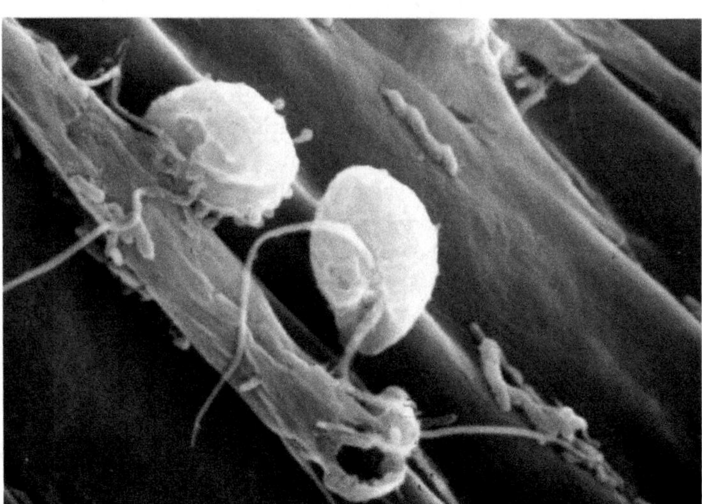

Figure 4.3 Image of two zoospores of *Phytophthora nicotianae* attracted to a tobacco root.
Photo by David Shew.

a host root it forms a **cyst** (Figure 4.2). The process of encysting requires the spore to generate new cell-wall material. The zoospores excrete little vacuoles filled with precursors of the cellulose cell wall through the cell membrane; those precursors then self-assemble around the zoospore, while the spore simultaneously either resorbs or sheds its flagella. The cyst is no longer motile, and often adheres to the root or soil particles. A zoospore that finds itself in unfavorable conditions (say, the soil is drying out) can encyst, even if it hasn't yet found a plant root, and the cell wall helps protect it from desiccation. A zoospore generally only has a matter of hours to swim before it must encyst or die. In contrast, a cyst can last for days or weeks, and once it has adequate moisture and nutrients around it, the cyst can germinate to produce a **hypha**. The hyphae can grow through substrate, and if near a host plant, can infect the plant.

Oomycete hyphae can continue to grow as coenocytic mycelium and reproduce asexually. If an oomycete finds itself with dwindling resources or other adverse conditions, it may produce **chlamydospores** (Figure 4.2). These are mitospores with thick cell walls that are able to survive for extended periods until conditions improve. Chlamydospores may be produced on mycelial side branches or even right in the middle of a hypha, like beads on a string. They are durable survival structures that are easy to produce quickly, but they do not disperse well. Chlamydospores germinate to resume growth when conditions are favorable.

When there is plenty of moisture, Oomycetes produce asexual sporangia. **Sporangia** are swollen structures filled with cytoplasm and a large number of nuclei (remember, as coenocytic organisms, oomycetes can easily move lots of nuclei to the same place). Especially when produced on above-ground plant parts, sporangia can serve in dispersal, breaking off from the hypha, and then washing away in the rain or down a stream. They can later germinate directly to produce more hyphae, infect a new host, or grow through other substrate. But when sporangia are produced with plenty of free water, for instance in saturated soils or water films, the real oomycete magic happens. Membranes begin to form within the cytoplasm, walling off each of many dozens of nuclei into small packets of cytoplasm and organelles. Each one then produces two flagella and becomes a zoospore. The sporangium bursts open, and the motile zoospores swim out into the free water!

Oomycetes are also capable of sexual reproduction by producing **oospores** (Figure 4.2). To make an oospore, an oomycete first makes two complementary structures specific to sexual reproduction: a narrow **antheridium** and a large, round **oogonium**. A septum is produced to separate those cells from the rest of the mycelium, isolating their nuclei. The nuclei of Oomycetes are **diploid** for nearly their entire life cycle (in the mycelium, in the zoospores, in the sporangia, and in the chlamydospores). The diploid nuclei in the antheridium and in the oogonium undergo meiosis, reducing the diploid (2N) nuclei to haploid (1N). Then the antheridium and oogonium fuse through **plasmogamy**, and the haploid nuclei from the antheridium migrate into the oogonium, creating an N+N cytoplasm. In some species, the antheridia and oogonia need to be produced by genetically different individuals of different sexual types in order to fuse; those species are **heterothallic**. Other species can reproduce sexually from a single genotype; they are **homothallic**. In either case, inside the fertilized oogonium, the nuclei fuse (**karyogamy**) to produce a diploid nucleus again. The oogonium then forms a thick cell wall with a diploid nucleus inside, creating an **oospore**. The process of sexual reproduction in Oomycetes is really different from that of Fungi, because Fungi have haploid nuclei, so they do not need a meiotic step before fertilization.

Oospores are often produced inside infected plants as the host plant is senescing or dying, and they are extremely tough structures. They can withstand desiccation, being eaten, bacterial attack, and more—surviving up to decades! Under appropriate conditions, chemicals excreted by the growing root of a susceptible host plant can induce the oospore to germinate, leading to the growth of coenocytic hyphae with diploid nuclei, beginning the cycle again (Figure 4.2).

4.2 Life-history strategies of plant parasitic oomycetes

The Oomycota include a diversity of organisms with different life-history strategies. Even within the Peronosporomycetes clade of Oomycota that contain the plant pathogens, there are two rather different strategies: the Pythiaceae are necrotrophs or hemibiotrophs, and the Peronosporaceae and Albuginaceae are obligate biotrophs.

The Pythiaceae include three important genera: *Pythium*, *Phytophthora*, and *Phytopythium*. *Phytophthora* means "plant destroyer," and it is a pretty good name. Most of the species in these genera infect plant roots, especially when soil is moist. Many cause "damping off," a dramatic disease of seeds and young seedlings (Section 8.4); in those cases the oomycete is an aggressive **necrotroph** that kills seedlings very quickly. Some species of *Phytophthora* attack above-ground parts of plants and may be **hemibiotrophs**, growing extensively in the plant tissue, perhaps producing sporangia, and killing host tissue later. *Phytophthora infestans* is one such pathogen (Box 4.1).

In contrast, the Peronosporaceae (commonly called **downy mildews**) are **obligate biotrophs** (Thines and Choi 2016). They infect leaf tissue and produce sporangia on branched **sporangiophores** on the undersides of infected leaves. The sporangia disperse to other leaves through the wind, where, given adequate moisture, they either release zoospores or germinate to produce hyphae and then infect a new host. As obligate biotrophs, downy mildews are not readily grown in pure culture. The most important genera of Peronosporaceae are *Peronospora*, *Plasmopara*, and *Bremia*.

Last comes *Albugo* in the Albuginaceae, commonly called **white rust** (but remember from Chapter 3 that true rusts are Basidiomycetes!). Like downy mildews, *Albugo* is an obligate biotroph, forming discolored galls on the upper surface of an infected leaf and white pustules filled with sporangia on the lower leaf surface. *Albugo* is particularly problematic on leafy vegetables like cabbage, lettuce, spinach, and beets, and it has a general life cycle (dispersal of sporangia which then produce zoospores or hyphae) very similar to the downy mildews.

Figure 4.4 Club root of cauliflower caused by the Rhizaria *Plasmodiophora brassicae*.
Photo by Rasbak, licensed under the Creative Commons Attribution-Share Alike 3.0 Unported license.

4.3 Other weird and wonderful SAR

We can't close without at least a brief mention of the Rhizaria, especially the genus *Plasmodiophora*, which causes club root of cabbages and other Brassicaceae (Figure 4.4) (Dixon 2009). *Plasmodiophora* needs to be here just because it is so fascinating and bizarre. Like the Oomycetes, it produces motile zoospores that swim through soil to new plant hosts, infecting wounds or root hairs. But instead of germinating and growing as hyphae, *Plasmodiophora* produces amoeba-like cells that create an aggregated form called a **plasmodium**. The plasmodium then generates yet more zoospores that infect other parts of the same plant or swim to nearby plant roots. Once inside the host plant, the plasmodium affects plant hormonal regulation, forcing the plant to create swollen root galls or "clubs," inside which *Plasmodiophora* grows. The plasmodium eventually is cleaved into millions of **resting spores**, which are desiccation resistant and can survive for decades until encountered by a new cabbage root. Unlike the Oomycetes, the cell walls of *Plasmodiophora* resting spores are comprised largely of chitin, proteins, and lipids, rather than cellulose.

Further reading

Lamour, K., editor. 2013. *Phytophthora: A Global Perspective*. CABI International, Wallingford, UK.

Rai, M., K. A. Abd-Elsalam, and A. P. Ingle. 2020. *Pythium. Diagnosis, Disease and Management*. CRC Press, Boca Raton, FL.

References

Berkeley, M. J. 1846. Observations, botanical and physiological, on the Potato Murrain. *Journal of the Horticultural Society of London* **1**:9–34 (reprinted as Phytopathological Classics number 38 by the American Phytopathological Society, 1948).

Boyle, P. P. and C. O. Gráda. 1986. Fertility trends, excess mortality, and the Great Irish Famine. *Demography* **23**:543–562.

Dixon, G. R. 2009. The occurrence and economic impact of *Plasmodiophora brassicae* and clubroot disease. *Journal of Plant Growth Regulation* **28**:194–202.

Goff, L. J. 1982. Symbiosis and parasitism: another viewpoint. *Bioscience* **32**:255–256.

Goss, E. M., J. F. Tabima, D. E. Cooke, S. Restrepo, W. E. Fry, G. A. Forbes, V. J. Fieland, M. Cardenas, and N. J. Grünwald. 2014. The Irish potato famine pathogen *Phytophthora infestans* originated in central Mexico rather than the Andes. *Proceedings of the National Academy of Sciences of the United States of America* **111**:8791–8796.

Large, E. C. 2003. *The Advance of the Fungi* (originally published 1940). APS Press, St. Paul, MN.

Thines, M. and Y.-J. Choi. 2016. Evolution, diversity, and taxonomy of the Peronosporaceae, with focus on the genus *Peronospora*. *Phytopathology* **106**:6–18.

CHAPTER 5

How to be a bacterium

Gregory S. Gilbert and Ingrid M. Parker

Bacteria are single-celled, **prokaryotic** organisms with a broad range of ecological strategies. Some are **autotrophic**, producing their own energy through **photosynthesis** (producing simple sugars using energy derived from light) or **chemosynthesis** (producing organic compounds using energy derived from chemical reactions, without light). However, all plant pathogenic bacteria are **heterotrophic**. Heterotrophic bacteria feed by secreting enzymes (exoenzymes) into the surrounding substrate to break down complex macromolecules into simple sugars and amino acids that can then be absorbed into the cell and metabolized. That means that such bacteria are **osmotrophic heterotrophs**—just like the Fungi and Oomycetes (Chapters 3 and 4).

5.1 What do Bacteria look like?

Morphologically, bacteria generally lack the charms and variety of the plants, fungi, and oomycetes we have looked at so far. Most plant pathogenic bacteria are simple rod-shaped cells called **bacilli**, generally between 1 and 4 μm in length (with a few bacteria as large as 20 μm) (Figure 5.1). A few have more interesting shapes. The **actinobacteria** form chains of filiform cells that superficially resemble very skinny fungal hyphae, and the **mollicutes** are an intriguing group of plant-pathogenic bacteria that includes wall-less bacteria such as **phytoplasmas** and some crazy corkscrew-shaped **spiroplasmas**. Section 5.6 explores the systematics and classification of bacteria.

As prokaryotes, Bacteria lack membrane-bound organelles (i.e., no mitochondria, chloroplasts, or defined nucleus). Each cell is surrounded by a plasma membrane (Figure 5.1). Usually, the cell membrane is in turn surrounded by a rigid, but porous, **peptidoglycan** cell wall. In Gram-negative bacteria (Section 5.6), the cell wall is then surrounded by a second, outer membrane. The cell wall and membranes together constitute the **bacterial cell envelope**. Some bacteria have an additional enclosing layer of polysaccharides that protect the cell and can help it adhere to surfaces and other bacteria. When the polysaccharide layer forms a semi-rigid structure, it is called a **capsule**; if more amorphous, it is called a **slime layer**.

Many bacteria are actively motile—they can move actively under their own power (Kearns 2010). Most commonly, motile bacteria have one to a few **flagella** (Figure 5.1), hollow tubes of the protein flagellin attached to small molecular motors that make the flagella rotate like corkscrews, allowing the bacterium to swim through water films. Bacterial flagella are functionally analogous to flagella in eukaryotes, including those of oomycete zoospores and mammalian sperm, but they are structurally quite different. Some motile bacteria move by twitching, where they extend thread-like, tubular **pili** (singular **pilus**) that attach to a surface ahead of the bacterium; the bacterium then retracts the pili, pulling the cell forward. Still others use focal-adhesion complexes that allow them to glide over a surface. Motile bacteria can move directionally in response to chemical gradients in their environment, a behavior called **chemotaxis**. Bacteria can detect gradients of nutrients and swim directionally toward the higher concentration (positive chemotaxis), or away from repellant chemicals (negative chemotaxis). Motility is important in active bacterial

The Evolutionary Ecology of Plant Disease. Gregory S. Gilbert and Ingrid M. Parker, Oxford University Press. © Gregory S. Gilbert & Ingrid M. Parker (2023).
DOI: 10.1093/oso/9780198797876.003.0005

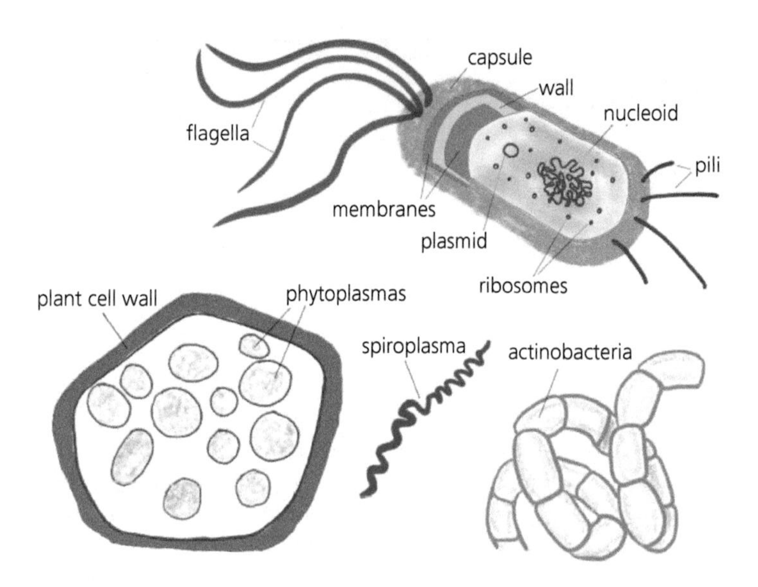

Figure 5.1 Morphology of plant-associated bacteria. Top shows the general ultrastructure of a motile, Gram-negative bacterial cell, such as a *Pseudomonas*. Wall-less mollicutes include phytoplasmas that grow as irregularly shaped cells within plant cells and cork-screw-shaped spiroplasmas often found in phloem cells. Actinobacteria form long chains that superficially resemble very fine fungal hyphae.
Drawing by Gregory S. Gilbert.

dispersal and collective behavior over short distances (μm to cm), but dispersal of bacteria over longer distances is mostly passive through the air, in water splash, on moving debris, soil, or tools, or by animal vectors.

5.2 How do Bacteria grow and reproduce?

Most bacteria are single-celled organisms, growing through a controlled process of cell division called **binary fission** (Figure 5.2). Bacteria, which are haploid, have a single circular chromosome (of double-stranded DNA) carrying several thousand genes, found in a region of the cell called the **nucleoid.** The bacterial cell may also include small, circular DNA molecules called **plasmids**. Plasmids can carry a handful of genes or hundreds.

In binary fission, the bacterial cell first elongates until it is nearly double the original length. The chromosomal DNA replicates and migrates toward opposite sides of the cell (Figure 5.2). Simultaneously, the bacterium produces a cell-division ring protein call **FtsZ**, or the **Z ring**, at the midpoint of the cell (Angert 2005). The FtsZ ring guides formation of cell wall growth, invaginating the cell envelope to form two roughly equal daughter cells with a single exact duplicate of the parent cell's chromosomal DNA.

With abundant nutrients and appropriate temperature and moisture, bacteria can complete binary fission in tens of minutes. Bacterial populations grow exponentially; each cell divides, and the two resulting daughter cells then each divide to produce two new progeny, those four into eight, and so on. The expected number of clonal bacterial descendants after n generations is then 2^n. That is, even for a slow-growing bacterium with a generation time of one hour, a single cell has the potential to produce 2^{24} = 16,777,216 descendant bacteria in one day. Because bacterial populations increase rapidly and exponentially (and because the distribution of bacterial population sizes is generally log-normal), we commonly refer to bacterial population sizes on a $\log_{10}$ scale. For instance, $\log_{10}(16{,}777{,}216)$ = 7.23. If our bacterial population continued to grow at the same rate for six more hours, we would reach 9.03 $\log_{10}$ cells. Bacterial abundance is often measured using dilution plating on culture media (Figure 9.13).

Plasmids, like chromosomal DNA, are passed on to daughter cells during binary fission, but plasmids can replicate independently of the chromosomal DNA, and a single bacterial cell may have multiple copies of a plasmid. Plasmids are also

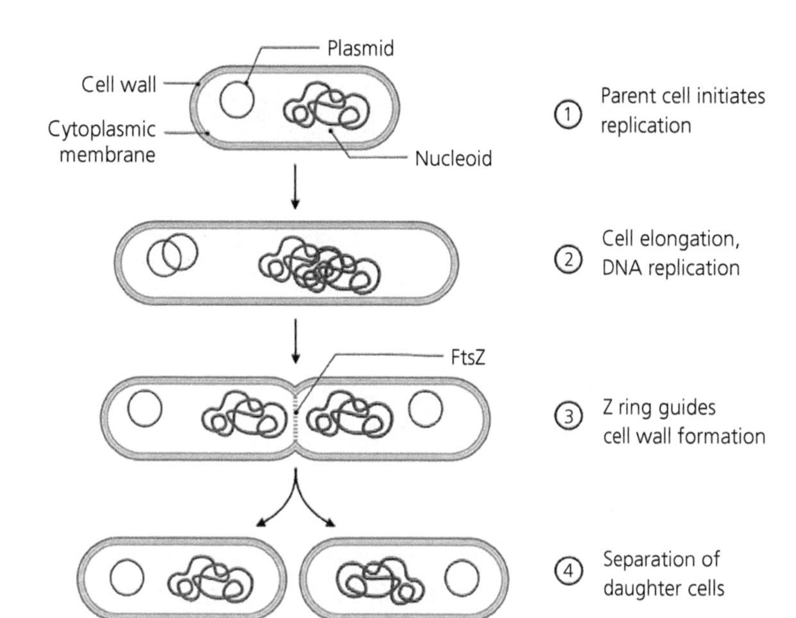

Figure 5.2 Binary fission in bacteria. Graphic by Ingrid M. Parker, created with BioRender.com.

sometimes lost from daughter cells. While the chromosomal DNA carries the genes necessary for normal cell processes, plasmids often carry auxiliary genes that affect the survival and ecological functions of the bacteria that carry them. For instance, genes that confer resistance to antibiotics are often found on plasmids, as are many genes associated with pathogenicity, or genes associated with the ability to use unusual kinds of substrates. Because the genes carried on plasmids are not essential for basic cell function, they show more evolutionary flexibility. The ease with which plasmids can be genetically manipulated and introduced into bacteria has allowed scientists to use plasmids as a biotechnological tool to study and modify individual genes.

5.3 Horizontal gene transfer

Whereas most genes in a bacterial genome are inherited from parent to offspring over many clonal generations of binary fission (**vertical gene transfer**), the evolution and ecology of bacteria have also been heavily influenced by **horizontal gene transfer**, in which individual genes or sets of genes move from one lineage into another. The primary modes of horizontal gene transfer involve conjugation, transformation, transduction, and transposable elements.

Plasmids can be transmitted from one bacterial cell to another through **bacterial conjugation** (Figure 5.3a). Importantly, while some plasmids may be restricted to a particular bacterial lineage, others are able to move between even distantly related bacteria. In conjugation, one cell (the donor) uses a pilus to initiate a bridge to another cell (the recipient). The donor cell contains a special plasmid, the **F-plasmid** (F for "fertility"), which is nicked, unwinds from the complementary strand, and passes into the recipient cell. Then, in both the donor and the recipient cell, DNA polymerase works to create the complementary copy of the single strand, so that the plasmid then returns to the full double-stranded DNA version. In this way, the plasmid has been introduced from the donor to the recipient; the recipient cell is converted from the F– state to the F+ state and can now act as a donor cell. Note that although bacterial conjugation is sometimes called "bacterial sex," it really has nothing to do with sex, since there is no formation of gametes and no diploid–haploid alternation, as would be the case in true sex. The genes on the F-plasmid will move with the plasmid into the recipient cell's lineage. Horizontal gene transfer via

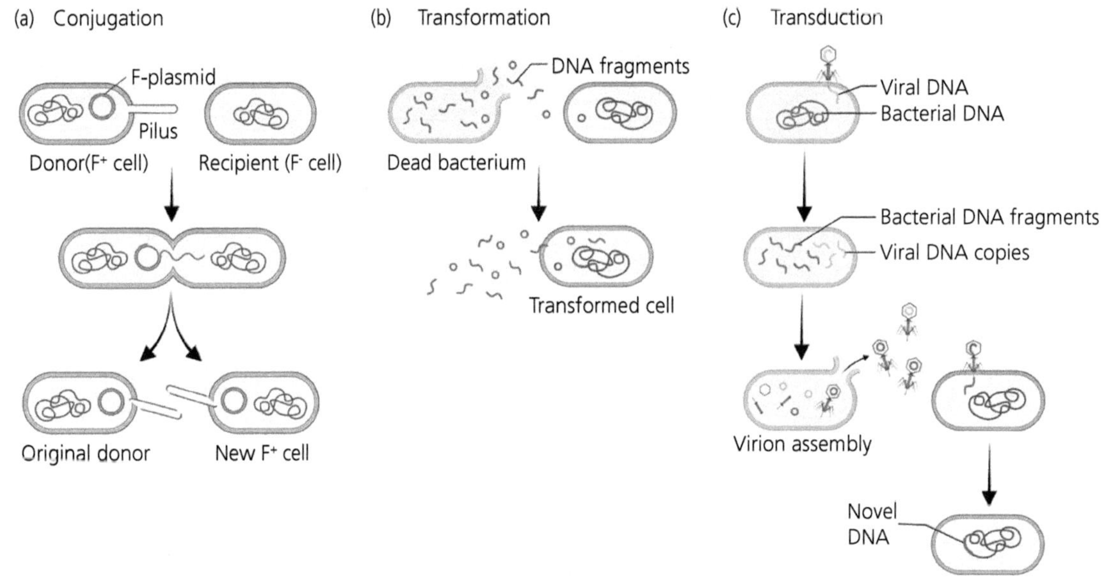

Figure 5.3 Horizontal gene transfer in bacteria may occur through (a) conjugation, in which an F+ donor cell provides a plasmid to an F− recipient cell, (b) transformation, in which fragments of DNA from dead bacterial cells are taken up directly, and (c) transduction, in which phage virus (the lunar-lander shaped structure attached to the cell) incorporates bacterial DNA inside its capsid during replication and then transmits it to a new bacterial cell along with its own viral DNA.
Graphic by Ingrid M. Parker, created with BioRender.com.

plasmids is an important mechanism in the spread of antibiotic resistance and other traits under strong selection.

A second important mode of horizontal gene transfer in bacteria is **transformation** (Figure 5.3b) (Thomas and Nielsen 2005). When bacterial cells die, fragments of the DNA from chromosomes and plasmids are released into the environment and are sometimes taken up directly by living bacterial cells. Transformation requires special proteins in the cell membrane that facilitate transporting the naked DNA into the bacterial cell, and usually only a small subset of bacterial cells express those proteins and are competent for transformation. Parts of the DNA fragments taken up by the bacterial cell are integrated into the chromosome or a plasmid during replication, "transforming" the bacterium with genes from a separate lineage.

Transduction occurs when **bacteriophage viruses (phage)** move DNA from one bacterium to another (Figure 5.3c). The phage attaches to the bacterial cell and injects its DNA into the bacterium, where viral DNA is replicated and viral genes (including those for the viral protein coat, or capsid) are expressed by the bacterial host cell (Chapter 6). When there are abundant copies of the viral DNA and coat protein in the bacterial cell, the viral DNA is packaged into the coat protein, creating new, functional phage virions that infect new bacteria after the host bacterium lyses. Sometimes bacterial plasmids are inadvertently packaged into the viral capsid, so when the virus injects its DNA into a new bacterial host, it also injects the plasmid DNA, effecting horizontal gene transfer.

Transposable elements, sometimes called "jumping genes" or transposons, are short stretches of DNA that can readily change positions within a genome, often through a "cut and paste" mechanism. Transposable elements are found in all kinds of organisms, not only bacteria; they were first described by Barbara McClintock in maize, a discovery that earned her a Nobel Prize! In bacteria, transposable elements can move between the chromosomal DNA and plasmids and back, or to the bacterial genome from fragments of DNA taken up from the environment via transformation. Like

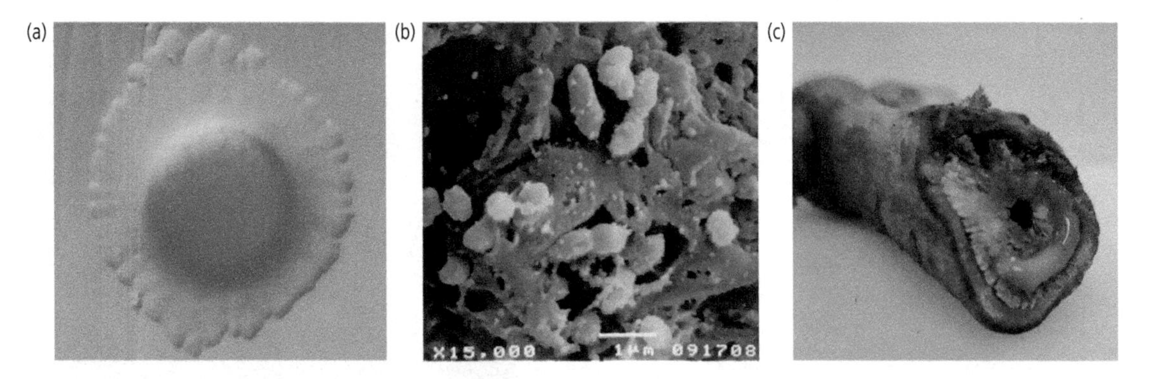

Figure 5.4 Many bacteria produce extracellular polysaccharides (EPS). (a) Bacterial colony producing abundant EPS while growing on agar culture medium. (b) Biofilm of bacteria embedded in EPS inside a wild fig fruit. (c) *Ralstonia solanacearum* produces copious EPS that plugs the vascular system of infected hosts, causing bacterial wilt.
Photos a and b by Gregory S. Gilbert. Photo c by Alejandra Huerta.

plasmids, transposons often include genes that provide the bacterial cell with traits that offer a fitness advantage, such as antibiotic resistance. In **horizontal transposon transfer (HTT)**, transposable elements can move genes not only between lineages of bacteria, but also between species of both plants and animals, which requires becoming integrated into the germ line.

5.4 How do Bacteria survive?

Bacterial cells are susceptible to desiccation, ultraviolet (UV) light, and heat. Bacteria have several mechanisms to protect against unfavorable conditions (Lebre et al. 2017). Many bacterial cells produce **extracellular polysaccharides** (often abbreviated as EPS)—or the more generic term, **extracellular polymeric substances** (Figure 5.4)—as a means to protect the cell from biotic and abiotic factors that can harm cells. A cell can increase production of these long-chain substances when threatened with desiccation, producing an EPS matrix in which the cells are embedded. Cells also upregulate the production of osmoprotectant compounds that protect DNA and proteins from damage. Bacterial cells can limit metabolic activities by becoming dormant, which conserves cell water content, helping them survive harsh conditions.

The EPS produced by bacteria helps individual cells stick together, and often contributes to their ability to adhere to their substrate and other species of bacteria, in what is called a **biofilm** (Bogino et al. 2013, Flemming et al. 2016) (Figure 5.4b). Biofilms are the normal structure of colonies of bacteria growing in most natural environments. If you've ever had a bacterial colony develop on leftover food, the shiny, sticky puddles that form are made of EPS containing millions of bacteria. The EPS and biofilms provide a protective and functional structure for bacterial colonies and can protect the bacteria against adverse conditions. Bringing many bacteria together in a small area facilitates cell-to-cell communication via chemical interactions called **quorum-sensing**. In many cases, quorum-sensing leads to gene expression changes (turning on or turning off transcription of particular genes) when bacteria detect neighboring cells through accumulation of signaling compounds called **autoinducers** (Von Bodman et al. 2003). Quorum-sensing is generally mediated by homoserine lactones in Gram-negative bacteria and by peptides in Gram-positive bacteria. When large numbers of aggregated bacteria produce enough autoinducers to reach a threshold level, quorum-sensing coordinates gene expression. Quorum-sensing can be important in disease development; in some bacteria, only when a threshold level of autoinducers and bacterial number is reached do the bacteria express genes key to pathogenesis. Quorum-sensing is also important in biofilm formation and horizontal gene transfer.

Some Firmicutes such as *Bacillus* and *Clostridium* produce **endospores** when resources are exhausted

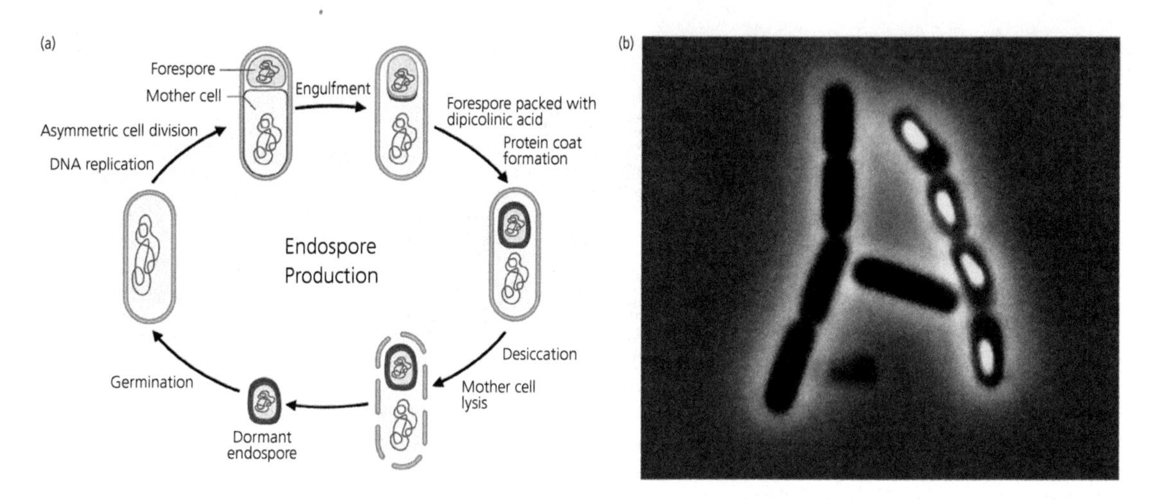

Figure 5.5 Endospores are produced by some bacteria and serve as long-term resting structures highly resistant to adverse environmental conditions. (a) Graphic of endospore production adapted by Ingrid M. Parker from "Sporulation," by BioRender.com (2022), retrieved from https://app.biorender.com/biorender-templates. (b) Photo of endospores forming within the biological control agent *Bacillus cereus* UW85 by Gregory S. Gilbert.

or unfavorable conditions develop (Figure 5.5) (Angert 2005). Endospores are dormant structures resistant to high temperature (you can boil them!), desiccation, and UV irradiation. Endospores are survival structures, not reproductive structures, because the cell is converted into a single spore instead of producing multiple progeny. The process of forming endospores begins like binary fission, with duplication of chromosomal DNA and the Z-ring initiation of cell division. In this case, though, the DNA migrates into unequal cells—a small **forespore** and a larger **mother cell**, which then completely engulfs the forespore. The mother cell packs the forespore with specialized proteins and dipicolinic acid to protect the DNA, together with necessary components for later rehydration and growth. The forespore then desiccates, the cytoplasm becomes mineralized, and a protective protein coat surrounds the final endospore. Finally, the mother cell disintegrates, leaving a highly resistant resting endospore. Endospores can lie dormant for many years, and then, under favorable conditions, pick right up where they left off, hydrating and germinating to begin normal growth and binary fission. Unfortunately (or fortunately?), the Proteobacteria and Mollicutes that include most plant pathogens do not produce endospores.

5.5 The environmental envelope of bacteria

Bacteria generally need humid conditions to be able to grow. Under dry conditions enzymatic activity is limited, as is diffusion of nutrients from the substrate into the cells. As a result, bacteria are generally limited to a narrower range of moisture than are fungi, usually requiring a water activity $a_w >> 0.9$. Unlike Fungi, which prefer more acidic conditions, Bacteria generally grow better under more basic conditions (Figure 5.6a).

Like Fungi, Bacteria can grow over a broad range of temperature from near freezing to well above human body temperature, but bacterial activity is greatest under moderate conditions, with optimum generally around 26–32°C for soil bacteria (Figure 5.6b). The details of minimum, maximum, and optimum temperatures vary among species, and these are not always easy to measure *in situ*. With bacteria in liquid culture, we use a microscope to count how many new cells are produced, but that is difficult inside a plant or in the soil. Instead, we track the metabolic signatures of bacteria; for example, Pietikåinen et al. (2005) measured bacterial activity in soil as the amount of radioactive methyl[^{3}H]thymidine taken up at different temperatures (Figure 5.6b).

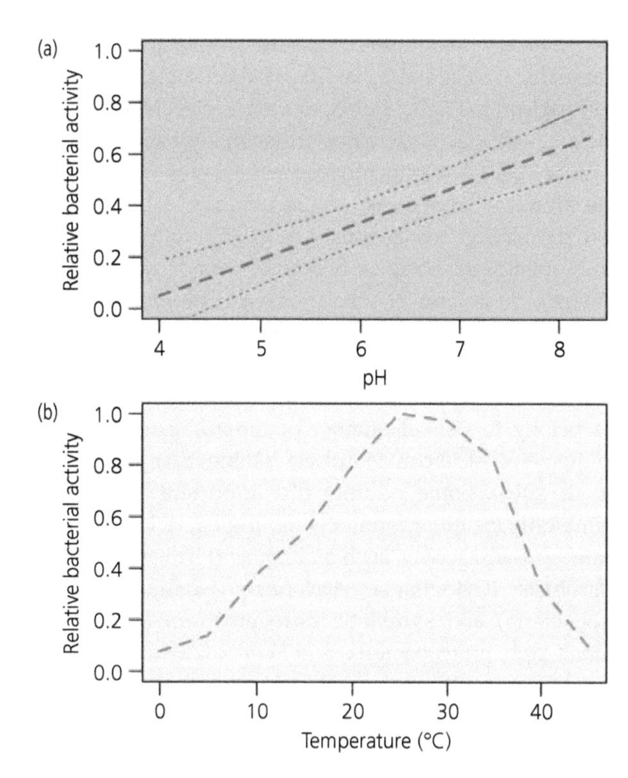

Figure 5.6 Bacterial growth is strongly sensitive to environmental conditions. (a) Bacteria grow poorly under acidic conditions, with activity (measured as [^{3}H]leucine uptake) increasing with soil pH. (b) Bacteria show optimal activity and growth at intermediate temperatures.

Figures by Gregory S. Gilbert, redrawn from (a) data in Rousk et al. 2009 and (b) Pietikåinen et al. 2005.

Bacterial species may be obligately **aerobic** (requiring oxygen to grow), obligately **anaerobic** (do not grow in the presence of oxygen), or facultatively anaerobic (grow best with oxygen but can grow without it); nearly all plant pathogenic bacteria are aerobic. However, anaerobic species can be harmful pathogens in flooded conditions or very humid storage conditions, such as when *Clostridium puniceum* causes slimy rot of potato tubers that have been stored wet. Oxygen diffusion is about 10,000-fold slower in water than in air, so even a thin film of water can promote anaerobic bacterial growth; wet conditions are a crucial part of the disease triangle for *C. puniceum*.

5.6 Bacterial classification

Morphology doesn't get you very far in telling bacteria apart, so traditionally they were classified by their chemical properties and physiological traits. Bacteria were divided into two groups based on how their cell walls react to a staining procedure using a crystal-violet stain followed by a pink counterstain: those that stain purple (because of their thick peptidoglycan layer) are called Gram positive, while those that stain pink are Gram negative (Gram, after the Danish bacteriologist Hans Christian Gram). Staining purple versus pink is a simple indicator of profound structural and chemical differences that predict behavior, including resistance to antibodies in human pathogens and evolution of resistance to antibiotics. Further classification was based on cell shape, motility, and a wide range of physiological tests (anaerobic growth, ability to use different carbon sources, etc.). Such trait-based grouping is still valuable because the traits often tell us something about the biology of the organisms, and because many traits map well onto modern molecular bacterial systematics.

Bacterial systematics is in a state of flux. Modern systematics of Bacteria is mostly based on comparison of genomic sequence data, with a special emphasis on 16S ribosomal RNA (small subunit) genes (Khaledian et al. 2020). However, the high rate of horizontal gene transfer in bacteria (Sections 5.3 and 13.7) creates patterns of reticulate evolution that complicate any vision of a hierarchical systematic classification of Bacteria in comparison to Plants, Fungi, or Oomycetes. Nevertheless, analysis of genomes across many bacterial lineages shows strong phylogenetic clades within the kingdom. Sequences of 120 ubiquitous bacterial proteins in the genomes of 94,759 culturable and uncultured bacteria reveal six major monophyletic clades and 99 subclades suggested for the rank of phylum (Parks et al. 2018). However, unlike plant-pathogenic fungi, which are found across hundreds of genera, most plant-pathogenic bacteria belong to a small number of genera, restricted to just six of those 99 phyla (Table 5.1) (Bull et al. 2010). Some additionally important plant symbionts include growth-promoting bacteria and pathogen antagonists such as *Bacillus* (Firmicutes), *Flavobacter* (Cytophagia), *Azotobacter* (Gammaproteobacteria) and symbiotic nitrogen-fixers *Rhizobium* (Alphaproteobacteria) and *Frankia* (Actinobacteria) (Chapter 15). Remarkably, of more than 3,500 currently described bacterial genera, about two dozen include nearly all the known bacterial symbionts of plants.

Bacteria from other clades (as well as Archaea) are likely to be important plant symbionts. Advances in high-throughput DNA sequencing (Section 9.8) have uncovered tremendous diversity among uncultured bacteria (that is, bacteria never grown on culture media), identified only by their DNA sequences (and occasionally through microscopy), and we can expect to find many hidden bacteria in important symbioses with plants in coming years.

This brings up the issue of the bacterial genera called *Candidatus* Phytoplasma. Phytoplasmas are insect-transmitted, wall-less bacteria that cause a variety of important plant diseases, but they have never been successfully grown in culture. Current rules of bacterial nomenclature are governed by the International Code of Nomenclature of Prokaryotes (ICNP, or the Prokaryotic Code) (Garrity et al. 2015), which does not permit the naming of new taxa based only on genetic sequences. What to do? As a stop-gap measure, the Prokaryotic Code erected the generic designation *Candidatus* followed by the "trivial" (commonly used) generic name to indicate a genus described only by its genome (IRPCM Phytoplasma/Spiroplasma Working Team—Phytoplasma taxonomy group 2004). Thus, the phytoplasma that causes ash yellows disease is named *Candidatus* Phytoplasma fraxini, often abbreviated as *Ca.* Phytoplasma fraxini. Note that only *Candidatus* is italicized, because the rest of the name is considered unofficial. The *Candidatus* structure is used for a number of other bacterial taxa that are recognizable only through genetic sequences, such as *Candidatus* Liberibacter.

Of course, bacterial diversity is much broader than this select list, and bacteria are critically important in biogeochemical cycling and as animal symbionts (including many mutualists and pathogens).

Table 5.1 Major genera of bacterial symbionts of plants are found in six phyla

Gram negative (Proteobacteria)		
Gammaproteobacteria	Betaproteobacteria	Alphaproteobacteria
Erwinia *Pantoea* *Pectobacterium* *Pseudomonas* *Xanthomonas* *Xylella*	*Acidovorax* *Burkholderia* *Ralstonia*	*Agrobacterium* *Ca.* Liberibacter
Gram positive		
Firmicutes	Actinobacteria	Mollicutes
Clostridium	*Clavibacter* *Streptomyces*	*Ca.* Phytoplasma *Spiroplasma*

5.7 Life histories of some plant-associated bacteria

Bacteria form many weird and wonderful symbioses with plants. Some are aggressive necrotrophs (e.g., *Pectobacterium* spp.), killing leaf tissue, flowers, fruits, roots, and tubers, and then living off the nutrients in the dead tissue. Some, like *Pseudomonas syringae* pv. *syringae*, are commonly found at low densities as epiphytes on leaves, and then under just the right conditions suddenly grow to high densities and cause foliar disease. Many bacteria are transmitted passively on the wind, in rain splash, or on farm equipment. The fire blight bacterium *Erwinia amylovora* is among them, but it also hitches a ride on the bodies of pollinating insects as they move among flowers collecting pollen and nectar, leading to infection of the bloom. The pollinators serve as **vectors**, transmitting the pathogen from one host to another.

Mollicutes such as phytoplasmas have intricate life cycles that require two very different kinds of hosts—a plant and an insect. Obligate pathogens, the phytoplasmas grow as blobbish, wall-less, triple-membrane-bound cells inside plant phloem. There they produce a variety of proteins that interfere with host plant development, causing disease. The phloem sap is a nutrient-rich food for a variety of phloem-feeding insects, like leaf hoppers and plant hoppers (Hemiptera). They stick their piercing-and-sucking mouthparts (called a **stylet**) into the phloem and suck up the sap, together with the phytoplasmas living there. The phytoplasmas pass through the stylet and intestine and enter the insect's hemolymph, which is its circulatory system. They make their way into the salivary glands, and when the insect moves to a new host plant to feed, the pathogen is transmitted through the stylet and into the phloem. This process is the same as persistent, propagative transmission of viruses (Figure 6.5b). The insects are both hosts and vectors.

A key aspect of bacterial life history is chemical interactions with other organisms. Bacteria have evolved a complex set of secretion systems to overcome the challenge of exporting proteins produced inside the cell across two or three lipid bilayer membranes (Green and Mecsas 2016). Bacteria share the highly conserved Sec and Tat secretory pathways found in all prokaryotic and eukaryotic organisms, but these are generally not able to move proteins out of bacterial cells without help from one of six important secretory systems with the clever names of Type 1 through Type 6 (Figure 5.7). Type 1 secretion system (T1SS; which

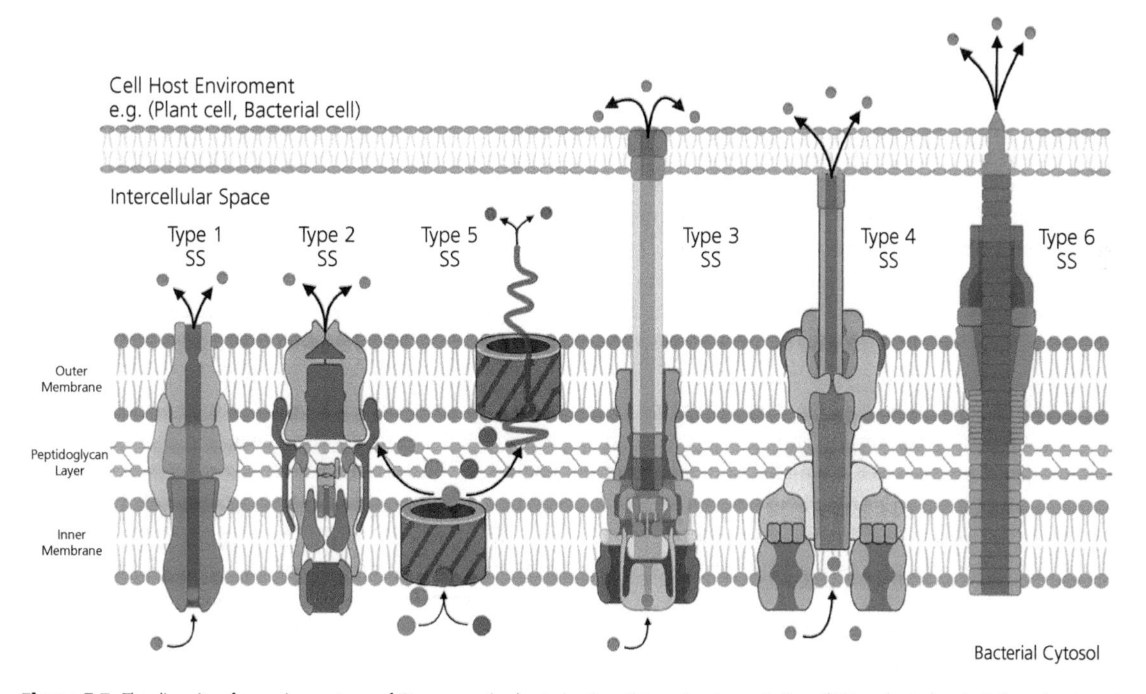

Figure 5.7 The diversity of secretion systems of Gram-negative bacteria. Secretion system types 1, 2, and 5 transport chemicals from the cytosol to the outside of the bacterial cell; these systems are essential in secreting digestive enzymes, signaling compounds, and some toxins. Secretion system types 3, 4, and 6 have needle-and-syringe-like structures that first cross the membrane to the outside of the bacterial cell and then into an adjacent host or bacterial cell where it can inject the chemicals. These types are particularly important in bacterial–plant symbioses and in bacterial conjugation.
Graphics by Andrea Gomez, created with BioRender.com.

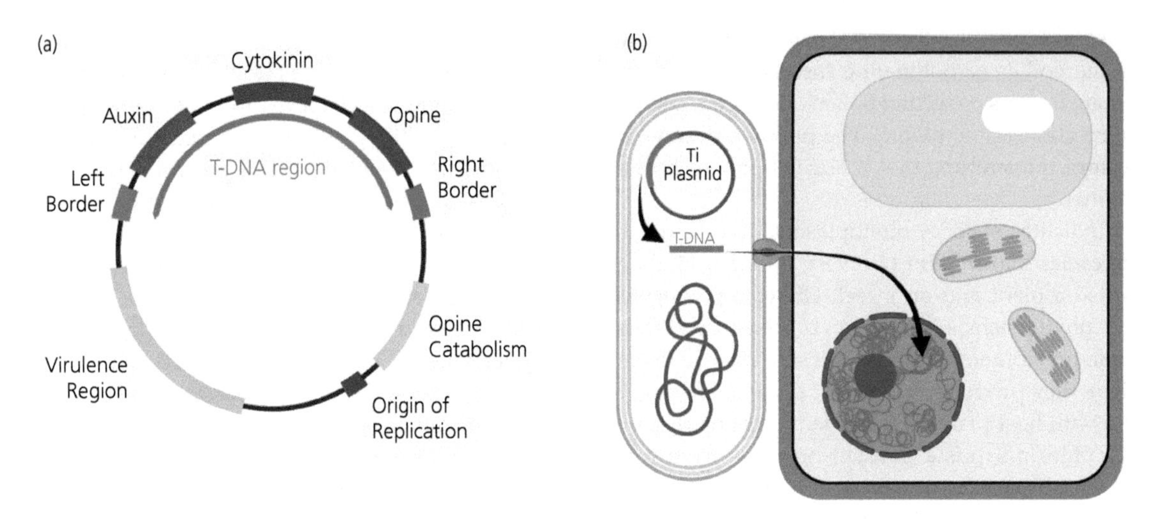

Figure 5.8 The Ti plasmid of *Agrobacterium radiobacter* allows the pathogen to genetically transform its plant host. (a) The T-DNA region of the Ti plasmid includes genes for the plant hormones auxin and cytokinin and for the bacterial energy source opine. The plasmid also includes genes that make the bacterium able to infect the plant and catabolize the opines produced by the transformed plant cells. (b) The bacterium infects the plant and transfers T-DNA into the plant cell using a Type 4 secretion system, where the T-DNA is taken up by the nucleus and integrated into the plant's genome.
Graphics by Ingrid M. Parker, created with BioRender.com.

works alone) and Type 2 (T2SS; which works in tandem with the Sec or Tat pathways) secrete a wide variety of digestive enzymes, adhesins, and some toxins; for plant pathogenic bacteria, this includes plant cell-wall degrading enzymes. Type 3 secretion systems (T3SS) are of particular importance to Gram-negative bacteria that are symbionts of plants because they are able to secrete a wide variety of proteins not only across the bacterial membranes but also across eukaryotic membranes in a two-step process to transport substrates directly into the plant host cell! T3SS have a structure reminiscent of a needle-and-syringe arrangement, and so are sometimes called "injectosomes," and can inject host cells with many types of effectors (Section 12.5) that are critical in plant–pathogen interactions. Type 4 secretory systems can transport both proteins and DNA across cell membranes, and are involved in bacterial conjugation (Section 5.3). Gram-positive bacteria have analogous secretory systems, that are structurally unrelated to those of Gram-negative bacteria.

Whereas many pathogens manipulate their host plants by producing toxins, enzymes, hormones, and polysaccharides, one bacterial pathogen requires special mention for the way it manipulates its host by hijacking the plant's own genome. *Agrobacterium radiobacter* (commonly referred to by its previous name of *A. tumefaciens*) causes crown gall on a wide range of woody dicotyledonous plants, causing problematic tumors on stems (Section 8.3). *A. radiobacter* has a large plasmid, called the Ti plasmid (Ti for Tumor-inducing) that includes a stretch of DNA called the T-DNA (for transfer DNA), with genes for the plant hormones auxin and cytokinin as well as genes to produce opines, which serve as bacterial food (Figure 5.8a). Amazingly, when the bacterium infects the host plant, it transfers its T-DNA into the plant cell, where it is then shuttled into the nucleus of the host plant cell and integrates directly into the plant DNA (Figure 5.8b); *Agrobacterium* genetically transforms the plant with its own DNA! The transformed plant expresses those genes as if they were its own, making extra auxin and cytokinins. Those two plant hormones work together to promote cell division, elongation, and differentiation (Table 2.1), which when over-expressed create a tumor. The bacteria live and grow inside the tumor, happily eating the opines provided by the transformed plant. In Section 17.3 we take a deeper look at *Agrobacterium* as a tool for creating transformed plants in biotechnology.

Bacterial symbionts of plants play a wide diversity of roles: as pathogens, as mutualists, and as antagonists of pathogens that can be useful in biological control. In Chapter 15, we explore the ecology of the plant microbiome, including symbiotic nitrogen fixation and all sorts of microbe–microbe interactions.

Further reading

Madigan, M., K. Bender, D. Buckley, W. Sattley, and D. Stahl. 2017. Brock Biology of Microorganisms, 15th edition. Pearson.

Bergey's Manual of Systematics of Archaea and Bacteria. 2015. Online book available at https://onlinelibrary.wiley.com/doi/book/10.1002/9781118960608.

References

Angert, E. R. 2005. Alternatives to binary fission in bacteria. Nature Reviews Microbiology **3**:214–224.

Bogino, P. C., M. M. Oliva, F. G. Sorroche, and W. Giordano. 2013. The role of bacterial biofilms and surface components in plant-bacterial associations. International Journal of Molecular Sciences **14**: 15838–15859.

Bull, C. T., S. De Boer, T. Denny, G. Firrao, M. F.-L. Saux, G. Saddler, M. Scortichini, D. Stead, and Y. Takikawa. 2010. Comprehensive list of names of plant pathogenic bacteria, 1980–2007. Journal of Plant Pathology **92**: 551–592.

Flemming, H.-C., J. Wingender, U. Szewzyk, P. Steinberg, S. A. Rice, and S. Kjelleberg. 2016. Biofilms: An emergent form of bacterial life. Nature Reviews Microbiology **14**:563–575.

Garrity, G. M., C. T. Parker, and B. J. Tindall. 2015. International code of nomenclature of prokaryotes. International Journal of Systematic and Evolutionary Microbiology **90**.

Green, E. R. and J. Mecsas. 2016. Bacterial secretion systems: an overview. Microbiology Spectrum **4**(1):10.1128/microbiolspec.VMBF-0012-2015.

IRPCM Phytoplasma/Spiroplasma Working Team—Phytoplasma taxonomy group. 2004. '*Candidatus* Phytoplasma', a taxon for the wall-less, non-helical prokaryotes that colonize plant phloem and insects. International Journal of Systematic and Evolutionary Microbiology **54**:1243–1255.

Kearns, D. B. 2010. A field guide to bacterial swarming motility. Nature Reviews Microbiology **8**:634–644.

Khaledian, E., K. A. Brayton, and S. L. Broschat. 2020. A systematic approach to bacterial phylogeny using order level sampling and identification of hgt using network science. Microorganisms **8**:312.

Lebre, P. H., P. De Maayer, and D. A. Cowan. 2017. Xerotolerant bacteria: surviving through a dry spell. Nature Reviews Microbiology **15**:285–296.

Parks, D. H., M. Chuvochina, D. W. Waite, C. Rinke, A. Skarshewski, P.-A. Chaumeil, and P. Hugenholtz. 2018. A standardized bacterial taxonomy based on genome phylogeny substantially revises the tree of life. Nature Biotechnology **36**:996–1004.

Pietikåinen, J., M. Pettersson, and E. Bååth. 2005. Comparison of temperature effects on soil respiration and bacterial and fungal growth rates. FEMS Microbiology Ecology **52**:49–58.

Rousk, J., P. C. Brookes, and E. Bååth. 2009. Contrasting soil pH effects on fungal and bacterial growth suggest functional redundancy in carbon mineralization. Applied and Environmental Microbiology **75**: 1589–1596.

Thomas, C. M. and K. M. Nielsen. 2005. Mechanisms of, and barriers to, horizontal gene transfer between bacteria. Nature Reviews Microbiology **3**:711–721.

Von Bodman, S. B., W. D. Bauer, and D. L. Coplin. 2003. Quorum sensing in plant-pathogenic bacteria. Annual Review of Phytopathology **41**:455–482.

CHAPTER 6

How to be a virus

Gregory S. Gilbert and Ingrid M. Parker

6.1 What are viruses?

Viruses are non-cellular pathogens. Unlike bacteria, fungi, and oomycetes, viruses do not have a cellular structure—they have no plasma membrane surrounding cytoplasm within which metabolism can occur, and no ribosomes to synthesize proteins. They lack the ability to self-replicate; they rely entirely on the cellular machinery of their host cells to reproduce, so they are all obligate parasites. They can persist in a dormant state but cannot grow or reproduce outside their hosts. Viral replication competes with the host for cellular resources, disrupting host physiology (Pallas and García 2011). Viral products can interfere with host gene expression, cell cycle control, and hormonal regulation, leading to disease symptoms (Figure 6.1). All living organisms have viruses—people, plants, fungi, bacteria, you name it.

At their most basic, viruses are composed of a genome (strands of **nucleic acid**, either DNA or RNA) surrounded by a protective **protein coat**, or **capsid**. A capsid together with its nucleic acid core make up a virus particle called a **virion**. Viruses are too small to be seen with a light microscope because they are smaller than half the wavelength of visible light; we have to use an electron microscope

Figure 6.1 Symptoms caused by Grapevine leafroll-associated virus on grape (*Vitis* sp.).
Photo by Gregory S. Gilbert.

The Evolutionary Ecology of Plant Disease. Gregory S. Gilbert and Ingrid M. Parker, Oxford University Press. © Gregory S. Gilbert & Ingrid M. Parker (2023).
DOI: 10.1093/oso/9780198797876.003.0006

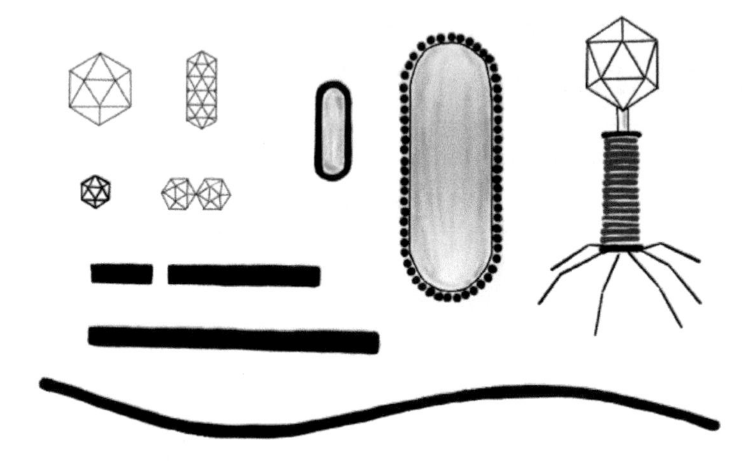

Figure 6.2 Some common shapes of plant viruses. At top left are four versions of polyhedrons. Gemini viruses comprise two 20-sided icosahedrons, with different parts of their genome in each icosahedron. Oval-shaped bacilliform viruses may be either simple or enveloped with glycoprotein appendages. Complex viral structures such as bacteria-infecting phage can look like a lunar-landing module. What look like straight or flexuous rods are actually long tubes formed by many coat-protein subunits strung together in a helix, with the genomic material inside. The sizes of the virus particles are approximately to scale, with the large, enveloped bacilliform rhabdovirus measuring about 50 nm in diameter.
Drawing by Gregory S. Gilbert.

to visualize them. Most viruses range from 20 to 400 nm in diameter, so the very biggest virus is about the same size as the very smallest bacterium (400 nm = 0.4 μm). Viruses have many different shapes, including spherical, helical (filamentous), and polyhedral, as well as the "lunar lander" structure of some complex **phages** (viruses that infect bacteria) (Figure 6.2). Some filamentous viruses, although only a few nm wide, can reach length of 2 μm! For some viruses, the genome is divided among two or more separate virions, requiring each of the component particles to be in the same host cell for the virus to cause disease.

For viruses to reproduce inside their host cells, they must accomplish two things: (1) convert their encoded functional genes as **messenger RNA (mRNA)**, so that the host produces the enzymes and coat proteins they need, and (2) replicate their nucleic acids so they can be packaged into complete virions. Each process is provided by the host's normal cellular machinery; the viral life cycle is essentially hijacking the host cell's basic functions of gene **transcription** and **translation** (Box 6.1). Let's look at the diversity of types of viral genomes, how their genes are expressed in the host plant, and how different types of viruses replicate. Then we'll look at how viruses affect their hosts, and how they move from host to host.

6.2 Genomes, replication, and types of viruses

Viral replication depends on gene expression, which entails the production of mRNA that can be translated into proteins (Box 6.1). Viruses are grouped according to their nucleic acid makeup and how they generate mRNA. They can be either DNA or RNA, and either double- or single-stranded. Single-stranded RNA viruses can either be positive sense (+) or negative sense (–).

Thus, plant viruses include six groups:

ssDNA:	Single-stranded DNA viruses (e.g., *Geminiviridae*)
dsDNA-RT:	Double-stranded, reverse-transcribing viruses (e.g., *Caulimoviridae*)
dsRNA:	Double-stranded RNA viruses (e.g., *Partitiviridae*)
ssRNA(+):	Positive-sense, single-stranded RNA viruses: (e.g., *Potyviridae*)
ssRNA(–):	Negative-sense, single-stranded RNA viruses (e.g., *Rhabdoviridae*)
Viroids:	Non-protein coding, circular, single-stranded RNA (e.g., *Avsunviroidae*)

Most plant-infecting viruses are ssRNA(+) viruses; their genome is functional mRNA and encoded genes can be directly translated into proteins by host ribosomes. For example, in the ssRNA(+) virus *Tobacco mosaic virus* (Box 6.2), ribosomes translate the viral RNA to generate each of its four gene products, including RNA replicase. The RNA replicase drives the transcription of complementary negative-sense ssRNA from the positive-sense viral RNA; the ssRNA(–) then acts as a template for RNA replicase to produce lots of copies of the original viral ssRNA(+), from which more coat proteins and replicases can be produced, and these auto-assemble to create virions.

Box 6.1 A brief review of transcription, translation, and DNA replication

There are two types of nucleic acids within which genes are encoded: DNA and RNA. Genes are encoded into DNA as sequences of four nucleotides: adenine (A), thymine (T), cytosine (C), and guanine (G). In RNA, the thymine is replaced by uracil (U). Sets of three nucleotides comprise a **codon**; there are 64 combinations of 3 of the 4 nucleotides (4^3): each of 61 of them codes for a single amino acid. Amino acids are the component parts of proteins. The remaining 3 codon combinations are "stop" sequences that terminate translation of RNA sequences into proteins. There are twenty usual **amino acids** used to create all manner of proteins, including most of the enzymes critical in cell functioning.

The double-stranded DNA found in cellular organisms includes a positive-sense (+) and a negative-sense (–) strand; the strands have complementary nucleotides (A paired with T, C with G). To produce a protein from the genes encoded in DNA, the first step is to **transcribe** an RNA strand as a complement of the negative-sense strand (using U instead of T). A **DNA-dependent RNA polymerase** enzyme does this job by first separating the DNA strands, and then guiding the assembly of the complementary RNA strand. This positive-sense RNA strand is called **messenger RNA (mRNA)** (Figure 6.3). **Ribosomes** (macromolecular complexes consisting of RNA and associated proteins) are abundant in the cytoplasm of all living cells and serve as platforms that guide the **translation** of the nucleotide codons into amino acids, stringing them together into proteins. Ribosomes bring together the mRNA and its corresponding **transfer RNA (tRNA)**, which carries the appropriate amino acid into place in a growing protein chain. Translation usually

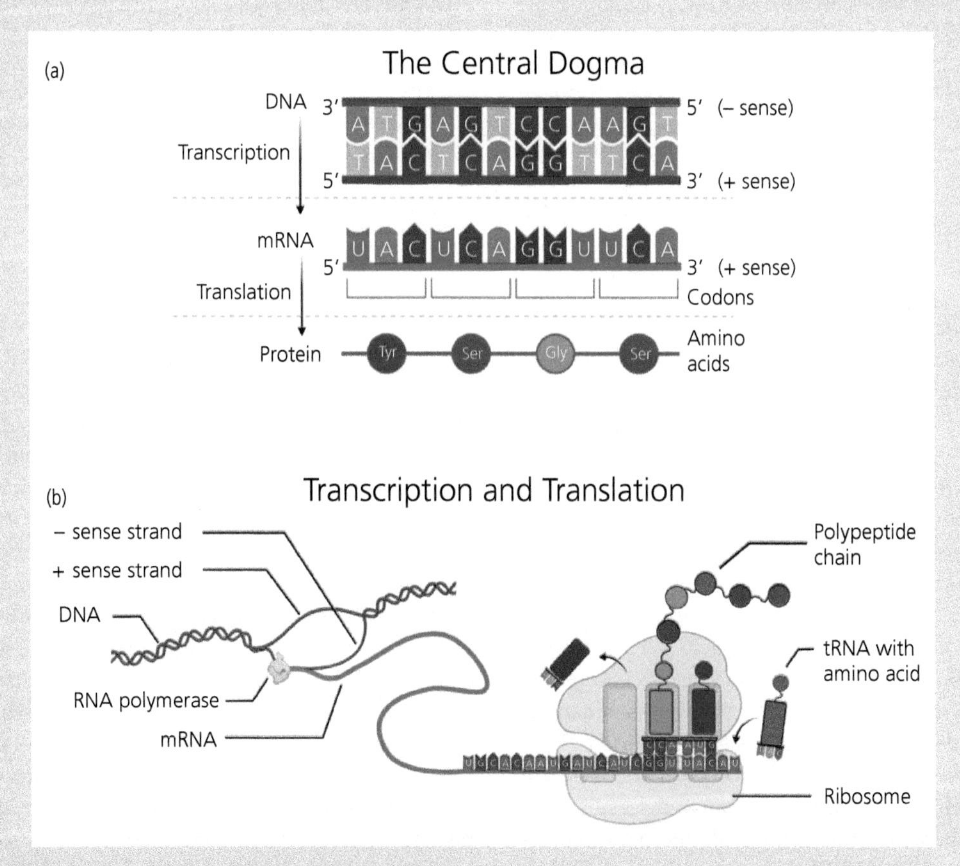

Figure 6.3 Basic process of gene expression. (a) The **Central Dogma** of molecular biology is that DNA is transcribed to mRNA, which is translated into amino acids, which are combined into proteins. (b) The negative-sense (antisense) strand of DNA is transcribed into mRNA by DNA-dependent RNA polymerase. Ribosomes in the cytoplasm guide translation of the mRNA sequence into chains of amino acids to create proteins. Ribosomes guide tRNAs that match the 3-nucleotide codons in the mRNA to place the appropriate amino acids in order to add to the growing polypeptide chain.

Graphics by Ingrid M. Parker, created with BioRender.com; panel (a) adapted from "Central Dogma," retrieved from https://app.biorender.com/biorender-templates.

Box 6.1 *Continued*

begins at an AUG start codon and continues until a stop codon is reached, where the tRNA brings a release factor that ends translation and releases the complete protein from the ribosome.

DNA is replicated (making exact copies of the double-stranded DNA) with the support of three enzymes: DNA helicase (which unzips the DNA into separate strands), DNA polymerase (which makes partial-length DNA strands with nucleotides complementary to each of the unzipped template DNA strands), and DNA ligase (which binds the partial-length DNA strands together into the full length). The product is two double-stranded DNA molecules, each containing one of the template strands and one new, complementary strand.

Whereas DNA is usually replicated from a DNA template and mRNA is transcribed from DNA using fundamental cellular processes universal to most organisms, making copies of RNA from RNA is not. For viruses with RNA genomes, RNA-to-RNA replication is essential, and there are two primary pathways: direct RNA replication, and replication with a DNA intermediary.

Viral RNA is usually single-stranded—either positive- or negative-sense (sometimes called sense and antisense). Direct RNA replication relies on an **RNA-dependent RNA polymerase**, also called **RNA replicase** or **RdRp**. To make a copy of positive-sense single-stranded RNA (ssRNA(+)), the RNA replicase first makes a complementary negative-sense strand of RNA (ssRNA(–)); the RNA replicase then makes a complementary positive-sense strand from that ssRNA(–) template. The process is similar to replicate ssRNA(–); RNA replicase first makes an mRNA template (which is ssRNA(+)), and then a complementary ssRNA(–) copy of that template. Plant cells do not normally produce RNA replicase, so RNA-based viruses encode a gene for RNA replicase in their genome.

The second approach, used by retroviruses, generates double-stranded **complementary DNA (cDNA)** from an RNA template through a process called **reverse transcription**. The process depends on a **reverse transcriptase** enzyme that includes RNA-dependent DNA polymerase activity (making a DNA copy from the single-stranded RNA) and then a DNA-dependent DNA polymerase to produce the double-stranded cDNA. Such cDNA can incorporate into the host genome, or DNA-dependent RNA polymerase can then make complementary RNA molecules, which are replicates of the original RNA viral genome.

In DNA-based viruses, genes are transcribed into mRNA using the standard gene-expression machinery of the host plant. In ssRNA(–) viruses, the genome is first transcribed into mRNA, because negative-sense RNA is complementary to mRNA, and then the mRNA is translated into proteins. Since RNA replication from RNA template is not a regular part of host cell function, all ssRNA(–) viruses have genes that code for RNA-dependent RNA polymerase enzymes (RNA replicase), and their virions come packed with RNA replicase to get the process started (Box 6.1).

Plants are also susceptible to diseases caused by **viroids**—short, circular, single-stranded RNA molecules that do not code for proteins; they are replicated by plant cells and cause disease symptoms probably by interfering with host gene expression. Some viroid RNAs themselves have ribozyme activity involved in viroid replication, suggesting that viroids may represent relics, or at least models, of pre-cellular evolution of life (Diener 1989).

The virus groupings described above appear to correspond somewhat with evolutionary relationships and histories. But systematic classification in viruses is tricky, because viruses have emerged many independent times, and they evolve rapidly because of high rates of mutation and reproduction. In addition, viruses experience substantial **horizontal gene transfer**, which creates reticulate rather than tree-like evolutionary relationships (Section 13.7). Mutation rates of RNA viruses are especially fast because RNA polymerases lack proofreading ability (Rubio et al. 2020). Per-site mutation rates of RNA viroids and viruses are in the range of 10^{-3} to 10^{-5} (as much as one mutation per 400 nucleotides), compared to 10^{-7} to 10^{-9} in bacteria, plants, and animals, as well as DNA viruses (Gago et al. 2009).

Box 6.2 Tobacco mosaic virus

Tobacco mosaic virus (TMV) is an ssRNA(+) virus of tremendous importance both for its impacts on many agronomic crops (not just tobacco), and for its contributions to our basic understanding of how viruses work (Scholthof 2004). Back in 1879, Adolf Mayer first showed that the mosaic disease of tobacco could be transmitted with plant juice extracts. A couple decades later in 1898, the celebrated microbiologist Martinus Beijerinck conducted filtration experiments to exclude all bacteria or fungi from the plant juice extracts, while retaining infectivity. He coined the term *virus* to describe the pathogenic agent. TMV continued to be the "first" in a series of events in virology. Wendell Stanley crystalized the virus (Stanley 1935) and showed that it retained pathogenicity, work for which he won a Nobel Prize in 1946. Fraenkel-Conrat and Williams (1955) showed that the component parts of TMV (the coat protein subunits and RNA) spontaneously self-assemble into virions when in solution. And finally, Tsugita et al. (1960), working in the Stanley lab, completed the first complete coat-protein sequence of any virus.

The TMV genome contains 6400 nucleotides that code for only four genes (Figure 6.4): a coat protein, two replicase-associated proteins (a full 183 kDA protein and a shorter 126 kDA version), and a movement protein. (In comparison, humans have 3.05 billion nucleotides that code for about 20,000 genes.) Each coat protein is composed of 158 amino acids. Once there are lots of copies in the cell along with strands of ssRNA(+), the virions auto-assemble, with one copy of the genome surrounded by many coat proteins arranged in a tube-like helix (Figure 6.4). Those TMV virions are now ready to infect a new plant!

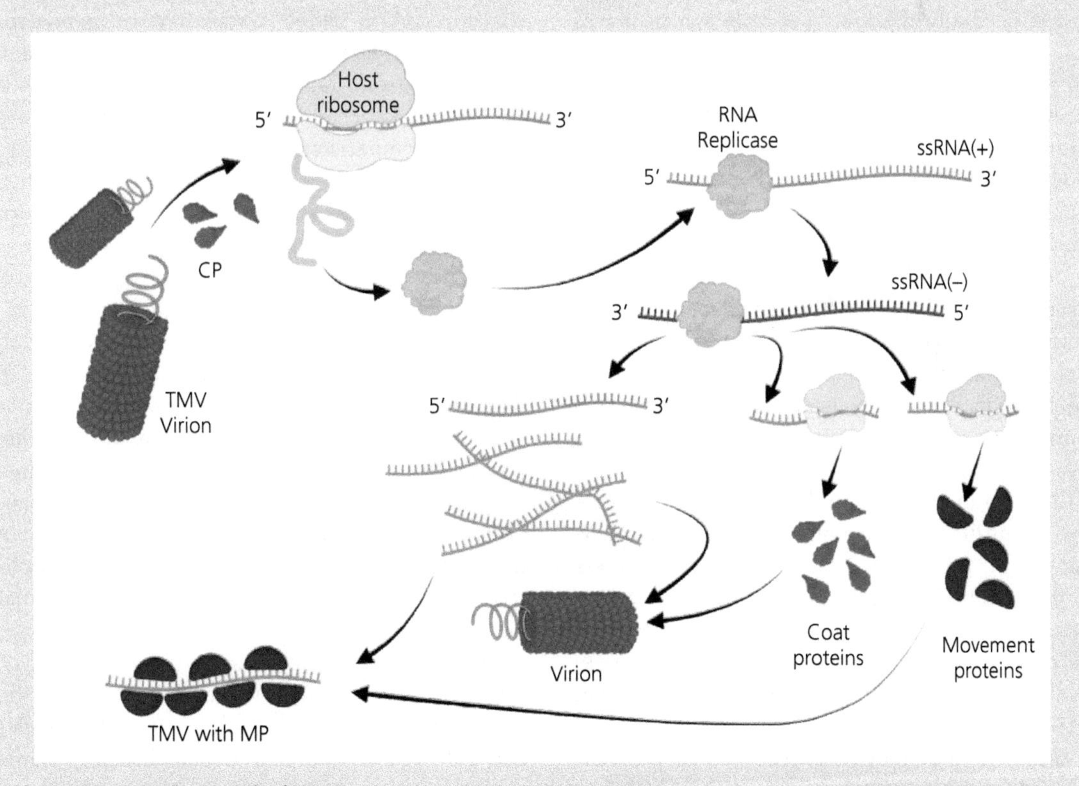

Figure 6.4 The replication cycle of *Tobacco mosaic virus* (TMV). The TMV particle (virion) enters a plant cell, and the ssRNA(+) strand dissociates from the coat protein (CP) molecules. The host ribosomes translate the RNA replicase proteins, which generate negative-sense RNA (ssRNA(–)) template strands. From this strand, the replicase generates many copies of the complete virus genome, as well as shorter RNA strands that are translated into coat proteins and movement proteins (MP). The TMV RNA strands are either encapsidated by CP to produce new virions, or they are wrapped with MP, which enables movement within or between cells.
Graphic by Ingrid M. Parker, created with BioRender.com.

Only in 2020 was general agreement reached for a hierarchical taxonomic framework for viruses that corresponds structurally to the hierarchical taxonomic structure in use for cellular organisms (International Committee on Taxonomy of Viruses Executive Committee 2020). It includes 5560 recognized species in 1019 genera, organized into 150 families, 14 orders, and 6 classes. All viruses are included in a single phylum within the realm (rather than domain) Riboviria. Species names for viruses do not follow the *Genus species* binomial convention used in the Linnean taxonomic system of cellular organisms. In fact, there is still no standard structure for the names other than that the name must be short and include more words than just the name of the host and the word "virus." In general, plant virus names take the form *Host symptom virus*, e.g., *Tobacco mosaic virus* or *Bean golden mosaic virus*. The host is generally the first host from which it was isolated, not necessarily the host in which it is causing disease. Common symptom descriptors used in the names of plant viruses include mosaic, streak, chlorosis, yellows, fleck, stunt, dwarf, necrosis, and spot (Chapter 8).

6.3 Viral transmission

Viruses generally enter the plant cell through wounds, which may be caused by the action of a vector organism or through physical abrasions or damage. Once inside the host cell, the coat protein components are stripped off the nucleic acid, which allow the transcription (DNA to RNA, or RNA(−) to RNA(+)) and translation (RNA to protein) to proceed. Some plant viruses enter the plant cell nucleus and complete their life cycle there, whereas others go through their cycle in the cytoplasm. Within a plant host, viruses can move between cells through the plasmodesmata (Section 2.1), aided by movement proteins coded in the viral genome (Heinlein 2015).

Viruses cannot move between plants on their own. Most are actively transmitted from plant to plant by a living vector—often an arthropod, nematode, fungus, or even a parasitic plant (Ng and Perry 2004). Animals such as aphids or nematodes that have piercing-and-sucking type mouthparts, called a **stylet**, are particularly good at transmission because they tap straight into living plant cells (Figure 6.5). Most virus transmission is **noncirculative**, or **nonpersistent**, in which the virus gets stuck to the outside of the stylet or is retained inside the stylet or in the foregut of the insect, and then is dropped off at the next plant that vector bites into (Figure 6.5a). Originally the term **stylet-borne** was considered a synonym for noncirculative transmission, before it was discovered that noncirculative viruses are sometimes carried inside the digestive system of the vector and not just on the outside. Noncirculative virus transmission usually involves rapid virus acquisition and transmission, with viruliferous periods in the vector that are short (minutes, not days).

By contrast, in some cases the virus enters the vector's circulatory system and persists there; this transmission is called **circulative** or **persistent** (Figure 6.5b). The virus enters the digestive system and moves from there into the hemocoel, which is the general body cavity, and finally into the salivary gland. In comparison to the noncirculative viruses, it takes longer for the aphid to acquire a circulative virus, and the virus may persist inside its vector for weeks, rather than minutes. Some viruses are able to reproduce inside the vector as well as in the plant; such transmission is persistent and also **propagative**, because the virus propagates inside the vector. This is similar to how phytoplasmas propagate in their insect vectors (Section 5.7). The ability to replicate in both plants and insects is relatively rare, and all the propagative viruses come from the same family, the Rhabdoviridae. Interestingly, the Rhabdoviridae also includes a number of animal viruses transmitted by insects, in which the virus can replicate within both its insect host and its mammal host.

Although most viruses require a living vector, some are **mechanically** transmitted. For instance, plants infected with *Tobacco mosaic virus* (TMV) that rub against and abrade neighboring plants can transmit the virus. Smokers who handle dried tobacco from an infected plant can transmit TMV to plants they touch, just by abrasion of the leaf surfaces. Although one might say the smoker is a vector, the transmission is mechanical. Research labs

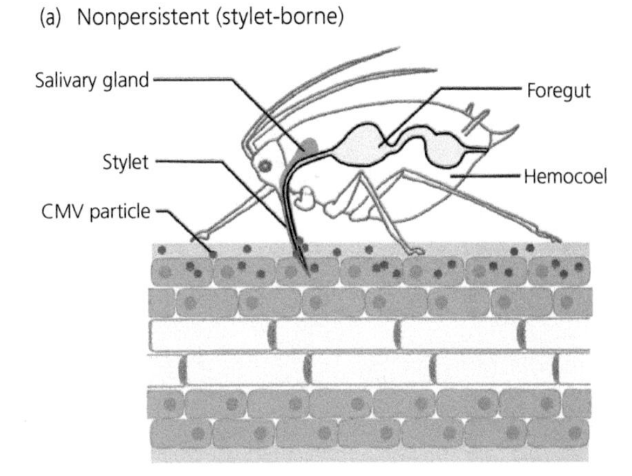

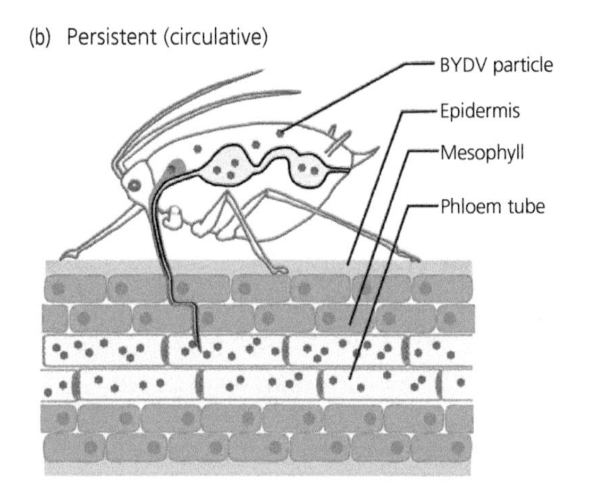

Figure 6.5 Virus transmission by aphids can be either nonpersistent (noncirculative) or persistent (circulative). (a) In nonpersistent transmission, as in the example of *Cucumber mosaic virus* (CMV), the aphid picks up the virus particles (in purple) either on the outside of the stylet (called stylet-borne), or inside the anterior parts of the digestive system (not shown). (b) In persistent transmission, as in the example of *Barley yellow dwarf virus* (BYDV), the aphid generally picks up the virus particles (in orange) by ingesting them along with phloem sap. The virus enters the digestive system and moves from there into the hemocoel, eventually ending up in the salivary gland, from which it can be transmitted to a new host plant.
Graphics by Ingrid M. Parker, created with BioRender.com.

that study TMV must follow extremely strict protocols in the greenhouse in order to avoid spreading the virus among their experimental plants.

Further reading

Hull, R. 2014. *Plant Virology*, 5th edition. Academic Press, New York.

References

Diener, T. 1989. Circular RNAs: relics of precellular evolution? *Proceedings of the National Academy of Sciences of the United States of America* **86**:9370–9374.

Fraenkel-Conrat, H. and R. C. Williams. 1955. Reconstitution of active tobacco mosaic virus from its inactive protein and nucleic acid components. *Proceedings of the National Academy of Sciences of the United States of America* **41**:690.

Gago, S., S. F. Elena, R. Flores, and R. Sanjuán. 2009. Extremely high mutation rate of a hammerhead viroid. *Science* **323**:1308–1308.

Heinlein, M. 2015. Plasmodesmata: channels for viruses on the move. *Plasmodesmata* **1217**:25–52.

International Committee on Taxonomy of Viruses Executive Committee. 2020. The new scope of virus taxonomy: Partitioning the virosphere into 15 hierarchical ranks. *Nature Microbiology* **5**:668.

Ng, J. C. and K. L. Perry. 2004. Transmission of plant viruses by aphid vectors. *Molecular Plant Pathology* **5**:505–511.

Pallas, V. and J. A. García. 2011. How do plant viruses induce disease? Interactions and interference with host components. *Journal of General Virology* **92**:2691–2705.

Rubio, L., L. Galipienso, and I. Ferriol. 2020. Detection of plant viruses and disease management: relevance of genetic diversity and evolution. *Frontiers in Plant Science* **11**:1092.

Scholthof, K.-B. G. 2004. Tobacco mosaic virus: a model system for plant biology. *Annual Review of Phytopathology* **42**:13–34.

Stanley, W. M. 1935. Isolation of a crystalline protein possessing the properties of tobacco-mosaic virus. Science **81**:644–645.

Tsugita, A., D. T. Gish, J. Young, H. Fraenkelconrat, C. A. Knight, and W. M. Stanley. 1960. The complete amino acid sequence of the protein of Tobacco Mosaic Virus. *Proceedings of the National Academy of Sciences of the United States of America* **46**:1463–1469.

CHAPTER 7

How to be a plant macroparasite

Gregory S. Gilbert and Ingrid M. Parker

7.1 Why a chapter on macroparasites?

Most plant pathogens are microscopic or even submicroscopic parasites, but we would be remiss if we did not introduce two **macroparasites** that are usually thought of as plant pathogens: parasitic plants and nematodes. We bundle them here in one chapter as misfits that often fall between the cracks in university courses, that differ from all the other parasites in being multicellular organisms that are larger than the plant cells they parasitize.

Whereas most plants produce their own sugars as photosynthetic autotrophs and get their water and mineral nutrients through a well-developed root system and mycorrhizal associates, some plants are heterotrophs that have evolved mechanisms to meet their needs by parasitizing other plants (Press and Phoenix 2005, Nickrent 2020). Some parasitic plants have gone all-in and lost their ability to photosynthesize, receiving all their nutrition directly from their host plant. For instance, *Rafflesia arnoldii*, also known as the corpse flower, is a rare plant native to Borneo and Sumatra that produces the largest (and probably worst-smelling) flower in the world (Figure 7.1E and Nikolov and Davis 2017). The plant is actually an endoparasite on the woody vine *Tetrastigma*, from which it derives all its nutrition and water, until it produces a single flower that can grow to 1 m in diameter! While all the plants that make a living as parasites are fascinating, a handful of them can be significant as plant pathogens, and those will be our focus.

Nematodes, or roundworms, are animals just barely big enough to see with the naked eye (though you really need a stereoscope or microscope to see them well). Nematodes are familiar because they can be parasites on humans and our pets (e.g., guinea-worms, roundworms, and liver flukes) (Jourdan et al. 2018). Most university students encounter the most famous nematode *Caenorhabditis elegans* in courses on genetics, developmental biology, and toxicology (Hunt 2017). However, those classes generally ignore nematodes that attack plants. Since they are not insects, nematodes are not usually covered in entomology courses, and they are neglected in most discussions of herbivory. Nematologists who study plant parasites most often find themselves in departments of plant pathology—and usually there is only one nematologist among a large group of people who study fungi, bacteria, and viruses. Particularly in tropical and subtropical settings, nematodes cause tremendous damage to a diversity of crops. Curiously, there has been almost no exploration of the impact or diversity of plant parasitic nematodes in wild systems outside of grassland and sand dune systems (Van der Stoel et al. 2002, Todd et al. 2006). But such neglect is not deserved! Future studies may reveal that they are very important in wild ecosystems. In this chapter we introduce the general life-cycle features of plant parasitic nematodes, representing each of several common life-history strategies.

7.2 Parasitic plants

There are about 4,700 species of flowering plants (about 1.6% of all angiosperms) that are parasitic on other plants (Nickrent 2020). They are found throughout tropical and temperate environments,

The Evolutionary Ecology of Plant Disease. Gregory S. Gilbert and Ingrid M. Parker, Oxford University Press. © Gregory S. Gilbert & Ingrid M. Parker (2023).
DOI: 10.1093/oso/9780198797876.003.0007

Figure 7.1 A menagerie of parasitic plants. (a) Orange stems of holoparasitic dodder (*Cuscuta gronovii*). (b) Holoparasitic western dwarf mistletoe *Arceuthobium campylopodum* on ponderosa pine (*Pinus ponderosa*). (c) Dark green clumps of the hemiparasitic bigleaf mistletoe (*Phoradendron tomentosum*) in branches of a western sycamore (*Platanus racemosa*). (d) The root parasite witchweed (*Striga hermonthica*) on pearl millet (*Cenchrus americanus*) starts as a root holoparasite and then transitions to a hemiparasite. (e) Stinking corpse flower *Rafflesia arnoldii* is a root holoparasite of the liana *Tetrastigma leucostaphylum* and produces the largest flower in the world. (f) The mycoheterotrophic snowplant, *Sarcodes sanguinea*, obtains nutrition from nearby trees bridged through the ectomycorrhizal fungus *Rhizopogon ellenae*.

Photos a, b, c, and f by Gregory S. Gilbert. Photo d by Muhammad Jamil. Photo e by Luke Triton under the Creative Commons Attribution-Share Alike 4.0 International license.

in wildlands, and in agriculture. Some parasitic plants specialize in parasitizing the above-ground stems of host plants (Figure 7.1a, b, and c) and others the root systems (Figure 7.1d, e, and f). In this chapter we look at the traits, evolution, physiology, and impact of parasitic plants with three different life-history strategies: holoparasites, hemiparasites, and mycoheterotrophs (Figure 7.1). **Holoparasites** such as *Rafflesia* acquire sugars (carbon) as well as mineral nutrients and water from their host plants (Figure 7.1a, b, and e). Holoparasites are either non-photosynthetic or have greatly reduced ability to photosynthesize. Holoparasites tap into both the xylem and phloem of their host plants. **Hemiparasites** are able to do their own photosynthesis to fix carbon from the atmosphere, but they tap into the xylem of the host plant to get water and mineral nutrients [*Holo* = entire and *Hemi* = half] (Figure 7.1c and d). **Mycoheterotrophs** are epiparasites that get their nutrients from neighboring plants, facilitated by a mycorrhizal fungal network (Figure 7.1f). Hemiparasitic plants are autotrophs like other plants, but holoparasites and mycoheterotrophs are heterotrophs, getting their carbon from another organism. Some parasitic plants start off as holoparasites and then later become hemiparasites (e.g., *Striga*, Section 7.2.3).

Parasitism is a derived trait in plants, having evolved independently from lineages of autotrophic plants a dozen times (Nickrent 2020). For instance, *Cuscuta* (dodder; Figure 7.1a and Section 7.2.1) has evolved non-photosynthetic vining stems that bind together hosts and penetrate the host vascular system. The genus *Cassytha* (laurel-dodder) evolved

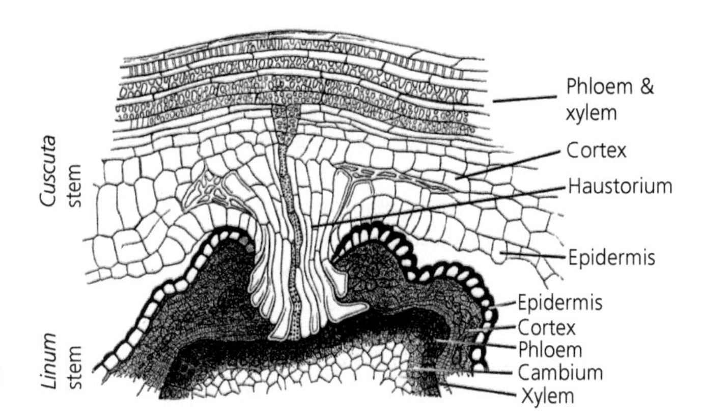

Figure 7.2 Haustorium of holoparasitic dodder (*Cuscuta epilinum*) parasitizing a stem of flax (*Linum usitatissimum*). Drawing modified by Gregory S. Gilbert from original that appeared in Sachs (1887).

a habit and structures nearly identical to *Cuscuta*, even though *Cassytha* is in the family Lauraceae and *Cuscuta* is in the family Convolvulaceae, and these families are separated by at least 100 million years of independent evolution!

Most parasitic plants are biotrophs, relying on their host plants to stay alive to be able to complete their life cycle. A key anatomical feature of holoparasitic and hemiparasitic plants is the production of **haustoria**: multicellular, root-like organs that absorb carbon, nutrients, and water from their host vascular tissue (Figure 7.2 and Teixeira-Costa 2021). Note that this is the same term we use for an analogous modified hyphal structure produced by many biotrophic fungal symbionts (e.g., rusts) to transfer nutrients from the plant host.

7.2.1 Dodder

A holoparasitic plant genus is dodder, including more than 200 species in the globally distributed genus *Cuscuta*, in the morning glory family (Convolvulaceae) (Figure 7.1a). Many species of dodder are highly host specific, while some have broad host ranges. Some species are serious agricultural pathogens, whereas others are influential and beneficial players in wild ecosystems (Grewell 2008). Dodder grows as long, tangled, masses of stems that are yellow, orange, or red. They are leafless and (for most species) nonphotosynthetic. The dodder stems wrap around and penetrate host stems with haustoria that tap into the vascular system to extract sugars, water, and mineral nutrients. They have remarkable methods for finding their host plant; if a dodder stem is growing free, it will actively grow towards a susceptible host and away from a non-host (Runyon et al. 2006)!

An additional importance of *Cuscuta* in plant pathology is that it can serve to transmit viruses and phytoplasmas from one host plant to another (Hosford 1967, Zhou et al. 2007). A long, twisty, stringy stem of a dodder individual can connect simultaneously to numerous host individuals. This effectively creates a conduit among the vascular systems of all those plants—the dodder serves as a parasitic bridge. Viruses and phytoplasmas move through the conduit, and so can physiologically important signaling compounds including chemical signals produced by one host plant in response to attack from an herbivore (Hettenhausen et al. 2017, Banerjee 2020).

7.2.2 Mistletoes

Mistletoes are above-ground parasites on woody plant species. All mistletoes are members of the sandalwood order (*Santalales*), which includes many clades of parasitic plants (Nickrent et al. 2010). There are two strikingly different life histories within the mistletoes: the leafy mistletoes (genus *Phoradendron* in the Americas and *Viscum* in Europe) and the dwarf mistletoes (genus *Arceuthobium*).

Leafy mistletoes (Figure 7.1c) are familiar in western culture as good-luck ornaments for the home; hanging mistletoe is a custom that dates back to ancient Druid and Norse cultures and was later associated with Christmas and kissing. But the name mistletoe derives from the Anglo-Saxon *mistel* (dung) and *tan* (twig): the poop-on-a-stick name refers to how the red, white, or yellow berries are eaten by birds, which then defecate sticky seeds a short time later (Aukema and Martínez del Rio 2002). The pooped seeds retain a sticky coating, and those that land on a tree branch will stick to it and later germinate. The mistletoe plant then forms a haustorium that pushes into the host plant and taps into the xylem, from which it obtains all its water and mineral nutrients. The shoot grows and produces green leaves, capable of photosynthesis. Thus, the leafy mistletoes are obligate hemiparasites.

Leafy mistletoe species are biotrophs that colonize a wide variety of (mostly) angiosperm woody plants. Some of them have a rather broad host range, and others are specialized on just one or a few hosts. They can cause branch swelling and witches' broom formation but generally do not cause severe disease impacts on their hosts. They sometimes facilitate the entry of destructive secondary pathogens into the host, which can lead to branch dieback or reduced host growth (Carnegie et al. 2009). A dramatic sight in winter is the profusion of parasitic shrubs of the evergreen leafy mistletoe scattered throughout the crown of a deciduous host tree.

In contrast to the leafy mistletoes, dwarf mistletoes (Figure 7.1b) have greatly reduced or no photosynthetic ability. *Arceuthobium* species attack gymnosperms in the Pinaceae in North America and Cupressaceae across Europe, Asia, and Africa. Unlike their leafy cousins, they can cause severe reductions in host-tree growth and survival and can be significant problems in forestry (Shaw et al. 2008). *Arceuthobium* has sticky (viscous) seeds, but instead of being bird-dispersed, seeds are explosively shot from the plant (ballistic dispersal) through a buildup of hydrostatic pressure, launching at 50 km/h and traveling for meters (Hawksworth 1959). The seeds stick to host tree needles and then, following a rain, swell and slide down to the young branch. There, the seeds germinate, producing a radicle that pushes into the host and forming a haustorium that taps into the phloem and xylem. Living entirely inside the host branch, the dwarf mistletoe is dependent on the host for all its water and nutrition for a few years. Eventually, the mistletoe erupts from the host branch and makes branches of its own. Its reduced leaves are like small scales, and while it has functional chloroplasts, its ability to photosynthesize is perhaps a tenth that of its host. Growth of dwarf mistletoe can cause branch swelling or witches' broom, lead to secondary infections, and reduce the growth of the host tree. Particularly when the infection happens on the trunk of a smaller tree, the swelling and distortion of the cambial tissue can girdle the host, killing it. This of course is not good for the mistletoe, because without a living host, the dwarf mistletoe dies as well.

7.2.3 Witchweed

Let's look at perhaps the most damaging of all parasitic plants, witchweed. Witchweed includes 33 species in the genus *Striga*, which is in the family Orobanchaceae (Figure 7.1d). They are beautiful flowering plants native to Africa and Asia, where some species can be devastating agronomic pathogens. *Striga* seeds are extremely tiny (~0.2 mm diameter), and can survive in the soil for many years. They germinate in response to sensing a nearby growing root of a host plant (Saucet and Shirasu 2016). The seedling radicle first attaches to the host root with specialized hairs and then penetrates into the root via a haustorium. The haustorium taps into the vascular tissue of the host plant, obtaining carbon from the phloem. In addition, finger-like tubular projections called oscula grow out of the haustorium and into the xylem cells, where they take up water and mineral nutrients from the host. At this stage, *Striga* is a holoparasite, dependent on the host plant for all its sustenance and growth, and it drains significant carbon and water resources from the host until it emerges above ground. It also affects the host by distorting its hormonal balance. Once above ground, *Striga* transitions to being a hemiparasite, able to do some (limited) photosynthesis but is completely reliant on the host for water and mineral nutrients because it does not produce its own root system. Eventually, *Striga* produces beautiful, brightly colored flowers (Figure 7.1d) and then thousands of tiny, tough seeds that infest the soil and are easily spread around when soil is moved on wind, water, or machinery.

The cereal crops it attacks—maize, sorghum, pearl millet, rice, and more—can be so dramatically affected by the rapidly growing and transpiring *Striga* that they are severely stunted and fail to reproduce (Watling and Press 1997). In heavily infested areas, *Striga* can limit the ability to grow certain crops (Dugje et al. 2006), and phytosanitary programs are serious about limiting its spread. *Striga* was accidentally introduced into the southeastern United States, where it spread across an area of 174,000 ha; after an intensive 50-year eradication effort, the infested area has been reduced to just 1% of that (Tasker and Westwood 2012). Other similar holoparasitic plants have less of an impact on their hosts. Another species in the Orobanchaceae, *Epifagus virginiana*, commonly called beechdrops, is an obligate root holoparasite of American beech trees (*Fagus americana*) (Tsai and Manos 2010). While the association with its host is very similar to that of *Striga*, *Epifagus* is so much smaller than its host plant that any impact on host growth is negligible.

7.2.4 Mycoheterotrophs

Another group of parasitic plants that we can't resist mentioning are the mycoheterotrophs. These are plants that are nonphotosynthetic and acquire their carbon, water, and mineral nutrients as parasites of fungi. The fungal associates acquire their carbon either through mycorrhizal associations with plant hosts or as saprotrophs, breaking down soil organic matter. Many mycoheterotrophic plants have completely lost their ability to photosynthesize—such as the spectacular red *Sarcodes sanguinea* or the ghostly white *Monotropa uniflora* (both Ericaceae). Such plants form root symbioses with mycorrhizal fungi, often with great specificity (Bidartondo et al. 2002). However, unlike other mycorrhizal associations, the plants do not provide their fungi with sugars in exchange for soil nutrients and water, since they are non-photosynthetic. Instead, the fungi form mycorrhizal associations with other, neighboring photosynthesizing plants (often trees) from which they obtain sugars. The mycoheterotroph then parasitizes the fungus, stealing the sugars that the fungus had acquired from the legitimate mycorrhizal host plants; this makes the parasitic plant an **epiparasite**. While this three-way interaction allows the mycoheterotrophic plants to thrive at the expense of neighbors, the neighboring plants do not seem to suffer significantly from their epiparasites (Bidartondo 2005).

7.3 Nematodes

Plant parasitic nematodes are just one trophic group within the diverse phylum Nematoda, which also includes many notable parasites of insects and vertebrates, fungal feeders (they suck out the innards of fungal hyphae), and grazers of bacteria, algae, and soil micro-invertebrates. The life history of a given nematode can usually be guessed by looking at anatomical features, especially the mouthparts, and the details of their life cycles vary according to their feeding practices and habitats. Here we confine ourselves to describing the anatomy and life cycles of plant parasitic nematodes.

Plant parasitic nematodes are the only animals that are traditionally thought of as plant pathogens, rather than herbivores. (Just go with it . . . it's what everyone does.) Most plant parasitic nematodes are soil-borne pathogens of plant roots, although there are some notable exceptions that attack other plant parts, like the pine-wood nematode, *Bursaphelenchus xylophilus* (Futai 2013). They range in size from about 0.3 mm to 2 mm in length and differ from microbial pathogens in that they ingest, rather than absorb, their food. They feed on plant cells using distinctive piercing–sucking mouthparts called **stylets** that function like retractable hypodermic syringes (Figure 7.3). In this respect, they may seem to have more affinity with piercing–sucking insects like mosquitos or aphids (Figure 6.5) than with microbial plant pathogens. They also differ from microbial pathogens by being multicellular, eukaryotic, unitary organisms that behave as individuals. Unlike Fungi, not every nematode cell is totipotent, able to grow independently to complete its life cycle. As **opisthokonts** (Eukaryotes that include Animals and Fungi, with the shared characteristic of having motile cells with a single posterior flagellum, like sperm; *opistho* = behind, *kont* = punting pole), they are more closely related to fungi than any other organisms we consider. Yet, as animals they lack cell walls. Their bodies are surrounded by a tough cuticle made of chitin, similar to an insect exoskeleton. In fact, within the animal kingdom, they are much more closely related to insects than they are to earthworms, mollusks, or people. Well, that description mostly says what they are not—so what are they?

We start with a general overview and then look at the four possible intersections of two key life-history traits of nematodes: plant parasitic nematodes may be **ectoparasites** (stay outside the plant) or **endoparasites** (enter inside the plant), and within each category they may be **migratory** (an individual moves around while feeding) or **sedentary** (stays put once it finds a host). Finally, we briefly describe the roles of nematodes as vectors or facilitators of other plant pathogens.

7.3.1 Anatomy of a plant parasitic nematode

A nematode is a multicellular animal (around 1,000 cells), but is basically a long, one-way tube designed for feeding and reproducing. For a plant-parasitic nematode, food is plant cellular cytoplasm, sucked in through the stylet, passed through the esophagus and intestines, and out the anus at the tail end (Figure 7.3). (Curiously, the shape and cuticular patterns around the anus are often important features to tell species apart.) The stylets of plant parasitic nematodes vary widely in size and shape, generally in relation to their life-histories, with ectoparasites generally having longer and more slender stylets than endoparasites. Fungal-feeding nematodes may also have a stylet, but they are shorter and finer than the stylets of most plant parasites. In plant parasitic nematodes the **median esophageal bulb** functions like a pump and serves to keep food moving into the intestine where it is digested, and the **basal esophageal bulb** is where digestive enzymes are generated and secreted into the digestive tract. Not surprisingly, the shapes and sizes of the stylets and esophageal bulbs are important criteria in the taxonomy and identification of plant parasitic nematodes (Figure 7.3). Nematodes are transparent and large enough that it is possible to see the stylet and esophagus of a plant parasitic nematode with a stereoscope or even a hand lens, which allows you to distinguish them. Nematodes have an unusual epidermis that is not composed of individual cells but is rather a mass of cell-like material with many nuclei and without discrete plasma membranes. The epidermis is covered with a tough but flexible cuticle made of chitin, similar to that found in arthropods.

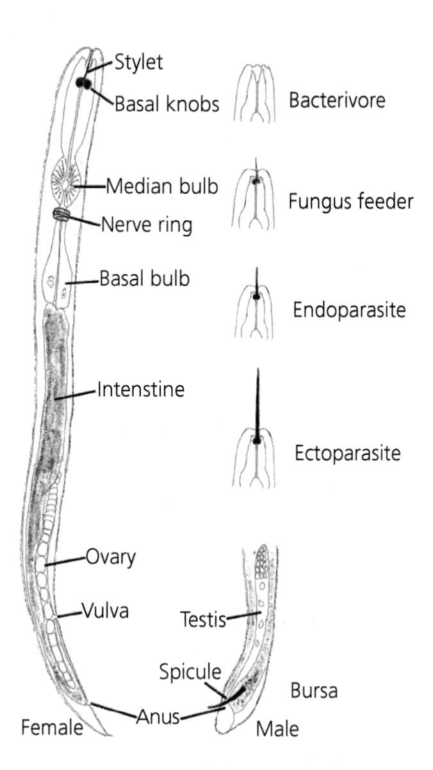

Figure 7.3 Anatomy of a plant parasitic nematode. Male and female nematodes show the characteristic retractable stylet at the mouth end and the enlarged esophageal bulb, which moves food through the digestive system and out the anus. The stylet is an easily seen feature that distinguishes plant parasites from bacteria-feeding nematodes. Fungal-feeding nematodes have fine, short stylets; endoparasites such as *Pratylenchus* penetrate inside the plant root, feeding on and killing individual root cells by consuming cell contents with a short stylet; ectoparasites such as *Belonolaimus* have very long stylets allowing them to penetrate deep within the plant root.
Drawing by Gregory S. Gilbert.

7.3.2 Nematode life cycle

Most nematodes have diploid nuclei, making haploid gametes in an ovary (female) or a testis (male). However, they are nearly as diverse as fungi or plants in their reproductive options, including **dioecy** (male and female individuals), **hermaphroditism** (an individual is both male and female), and **parthenogenesis** (asexual reproduction). Some nematodes are even polyploid, which

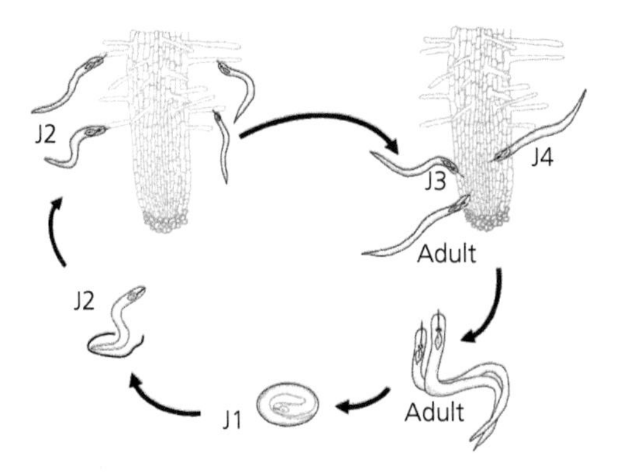

Figure 7.4 Life cycle of the migratory ectoparasite *Belonolaimus* (sting nematode). Sting nematodes follow a typical life cycle for most plant parasitic nematodes. The first stage juvenile (J1) develops in the egg; J2 emerges from the egg, migrates through soil, and feeds on root hairs. Nematodes grow through three more molts to become J3, J4, and finally adults, each of which feed in the zone of elongation. The sting nematode is dioecious, with separate male and female adults that mate to produce fertile eggs. Many nematode species can instead (or additionally) lay eggs parthenogenically (with males absent or rare), and some are hermaphrodites, able to self-fertilize their own eggs.
Drawing by Gregory S. Gilbert.

is rather unusual for an animal. Let's start off by discussing a generalized life cycle for a generic plant parasitic nematode, using the sting nematode *Belonolaimus* as a model (Figure 7.4), and then explore some of the variations. Different species have different basic cycles, and some species can choose different pathways depending on the circumstances.

Plant-parasitic nematodes lay eggs, usually in the soil or on plant roots. The first-stage juvenile (called the J1) develops while still inside the egg. It is a fully formed, but very small, nematode inside the egg. The J1 then molts (sheds its cuticle and grows larger) to become a J2 (second-stage juvenile), which hatches from the egg and can move freely through the soil. Nematodes can't swim, but they can wiggle their way through soil pores. The J2 stage is when plant parasitic nematodes often act like "troublemaking kids;" that is when they move around, looking for a host plant, and cause initial infections. Because they must move through the soil with its complex three-dimensional structure, nematode dispersal is often limited to distances measured in centimeters or meters. Once they find a host, they feed on the root cells using their stylets. The nematodes grow, molt again to the J3 stage, eat some more, grow some more, and molt again to enter the J4 stage, the last juvenile stage. With the next molt, they become mature adults. Many nematodes can reproduce through **parthenogenesis**, with female adults producing unfertilized but viable eggs that are genetic clones of the female.

In addition to asexual parthenogenesis, there are two versions of sexual reproduction. In the **dioecious** version, J4s molt to become either adult males (with a testis that produces sperm) or adult females (with an ovary that produces eggs). In many plant-parasitic nematodes, females can produce both parthenogenic diploid eggs and sexual haploid eggs, and they are facultatively sexual. In some species, adult males are extremely rare, and sexual reproduction is an infrequent event. In these situations, the ecological dynamics are driven by parthenogenic reproduction, although sexual reproduction is still evolutionarily important because it allows for genetic recombination and outcrossing between individuals. The second form of sexual reproduction in nematodes is **hermaphroditism**. In some nematode species, J4s molt to produce adults with both ovary and testis within the same individual. In such cases, sperm from the testis can fertilize eggs in the ovary of the same nematode, which is self-fertilization, an extreme form of inbreeding. This still involves genetic recombination, because it requires meiosis followed by fertilization and karyogamy, but it does not involve outcrossing. This is analogous to the common phenomenon of selfing in plants. Hermaphrodites can also fertilize the eggs of other hermaphrodites, leading to outcrossing. Note that whereas eggs produced through parthenogenic reproduction as well as the eggs produced through hermaphroditic reproduction are both the offspring of a single nematode, the former is asexual reproduction and the second is sexual, involving recombination.

Plant-parasitic nematodes can survive periods without a host in a number of different ways, including producing cryo-protectants that allow them to desiccate or freeze (anhydrobiosis) without cell damage, and then rehydrate when temperature and moisture are adequate (Forge and MacGuidwin 1992, Wharton 1995). Some nematodes such as *Heterodera* spp. produce egg-packed cysts that can survive freezing and drying and can remain viable in soil for decades (Krusberg and Sardanelli 1989). Eggs of the more specialized cyst and root-knot nematodes often require specific chemical cues from their preferred hosts before they will hatch, ensuring that the J2 do not end up wandering around looking for food when the snack bar is not yet open.

7.3.3 Nematode life-history strategies

Plant-parasitic nematodes generally fall into one of four life-history strategies, based on whether they parasitize from inside or outside the host and whether they move around or stay in one place in the host (Table 7.1). Migratory nematodes are generally necrotrophic, killing host cells by consuming their contents, and then moving on to attack other cells. They are analogous to grazing herbivores, which consume a different part of the plant with each bite. Sedentary nematodes generally have a more intimate, biotrophic symbiosis with their host plants, establishing a stable feeding relationship that benefits the nematode but deforms and impedes the normal functions of the host. All the species within a nematode genus generally share a life-history strategy, so we will usually just use the generic names here.

A typical migratory ectoparasite is *Belonolaimus*, the sting nematode (Figure 7.4). A J2 juvenile moves through the soil until it encounters a susceptible host and begins feeding on small roots (usually on root hairs, at the zone of elongation, or at the apical meristem) with its stylet. *Belonolaimus*, like nearly all migratory ectoparasites, have particularly long stylets that enable them to access cortical and meristem cells well beyond the epidermal cells. It continues to grow and passes through successive molts, all while moving around the plant root feeding on (and killing) root cells as it goes. The adult nematodes mate in the soil and lay eggs, which begin the cycle again. Sting nematodes are among the most destructive of plant-parasitic nematodes, attacking a wide diversity of plant hosts and stopping the growth of root systems by killing the root meristem.

Table 7.1 Intersecting life-history strategies of some plant parasitic nematodes

Life history	Ectoparasites	Endoparasites
Migratory	*Xiphinema* (dagger) *Belonolaimus* (sting)	*Pratylenchus* (lesion) *Rodopholus* (burrowing)
Sedentary	*Tylenchulus* (citrus) *Rotylenchulus* (reniform)	*Heterodera* (cyst) *Meloidogyne* (root knot)

Sedentary ectoparasites, such as *Rotylenchulus* (the reniform nematode), start the infection process when an adult female (rather than the usual J2, which in *Rotylenchulus* actually molt to adult stages without feeding!) partially penetrates the host root and produces secretions that induce the production of a stable feeding site in the vascular cylinder of the root (Wubben et al. 2010). The feeding site, called a **syncytium**, is formed by the dissolution of plant cell walls to form large, multinucleate structures on which the nematode feeds. The rest of the female nematode's body remains outside the plant root. Males live in the soil and do not parasitize plants; they mate with the embedded females.

A migratory endoparasitic counterpart to *Belonolaimus* would be *Pratylenchus*, the lesion nematode. It has a life history and pathogenic impacts similar to *Belonolaimus*, with an important behavioral distinction. *Pratylenchus* actually forces itself inside a living plant root, and then, while moving through the root cortex, uses its short stylet to sequentially perforate cell walls so it can feed on (and kill) plant cells in the root as it goes. This type of feeding causes lesions composed of dead cells in the roots. The nematodes are able to enter and leave the roots, with all stages from J2 to adult being infectious; they lay their eggs inside and outside of the roots (Jones and Fosu-Nyarko 2014).

The sedentary endoparasites are typified by *Heterodera* (cyst nematodes) and *Meloidogyne* (root-knot nematodes). These nematodes enter inside a plant root, but instead of killing cells by sucking out their contents, they manipulate host cell growth and division to provide a steady supply of food for nematode development. In the process they cause deformations and reduced function in the root system. The life cycle of *Meloidogyne* follows that of a generic plant-pathogenic nematode (Figure 7.4) until after it penetrates the zone of elongation. Once

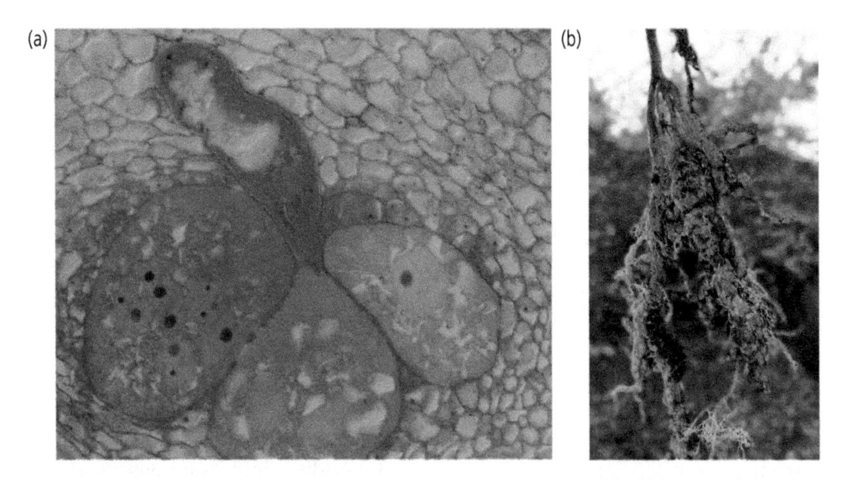

Figure 7.5 Root-knot nematodes. (a) *Meloidogyne* spp. are sedentary endoparasites that manipulate plant hormones to induce the formation of giant cells from which a female nematode is feeding (note the stylet). The giant cell at left has multiple nuclei visible. Cross-section of infected rice root shown 7 days after inoculation. (b) Hormonal imbalance leads to gall formation that gives the disease its root-knot moniker.
(a) Image provided by Dr. Diana Fernandez, Institut de recherche pour le développement (IRD), France. (b) Photo by Gregory S. Gilbert.

inside, it injects secretions from esophageal glands into plant cells to alter their hormonal balance, causing them to grow without dividing, and creating **giant cells** (Figure 7.5a and Bellafiore et al. 2008). The plant root responds to the nematode feeding and giant-cell production with gall deformation of the root system, called root knot (Figure 7.5b). Galled roots do not function well for acquiring water or nutrients for the plant. The nematodes continue to grow and molt, all while feeding from the giant cells (but without killing them). Reproduction is usually parthenogenic, and adult females, while feeding from a giant cell, swell dramatically so that their tail end pushes through to the outside of the root, where they lay a mass of eggs in a gelatinous matrix, beginning another cycle.

7.3.4 Nematodes in disease complexes

Nematodes are also important in plant diseases because of their important interaction with other pathogens. First, by creating wounds and damaging root cells, they can have synergistic interactions with soil-borne fungi and oomycetes, together causing more disease than each would on its own (Mai and Abawi 1987). Second, a particular group of migratory ectoparasites (order Dorylaimida) can be important vectors of plant viruses (Andret-Link et al. 2017). Much like the piercing-and-sucking insects that are common vectors of plant viruses, nematodes insert their stylets into living plant cells, extract the cell contents, and then move on to the next cell or plant while carrying the virus on their stylet or in their esophagus. The virus can persist in the alimentary tract of the nematode and be transmitted to additional hosts until the nematode molts, at which time the virus is lost. At least 30 nematode species are known to transmit viruses, generally in the nepovirus and tobravirus groups, which cause some economically important diseases including grapevine fanleaf virus and tomato ringspot.

7.4 Conclusions

Macroparasites of plants fall into a rather special category of plant pathogens because their size and feeding habits differ from all the other pathogens we have discussed. But they share much in how they affect host individuals and populations, and their peculiarities open intriguing possibilities to explore the evolutionary ecology of plant diseases. They will make occasional appearances throughout this book, and we hope they are increasingly brought into the mainstream of plant disease ecology research.

Further reading

Jones, J. T., A. Haegeman, E. G. Danchin, H. S. Gaur, J. Helder, M. G. Jones, T. Kikuchi, R. Manzanilla-López, J. E. Palomares-Rius, and W. M. Wesemael. 2013. Top 10 plant-parasitic nematodes in molecular plant pathology. *Molecular Plant Pathology* 14:946–961.

Kuijt, Job. 1969. *The Biology of Parasitic Flowering Plants*. University of California Press, Oakland, CA.

References

Andret-Link, P., A. Marmonier, L. Belval, K. Hleibieh, C. Ritzenthaler, and G. Demangeat. 2017. Ectoparasitic nematode vectors of grapevine viruses. In: B. Meng, G. P. Martelli, D. A. Golino, M. Fuchs, editors. *Grapevine Viruses: Molecular Biology, Diagnostics and Management*, pp. 505–529. Springer, Berlin.

Aukema, J. E. and C. Martínez del Rio. 2002. Where does a fruit-eating bird deposit mistletoe seeds? Seed deposition patterns and an experiment. *Ecology* **83**:3489–3496.

Banerjee, A. 2020. Inter-plant communication via parasitic bridging. *Journal of Experimental Botany* **71**:749–750.

Bellafiore, S., Z. Shen, M.-N. Rosso, P. Abad, P. Shih, and S. P. Briggs. 2008. Direct identification of the *Meloidogyne incognita* secretome reveals proteins with host cell reprogramming potential. *PLoS Pathog* **4**: e1000192.

Bidartondo, M. I. 2005. The evolutionary ecology of myco-heterotrophy. *New Phytologist* **167**:335–352.

Bidartondo, M. I., D. Redecker, I. Hijri, A. Wiemken, T. D. Bruns, L. Dominguez, A. Sersic, J. R. Leake, and D. J. Read. 2002. Epiparasitic plants specialized on arbuscular mycorrhizal fungi. *Nature* **419**:389–392.

Carnegie, A. J., H. Bi, S. Arnold, Y. Li, and D. Binns. 2009. Distribution, host preference, and impact of parasitic mistletoes (Loranthaceae) in young eucalypt plantations in New South Wales, Australia. *Botany* **87**: 49–63.

Dugje, I., A. Kamara, and L. Omoigui. 2006. Infestation of crop fields by *Striga* species in the savanna zones of northeast Nigeria. *Agriculture, Ecosystems & Environment* **116**:251–254.

Forge, T. and A. MacGuidwin. 1992. Effects of water potential and temperature on survival of the nematode *Meloidogyne hapla* in frozen soil. *Canadian Journal of Zoology* **70**:1553–1560.

Futai, K. 2013. Pine wood nematode, *Bursaphelenchus xylophilus*. *Annual Review of Phytopathology* **51**:61–83.

Grewell, B. J. 2008. Parasite facilitates plant species coexistence in a coastal wetland. *Ecology* **89**:1481–1488.

Hawksworth, F. G. 1959. Ballistics of dwarf mistletoe seeds. *Science* **130**:504.

Hettenhausen, C., J. Li, H. Zhuang, H. Sun, Y. Xu, J. Qi, J. Zhang, Y. Lei, Y. Qin, and G. Sun. 2017. Stem parasitic plant *Cuscuta australis* (dodder) transfers herbivory-induced signals among plants. *Proceedings of the National Academy of Sciences of the United States of America* **114**:E6703–E6709.

Hosford, R. M. 1967. Transmission of plant viruses by dodder. *Botanical Review* **33**:387–406.

Hunt, P. R. 2017. The *C. elegans* model in toxicity testing. *Journal of Applied Toxicology* **37**:50–59.

Jones, M. and J. Fosu-Nyarko. 2014. Molecular biology of root lesion nematodes (*Pratylenchus* spp.) and their interaction with host plants. *Annals of Applied Biology* **164**:163–181.

Jourdan, P. M., P. H. Lamberton, A. Fenwick, and D. G. Addiss. 2018. Soil-transmitted helminth infections. *Lancet* **391**:252–265.

Krusberg, L. R. and S. Sardanelli. 1989. Survival of *Heterodera zeae* in soil in the field and in the laboratory. *Journal of Nematology* **21**:347.

Mai, W. and G. Abawi. 1987. Interactions among root-knot nematodes and Fusarium wilt fungi on host plants. *Annual Review of Phytopathology* **25**:317–338.

Nickrent, D. L. 2020. Parasitic angiosperms: how often and how many? *Taxon* **69**:5–27.

Nickrent, D. L., V. Malécot, R. Vidal-Russell, and J. P. Der. 2010. A revised classification of Santalales. *Taxon* **59**:538–558.

Nikolov, L. A. and C. C. Davis. 2017. The big, the bad, and the beautiful: Biology of the world's largest flowers. *Journal of Systematics and Evolution* **55**:516–524.

Press, M. C. and G. K. Phoenix. 2005. Impacts of parasitic plants on natural communities. *New Phytologist* **166**: 737–751.

Runyon, J. B., M. C. Mescher, and C. M. De Moraes. 2006. Volatile chemical cues guide host location and host selection by parasitic plants. *Science* **313**:1964–1967.

Sachs, J. 1887. *Vorlesungun über Pflanzen-Physiologie*, 2nd edition. Engelmann, Leipzig.

Saucet, S. B. and K. Shirasu. 2016. Molecular parasitic plant–host interactions. *PLoS Pathogens* **12**:e1005978.

Shaw, D. C., M. Huso, and H. Bruner. 2008. Basal area growth impacts of dwarf mistletoe on western hemlock in an old-growth forest. *Canadian Journal of Forest Research* **38**:576–583.

Tasker, A. V. and J. H. Westwood. 2012. The US witchweed eradication effort turns 50: a retrospective and look-ahead on parasitic weed management. *Weed Science* **60**:267–268.

Teixeira-Costa, L. 2021. A living bridge between two enemies: haustorium structure and evolution across parasitic flowering plants. *Brazilian Journal of Botany* **44**: 165–178.

Todd, T., T. Powers, and P. Mullin. 2006. Sentinel nematodes of land-use change and restoration in tallgrass prairie. *Journal of Nematology* **38**:20.

Tsai, Y.-H. E. and P. S. Manos. 2010. Host density drives the postglacial migration of the tree parasite, Epifagus virginiana. *Proceedings of the National Academy of Sciences of the United States of America* **107**:17035–17040.

Van der Stoel, C., W. Van der Putten, and H. Duyts. 2002. Development of a negative plant–soil feedback in the expansion zone of the clonal grass *Ammophila arenaria* following root formation and nematode colonization. *Journal of Ecology* **90**:978–988.

Watling, J. and M. Press. 1997. How is the relationship between the C4 cereal *Sorghum bicolor* and the C3 root hemi-parasites *Striga hermonthica* and *Striga asiatica* affected by elevated CO_2? *Plant, Cell & Environment* **20**:1292–1300.

Wharton, D. A. 1995. Cold tolerance strategies in nematodes. *Biological Reviews* **70**:161–185.

Wubben, M. J., S. Ganji, and F. E. Callahan. 2010. Identification and molecular characterization of a β-1,4-endoglucanase gene (Rr-eng-1) from *Rotylenchulus reniformis*. *Journal of Nematology* **42**:342.

Zhou, L., D. Gabriel, Y. Duan, S. Halbert, and W. Dixon. 2007. First report of dodder transmission of huanglongbing from naturally infected *Murraya paniculata* to citrus. *Plant Disease* **91**:227.

CHAPTER 8

Types of plant diseases

Gregory S. Gilbert and Ingrid M. Parker

"There are two types of people in this world; those who think there are two types of people and those who don't." This quote from Robert Benchley in 1920[1] rings true because we have a natural tendency to classify the world around us, placing people, things, and situations into types. As long as we remember that these inferences are imperfect, they are useful because they allow us to form quick expectations from limited information. In disease ecology, classifications are useful because they provide reasonable first-pass guesses about the behavior, impacts, and critical control points for different kinds of diseases.

This chapter provides an overview of the major categories of plant diseases. We look at diseases on different parts of the plant, caused by the many different kinds of pathogens introduced in Chapters 3–7. To cause disease, the pathogen must gain access to a part of the host that can be invaded; plant pathologists call this an **infection court**, and it is specific to the type of disease. Our goal in this chapter is to introduce the diverse kinds of plant diseases, the symptoms that characterize them and the types of pathogenic organisms that cause them, and to consider how different kinds of diseases impact individual plants. Along the way, you will become familiar with some of the terminology used to describe plant diseases in the literature.

[1] Here paraphrasing a commentary by Robert Benchley in his satirical 1920 literary critique of the New York City telephone directory. "There may be said to be two classes of people in the world; those who constantly divide the people of the world into two classes, and those who do not. Both classes are extremely unpleasant to meet socially, leaving practically no one in the world whom one cares very much to know."

8.1 Major types of plant diseases

If you wanted to create your own classification of plant diseases, how would you organize them into groups? You might consider (1) the parts of the plant that are affected, (2) when the pathogen attacks during the lifecycle of the host, (3) the symptoms and impact of the pathogen on the host plant, (4) the mechanisms by which disease develops, or (5) the kind of organism that causes the disease. Each of these classification criteria is both useful and imperfect. Any of them provides insight into disease cycles, pathogen life histories, ecological patterns, and evolutionary processes.

Here we provide a brief tour of major kinds of plant diseases. We organize our disease typology according to the parts of the plant, the processes of disease development, and the major stages of the plant life cycle. We describe symptoms and provide a few representative examples of pathogens associated with each type (Table 8.1), with the caveat that many pathogens and diseases could be appropriately placed into multiple categories.

8.2 Foliar diseases

Leaves are the main site of photosynthesis, so foliar diseases have an impact on the plant by reducing photosynthesis, which affects all plant functions. In addition, growth of foliar pathogens suppresses plant performance by redirecting photosynthetic products into growth of fungal mycelium and spores at the expense of plant growth and reproduction. We divide foliar diseases into those

The Evolutionary Ecology of Plant Disease. Gregory S. Gilbert and Ingrid M. Parker, Oxford University Press. © Gregory S. Gilbert & Ingrid M. Parker (2023).
DOI: 10.1093/oso/9780198797876.003.0008

Table 8.1 A typology of major types of plant diseases and their characteristics

Disease type	Parts affected	Symptoms, development, impacts	Example pathogens
Foliar diseases			
Necrotrophic foliar diseases	Leaves	Active killing and degradation of infected leaf tissue; reduced photosynthesis, growth, and reproduction	Bacteria: *Pseudomonas*, *Xanthomonas*; Fungi (Ascomycetes): *Alternaria*, *Botryotinia*, *Cochliobolus*, *Glomerella*, *Sclerotinia*, *Phomopsis*; Fungi (Basidiomycetes): *Mycena*; Oomycetes: *Phytophthora*
Biotrophic foliar diseases	Leaves	Extraction of water, nutrients, and photosynthate from plant hosts; reduced growth and reproduction	Fungi (Ascomycetes): *Microsphaera*, *Rhytisma*; Fungi (Basidiomycetes): *Puccinia*, *Melampsora*; Oomycetes: *Peronospora*; Viruses: *Tobacco mosaic virus*, *Cucumber mosaic virus*
Developmental diseases			
Galls, knots, galls, curls, and witches' brooms	Branches, leaves, roots	Modified cell division, growth, and development through manipulation of plant hormones; reduced growth and reproduction	Bacteria: *Agrobacterium*, *Pseudomonas*; peanut witches' broom phytoplasma; Fungi (Ascomycetes): *Taphrina*; Fungi (Basidiomycetes): *Crinipellis*, *Exobasidium*; Nematodes: *Meloidogyne*, *Heterodera*; Parasitic plants: *Arceuthobium*, *Phoradendron*; Rhizaria: *Plasmodiophora*; Viruses: *Rose rosette virus*
Stunting and dwarfing	Leaves, roots, stems	Energy and water costs to support growth of pathogen; alterats host gene expression and developmental; reduced growth, competitive ability, and reproduction	Bacteria: *Candidatus* Phytoplasma, *Spiroplasma*; Parasitic Plants: *Striga*, *Orobanche*, *Cuscuta*, *Phoradendron*, *Arceuthobium*; Viruses: *Barley yellow dwarf virus*, *Tomato bushy stunt virus*
Root and vascular diseases			
Damping-off	Germinating seeds and young seedlings	Kills young growing tissue; lethal	Fungi (Ascomycetes): *Gibberella*, *Nectria* (*Fusarium*); Fungi (Basidiomycetes): *Thanatephorus* (*Rhizoctonia*); Oomycetes (*Pythium*, *Phytophthora*)
Vascular diseases	Xylem and phloem	Impedes function of the vascular system, limits access and movement of water, nutrients, and sugars; reduced growth, reproduction, survival	Bacteria: *Clavibacter*, *Erwinia*, *Pantoea*, *Ralstonia*, *Xanthomonas*, *Xylella*, *Candidatus* Phytoplasma, *Spiroplasma*; Fungi (Ascomycetes): *Ceratocystis*, *Verticillium*, *Ophiostoma*, *Fusarium*; Nematodes: *Bursaphelenchus*
Root rot and crown rot	Roots and, crowns	Decay of vascular or cambial tissue; reduced access to water and nutrients; slows growth and limits reproduction; makes more susceptible to physical damage; can be lethal	Fungi (Ascomycetes): *Gibberella* (*Fusarium*), *Gaeumannomyces*, *Sclerotinia*, *Thielaviopsis*, *Leptographium*; Fungi (Basidiomycetes): *Thanatephorus* (*Rhizoctonia*), *Sclerotium*; Nematodes: *Pratylenchus*, *Xiphinema*; Oomycetes: *Phytophthora*, *Aphanomyces*, *Pythium*
Diseases of woody stems			
Cankers	Stems and branches of woody plants	Damages vascular and cambial tissue; can girdle stem; dieback of branches, death of small trees	Bacteria: *Erwinia*, *Pseudomonas*, *Xanthomonas*; Fungi (Ascomycetes): *Botryosphaeria*, *Ceratocycsits*, *Endothia*, *Eutypa*, *Hypoxylon*, *Nectria*; Oomycetes: *Phytophthora*

continued

Table 8.1 *Continued*

Disease type	Parts affected	Symptoms, development, impacts	Example pathogens
Heart rot and Butt rot	Trunks of woody plants	Decay of dead heartwood; breakdown of lignin and/or cellulose; decay of internal heartwood; can weaken to physical damage	Fungi (Basidiomycetes): *Armillaria*, *Fomes*, *Fomitopsis*, *Heterobasidion*, *Phellinus*, *Ganoderma*, *Laetiporus*, *Pleurotus*, *Phaoelus*
Diseases of reproductive structures			
Flower diseases	Flowers and developing ovaries	Infection through style, replacing development of embryo; reduced plant reproduction	Bacteria: *Erwinia*; Fungi (Ascomycetes): *Claviceps*, *Aspergillus*; Fungi (Basidiomycetes): *Microbotryum*, *Ustilago*
Fruit and post-harvest rot	Fruits and vegetables	Growth in fruits and plant parts for consumption by humans; reduced seed dispersal by frugivores; food waste for human consumers	Bacteria: *Pectobacterium*, *Pseudomonas*; Fungi (Ascomycetes): *Botrytis*, *Elsinoe*, *Glomerella*, *Monilinia*, *Penicillium*; Fungi (Zygomycetes): *Rhizopus*; Oomycetes: *Pythium*
Seed rot and seed-borne diseases	Seeds	Destruction of embryo or infestation of seed coat, on plant or post dispersal; reduced plant reproductive success; primary inoculum for disease in next generation	Bacteria: *Pseudomonas*, *Clavibacter*, *Ralstonia*; Fungi (Ascomycetes): *Alternaria*, *Aspergillus*, *Claviceps*, *Diplodia*, *Phomopsis*, *Pyrenophora*, *Hypocrea*, *Leptosphaeria*, *Nectria*; Viruses: *Bean common mosaic virus*, *Cucumber mosaic virus*

caused by necrotrophic pathogens and those by biotrophic pathogens.

8.2.1 Necrotrophic foliar diseases

Necrotrophic foliar pathogens (including fungi, bacteria, and oomycetes) actively kill leaf cells, sometimes by producing toxins that diffuse into surrounding cells or by penetrating directly into cells and disrupting cell integrity (van Kan 2006). In the process, they may produce symptoms of **chlorosis** and/or **necrosis**. In chlorotic leaf tissue, the leaf appears yellow or pale in color because chlorophyll activity is impeded, but the cells are not killed. In contrast, cells in necrotic tissue are dead (Figure 8.1a). Often, necrotic areas are outlined by a chlorotic halo, where toxin production by the pathogen has interrupted photosynthetic activity in advance of killing the cells. After weakening or killing host cells, the pathogen can grow into and consume the damaged plant tissue, while the impaired host is unable to mount a local defense. Sometimes, the necrotic tissue is surrounded by red or purple borders where the host is producing anthocyanins (Figure 8.1b and Section 12.2).

Necrotrophic diseases of leaves can be called **spots** (small, defined, often circular), **blotches** (larger, often irregular, sometimes delineated by leaf veins), or **blights** (rapidly consuming entire leaves and sometimes branches). The term anthracnose is used for foliar diseases characterized by discrete, blackened areas of necrosis (*anthrac* = coal) caused by several genera of ascomycetes that produce mitospores in acervuli (Figure 3.6). Unless very severe, foliar diseases seldom kill a host plant directly. However, they can inhibit plant growth, which in turn can reduce plant competitive ability, or make the plant vulnerable to other environmental stressors. For instance, the agricultural weed velvetleaf (*Abutilon theophrasti*) is a strong competitor that inhibits the growth of soybean plants (*Glycine max*). Ditommaso and Watson (1995) showed that inoculating velvetleaf with the foliar nectrotroph *Colletotrichum coccodes* reduced the competitive ability of velvetleaf, allowing the soybeans to flourish.

8.2.2 Biotrophic foliar diseases

In addition to necrotrophic foliar pathogens, biotrophic pathogens of leaves include two major groups of fungi: **rusts** (which are basidiomycetes;

Figure 8.1 Foliar diseases create a diversity of signs and symptoms. (a) Guignardia leaf blotch caused by *Guignardia aesuli* on California buckeye (*Aesculus californica*) is recognized by broad patches of brown necrosis with broad yellow chlorotic margins. (b) Often, host responses to infection include the production of anthocyanins that create red, maroon, or black rings around necrotic areas, such as surrounding the leaf spots on this *Corymbia ficifolia*. (c) Rusts form pustules with masses of characteristic orange to red spores on the undersides of leaves or on stems; shown is *Coleosporium madiae* on coast tarweed (*Madia sativa*). (d) White mycelium and conidia covering the leaf surface are signs of powdery mildew, here showing *Phyllactinia guttata* on California hazel (*Corylus californica*).
Photos by Gregory S. Gilbert.

Lorrain et al. 2019) and **powdery mildews** (which are ascomycetes; Glawe 2008). They infect the host and grow as parasites by taking water, nutrients, and sugars directly from the host. In addition, rusts cover leaf tissue with spore-producing pustules (Figure 8.1c). Powdery mildews extract nutrients and sugars from the host through haustoria that penetrate the leaf surface, and produce a powdery layer of mitospores that covers the leaf surface and sometimes deforms the leaves (Figure 8.1d). Because parasitizing a living host requires greater biological finesse than the kill-and-consume approach of necrotrophic pathogens, most biotrophs tend to be rather specialized on one or a few closely related host species (Section 13.5). Heteroecious rust species have the unusual feature of requiring two separate host species to complete their life cycle, and the two hosts are typically not closely related at all (e.g., five-needle pines (Pinaceae) and currants (*Ribes* spp., Grossulariaceae) are the obligate alternate hosts of *Cronartium ribicola*).

In addition to rusts and powdery mildews, there are a number of other biotrophic pathogens of leaves. One of the prettiest foliar diseases is caused by tar-spot fungi such as *Rhytisma* spp. that live as foliar biotrophs and then produce ascospore-producing stroma on the leaf surface (see cover image!). **Downy mildews** are biotrophic Oomycetes (e.g., *Peronospora*) (Section 4.2) that are devastating pathogens of many crops. Plant viruses are also biotrophs, because they depend entirely on host cell processes to reproduce (Chapter 6). On leaves, viruses often induce patterns of chlorosis like yellowing, mottling, or mosaics that reduce photosynthesis, plant growth, and reproduction (Figure 6.1).

8.3 Developmental diseases

8.3.1 Galls, knots, curls, and witches' brooms

Some pathogens induce changes in plant development that lead to the formation of galls, knots, curls, or other swellings and deformations. They often do this by wielding exquisite hormonal control over their plant hosts (Section 2.7), either by producing analogous plant hormones themselves or by manipulating the plant to produce unusual concentrations of hormones that alter cell division and elongation processes. The pathogen stimulates hormonal imbalances in the plant that lead to the loss of apical dominance, promoting exuberant growth of lateral buds on the branches to create clusters of shoots. Root-knot nematodes (*Meloidogyne*) alter plant gene expression and deform root systems (Section 7.3). The primitive ascomycete *Taphrina deformans* causes Peach leaf curl disease in part by producing the plant hormone auxin (Cissé et al. 2013). *Exobasidium* spp. create striking discolorations and leaf galls that then get covered with fungal mycelium and basidia (Figure 8.2a). The coffee-leaf-infecting mushroom *Mycena citricolor* produces a potent oxidase of auxin which causes the coffee bush to prematurely drop all its leaves; the fungus grows saprotrophically on the fallen leaves—while glowing in the dark! (Sequeira and Steeves 1954, Oliveira et al. 2012). The most elegant of examples of hormonal regulation is crown gall, caused by *Agrobacterium radiobacter*, which genetically transforms the infected host cells to over-produce auxin and cytokinins, leading to run-away growth of plant cells that form galls (Figure 8.2b; Section 5.6). Witches' brooms are dense, bushy, clusters of dwarfed shoots on branches that can be caused by fungi, phytoplasmas, viruses, and mistletoes (Figure 8.2d) (Chapters 3, 5, 6, and 7). Each of these deformations interferes with water transport, photosynthesis, growth, and reproduction, but is not usually directly lethal for the host.

8.3.2 Stunting and dwarfing

The symptoms of many common diseases lack dramatic death or decay, but they are no less problematic for plant performance. The impact of many pathogens is to reduce growth of the host, causing stunting or dwarfing. Sometimes this alters the shape of the plant, but sometimes it simply means that infected plants are smaller. Wild grasses infected with *Barley yellow dwarf virus* often look healthy; one only discovers that they are severely stunted when they are side-by-side with noninfected plants; biomass of BYDV-infected grasses can be 22–89% smaller than noninfected plants (Malmstrom et al. 2005). Many viruses, spiroplasmas, and phytoplasmas cause such stunted growth and dwarfing, due to the costs to the host of supporting pathogen replication and to impacts on host gene expression and developmental processes. Some parasitic plants (Figure 7.1) cause severe stunting of the host because they produce a drain on water and photosynthate from the host plants.

There are two important effects of the reduced growth. First, smaller plants usually produce fewer seeds. Second, in a world where plants compete with neighbors, smaller plants generally have lower survival and fitness than their larger neighbors, creating feedback that makes them even less likely to thrive (Chapter 14).

Figure 8.2 Pathogens sometimes change developmental processes in hosts creating disruptive patterns of growth. (a) *Exobasidium vacinii* causes leaf galls on manzanita (*Arctostaphylos*) that are brightly colored from anthocyanin production, and deformed. A layer of white mycelium and basidia form on the underside of the leaf. (b) A massive crown gall caused by *Agrobacterium radiobacter* forms at the base of the trunk of a walnut tree, reducing host growth and productivity. (c) The fungus *Dibotryon morbosum* causes black knot of cherry (*Prunus serrotina*), forming gnarly deformation on branches and the trunk. (d) Many kinds of pathogens disrupt apical dominance, leading the host to generate highly branched witches' brooms, like this one on a species of *Inga*.
Photo a by Jon Detka. Photo b by Cole Margarite. Photos c and d by Gregory S. Gilbert.

8.4 Root and vascular diseases

8.4.1 Damping-off of seedlings

Usually, the most vulnerable stage in a plant life cycle is a young seedling. The seed has just germinated, and all growth energy comes from the cotyledons. With a rapidly growing root the tissues must be soft and pliable, so there is no opportunity for hardened cell walls, thick cuticles, or suberized tissue; this creates a particularly vulnerable infection court for a variety of pathogens. The root (Figure 2.6) leaks nutritious chemicals that attract motile zoospores of Oomycetes like *Pythium* and *Phytophthora*, and growing hyphae of Fungi like *Gibberella*, *Nectria*, and *Thanatephorus*. When these necrotrophic pathogens colonize and kill the plant tissue, they disrupt root or shoot functions and rapidly kill the seedling—a set of diseases collectively called **damping-off** (Figure 8.3a). The pathogen (with the help of secondary invaders) can then quickly clean up the mess, turning the seedling into a rapidly vanishing pile of mush. Damping-off usually happens after the seed germinates but before the seedling emerges from the soil (**pre-emergence damping-off**) or just after the seedling pops out of the ground (**post-emergence damping-off**) (Lamichhane et al. 2017). In general, once a seedling sets its first true leaves it has passed through the high-danger zone and damping-off becomes less of a problem. In situations with high seedling density, damping-off can spread quickly through a population of seedings, dramatically reducing the host population. Damping-off is particularly problematic in nurseries, but can also cause major losses in field-planted seed and in wild plant systems (Sections 11.5, 14.4, and 17.6).

8.4.2 Vascular diseases

The distinctive structures and vascular functions of xylem and phloem (Chapter 2) lead to divergent interactions with pathogens. Pathogens that disrupt the xylem cause wilting, reduced growth, and sometimes death. **Vascular wilt** pathogens like the fungus *Verticillium* or the bacteria *Ralstonia* and *Xylella* can grow inside the very low nutrient environment of xylem, scavenging scarce nutrients from the xylem sap or secreting phytotoxins to induce nutrient leakage from surrounding cells (Yadeta and Thomma 2013). Vascular wilt pathogens block xylem vessels with their cells and polysaccharide exudates (Section 5.4). Host plants block the spread of pathogens by sealing off xylem vessels with **gums** and **tyloses**. Both the pathogen activities and the host response can induce xylem occlusion and cavitation—the disruption of water flow by air bubbles (Sabella et al. 2019). Reduced transpiration limits photosynthesis, causing wilting and stunted growth. Plants affected by vascular wilt effectively suffer from drought, even if they are in moist soil (Figure 8.3b).

In comparison to xylem, phloem is a very high nutrient environment. Fastidious bacteria that live in phloem, such as *Spiroplasma citri*, *Candidatus* Liberibacter asiaticus, and *Candidatus* Phytoplasma asteris, can lead to an accumulation of photosynthates in source organs (usually leaves) and starve the sink organs (usually roots) of needed energy. They also produce chemicals that manipulate host physiology and interfere with defense responses. Symptoms of these phloem-associated diseases include scorching, stunting, and dieback of the host (Bendix and Lewis 2018).

8.4.3 Root rot and crown rot

Roots are the literal foundations of the plant, and they usually grow in soil. Soil is a moist environment protected from UV radiation, which makes it a great place for soil-borne fungi, oomycetes, bacteria, and nematodes that attack plant roots. Disease or death of root tissue interferes with water and nutrient uptake; the impact on the plant will depend on how much of the root system is affected by the pathogen. Rot indicates active destruction, decay, and decomposition of plant material—a kind of necrotrophy. Root rot may occur as localized infections at the vulnerable zone of elongation near the growing tips of individual roots (Figure 2.6). Such distributed infections reduce plant growth and vigor, and may lead to foliar discoloration, wilting, or stunting. Some root-rot pathogens grow upward into the **crown**—the part of the plant where the roots and stem meet near the soil surface (Figure 8.3c). When rot penetrates the main root system or crown, it can kill the host outright by shutting down water flow, and rapid death of above-ground parts is often a sign of root rot (Figure 8.3d). In trees, the effects of root rot may be seen as reduced growth, thinning canopies, and senescence of distant leaves or branches.

Root-rot is caused by a diversity of oomycetes, ascomycetes, and basidiomycetes. Nematodes are also important in causing damage to root systems when they feed on root cells by sucking out their contents or penetrate roots and feed internally. In

Figure 8.3 Symptoms of root and vascular diseases are associated with disrupting the transpiration stream. (a) Damping-off kills newly germinated seedlings, collapsing the rapidly growing root or shoot tissues. (b) Bacterial wilt of tomato caused by *Ralstonia solanacearum* clogs the vascular system and prevents transpiration, causing the plant to wilt. (c) Root-rot and crown-rot pathogens rapidly destroy vascular tissue, visible as discolored tissue in longitudinal section. (d) Macrophomina crown rot (*Macrophomina phaseolina*) disrupts normal transpirational flow in strawberries, leading to wilting and death of older leaves, while younger leaves can remain green.
Photo a by Daren Mueller, Iowa State University, Bugwood.org. Photo b by Alejandra Huerta; Photo c by Gregory S. Gilbert. Photo d by Joji Muramoto.

some cases, soil-borne fungi and nematodes act synergistically to damage root systems (Section 7.3)

8.5 Diseases of woody stems

The stems of perennial woody plants are susceptible to two major kinds of diseases: cankers that affect the living sapwood tissue (cambium, phloem, and active xylem) and heart rot that decays the dead xylem tissue of the heartwood.

8.5.1 Cankers

The stems and branches of woody plants can be damaged or destroyed by a diversity of bacteria, oomycetes, and fungi that cause **cankers**—dead sections of bark and underlying tissue on trunks and branches of trees. Infection usually begins through infection courts created by a wound or break in the bark, followed by colonization of the sapwood. Canker pathogens may kill the tissue quickly, causing sunken, sometimes oozing lesions beneath the bark that interfere with growth and vascular function (Figure 8.4a) (Lynch et al. 2013). In other cases, the pathogen and host may be locked in a long-lasting, back-and-forth battle in which the plant tries to wall off the pathogen by producing physical and chemical barriers, including masses of rapidly dividing cells called **callus tissue**; the pathogen then breaches the barriers and colonizes more host tissue, followed by more barriers, and more breaching, creating a slowly expanding, and often swollen, messy, and rough canker (Figure 8.4b) (Ward et al. 2010). Symptoms like wilt and dieback occur on branches distal to the canker. When a canker encircles enough of the trunk to completely prevent the movement of water and sap between the above- and below-ground parts, the tree is girdled and eventually dies.

8.5.2 Heart rot

The inner part of the trunk of most trees—the heartwood—is non-conductive, plugged xylem tissue that only provides structural support. Heartwood tissue is heavily reinforced with lignin, which makes it highly resistant to attack by most organisms (Section 12.2). A special group of basidiomycete polypore fungi are able to tolerate and degrade lignin. Those that aggressively degrade lignin cause **white rot**; removing the dark-colored lignin leaves behind mostly the light-colored cellulose and hemicellulose (Figure 8.4c). **Brown cubical rot** is caused by fungi that tolerate the lignin but preferentially degrade the cellulose and hemicellulose, leaving much of the darker lignin behind (Figure 8.4d). Heart-rot fungi infect through wounds in the bark, through branch scars, or by growing up from the root and into the trunk. After lignin-degrading (lignolytic) fungi have colonized and removed the lignin, heartwood is additionally susceptible to ascomycete fungi, bacteria, termites, and ants that cannot tolerate lignin but are voracious consumers of the remaining cellulose (Section 15.11).

Most heart-rot fungi are not able to invade and kill the living sapwood, so they do not interfere with growth or vascular function. Trees hollowed by heart-rot fungi can continue living and growing for decades (Figure 8.4e). Cavity-nesting birds and mammals take advantage of trees hollowed out by heart-rot fungi (Hennon and Mulvey 2014). Are heart-rot fungi pathogens or clever saprotrophs?

Heart-rot decay can structurally weaken the tree so that it is more susceptible to snapping-off or falling down in a storm (Larson and Franklin 2010). Where the trunk meets the roots is usually called the butt of the tree, because for trees, "crown" is used to refer to the branches and leaves. If you imagine a tree swaying in the wind with roots firmly stuck in place, the crown/butt is the fulcrum of a long lever. Wet, windy conditions lead to the destruction of many trees in storms. If the roots are weakened from root-rot, a living tree can be ripped out of the soil, left lying in a heap with its roots exposed; this is called **tip-up mortality**. If the tree is weakened right at the fulcrum from butt-rot, it can break; this is called **snap-off mortality** (Figure 8.4f). When a tree snaps-off or tips-up in a windstorm, was it killed by the wind or by the fungi that decayed its roots and butt?

8.6 Diseases of reproductive structures

8.6.1 Flower diseases

The reproductive parts of plants are vulnerable to attack by pathogens. Bacteria such as the fire-blight pathogen *Erwinia amylovora* infect and destroy flowers, then go on to colonize stems and cause additional damage as canker pathogens (Farkas et al. 2012). Smut fungi infect flowers, and some, like *Microbotryum* spp. (cause of anther smut on plants in the Caryophyllaceae; Section 2.5 and Figure 2.10) sterilize the plant by suppressing ovary development and replace the pollen on the anthers with teliospores that are dispersed by pollinators to new plants. Some smut fungi grow down the style into the ovary like a pollen tube, where the fungus consumes and replaces the developing embryo. For example, the corn smut fungus (*Ustilago maydis*) replaces corn kernels with large, swollen fungal

Figure 8.4 Cankers are created when pathogens disrupt tissue growth in the host plant. (a) Dark, sunken, annual cankers (white arrow) of beech bark disease develop when the beech scale insect (white tufts) allows the fungus *Neonectria faginata* to colonize the bark of the tree and kill the sapwood beneath. As new cankers are created each year, the dead areas eventually encircle the trunk, girdling and killing the tree. (b) Perennial target canker caused by *Neonectria ditissima* on wild cherry (*Prunus serrotina*); rings near the bottom of the exposed area of the canker (white arrow) show the recurring battle between the fungus advance and temporary walls of callus tissue created by the host. Over years, the canker deformation becomes massive and can girdle the tree. Wood decay is mostly caused by a group of poroid basidiomycete fungi that are able to tolerate and degrade lignin. (c) Lignolytic fungi that aggressively remove dark-colored lignin create white rot, leaving the oak (*Quercus* sp.) wood stringy and light colored. (d) Wood decay fungi that preferentially remove the light-colored cellulose and hemicellulose and leave lignin behind create brown cubical rot, with the pine (*Pinus* sp.) wood becoming more darkly colored. (e) Cross-section of a fallen redwood tree (*Sequoia sempervirens*) shows circular patterns of pocket decay (a brown cubical rot) caused by *Poria sequoia* in the heartwood. (f) Heart rot made this still-alive oak tree (*Quercus rubra*) susceptible to snapping off during a storm.
Photos by Gregory S. Gilbert.

masses known as *huitlacoche* (Figure 8.5a), a traditional food in Mexico since the time of the Aztecs (Juárez-Montiel et al. 2011).

8.6.2 Fruit rot and post-harvest rots

As the ovules develop into seeds, the ovaries and supporting tissues give rise to fruits that are nutritious and often sweet. Pod rot of cacao, caused by *Moniliophthora roreri*, is a serious threat to the supply of high-quality chocolate (Cubillos 2017)! *Botrytis cinerea* attacks grapes both on the vine and on the table; under the right conditions, grapes fermented by *Botrytis* on the vine (called noble rot) are then used to produce very special sweet botryotized wines (Magyar 2011). Some fungi (like *Mucor* and *Rhizopus*) are particularly adapted to growing at the low water activity created by high concentration of sugar in fruits (Figure 8.5b) (Box 3.2).

Diseases of fruits can have important ecological impacts on plant populations by reducing palatability of fruit to animals that disperse seed. Fruit-rot microbes may make their fruit substrates putrid in part to better compete with macro-frugivores (Janzen 1977, Ruxton et al. 2014). On the other

Figure 8.5 Pathogens affect reproductive parts of plants in several ways. (a) Smut fungi such as the corn smut pathogen *Ustilago maydis* colonize flowers and convert reproductive organs into spore-producing structures, such as these masses of *huitlacoche* that replace corn kernels. (b) Some rapidly growing fungi such as *Mucor pyriformis* are able to tolerate high sugar concentrations of fruits. (c) Soft rot of potato (*Solanum tuberosum*) caused by the bacterium *Pectobacterium tuberosum* when tubers are stored wet or at warm temperatures. (d) Seeds are often contaminated with microbes that can grow on them in storage or when planted. Shown are soybean seeds infected by *Phomopsis* sp.
Photo a by Shannon Lynch. Photo b by Gregory S. Gilbert. Photo c by Alejandra Huerta. Photo d by Daren Mueller, Iowa State University, Bugwood.org.

hand, plants need those macro-frugivores to disperse their seeds. Consequently, plants that rely on palatable fruit to attract animal dispersers often invest in defensive chemicals that protect their fruits against fruit-rot microbes (Cipollini and Stiles 1993). There is a fine line between too much and too little defense, because the defensive compounds produced to protect against micro-frugivores sometimes deter macro-frugivores, too. Evolutionary dynamics involving plants, frugivores, and fruit-rot microbes have undoubtedly shaped the flavors of our cuisine in every part of the world.

Microbes that ferment fruits are one example of post-harvest rots, which are a major battleground between microbes and humans. Most plant parts have been colonized by harmless bacteria and fungi, but given a little time and the right environment, they make short order of the moist, nutrient-rich plant parts that make up our fruits and vegetables. For instance, potato tubers that are not stored cool and dry are susceptible to developing bacterial soft rot (Figure 8.5c). All the fruits, tubers, stems, and leaves we like to eat are susceptible to post-harvest decay from bacterial and fungal pathogens. That is why we have refrigerators—to keep temperatures below the range at which microbes grow well (Figures 3.9 and 5.6).

8.6.3 Seed decay

The surface or interior of seeds are often colonized by fungi (Section 15.9). Seed-borne fungi may infect the seed while it is still on the plant or after they have dispersed. Because seeds have extremely low water activity (5–10% water content) compared to other plant materials (30–70% in leaves), seed-colonizing pathogens must be adapted to grow on dry, or "xeric," substrate and are called **xerotolerant** (Box 3.2).

Some seed-borne fungi produce **mycotoxins** that are acutely toxic or highly mutagenic to humans and other animals. The ergot fungus *Claviceps* replaces the embryo of a developing seed with a sclerotium, while *Aspergillus* can colonize the developed seed. Both can lead to contaminated grain harvests and can result in toxic animal feed and foods, with serious health effects on people and livestock (Section 15.9).

Pathogen-infested crop seed can be an important source of primary inoculum when those seeds are later planted out into a clean field (Figure 8.5d) (Section 17.2). In wild populations, a soil seed bank of dormant seeds is an important part of the life cycle of many plant species, allowing plant populations to persist through periods of inhospitable environmental conditions (Section 14.4). The impacts of pathogens on buried seeds varies considerably among plant species, depending on a suite of defensive traits, with annual rates of seed mortality in the soil ranging from less than 1% in some temperate species to up to half of buried seeds of some pioneer tree species in lowland tropical forest (Dalling et al. 2020).

Further reading

For a more encyclopedic coverage of plant diseases, we recommend the compendium series produced by the American Phytopathological Society for diseases of economically important crops and the many regional online guides to crop diseases produced by agricultural extension services in addition to:

Agrios, G. N. 2004. *Plant Pathology*, 5th edition. Elsevier, Amsterdam.

Sinclair, W. A. and H. H. Lyon. 2005. *Diseases of Trees and Shrubs*, 2nd edition. Comstock Publishing, Sacramento, CA.

References

Bendix, C. and J. D. Lewis. 2018. The enemy within: phloem-limited pathogens. *Molecular Plant Pathology* **19**:238–254.

Cipollini, M. L. and E. W. Stiles. 1993. Fruit rot, antifungal defense, and palatability of fleshy fruits for frugivorous birds. *Ecology* **74**:751–762.

Cissé, O. H., J. M. Almeida, Á. Fonseca, A. A. Kumar, J. Salojärvi, K. Overmyer, P. M. Hauser, and M. Pagni. 2013. Genome sequencing of the plant pathogen *Taphrina deformans*, the causal agent of peach leaf curl. *Mbio* **4**(3):e00055-13.

Cubillos, G. 2017. Frosty pod rot, disease that affects the cocoa (*Theobroma cacao*) crops in Colombia. *Crop Protection* **96**:77–82.

Dalling, J. W., A. S. Davis, A. E. Arnold, C. Sarmiento, and P.-C. Zalamea. 2020. Extending plant defense theory to seeds. *Annual Review of Ecology, Evolution, and Systematics* **51**:123–141.

Ditommaso, A. and A. K. Watson. 1995. Impact of a fungal pathogen, *Colletotrichum coccodes* on growth and competitive ability of *Abutilon theophrasti*. *New Phytologist* **131**:51–60.

Farkas, A., E. Mihalik, L. Dorgai, and T. Bubán. 2012. Floral traits affecting fire blight infection and management. *Trees* **26**:47–66.

Glawe, D. A. 2008. The powdery mildews: a review of the world's most familiar (yet poorly known) plant pathogens. *Annual Review of Phytopathology* **46**:27–51.

Hennon, P. E. and R. L. Mulvey. 2014. Managing heart rot in live trees for wildlife habitat in young-growth forests of coastal Alaska. General Technical Report PNW-GTR-890. Portland, OR.

Janzen, D. H. 1977. Why fruits rot, seeds mold, and meat spoils. *American Naturalist* **111**:691–713.

Juárez-Montiel, M., S. R. de León, G. Chávez-Camarillo, C. Hernández-Rodríguez, and L. Villa-Tanaca. 2011. Huitlacoche (corn smut), caused by the phytopathogenic fungus *Ustilago maydis*, as a functional food. *Revista Iberoamericana De Micologia* **28**:69–73.

Lamichhane, J. R., C. Dürr, A. A. Schwanck, M.-H. Robin, J.-P. Sarthou, V. Cellier, A. Messéan, and J.-N. Aubertot. 2017. Integrated management of damping-off diseases. A review. *Agronomy for Sustainable Development* **37**:10.

Larson, A. J. and J. F. Franklin. 2010. The tree mortality regime in temperate old-growth coniferous forests: the role of physical damage. *Canadian Journal of Forest Research* **40**:2091–2103.

Lorrain, C., K. C. Gonçalves dos Santos, H. Germain, A. Hecker, and S. Duplessis. 2019. Advances in understanding obligate biotrophy in rust fungi. *New Phytologist* **222**:1190–1206.

Lynch, S. C., A. Eskalen, P. J. Zambino, J. S. Mayorquin, and D. H. Wang. 2013. Identification and pathogenicity of Botryosphaeriaceae species associated with coast live oak (*Quercus agrifolia*) decline in southern California. *Mycologia* **105**:125–140.

Magyar, I. 2011. Botrytized wines. *Advances in Food and Nutrition Research* **63**:147–206.

Malmstrom, C., C. Hughes, L. Newton, and C. Stoner. 2005. Virus infection in remnant native bunchgrasses from invaded California grasslands. *New Phytologist* **168**:217–230.

Oliveira, A. G., D. E. Desjardin, B. A. Perry, and C. V. Stevani. 2012. Evidence that a single bioluminescent system is shared by all known bioluminescent fungal lineages. *Photochemical & Photobiological Sciences* **11**: 848–852.

Ruxton, G. D., D. M. Wilkinson, H. M. Schaefer, and T. N. Sherratt. 2014. Why fruit rots: theoretical support for Janzen's theory of microbe–macrobe competition. *Proceedings of the Royal Society of London B: Biological Sciences* **281**:20133320.

Sabella, E., A. Aprile, A. Genga, T. Siciliano, E. Nutricati, F. Nicolì, M. Vergine, C. Negro, L. De Bellis, and A. Luvisi. 2019. Xylem cavitation susceptibility and refilling mechanisms in olive trees infected by *Xylella fastidiosa*. *Scientific Reports* **9**:1–11.

Sequeira, L. and T. A. Steeves. 1954. Auxin inactivation and its relation to leaf drop caused by the fungus *Omphalia flavida*. *Plant Physiology* **29**:11–16.

van Kan, J. A. L. 2006. Licensed to kill: the lifestyle of a necrotrophic plant pathogen. *Trends in Plant Science* **11**:247–253.

Ward, J. S., S. Anagnostakis, and F. J. Ferrandino. 2010. Nectria canker incidence on birch (*Betula* spp.) in Connecticut. *Northern Journal of Applied Forestry* **27**:85–91.

Yadeta, K. and B. Thomma. 2013. The xylem as battleground for plant hosts and vascular wilt pathogens. *Frontiers in Plant Science* **4**:97.

CHAPTER 9

How to do disease ecology

Gregory S. Gilbert and Ingrid M. Parker

9.1 Diagnosing and measuring disease

As disease ecologists, we generally want to know how much disease there is, where it is, what it is doing to the plants, and how it changes over time. In this chapter we explore how to measure **disease intensity**—a general term for the amount of disease—which can be assessed on individual plants or whole populations and in a number of different ways. Measuring disease begins with accurate **disease diagnosis**, where observed signs and symptoms on diseased plants are associated with a particular causal pathogen. Visual assessments are usually confirmed by isolation or detection of a known pathogen associated with those symptoms on a particular host; for unfamiliar symptoms, completing Koch's proof of pathogenicity is used to establish the cause (Box 1.1, Figure 1.2). For well-studied diseases, visual assessments can be used to determine which plants in a population are diseased; this allows us to measure **disease prevalence** (the proportion of individuals in a population that are diseased), as well as spatial patterns of disease. Repeated measurements of disease prevalence over time provide an assessment of **disease progress**—how disease spreads through populations over time and across space. The quantitative measure of the disease intensity on individual plants is called **disease severity**, and the effects of disease on host growth, survival, and reproduction are thought of as **disease impact.**

This chapter provides an overview of some of the tools use to diagnose disease and to measure the various components of disease intensity that help to answer questions in disease ecology. We include descriptions of serological and DNA-based methods for detection and identification, techniques for culturing pathogens, and approaches to quantifying, sampling, and assessing disease intensity. In later chapters we explore how these tools are used to understand temporal patterns of disease spread (Chapter 10) and spatial spread of pathogens and disease (Chapter 11).

9.2 Components of disease intensity

The terminology used in plant pathology has not always been consistent with how the same terms are used in medical or zoological epidemiology. The term **disease incidence** is used in much of the plant pathology literature to mean the proportion of plants in the population that are diseased (Madden et al. 2007). In contrast, the medical (or other animal) disease literature uses disease incidence to mean "the number of new health-related events in a defined population within a specified period of time ... measured as a frequency count, a rate [risk], or a proportion" (Porta 2016). In that sense, incidence is a probability of *new* occurrences over time (e.g., as a frequency count: 268,600 new cases of invasive breast cancer are expected among US women in 2019; or as a rate or risk: 130 per 100,000 women are expected to develop breast cancer in 2019 (American Cancer Society 2019)). Medical epidemiologists (and many plant pathologists) instead use the term **disease prevalence** (sometimes called point prevalence) to describe a "snapshot" proportion of individuals with the disease at a particular time (Noordzij et al. 2010). Prevalence can be calculated with a variety of different units (e.g., # of infested fields in a landscape, # of diseased leaves

The Evolutionary Ecology of Plant Disease. Gregory S. Gilbert and Ingrid M. Parker, Oxford University Press. © Gregory S. Gilbert & Ingrid M. Parker (2023).
DOI: 10.1093/oso/9780198797876.003.0009

per plant), but most often prevalence refers to the proportion of plants that are diseased, given as:

$$Disease\ prevalence = \frac{\#\ diseased\ plants}{\#\ plants\ examined} \quad \text{[Eq. 9.1]}$$

We use the term disease prevalence for observed measures of occurrence at a particular time and restrict the use of the term disease incidence to the sense used in the broader disease literature, to mean the frequency or rate of new occurrences of disease.

Importantly, prevalence does not include the extent or impact of disease on individual plants. Prevalence is based on the binary response of being diseased or not; what proportion of plants in a field are diseased; what proportion of fields in a landscape are **infested**; what proportion of leaves are **infected**. A field of tomatoes where half the examined plants each have one lesion of late blight on a single leaf would have the same disease prevalence as a field where half the plants were already nearly dead from extensive late blight symptoms; the prevalence would be 0.5 in both cases.

So to more fully capture variation in disease intensity we also need to measure disease severity, most commonly expressed as a proportion of the relevant host tissue showing disease symptoms (Madden et al. 2007). This can be the proportion of leaf area that is necrotic, the fraction of plant surface covered by pustules, the volume of wood decayed, the proportion of xylem vessels occluded, or other quantitative measures. Finally, we often want to measure the impact of disease on host performance. There are many possible ways that disease can weaken host performance, including the reduction in transpiration or photosynthesis, reduced growth rate, dieback, loss of crop yield or plant fecundity, or shortened survival time, and changes in host population dynamics. In studies of wild plants, we are particularly interested in the impact of disease on **lifetime fitness** of the host plants (Section 13.1).

9.3 Measures of disease intensity

Measuring disease prevalence relies on the qualitative, binary measure of whether a plant has a particular disease or not, whereas disease severity requires a quantitative measure of how much a plant is affected by the disease. Choosing to

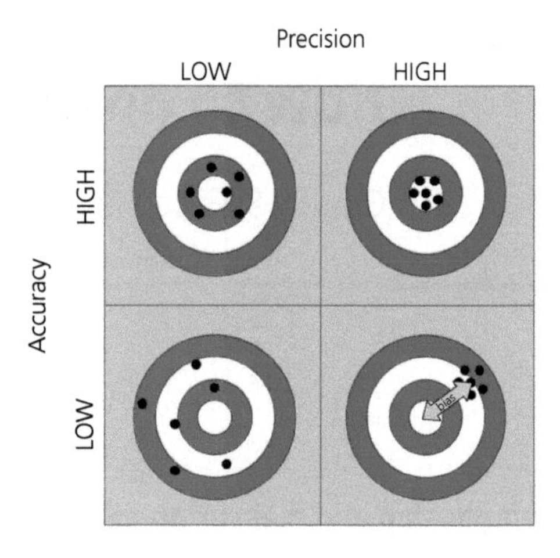

Figure 9.1 Relationship between accuracy and precision in producing valid measurements. Imagine that the center of the target is the true value. Ideally, we strive for a measure with high accuracy (at the center of target) and high precision (multiple measures give similar values). Measurements with a high accuracy but low precision still have an average near the center of the target, and with a large enough sample size will reveal a valid estimate of the true value. Measurements with high precision but a known bias that reduces accuracy can be adjusted to provide a valid estimate of the true value. Graphic by Gregory S. Gilbert.

measure prevalence or severity (or both) depends on the question at hand. In either case, selecting an approach to measure disease intensity involves trade-offs between the effort required, the accuracy of the determination, and the precision (or reliability) of the measurements (Figure 9.1). **Accuracy** is how close a measurement is to the true value. **Precision** is how similar measures are to each other when repeated under similar circumstances. Precision is a measure of statistical random error, and can be increased by using tools with finer resolution and with observer training to improve repeatability. Accuracy is a measure of systematic (non-random) error, also called **bias**, and can be improved through training and calibration of both equipment and the people making the measurements. When the scale and direction of bias is known, it can be adjusted to improve accuracy. Plant pathologists have combined statistical analyses with empirical data to develop advice about how to handle different kinds of bias (Bock et al. 2010). Ideally, of course, we aim for measurements that are both accurate and precise. The **validity** of a measurement is how well you are measuring what you intend to measure; a measure is valid if it is precise (reliable) and accurate (unbiased).

For binary measures such as disease prevalence, there are two kinds of error: **false positives** and **false negatives**. False positives occur when we diagnose a plant with a particular disease when in fact it does not have it; false negatives occur when we declare a plant healthy when it is actually diseased. Tests that have high **sensitivity** to ensure they do not miss infected individuals may sometimes generate false positives in plants that are not really infected. Similarly, tests that avoid false positives by increasing **stringency** may lack the sensitivity to detect low levels of infection, generating false negatives in plants that are really infected. The relative consequences of false positives and false negatives must be carefully considered and depend on the context. A false-negative test for infestation with the seed-borne fungus *Gibberella fujikuroi* on rice seed could lead a farmer to confidently plant a field only to later lose the crop to the Bakanae disease it causes (Carneiro et al. 2017). A false-positive test for an emergent disease like Huanglongbing (aka Citrus Greening) caused by the bacterium *Candidatus* Liberibacter asiaticus could result in costly and contentious eradication efforts employed for no reason (Parnell et al. 2017). For pathogens that represent significant threats, a highly sensitive test with numerous false positives might be preferred to avoid missing critical cases, but verifying positive tests from pathogen surveillance is important prior to taking high-stakes management actions.

9.4 Disease diagnosis: is this plant sick?

The first step in measuring disease prevalence is to reliably determine which individual plants are diseased and which are not. Pathologists assess disease status either by looking for the presence of symptoms of the disease or by detecting the pathogen itself. Both approaches are necessary because multiple pathogens can cause similar symptoms on the same host, and because a single pathogen may cause different symptoms on different host species or be present in a latent infection. This means we need to recognize when a plant is diseased and to be able to detect and identify the pathogen that causes it.

What is the difference between **identification** and **recognition**? When you first meet someone in an official capacity, you might ask for some kind of certified identifying documents (ID) like a passport or driver's license to identify them. Once you have identified them, you will use characteristic features (their face, or their email address or telephone number) that allow you to recognize that person in the future without having to check their ID each time. Similarly, a birdwatcher will do a careful comparison of field markings and song characteristics to confidently identify a bird species the first time, but will soon develop a set of features for quick recognition of that species. For plant diseases, once we have gone through the careful process of cataloging the signs and symptoms associated with infection by a particular pathogen (one that has been identified using appropriate morphological and molecular approaches) on a particular host, we can often use a small set of reliable traits—symptoms on the host and signs of the pathogen—to recognize diseases.

9.5 Visible and nearly visible assessments: symptoms and signs

Disease assessment usually begins with examining **symptoms** such as patterns of necrosis, chlorosis, changes in pigmentation, wilting, death, decay, cankers, and deformations (Chapter 8). Symptoms are host expressions of infection by a pathogen, and, because they are also shaped by environmental conditions, they are naturally variable. Nevertheless, a practiced eye, familiarized with the local range of variation in diseased and non-diseased plants, can often provide rapid and valid disease diagnosis, and measurement of disease prevalence. For instance, by looking through binoculars while bouncing up and down in a 6-m long boat with an outboard motor, we categorized disease symptoms on some 167,000 palms along 392 km of coastline and 306 islands in the indigenous comarca of Guna Yala, Panama, documenting the prevalence and spatial spread of a new, phytoplasma-caused disease of coconut palms (Gilbert and Parker 2008).

What if you don't have "a practiced eye"? Visual symptoms are distinguished based on characteristic shapes and color patterns (Chapter 8), and can be assessed with plant in hand, through binoculars, or remotely using cameras mounted on drones or even satellites (Oerke 2020). Web-based tools have transformed the way we identify unfamiliar disease symptoms. When faced with an unfamiliar disease, the best way to find information is to start with a web search on your plant host or close relatives of your host. For example, search "oak diseases in <my state>" on the Internet. Many government agencies, agricultural extension service offices, universities, and even nonprofit organizations have websites dedicated to helping you identify diseases. Some of these organizations also provide diagnostic services. The websites of the American Phytopathological Society and similar associations across the globe serve as useful references for many well-studied diseases. These resources can help assess disease status by careful comparison to published descriptions and photos of disease symptoms.

Diseased plants may undergo a number of physiological changes that are not readily detectable by the human eye. Diseases that affect water relations can be detected by sensors that measure plant temperature through infrared thermography (Lindenthal et al. 2005); diseases that interfere with the plant's photosynthetic apparatus can be detected using hyperspectral sensors that measure leaf fluorescence; and volatile chemicals produced by infected plants can be detected by e-noses (gas chromatographs) (reviewed in Oerke 2020). Disease detection using such invisible symptoms tends to require specialized equipment and training, but sometimes has the advantage of either larger scale or earlier detection than is possible using visual assessment. For instance, drone-based hyperspectral images and a bit of post-flight image analysis can assess disease prevalence across an entire field of corn in a fraction of the time it would take to do visually, or it can allow measurement of disease prevalence across forested landscapes that are difficult to access (Zhang et al. 2018). Fancy electronics are not always required; dogs have been trained to sniff out the volatile chemicals produced by citrus trees infected with huanglongbing disease caused by two *Candidatus* Liberibacter bacteria and spread by the psyllid *Diaphorina citri* (Gottwald et al. 2020) (Figure 9.2).

Figure 9.2 Detector canine 'Bello' works in a citrus orchard in Texas, searching for huanglongbing disease on citrus trees. Bello can detect the volatiles produced directly by the bacterial pathogen with more than 99% accuracy.
Photo provided by USDA-ARS-USHRL, Fort Pierce, FL.

Signs of the pathogen—structures of the pathogen or pathogen products that are visible to the naked eye, with a hand lens, or through a microscope—complement symptoms in the diagnosis of plant diseases. Signs include ascomata, basidiomata, acervuli, pycnidia, pustules, stroma, telia, mycelia, mitospores, mistletoe shoots, and slimy polysaccharide exudates (see Chapters 3–7). Some pathogens (e.g., powdery mildews, rusts, smuts), produce few visible symptoms in their host plants, with signs of the pathogen itself the main visible diagnostic of a disease. However, the structures of many pathogen species can be morphologically difficult to tell apart.

9.6 Culture-based approaches

Signs of the pathogen are often not readily visible on diseased plants, but we can use some simple techniques to encourage their production and make them easier to observe. The simplest technique is to put a sample of a diseased plant in a **moist chamber** for a few days to weeks to encourage pathogen growth and reproduction. A low-tech approach is to wash off a diseased leaf or stem and place it together with a moist paper towel in a plastic bag, airtight storage box, or Petri dish. Check it regularly with a hand lens or under a stereoscope for fungal structures.

It is often useful to isolate the pathogen into pure culture on **agar growth medium** (Figure 9.3, Box 9.1). In some cases, reproductive structures of the pathogen are readily visible on the plant (basidiomata, ascomata, pycnidia, mycelium), and small amounts of fungal mycelium or spores can be transferred directly to agar medium to grow in pure culture. To isolate putative fungal pathogens from diseased plant tissue, place very small pieces of plant tissue on agar medium, and allow the pathogen to grow into the agar (Figure 9.3). Because all plants are always covered with many bacteria and fungi that have nothing to do with the disease, you must first **surface sterilize** the plant tissue. This kills whatever microbes are living on the outside, allowing only those on the inside to grow into the growth medium. Once a fungus or bacterium is growing into the medium, a small number of cells can be transferred to a new plate to grow in pure culture. Pure cultures can be stored (by freezing, freeze-drying, or under water or oil), used for pathogen identification, or grown for further study.

Once you have a fungus or bacterium in pure culture, how do you identify it? Sometimes fungi produce spores that can be identified

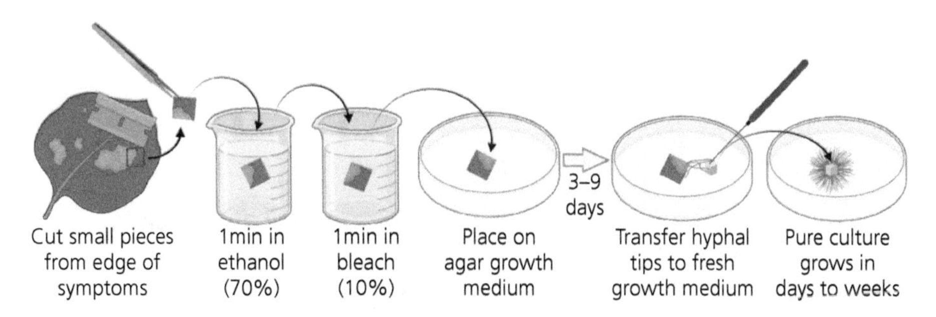

Figure 9.3 Process for surface sterilization and isolation of pathogens in pure culture. We usually isolate pathogens from diseased plant tissue (leaves, stems, roots, wood) by cutting small pieces of tissue (2–3 mm on a side) right from the edge of the diseased area, using fine scissors, a hole punch, a knife, or a razor blade. Fragments need to be small to increase the chance of isolating only one kind of microbe from a plant fragment. Submerse fragments for 1 min in 70% ethanol, followed by 1 min in 10% commercial chlorine bleach (0.6% sodium hypochlorite). Placing fragments into a tea strainer can make submersion easier. Then use forceps that were sterilized in the bleach to place the surface-sterilized plant fragments on agar growth medium in a Petri plate (Box 9.1). Fungi should grow out of the infected plant material in several days (bacteria and some oomycetes somewhat more quickly). Very small chunks of mycelium (or bacterial cells) can then be transferred with a sterile needle to a fresh plate of agar medium to allow the pathogen to grow in pure culture, from which it can then be identified or used in experiments.
Graphic by Gregory S. Gilbert, created with BioRender.com.

Box 9.1 Kinds and preparation of growth media

General growth media for fungi, bacteria, and oomycetes have four basic ingredients: water, a carbon source, a nitrogen source, and a mix of other necessary nutrients (minerals, vitamins, etc.). These ingredients can be combined into a liquid broth and poured into test tubes or flasks to grow the microbes in liquid culture. Alternatively, agar (an extract of brown algae) can be added as a gelling agent to make a semi-solid substrate that can be dispensed into Petri plates (Petri is capitalized in honor of the inventor, Julius Petri). The agar itself cannot be consumed by most fungi and bacteria, so it provides a stable structure for the needed water and nutrients (you could try gelatin, but many microbes digest the gelatin and turn the growth medium to mush). Carbon and nitrogen sources can be simple organic molecules (e.g., glucose and ammonium nitrate) or complex ones (e.g., starch, cellulose, proteins). An easy way to get all the carbon, nitrogen, and other nutrients plant pathogens need to grow is to extract it from plant material. Four common sources for fungal growth media are extracts of malt (from malted barley), potatoes, corn meal, and oatmeal. Many plant pathogenic bacteria grow well on trypticase soy extract. Sometimes media will be supplemented with simple sugars to enhance microbial growth.

The media ingredients are mixed together in a flask and then sterilized in an autoclave. An **autoclave** is a pressure cooker that allows the medium to be heated to a temperature high enough (121°C for about 15 minutes) to kill heat-resistant bacterial spores that would survive simple boiling. When agar (which includes a long, unbranched, but coiled polysaccharide called agarose mixed with shorter agaropectins) is heated above about 85°C, it dissolves in water and uncoils. After autoclaving, the medium can be slowly cooled to about 52–56°C, where it is cool enough to handle, but the agar remains liquid. At that temperature the agar can be poured into sterile Petri dishes. As the agar cools to below about 40°C, the agar coils up and traps water, forming a gel. The agar gel will stay solid up to its melting temperature of 85°C. The property of melting at a higher temperature (85°C) than the temperature for solidifying (40°C) is called **hysteresis**, and together with being transparent and inert, it is the feature that makes agar useful in microbiology.

Mycologists, bacteriologists, and plant pathologists have developed a tremendous array of **selective media** to help isolate particular pathogens of interest (Bills and Foster 2004). Such selective media can also be used to detect and quantify particular pathogens in the air, water, or soil. We can make growth media selective by adding substances that prevent the growth of many microbes but permit the growth of those of interest. For instance, when trying to isolate fungi, we often add broad-spectrum antibacterial antibiotics such as chloramphenicol or tetracycline to the agar. To prevent fungal contamination when trying to isolate bacteria, we might add the antifungal antibiotic cycloheximide. To discourage ascomycetes but allow slower-growing (and resistant) wood-decay basidiomycete fungi to grow, the selective fungicide benomyl can be used to create a selective medium. If *Phytophthora* is suspected to be causing unusual cankers on a tree, small chunks of wood from the edge of the canker can be placed onto PARP medium (Jeffers and Martin 1986), which has a mixture of antifungal and antibacterial additives that make the medium highly selective for oomycetes. Add a bit of hymexazol to the PARP, and it further prevents the growth of most species of the oomycete *Pythium*, while still allowing the growth of *Phytophthora*.

morphologically (Seifert et al. 2011). Bacteria have been traditionally identified by their metabolic profiles (e.g., which carbon sources they can break down) and chemical attributes (e.g., Gram staining of cell walls) (Section 5.6 and Bergey's Manual Trust 2015). However, current identification of most microbes relies on comparing sequences of specific genes (DNA barcodes; Section 9.8) to global databases.

9.7 Serological approaches

There are two important limitations to culture-based approaches: (1) many pathogens cannot be readily grown in culture and (2) it can take days to weeks to see results. For many plant pathogens, there are rapid tests available that depend on either serological (antibody-based) or molecular (DNA- or RNA-based) approaches.

Serological, or immunological, approaches are rapid assays that use **antibodies** that bind to specific **antigens** (usually surface proteins on cells) to detect pathogens within diseased plant tissue. Antibodies bind to specific antigen **epitopes** on the pathogen (epitopes are the specific protein segment on the antigen recognized by the antibody). They are produced commercially and are used either as test strips (qualitative measure of presence/absence) or in enzyme-linked immunosorbent assays (which allow quantification of the amount of pathogen present) (Box 9.2). In either case, if a pathogen is present in an extract of diseased plant tissue, a color change in the test indicates whether the pathogen is present in the sample, and results are ready in minutes or hours. Such tests are relatively inexpensive, require minimal training and equipment, and can sometimes be done in the field.

There are two main limitations of serological testing. First, antibodies are produced commercially in response to market demand, so tests are generally only available for common pathogens of economically important crops. Second, controlling the proportion of false-negative and false-positive results can be challenging because antibodies must be sensitive enough to bind to antigens that are reliably present on the pathogen of interest (to reduce false negatives) but they must not bind to similar antigens on related, but non-target microbes (which would create false positives). There is often a trade-off between these requirements, because increasing the specificity of the test can lead to a reduction in

Box 9.2 Serological testing using ELISA or test strips to detect plant pathogens

ELISA (enzyme-linked immunosorbent assay) tests are usually based on what is called a double-antibody sandwich test. First, "capture antibodies" are stuck to a solid surface, such as the small wells in a microtiter plate. Then a plant extract, which is expected to contain the pathogen, is made by macerating plant material in a buffer, sometimes just by mushing it around in a plastic bag. The plant extract is then washed over the capture antibodies, which bind to corresponding antigen epitopes on the pathogen, forming a conjugate. The remaining extract is washed away. Next, a solution containing a detection antibody is added; the detection antibody is specific to a different epitope on the antigen, and it has been modified with the attachment of a small reporter enzyme (usually horseradish peroxidase or alkaline phosphatase). The detection antibody binds to epitopes on the pathogen antigens that are stuck to the original capture antibody, creating an antibody–pathogen–antibody sandwich. Any excess detection antibody is then washed away. Finally, a solution with the substrate for the reporter enzyme is added. Usually this is a substrate that starts off colorless but changes color when the enzyme acts on it. Only those wells that received the pathogen turn color, because only they retained the detection antibody. The intensity of color change can be measured with a spectrophotometer to estimate the amount of pathogen in the plant extract.

Strip tests are modifications of the same principle (Figure 9.4). A strip test is made of a wicking material that pulls a liquid sample from one end of the strip to the other using capillary action. The sample is placed at one end of the strip. The wicking solution first passes through an antibody conjugate pad, where loose antibodies with colored labels are waiting—antibodies that are specific to antigen epitopes on a particular pathogen. The fluid continues to be wicked along (carrying with it any antigen–antibody conjugate) until it reaches a short strip (test line) coated with capturing antibodies, specific to different epitopes on the same antigens (these capturing antibodies are stuck firmly to the strip). Any antigen-labeled-antibody conjugates in the sample will get captured by the capturing antibodies and be stuck there. If the antigen was present in the sample, this creates a concentrated patch of labeled antibody sandwiches that can be seen as a colored line for positive samples. The rest of the sample fluid (including excess labeled antibodies) continues to be wicked along. A little beyond the test line is another short patch of capturing antibodies, but this time they are antibodies to the labeled antibodies (that is, the labeled antibodies are now the antigens)! That captures any of the original labeled antibodies that did not form conjugates with pathogen antigens, and forms another colored line, indicating that the test worked. If there is no pathogen present, only the control line will be visible; if the pathogen is present, both lines appear. By the way, this is how a human pregnancy test works, with monoclonal antibodies that detect the hormone human chorionic gonadotropin (hCG).

Box 9.2 *Continued*

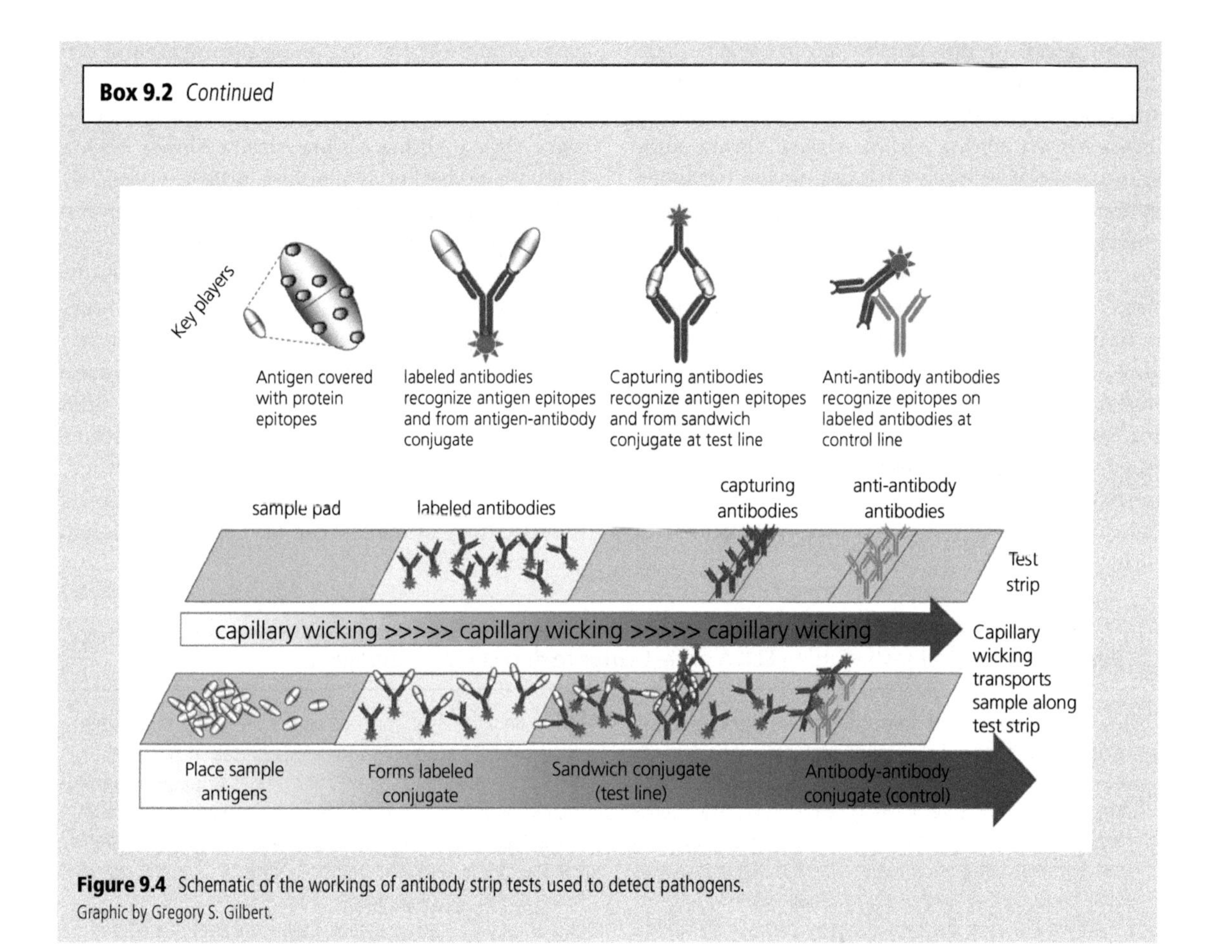

Figure 9.4 Schematic of the workings of antibody strip tests used to detect pathogens.
Graphic by Gregory S. Gilbert.

sensitivity. Finding the sweet spot remains a significant challenge for serological testing for plant pathogens. However, the low cost and ease of use make them valuable diagnostic tools for detecting many important plant pathogenic viruses, bacteria, fungi, and oomycetes (Lacroix et al. 2016, Wakeham and Pettitt 2017).

The kind of antibodies used influences the rate of false negatives and false positives. Antibodies are produced by injecting animals (rabbits, goats, chickens, rats, llamas) with the pathogen or parts of the pathogen to stimulate a response in the animal's immune system. **Polyclonal antibodies** are produced in the blood by plasma cells, and they are then purified from blood extracted from the animal; polyclonal antibodies may include a mixture of antibodies produced by different plasma cells that recognize different epitopes on the antigens. Polyclonal antibodies can be effective at consistently recognizing a pathogen (low false negatives) but because some of the antigen epitopes might be shared with other pathogens, they tend to have an annoyingly high false-positive rate. In contrast, **monoclonal antibodies** each recognize a single antigen epitope, and thus provide high specificity. To create monoclonal antibodies, a mouse is injected with the pathogen or parts of the pathogen to stimulate an immune response. Spleen cells that produce antibodies are harvested and mixed with plasma myeloma cells to create **hybridomas**, a fusion of an antibody-producing spleen cell and a plasma myeloma cell, which is an immortal cancerous cell that can grow indefinitely in laboratory culture. Each hybridoma cell line is grown in laboratory culture, where it produces monoclonal antibodies that bind to a single antigen epitope. With careful screening, cell lines that produce monoclonal antibodies only against the pathogen of interest are selected, leading to a low false-positive rate when they are used in diagnostic tests for a plant pathogen. However, any natural variation in the antigen in the pathogen population can lead to false negatives, because the high specificity of the monoclonal antibodies may not allow them to recognize the variant.

9.8 DNA-based approaches

The tools of molecular biology have transformed our ability to distinguish among species and understand their evolutionary relationships, beginning with nucleic acid hybridization techniques in the 1970s. In 1977, Frederick Sanger invented the chain termination method of **DNA sequencing** and used it to sequence the genome of bacteriophage ϕX174 (Heather and Chain 2016). The development of the **polymerase chain reaction** (**PCR**) revolutionized the tools available to detect and identify plant pathogens (Box 9.3). PCR allows us to inexpensively make millions of copies of pathogen DNA, which can be visualized, measured, or sequenced. PCR facilitates pathogen detection, quantification, and identification; it is a flexible tool that can be used in conjunction with culture-based approaches or in rapid tests directly from plant material. Whereas serological approaches rely on variation in surface proteins or other antigens that are phenotypes of the pathogens, DNA-based approaches look directly at variation in the DNA (or RNA) sequences in the pathogen genome. Here we will look at a few of these approaches.

Box 9.3 The basics of PCR

Polymerase chain reaction (**PCR**) is a process to quickly make millions of copies of specific stretches of DNA. Since Kary Mullis invented PCR in the 1980s (for which Mullis received the 1993 Nobel Prize in Chemistry—see Saiki et al. 1985), PCR has transformed all fields of life sciences.

PCR relies on two short sequences of single-stranded DNA (**primers**) that match sequences found at either end of the stretch of double-stranded DNA to be copied (**template**); one primer matches to one strand of DNA (the 5′ to 3′ strand) and the other primer to the complementary DNA strand (the 3′ to 5′ strand), such that the target region for amplification is sandwiched between the two primers. In addition to the primers and template DNA (extracted from the organism of interest, or from an environmental sample), a PCR reaction requires abundant free bases (ATCG), and a DNA polymerase enzyme. Usually this is the DNA polymerase called Taq, which is produced by the hot-springs-loving bacterium *Thermus aquaticus* (discovered by Alice Chien et al. 1976). Taq is special because its optimum temperature for copying DNA is a warm 75°C (it barely does anything at room temperature), and it does not break down at high temperatures close to boiling, like 95°C, where double-stranded DNA becomes denatured (separated into its two complementary strands).

PCR reactions are usually done in machines called thermocyclers, which allow rapid and controlled heating and cooling of samples in repeated cycles (Figure 9.5). In a PCR reaction, the template double-stranded DNA is first denatured by heating to about 95°C. The temperature is then quickly lowered to about 50°C, where the primers automatically anneal (bind) to their corresponding sites on the template DNA—one on each of the separated strands. With warming to about 72°C for a few minutes, the DNA polymerase extends the DNA sequence from those primers, placing the appropriate matching base pairs (A matches T, C matches G) in the correct order to make a new complementary strand for each strand of the template DNA; this returns the single-stranded denatured DNA to its double-stranded condition. The temperature is then raised again to 95°C to denature all the new double-stranded copies of DNA (which now serve as template), and the cycle begins again. PCR generally runs through 20–30 such cycles in an hour or so. Because each template produces two copies, and those copies another two copies each, the process is exponential; 20 cycles would make 2^{20} copies (a little over a million) and 30 cycles would make 2^{30} (over a billion) copies of a single initial strand of DNA.

Box 9.3 *Continued*

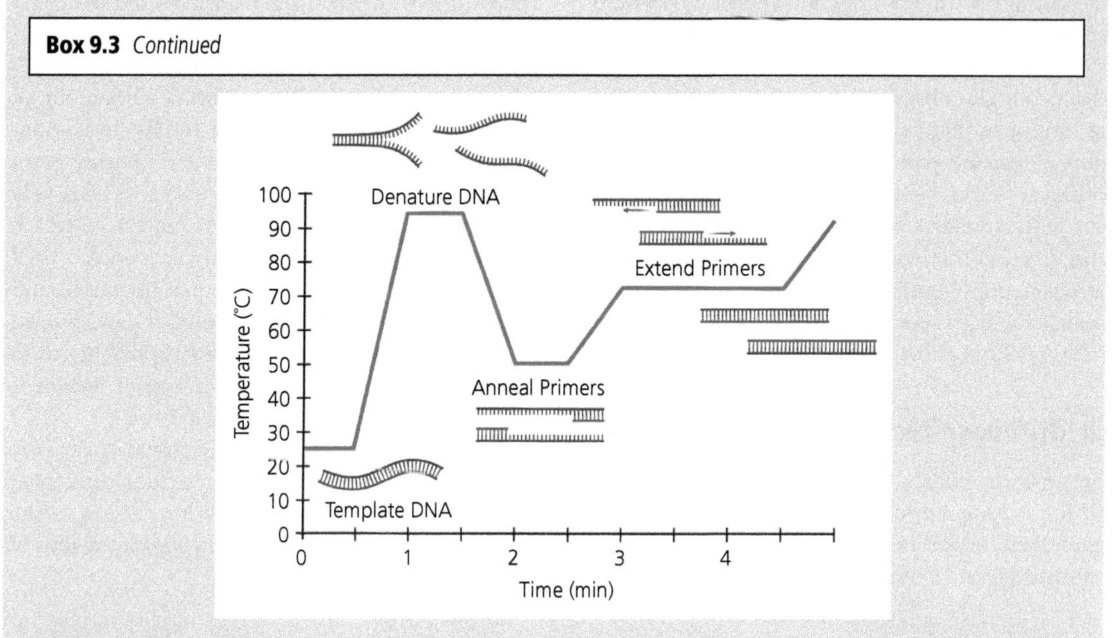

Figure 9.5 Temporal scheme of the temperature changes (orange line) that lead to copies of template DNA in a single cycle of PCR as controlled in a thermocycler.

Graphic adapted by Ingrid M. Parker from "PCR stages relative to temperature," by BioRender.com (2022), retrieved from https://app.biorender.com/biorender-templates.

9.8.1 Barcodes

Identification of most microbes relies on using "**barcode**" sequences for comparison to global databases of microbes and plants. They are called barcodes as analogies to the barcodes used at supermarket checkouts that provide unique identifiers for different products. The DNA sequences used as barcodes are stretches of several hundred base pairs in regions that undergo just the right rate of evolution to distinguish among taxa. When the rate of evolution is too slow, such as with genes so essential to core cell functions that most changes are fatal, then sequences are not useful because fixed mutations are too rare to distinguish among species. When the rate is too fast, such as with genes under strong selection in populations (such as virulence genes), then sequences are not useful because they are highly variable within a species (Section 13.1). In order to identify plant pathogens easily and cheaply from their DNA sequences, we all need to agree to look at the same stretch of DNA. For fungi, that sweet spot where sequences distinguish among species but are not so variable that every population looks different, is the **internal transcribed spacer (ITS) region** (Schoch et al. 2012).

The ITS region includes the spacer DNA within a larger section of nuclear DNA that codes for ribosomal RNA (Figure 9.6). This larger DNA region has several features that are useful for different applications. It is divided into three components: a region coding for the 18S small ribosomal subunit (18S SSU rRNA), a region that codes for the 28S large ribosomal subunit (28S LSU rRNA), and the ITS, which itself has three parts. The 18S SSU rRNA and 28S LSU rRNA components generate critical functional parts of the ribosome, so evolution of those components (observed as changes in the DNA sequences) is rather slow. Evolution of the SSU is even a bit slower than the LSU (Raja et al. 2017). Because they evolve slowly, the sequences of base pairs in the 18S SSU rRNA and 28S LSU rRNA regions have low levels of variation among taxa. They are most useful in reconstructing deeper, more ancient phylogenetic

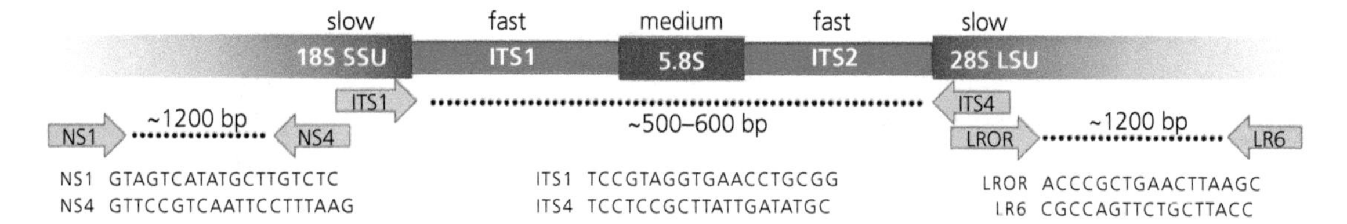

Figure 9.6 Structure of the nuclear ribosomal DNA regions used for fungal barcodes. The DNA that codes for the ribosomal 18S small subunit (18S SSU) is separated from the 28S large subunit (28S LSU) by the internal transcribed spacer (ITS) region. The ITS1 and ITS2 regions are cleaved out and are subject to fast rates of evolution; the 18S, 5.8S, and 28S regions code for ribosomal components and evolve more slowly. Three pairs of universal primers (NS1 & NS4, ITS1 & ITS4, and LROR & LR6; shown as arrows indicating direction of priming) are used for sequencing part of the 18S SSU, the ITS region, and part of the 28S LSU, respectively (not to scale). Commonly used primer sequences are shown, 5′ → 3′.
Graphic by Gregory S. Gilbert.

relationships among fungi, like divergences at the order or family level (Raja et al. 2017).

In contrast, greater variation in the internal transcribed spacer region tends to coincide roughly with recognized differences among fungal species. The 500–600 base pair ITS region includes three sections (Figure 9.6): ITS1, ITS2, and a 5.8S region between them that contributes rRNA to the large subunit of the ribosome. ITS1 and ITS2 are cleaved away during rRNA maturation, such that neither region contributes to functional parts of the ribosome, and they are freer to evolve through genetic drift (Section 13.2).

Because it hits the sweet spot of just the right amount of variation among fungal lineages, the ITS region was chosen by a consortium of mycologists as the "official" molecular barcode for identifying fungi, and the research community continues to build extensive public reference databases of ITS sequences for fungal identification (Schoch et al. 2012). The benefit of agreeing on a single barcode is that a standardized set of primers can be used to get a sequence of the same genome region for any fungus, allowing ready comparison and contribution to global databases. The complete ITS sequence can be read from a fungal sample with a few hours' work and a few dollars in supplies and sequencing costs.

While ITS sequences are widely and effectively used to identify fungal species, there are also some important limitations. First, ITS sequences do not always evolve rapidly enough to differ consistently between closely related species, so to distinguish among them sometimes requires sequencing more rapidly evolving barcode regions. There is not yet consensus on which other regions are optimal, and a number of candidate genes are being tested (Stielow et al. 2015). Second, ITS sequences of distantly related fungi may show the opposite problem: they may be so different that it is not possible to align the sequences to understand their phylogenetic relationships. In some cases, complementing ITS sequences with those of more slowly evolving barcodes of the 18S SSU rRNA and 28S LSU rRNA regions is helpful (Tedersoo and Lindahl 2016). Holistically inspecting the ITS1 and ITS2 regions together with the 5.8S region can provide complementary insights into how two species are similar or different (Figure 9.7).

Other regions of DNA serve as standard barcodes for other groups of pathogens. For bacteria, it is the 16S ribosomal subunit (16S SSU rRNA) (Clarridge 2004); for nematodes, part of the 18S subunit is used (18S SSU rRNA) (Floyd et al. 2002); and for oomycetes a combination of ITS and the mitochondrial cytochrome c oxidase subunit 1 (*cox1*) or subunit 2 (*cox2*) genes are used (Robideau et al. 2011, Choi et al. 2015). For plants, two chloroplast genes, *rbcL* and *matK*, along with ITS, serve as the core barcodes (China Plant BOL Group et al. 2011). Viruses don't have universal barcode gene candidates, and instead the entire viral (DNA or RNA) genome is often sequenced (Campillo-Balderas et al. 2015).

Researchers who generate ITS sequences for an organism they have identified then upload that sequence data into databases such as **GenBank**, run by the International Nucleotide Sequence Database Collaboration (including the DNA DataBank of Japan (DDBJ), the European Nucleotide Archive

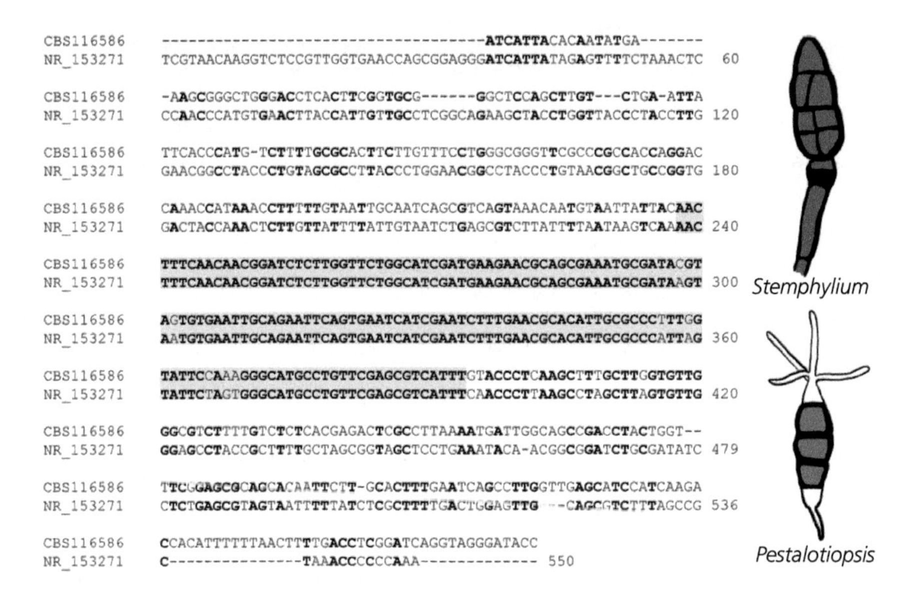

Figure 9.7 ITS1–5.8S—ITS2 DNA barcode sequences aligned for two plant pathogenic fungi, *Stemphylium solani* CBS116586 and *Pestalotiopsis sequoiae* NR_153271. Sequences are from carefully curated specimens associated with collections originally used to give a name to the fungus. Sequences have been aligned to the extent possible, and base pairs that match are shown in **bold**. Within the conserved 5.8S region (bases 237–394, grey background) only 8 of 158 base pairs differ. In contrast, both the lengths and the base sequences of the rapidly evolving ITS1 and ITS2 regions differ greatly between the species. More closely related species have more matching base pairs.
Graphic by Gregory S. Gilbert.

(ENA), and the US National Center for Biotechnology Information (NCBI). These barcode sequences are then available at www.ncbi.nlm.nih.gov/genbank/. The International Barcode of Life (iBOL; https://ibol.org/) consortium and their Barcode of Life Data Systems (BOLD; http://v4.boldsystems.org/) coordinate the collection and use of barcode data across the tree of life (www.barcodinglife.org). The UNITE community (https://unite.ut.ee/) has compiled a well-curated database of barcode sequences for Fungi, in particular (Nilsson et al. 2019).

It is important to recognize that DNA barcodes only work if they can be compared to a reliable database. The databases of verified ITS sequences are still very incomplete (only a fraction of described plant pathogen species have been sequenced, and only a fraction of all species have even been described (Blackwell 2011, Wu et al. 2019). In addition, significant issues in the reliability of accompanying taxonomic data are well known (Kõljalg et al. 2013). For example, searching the GenBank database for matches to fungal ITS DNA sequences can sometimes return close matches to DNA sequences from plant species. But upon careful inspection, it becomes clear that what had been registered as DNA sequence from the plant had actually come from a fungal endophyte living inside the plant tissue (Camacho et al. 1997)! The sequence was nothing like the DNA of its plant host.

Purified DNA extracted from a single species, as in the culture-based approaches described above, were traditionally identified with Sanger sequencing. But leaves, roots, wood, or soil often contain many different fungal species—wouldn't it be great to be able to get the ITS sequences for all the fungi in a plant sample at the same time? **High-throughput sequencing** to the rescue! Sometimes called "next-gen sequencing," high-throughput sequencing refers to a suite of recent technological advances by several different companies (Illumina, PacBio, and Oxford Nanopore) that sequence multiple stretches of DNA in parallel (Heather and Chain 2016). While Sanger sequencing can read a barcode gene of a single organism at a time, high-throughput sequencing enables sequencing of multiple taxa simultaneously. We can identify all the organisms present in a sample from their DNA; this is called **metabarcoding** or **metagenomics**. The particular details of the technology vary among the different platforms, and there are trade-offs among read length, accuracy, and cost per sample. But regardless of the particulars, high-throughput sequencing has already provided an unprecedented view into plant microbiomes—essentially allowing the study of whole assemblages of fungi or bacteria living inside plants (Copeland et al. 2015).

9.8.2 Pathogen-specific primers

The pairs of primers used to amplify the ITS barcode are useful because they are universal primers for the ITS regions of (nearly) all fungal species. A different approach is to find pairs of primers that each recognize DNA sequences that are unique to a particular taxon of interest; ideally, such pathogen-specific primers amplify a particular stretch of DNA from the target pathogen but not DNA from any other taxa.

Specific primer pairs need to be developed individually for each target pathogen. This is especially useful when a particular target plant is commonly attacked by a particular set of pathogens; a suite of primers specific to each target pathogen can be developed and then used in combination to detect each of the pathogens, simultaneously, in the same sample. Once primers are developed and validated, DNA extracts from many plant samples—as well as air, water, or soil samples—can be prepared (often using robots) and amplified for detection and quantification of those pathogens.

For instance, Sun et al. (2020) identified three different pairs of primers; each set was highly specific to just one of three common wheat pathogens (Figure 9.8a). PCR with each primer pair amplified DNA from only one target fungal species (at least among the fungi commonly found in the study area), and each PCR product was of a different length. PCR on DNA extracted from a diseased

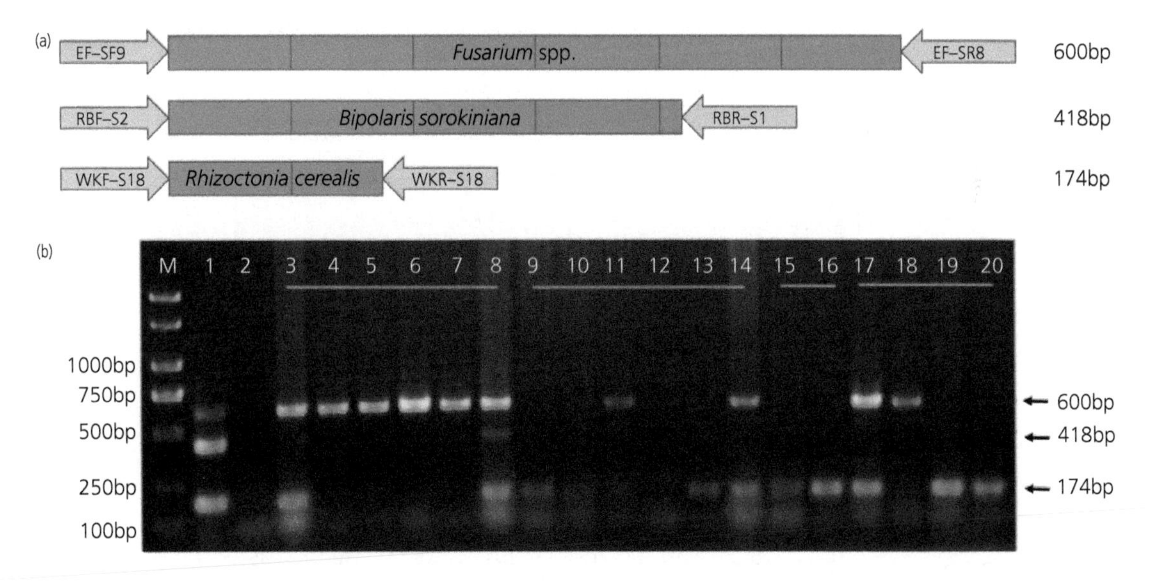

Figure 9.8 The use of pathogen-specific primers to detect the cause of disease on wheat (Sun et al. 2020). (a) Three pairs of primers (grey arrows) are specific to each of three wheat pathogens, and each produces a product of characteristic length. (b) DNA extracted from plant samples was amplified with the three primer pairs, and the DNA products were placed in wells at the top of the agarose gel. An electric current pulls the negatively charged DNA through the gel; smaller fragments move more quickly than larger fragments, so fragments end up separated in bands of similar size. Lane M is a DNA ladder with fragments of known lengths (number of base pairs shown at left); Lane 1 is a positive control with DNA from each of the three fungi; Lane 2 is a negative control without fungi; Lanes 3–20 are DNA extracted from wheat plants from four fields in Shandong Province, China. *Fusarium* was detected in samples 3–8, 11, 14, 17, and 18; *Bipolaris* was in sample 8.

Figure adapted with permission, from Table 2 and Figure 6 in Sun et al. (2020).

wheat plant, which included DNA from fungi living inside the plant, amplified the DNA of the target pathogen if it was present. For example, when DNA of *Rhizoctonia cerealis* was present, amplification using the primer pair WKF-S18 and WKR-S8 generated a PCR product 174 base pairs long. The other two primer pairs yielded DNA products of different lengths, specific to *Bipolaris sorokiniana* or to *Fusarium* species. Sun et al. took advantage of this specificity and used all three primer pairs at the same time on one sample; that is, they multiplexed the three primers, allowing them to simultaneously detect the presence of any of the three pathogens. An agarose gel of the PCR products, viewed under UV light, allowed visualization of whether 0, 1, 2, or all 3 of the pathogens were present (Figure 9.8b).

Such pathogen-specific primers can be used to quantify the amount of pathogen present in a sample using **real-time** or **quantitative PCR (qPCR)** (Luo et al. 2017). qPCR machines do PCR just like any other thermocycler, but are modified to measure the amount of PCR product (copies of DNA sequence) after each cycle. A detector compound (usually either SYBR Green® or TaqMan®) is included in the solution; the detector compound binds to DNA and then fluoresces (it only fluoresces when bound to double-stranded DNA). The more PCR product, the more fluorescence. qPCR machines include a fluorometer to measure fluorescence after each PCR cycle. The number of PCR cycles needed before fluorescence reaches a minimum detection threshold depends on how much target DNA was in the original sample; if there was a lot of target DNA (i.e., a lot of the target pathogen) then only a few PCR cycles are needed to reach the threshold, but if there was very little target DNA, many more cycles are required (Figure 9.9). The number of cycles needed to reach that detection threshold is called the cycle threshold, C_t. Standard curves are created by doing separate PCR reactions on sequential dilutions of a known amount of pathogen, establishing a C_t for each amount. The C_t value of a sample of unknown pathogen amount can then be compared to a standard curve of C_t values for known amounts to estimate the pathogen concentration.

A recently developed technology with pathogen-specific primers is called **loop-mediated isothermal**

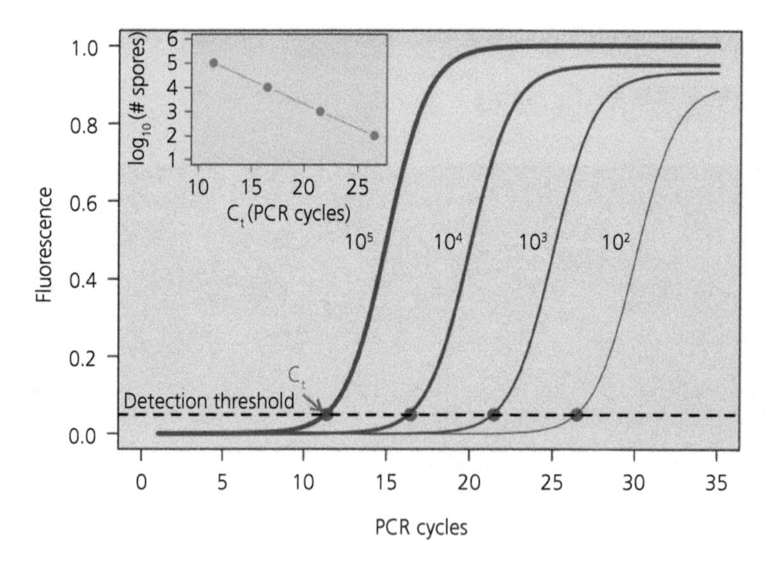

Figure 9.9 Time course of qPCR to measure the quantity of fungal spores in a sample. The thickest line shows observed fluorescence (measure of amplified DNA) with increasing numbers of PCR cycles for a sample with 100,000 (10^5) spores. Each successive curve represents results from PCR of a 10-fold dilution series. The number of PCR cycles at which time the fluorescence passes the detection threshold (C_t) is marked with a green circle. The inset graph shows the relationship between the number of spores in the sample and the C_t; this standard curve can then be used to estimate the number of spores in an unknown sample, based on the observed C_t value.
Figure by Gregory S. Gilbert.

amplification (LAMP). LAMP is a non-PCR method using a strand-displacing DNA polymerase and 4–6 primers that recognize 6–8 distinct sequences on the target DNA, producing a large number of strung-together copies of the target DNA in less than an hour, without the need for PCR (Notomi et al. 2000). LAMP produces very long strands of DNA product, with two of the primers creating loop structures that facilitate additional amplification. The combination of four distinct primers provides for high selectivity in amplifying only the target. The entire reaction takes place at a constant temperature, so there is no need for an expensive or bulky thermocycler as used in PCR. DNA product accumulates more quickly than in PCR (readable in as little as 15 minutes), and the DNA product can be read through either simple fluorescent or pH-sensitive colorimetric assays. These characteristics mean that LAMP is a portable and rapid technology suited for field diagnostics, so-called point-of-care uses (Aglietti et al. 2019).

9.9 How do we measure disease severity?

Disease severity is a measure of how badly individual plants are affected by the disease. Unlike assessments of disease prevalence, where each plant is rated for a qualitative, binary response (a plant is diseased or it is not), severity is measured as a matter of degree of disease on the plant. Some approaches for measuring disease severity are as old as plant pathology itself, while others are taking advantage of the most exciting new technologies in engineering and AgTech. As for prevalence, techniques to estimate severity are subject to trade-offs of time and cost as well as accuracy and precision, and the choice of which approach to use depends on the questions you want to answer and the resources available.

The most common estimates of disease severity are "by eye" (ocular or **visual assessment**). Unfortunately, people are hopelessly poor at visually assessing proportions without bias, and every study that quantifies disease severity via symptoms and signs must address this challenge openly. There are tools available to help, and training is essential (Bock et al. 2010). Mitigating the problems of rater variability and bias center on the concepts of precision and accuracy we discussed in Section 9.3. Whether low- or high-tech, tools for visual assessment are all designed to make data collection faster, easier, and more accurate; we introduce three approaches here: ordinal rating scales, visual aids for assessing proportions, and digital image analysis.

Estimates of severity may be either **continuous** (e.g., measurement of the percentage of plant tissue damaged on a continuous scale of 0 to 100) or **ordinal** (a series of categories or a ranked scale that reflect increasingly damaged states). It is sometimes easier and faster to assign samples to categories than to give a precise quantitative score, although a quantitative score is more informative. (For example, a term paper could be given a "B" grade, or assigned 83.5% of the possible points. It is easier to develop an intuition for the difference between an "A paper" and a "B paper" than to develop an intuition for what makes a paper 83.5%.) Unlike for continuous variables, the statistical analysis of ordinal variables requires an additional step of considering how the values from categories or rankings should be summarized.

In plant pathology, the simplest rating scales are **ordinal scales of disease severity** devised to reflect an ordered progression of symptoms and impact on the host plant. Such scales must be developed individually for each disease on each host species, depending on the kinds of symptoms associated with the disease and the structure of the plant host. For instance, *Okra enation leaf curl virus* can be scored based on the degree of leaf curling, plant stunting, and production of axillary branches (Table 9.1). Each plant in a sample is given a severity score (P) based on the rating scale. A **disease severity index** can

Table 9.1 Disease rating scale for symptoms caused by *Okra enation leaf curl virus*

Rating	Description
0	No symptoms
1	Top leaves curled
3	Top leaves curled and slight stunting of plant
5	All leaves curled, twisting of petiole and slight stunting of plant
7	Severe curling of leaves, twisting of petiole, stunting of plant, and proliferation of axillary branches

Adapted from Yadav et al. (2018).

then be calculated as:

$$DSI\,(\%) = \frac{\sum_{i=1}^{k}(P_i \times Q_i)}{(M \times N)} \times 100 \qquad \text{[Eq. 9.2]}$$

where Q_i is the number of plants with rating P_i, M is the maximum value of P, N is the total number of plants rated, and k is the number of rating classes.

A similar ordinal rating scale is sometimes used to reflect broad categories of the percent of plant tissue that is symptomatic. For instance, Benitez-Malvido et al. (1999) used a 6-category scale of disease severity based on visual estimates of the area affected by disease on the leaves of rainforest tree seedlings. Their categories were 0 = intact, 1 = 1–6%, 2 – 6–12%, 3 = 12–25%, 4 = 25–50%, and 5 = 50–100%. Leaf ratings were combined into a disease index for each plant as:

$$DI = \sum_{i=1}^{5} n_i C_i N \qquad \text{[Eq. 9.3]}$$

where n_i is the number of leaves in the ith category of damage, C_i is the midpoint of each severity category, and N is the total number of leaves on the plant.

Such ordinal scales are useful for rapid and inexpensive estimates of disease severity. They are often used for screening of a large number of plants, such as in breeding trials or landscape-level assessments (Tahi et al. 2000). Raters must be trained to provide accurate (reliable and repeatable) estimates; for example, a common bias is that low disease intensity tends to be overestimated (Bock et al. 2010). The coarseness of ordinal scales results in data with low **resolution** (low ability to differentiate between values that are slightly different), which compromises the precision of the measurements and limits their usefulness for many applications.

Visual aids are useful tools to improve accuracy and precision of severity estimates, especially for foliar diseases. A series of images with representative patterns of different proportions (say, 5, 25, 50, 75, 99%) of foliar disease can be taken to the field as normalizing reference points for visual assessments (James 1971). The assessor compares leaves to images and interpolates between illustrated reference points to record more continuous estimates of disease severity. With training and practice, the use of such visual keys improves both accuracy and precision of estimates (Bock et al. 2010).

The advent of digital photography and (free!) image processing software provides a whole new way to measure foliar disease severity using **digital image analysis**. A digital image of diseased leaves can be taken on live plants in the field or greenhouse, or leaves can be collected and brought back to the lab for imaging. Image analysis is done using software packages such as *ImageJ*, which is available free from the National Institutes of Health (https://imagej.nih.gov). First the whole leaf area is measured, then color filters allow delineation and measurement of symptomatic tissues (Figure 9.10a–c). With training to calibrate the setting of color filters, such image analysis provides an accurate and precise estimate of severity as percent diseased leaf tissue. It is also easy to create your own normalizing key for field visual estimation of disease severity customized for the host and pathogen you want to study (Figure 9.10d). Using digital image analysis to support your own field data collection can be a "best of both worlds" approach to rapid symptom assessment.

As with disease prevalence, disease severity can be estimated with exciting new technologies that unveil disease symptoms invisible to the naked eye. We see colors in the visible spectrum, and that allows us to recognize symptoms like chlorosis, anthocyanin production, and necrosis as disease symptoms (Chapter 8). But physiological symptoms in stressed and diseased plants may not be perceptible via visible wavelengths yet be detected as changes in other parts of the spectrum. Lindenthal et al. (2005) used **thermal imaging (thermography**, which detects long-infrared wavelengths) to create heat maps of individual leaves showing that highly localized temperature patterns reflected disease severity even before the onset of visible symptoms (Figure 9.11a and b). **Hyperspectral imaging** collects data for every pixel in an image from across the electromagnetic spectrum, from infrared to ultraviolet, uncovering otherwise

Figure 9.10 Digital analysis of percent of leaf tissue covered by lesions of blackberry leaf spot caused by *Mycosphaerella rubi* using NIH *ImageJ* software. (a) Original photo taken against a light blue background ("bluescreen" works better than traditional "greenscreen" for plants). (b) Area of the entire leaf selected and measured by filtering out blue background. (c) Tan lesions with a maroon border selected using color threshold filters and area of lesions measured. Severity = area of lesions/total leaf area *100. (d) Estimates of leaf area covered by blackberry anthracnose lesions for a range of disease severity. A set of such images can be used in the field to standardize visual severity estimates.
Images by Gregory S. Gilbert.

invisible disease symptoms with high sensitivity. Hyperspectral imaging of above-ground parts of sugar beets was three to five times better at detecting plants suffering from Rhizoctonia root rot than traditional visual assessment (Barreto et al. 2020). Hyperspectral imaging systems can be used on individual plants or they can be mounted on flying drones (**UAVs**: **unmanned aerial vehicles**) to capture images at large scale or in difficult terrain.

Acoustic tomography uses differences in the speed of sound through solid and decayed wood to create non-invasive images of living tree trunks in minutes to measure the severity of heart-rot decay (Figure 9.11c). Commercial units are widely used by arborists as well as researchers. One of us has used tree tomography to study internal decay of trees in tropical rainforests (Gilbert et al. 2016), and then collaborated with park and city managers in Panama to help evaluate the safety risk of potential hazard trees. Acoustic tomography can avoid disasters by identifying trees that are no longer structurally sound, and it can also save trees that look suspicious

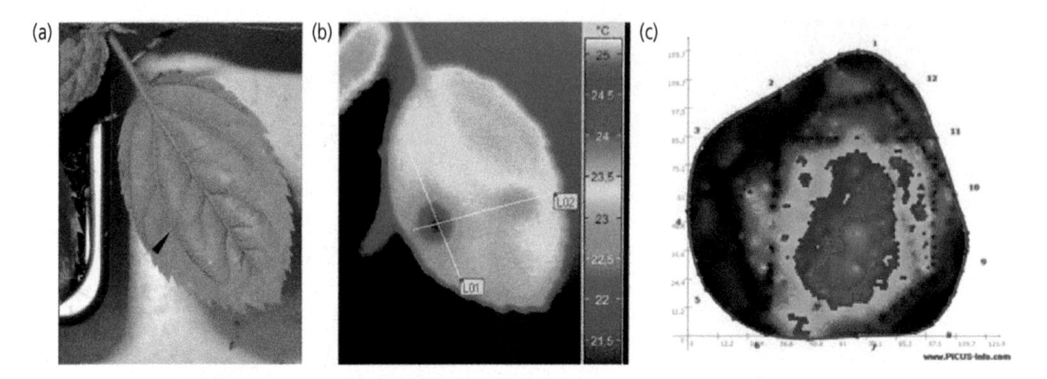

Figure 9.11 Measuring disease severity with high-tech methods to visualize the not-so-visible. (a) Extent of infection following inoculation (black arrow) of an apple leaf with the scab fungus *Venturia inaequalis* indicated by (b) characteristic cooler leaf temperatures made visible through thermal imagery. (c) Acoustic tomography shows heart-rot decay (bright colors) in the interior of a tree trunk; brown colors around the perimeter reflect healthy wood, with no exterior signs of decay.
Images a and b reprinted by permission from Springer Nature, Mahlein et al. (2012); Image c by Gregory S. Gilbert.

from the outside but in fact have a healthy trunk and just need a little judicious pruning. New and exciting biosensor technologies to measure disease severity are on the horizon and will no doubt revolutionize disease assessment in the future (Oerke 2020).

9.10 How do we measure impact on the host?

Diseases can affect photosynthesis, water relations, leaf lifespan, access to nutrients, and other physiological processes. Which of these things are most important to the plant? *Are* they important? How do we know? We need measures of overall impact in order to compare different pathogens, compare different crop lines or host species, or evaluate the importance of pathogens in agricultural and wild systems.

For crops, there is a simple bottom line through which all impacts can be interpreted: yield. If disease results in a 10% decline in water use efficiency but does not reduce yield, then we are less concerned about that pathogen than if yield were also reduced by 10%. For pathogen effects on wild plants, what constitutes a meaningful impact on the host is less clear. The bottom line for wild species (instead of yield) is fitness. We discuss fitness at length in Chapter 13. Plant survival, growth, and reproduction can all contribute to plant fitness, which, in brief, can be thought of as total reproductive success, or the contribution that an individual makes to the gene pool of the next generation. Pathogens that reduce plant growth rate may affect fitness because smaller plants make fewer seeds, or because smaller plants make weak seeds that are vulnerable to desiccation, or because smaller plants get outcompeted by larger neighboring plants. Diseased plants may more often die before reproducing, or they may die sooner than healthy plants; here we quantify the impact of pathogens on host fitness using survival analysis (Scherm and Ojiambo 2004).

An important aspect of measuring disease impact on a host happens before there is even anything to measure: that aspect is thinking through your **study design**. Here we consider two types of approaches to quantifying impact: experimental studies and observational studies.

Ideally, impacts of disease are measured with **experimental studies**, comparing performance of plants infected with pathogens to those that are not, in randomized, controlled trials. There are three basic experimental designs for creating such infected versus non-infected comparisons.

(1) Inoculate plants with a pathogen and compare the effects to non-inoculated controls. For instance, Wheeler et al. (2016) filled pots with potting soil that was mixed with several strains of *Verticillium dahliae* to a concentration of

10 cfu g^{-1} soil. They sowed seeds of eight different crop species into the infested soils, as well as into non-inoculated soil as controls. They measured plant height over time, infection rates, and the final biomass of the roots and the shoots. Comparison of outcomes in infested soils to control soil allowed them to experimentally measure how various pathogen strains affected different crop species.

(2) Excluding natural infection from some plants through the use of protectant fungicides, bactericides, or insecticides. For instance, randomized field trials with seven wheat cultivars that were sprayed with fungicide or left non-sprayed showed increases in grain yield of 6% to 48% associated with protection from rust infection (Soko et al. 2018). Some fungicides can have direct beneficial or negative impacts on plant growth even in the absence of disease, so fungicide in the absence of the pathogen should be included as a control treatment.

(3) Comparison of resistant vs susceptible genotypes of the same plant species. For instance, a large replicated field study of susceptible and resistant cultivars of soybean across fields infested with soybean cyst nematode (*Heterodera glycines*) showed a 14.9 kg ha^{-1} yr^{-1} yield benefit from genetic resistance (De Bruin and Pedersen 2008). Resistant and susceptible host genotypes may differ in growth and fitness attributes not related to impacts on disease, so comparing the genotypes in the absence of the pathogen should be included as an appropriate control.

Combining the second and third approach into one study, Gortari et al. (2019) used fungicide treatments to experimentally exclude natural infection on both susceptible and resistant clones of poplar trees (*Populus deltoides*) by the rust fungus *Melampsora medusae*. The authors grew clones in pots outdoors in an area heavily infested with the rust, and sprayed half the plants every 15 days with a systemic fungicide that prevented infection. None of the sprayed plants developed rust infections, but all the non-sprayed plants became infected. Severity of infection in the resistant cultivars was about 30%, while severity in the susceptible clone was more than double that. Resistant clones produced about 30% more leaves than the susceptible clones when protected with fungicides, whereas the susceptible clones had double the growth benefit from fungicide treatment.

Unfortunately, controlled, randomized experiments are not always possible. To measure the impacts of disease on the growth of large, long-lived species like trees, or for studies across large landscapes, we almost always use **observational studies**. For example, the rates of radial growth in diseased trees can be compared to growth of nearby trees that are not diseased, or rates can be compared before and after naturally occurring disease (Busby and Canham 2011). The problem with observational approaches to estimate disease impacts is that they do not readily control for factors such as small-scale variation in environmental conditions that could make some trees more susceptible to disease while also making them grow slower and experience higher mortality. It is difficult to disentangle how much growth reduction is caused by the disease, and how much disease development and growth reduction are jointly caused by those other factors.

9.11 How do we integrate disease intensity over time? AUDPC

Plants seldom recover after developing disease. That means that disease progress over time is cumulative. How quickly the disease spreads is often as important in determining the overall impact of disease as is the final amount of disease. For instance, a disease that spreads very quickly to infect 23% of the plants in a population in the first few weeks of the growing season is likely to have a bigger impact on host growth and reproduction than one that spreads very slowly but reaches a final prevalence of 23% at the end of the season (Figure 9.12). Similarly, because disease severity increases over time, newly symptomatic tissue is added to that already diseased. The cumulative nature of plant disease progress can be captured by calculating the **area under the disease progress curve (AUDPC)**, originally introduced by Shaner and Finney (1977) to compare the spread of powdery mildew on different cultivars of wheat.

Calculating the AUDPC requires repeated measures of disease intensity (Y, which could represent either prevalence or severity) over time (t). AUDPC can then be calculated using the formula:

$$AUDPC = \sum_{i=1}^{n} [(Y_{i+1} + Y_i)/2]\,[t_{i+1} - t_i] \quad \text{[Eq. 9.4]}$$

where Y_i is disease intensity at time t_i, and Y_{i+1} is the disease intensity at the next time step, time t_{i+1}. For each time-interval, the average disease intensity for that interval $[(Y_{i+1} + Y_i)/2]$ is calculated as the height of a rectangle (Figure 9.12), and the width of the rectangle is the length of the time interval $[t_{i+1} - t_i]$; the product of those two gives the area of the rectangle. The sum of the areas of all the rectangles gives an estimate of the AUDPC. This process is basically calculating the Riemann sum (an approximation for the integral) for the curve. One could also calculate the AUDPC as the definite integral of a curve fit to disease progress data, but variability in the shapes of disease progress curves and uncertainty in the fitting of functions to those shapes means the Riemann-sum approximation is more broadly applicable in real-world situations. AUDPC is used as an integrated measure of disease intensity in all areas of phytopathology including selection of resistant genotypes, studies of disease physiology, epidemiology, development of control methods, and evolution of virulence (Simko and Piepho 2012).

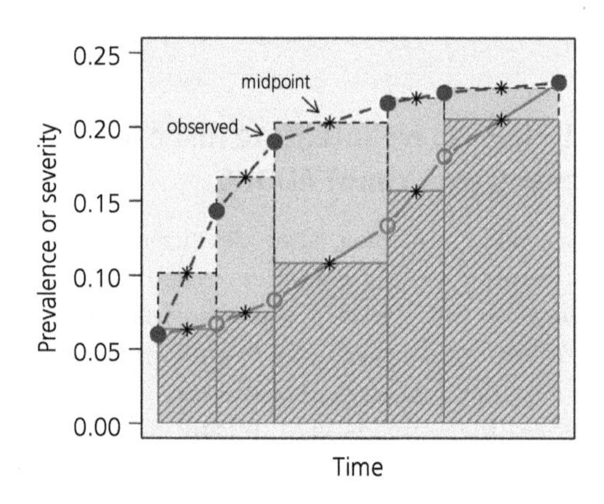

Figure 9.12 Area under the disease progress curve (AUDPC) for two epidemics. Disease prevalence (or severity) at each observation time-point is shown as an open orange circle (epidemic A) or filled purple circle (epidemic B). The midpoint disease values for each time-interval are marked with asterisks. Rectangles with lower and upper boundaries from zero to the midpoint, and from the start to end of a time interval, are shown in diagonal hatching for epidemic A and light blue for epidemic B. AUDPC is calculated as the sums of those rectangular areas, as given in Equation 9.4: 0.92 for epidemic A and 1.35 for epidemic B.
Figure by Gregory S. Gilbert.

9.12 How do we measure pathogens outside of their hosts?

Disease ecologists often want to measure the abundance and distribution of pathogens in the environment, independent of hosts. Resting spores of fungi and oomycetes in the soil, as well as spores dispersing through the air, rain-splash, or streams are of great interest in studies of temporal (Chapter 10) and spatial (Chapter 11) dynamics of plant diseases and in making management decisions (Chapter 17). Plant pathogens can be detected, identified, and quantified using all the same techniques described above for plant samples, including cultural approaches using selective growth media, serological testing, PCR using pathogen-specific probes, and high-throughput sequencing using barcodes. Plant pathologists have developed an array of methods for collecting environmental samples to be processed using those techniques.

Many pathogens have resting structures in the soil that serve as primary inoculum to start disease cycles. Farmers benefit from knowing the amount of pathogen inoculum in a field before deciding what to plant, and ecologists use information about spatial patterns of pathogens in studies of spatial dynamics of plant communities. Let's consider three approaches to quantifying pathogens in soil: PCR-based approaches, cultural approaches, and most-probable-number baiting.

Pathogens can be detected and measured in soil by extracting DNA from a sample and using PCR-based approaches on the soil extracts (Nowakowska et al. 2017). PCR-based methods have the disadvantage that they do not distinguish between live and recently dead organisms, so it is not always straightforward to relate results from soil extracts to potential for disease development.

Cultural methods can be used, but face a really big challenge of scale: microbe densities can range from 1 (10^0) to 1,000,000 (10^6) or more spores or cells per cm^3 of soil. Bacteria can reach 10^9 cells on the surface of a root! Pathologists and microbial ecologists often rely on dilution series to estimate the amounts of microbes of interest in environmental samples.

Two serial dilution approaches that have been widely used in plant pathology for decades are dilution plating and most-probable number approaches. **Dilution plating** relies on agar growth media that are selective for the microbes of interest, and allows the quantification of viable, culturable microbes in soil or other substrates, including infected plant material (Figure 9.13). Because

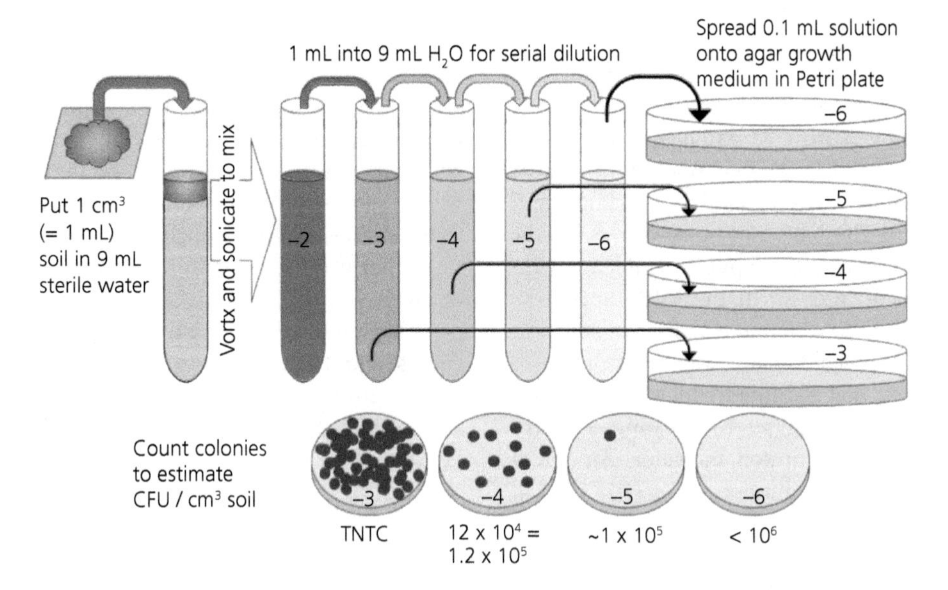

Figure 9.13 Dilution plating to quantify microbes from soil or other substrates. To quantify the number of culturable microbes in soil, we start by diluting 1 cm^3 (which is the same as 1 mL) in 9 mL sterile water to create a 1:10 suspension. To make sure the microbes are in solution, we often use sonication (high-frequency sound in a water bath) to shake apart soil particles and microbial cells, and then mix the suspension on a vortex mixer. The first tube is labeled "–2," for 10^{-2} dilution (technically this is a 10^{-1} dilution, but because later we will only quantify 0.1 mL of the solution, we label it –2 from the start). A 1-mL sample of that suspension (after vortexing to mix it) is then diluted 10-fold with 9 mL of sterile water (labeled –3 as the 10^{-3} dilution). A subsequent sample of that solution is diluted 10-fold, and so forth, up to serial dilution 10^{-6} or more. A small aliquot (0.1 mL) of each (we often skip –2, because it is too thick) is then placed on selective growth agar medium in separate Petri plates, and spread evenly across the surface with a bent glass rod. After an appropriate time (days to a week), growth of colonies characteristic of the target pathogen can be counted. If the soil is heavily infested, the first few dilutions may be so covered with microbial growth that colonies are too numerous to count (TNTC), while the final dilutions may have no growth at all. Somewhere in the series should be a countable number of discrete colonies. The concentration of the pathogen—described as colony-forming units (cfu)—in the original soil can then be calculated. The shortcut is just to use the exponent labeled on the plate. Here we see that 12 colonies formed on the 10^{-4} dilution plate. The inverse of 10^{-4} is 10^4, and we can say there were $12\times10^4 = 1.2\times10^5 = 120{,}000$ cfu per cm^3 of soil. The long version of the calculation is:

The 12 cfu on the –4 plate come from 12 cfu per 0.1 mL, or 120 cfu/mL, in the –4 solution.

Given the 10-fold dilution series, that means there would have been:

1,200 cfu/mL in the –3 solution, and
12,000 cfu/mL in the –2 solution.

The –2 solution is itself a 10-fold (1:9) soil to water dilution, so:

the original 1 cm^3 of soil would have contained 120,000 cfu, or 1.2×10^5 cfu per cm^3.

We often make an additional adjustment for the bulk density of the soil (say, 1.4 g/cm^3 soil) as: 1.2×10^5 cfu per cm^3/1.4 g soil per cm^3 = 8.6×10^4 cfu/g dry weight of soil.

Graphic by Gregory S. Gilbert.

it is not possible to know exactly what kind of propagule leads to a colony in a Petri plate (e.g., it could be a fungal spore, a hyphal fragment, or a combination), we usually refer to the number of **colony-forming units** (**cfu**) per volume or mass of soil. Dilution plating is a simple, inexpensive, and effective way to quantify viable populations of a target microorganism that can be reliably identified when growing on an effective selective medium. However, it is also slow, because it can take days or weeks for the colonies to develop.

Another important dilution-series technique for quantifying viable soil-borne or water-borne pathogens is by **baiting** of pathogens out of soil with plant material in combinations with a **most probable number** (**MPN**) analysis (Adams and Welham 1995). The baits can be seedlings, seeds, leaves, or fruits that are susceptible to infection with the pathogen of interest. In the MPN approach, baits are exposed to different dilutions of soil, and then scored as infected or not after an appropriate amount of time. Soil is diluted with sterilized sand or soil (for most fungi) or by creating a slurry of different amounts of soil in water (for many oomycetes) to create a dilution series (1:1, 1:2, 1:4, 1:8, etc.). Each dilution is distributed to a large number of containers (pots or cups), and a single bait placed in each. After days or weeks, the baits are examined visually for symptoms of infection or disease development. At the most concentrated dilutions, all baits may be symptomatic, whereas at the greatest dilutions, none may develop symptoms. At an intermediate dilution, where some, but not all the baits are symptomatic, the most probable number of pathogen propagules per unit volume of soil is then given by:

$$MPN = \frac{1}{v}\ln\left(\frac{n}{s}\right) \qquad \text{[Eq. 9.5]}$$

where v is the volume of soil in the container, s is the number of baits that were infected, and n is the total number of baits. Note that the MPN approach is not restricted to baiting from soil; it can be used to quantify pathogens (or other microbes) in other substrates like water or grain with any detection technique that tells you if a target pathogen or other microbe is present or absent in a sample, such as PCR with pathogen-specific primers (Lee et al. 2009).

Epidemiological studies are often informed by measuring the quantity and distribution of pathogen propagules in the air. Plant pathologists have developed methods to measure air-borne fungi and bacteria with both passive and active sampling (Figure 11.5) (West and Kimber 2015). Passive sampling can be done with microscope slides (Galante et al. 2011), passive dust collectors (Chaudhary et al. 2020), and settling and rainwater collectors (Peay and Bruns 2014); these methods collect spores that are deposited on surfaces over time. Active samplers include volumetric spore traps that use vacuums to force known quantities of air (say 15 L per minute) onto a sticky surface or collecting tube, where particles including fungal spores and bacteria are collected (Crandall and Gilbert 2017), or samplers with rotating arms or slides with sticky surfaces that impact a known volume of air over the period of sampling (Torfs et al. 2019). Active samplers have the advantage of allowing estimates of the concentration of propagules in the air, whereas passive samplers have the advantage of integrating deposited propagules over longer periods of time. Spores and bacteria collected in the samples can be identified and quantified either through microscopy or through any of the PCR-based techniques used to detect specific pathogens or to characterize the entire microbial community.

Keep an eye out as you read the rest of the book for where the methods and approaches described in this chapter are used to answer questions in all aspects of plant disease ecology.

Further reading

Boonham, N., J. Tomlinson, and R. Mumford, editors. 2016. *Molecular Methods in Plant Disease Diagnostics: Principles and Protocols*. CABI, Wallingford, UK.

Dhingra, O. D. and J. B. Sinclair. 1995. *Basic Plant Pathology Methods*, 2nd edition. CRC Press, Boca Raton, FL.

References

Adams, M. and S. Welham. 1995. Use of the most probable number technique to quantify soil-borne plant pathogens. *Annals of Applied Biology* **126**:181–196.

Aglietti, C., N. Luchi, A. L. Pepori, P. Bartolini, F. Pecori, A. Raio, P. Capretti, and A. Santini. 2019. Real-time loop-mediated isothermal amplification: an early-warning tool for quarantine plant pathogen detection. *AMB Express* **9**:50.

American Cancer Society. 2019. *Cancer Facts and Figures 2019*. American Cancer Society, Atlanta, GA.

Barreto, A., S. Paulus, M. Varrelmann, and A.-K. Mahlein. 2020. Hyperspectral imaging of symptoms induced by Rhizoctonia solani in sugar beet: Comparison of input data and different machine learning algorithms. *Journal of Plant Diseases and Protection* **127**:441–451.

Benitez-Malvido, J., G. García-Guzman, and I. D. Kossmann-Ferraz. 1999. Leaf-fungal incidence and herbivory on tree seedlings in tropical rainforest fragments: an experimental study. *Biological Conservation* **91**:143–150.

Bergey's Manual Trust. 2015. *Bergey's Manual of Systematics of Archaea and Bacteria*. https://onlinelibrary.wiley.com/doi/book/10.1002/9781118960608. John Wiley & Sons, Hoboken, NJ.

Bills, G. F. and M. S. Foster. 2004. Formulae for selected materials used to isolate and study fungi and fungal allies. In: G. M. Mueller, M. S. Foster, and G. F. Bills, editors. *Biodiversity of Fungi: Inventory and Monitoring*, pp. 595–618. Elsevier Academic Press, Cambridge, MA.

Blackwell, M. 2011. The Fungi: 1, 2, 3... 5.1 million species? *American Journal of Botany* **98**:426–438.

Bock, C. H., G. H. Poole, P. E. Parker, and T. R. Gottwald. 2010. Plant disease severity estimated visually, by digital photography and image analysis, and by hyperspectral imaging. *Critical Reviews in Plant Sciences* **29**:59–107.

Busby, P. E. and C. D. Canham. 2011. An exotic insect and pathogen disease complex reduces aboveground tree biomass in temperate forests of eastern North America. *Canadian Journal of Forest Research* **41**:401–411.

Camacho, F., D. Gernandt, A. Liston, J. Stone, and A. Klein. 1997. Endophytic fungal DNA, the source of contamination in spruce needle DNA. *Molecular Ecology* **6**:983–987.

Campillo-Balderas, J. A., A. Lazcano, and A. Becerra. 2015. Viral genome size distribution does not correlate with the antiquity of the host lineages. *Frontiers in Ecology and Evolution* **3**:10.3389/fevo.2015.00143.

Carneiro, G. A., S. Matić, G. Ortu, A. Garibaldi, D. Spadaro, and M. L. Gullino. 2017. Development and validation of a TaqMan real-time PCR assay for the specific detection and quantification of *Fusarium fujikuroi* in rice plants and seeds. *Phytopathology* **107**:885–892.

Chaudhary, V. B., S. Nolimal, M. A. Sosa-Hernández, C. Egan, and J. Kastens. 2020. Trait-based aerial dispersal of arbuscular mycorrhizal fungi. *New Phytologist* **228**:238–252.

Chien, A., D. B. Edgar, and J. M. Trela. 1976. Deoxyribonucleic acid polymerase from the extreme thermophile *Thermus aquaticus*. *Journal of Bacteriology* **127**:1550–1557.

China Plant BOL Group, D.-Z. Li, L.-M. Gao, H.-T. Li, H. Wang, X.-J. Ge, J.-Q. Liu, Z.-D. Chen, S.-L. Zhou, and S.-L. Chen. 2011. Comparative analysis of a large dataset indicates that internal transcribed spacer (ITS) should be incorporated into the core barcode for seed plants. *Proceedings of the National Academy of Sciences of the United States of America* **108**:19641–19646.

Choi, Y. J., G. Beakes, S. Glockling, J. Kruse, B. Nam, L. Nigrelli, S. Ploch, H. D. Shin, R. G. Shivas, and S. Telle. 2015. Towards a universal barcode of oomycetes—a comparison of the *cox1* and *cox2* loci. *Molecular Ecology Resources* **15**:1275–1288.

Clarridge, J. E. 2004. Impact of 16S rRNA gene sequence analysis for identification of bacteria on clinical microbiology and infectious diseases. *Clinical Microbiology Reviews* **17**:840–862.

Copeland, J. K., L. J. Yuan, M. Layeghifard, P. W. Wang, and D. S. Guttman. 2015. Seasonal community succession of the phyllosphere microbiome. *Molecular Plant–Microbe Interactions* **28**:274–285.

Crandall, S. G. and G. S. Gilbert. 2017. Meteorological factors associated with abundance of airborne fungal spores over natural vegetation. *Atmospheric Environment* **162**:87–99.

De Bruin, J. L. and P. Pedersen. 2008. Yield improvement and stability for soybean cultivars with resistance to *Heterodera glycines* Ichinohe. *Agronomy Journal* **100**: 1354–1359.

Floyd, R., E. Abebe, A. Papert, and M. Blaxter. 2002. Molecular barcodes for soil nematode identification. *Molecular Ecology* **11**:839–850.

Galante, T. E., T. R. Horton, and D. P. Swaney. 2011. 95% of basidiospores fall within 1 m of the cap: a field-and modeling-based study. *Mycologia* **103**:1175–1183.

Gilbert, G. S., J. O. Ballesteros, C. A. Barrios-Rodriguez, E. F. Bonadies, M. L. Cedeño-Sánchez, N. J. Fossatti-Caballero, M. M. Trejos-Rodríguez, J. M. Pérez-Suñiga, K. S. Holub-Young, and L. A. Henn. 2016. Use of sonic tomography to detect and quantify wood decay in living trees. *Applications in Plant Sciences* **4**:1600060.

Gilbert, G. S. and I. M. Parker. 2008. Porroca: An emerging disease of coconut in Central America. *Plant Disease* **92**:826–830.

Gortari, F., J. J. Guiamet, S. C. Cortizo, and C. Graciano. 2019. Poplar leaf rust reduces dry mass accumulation and internal nitrogen recycling more markedly under low soil nitrogen availability, and decreases growth in the following spring. *Tree Physiology* **39**:19–30.

Gottwald, T., G. Poole, T. McCollum, D. Hall, J. Hartung, J. Bai, W. Luo, D. Posny, Y.-P. Duan, and E. Taylor. 2020. Canine olfactory detection of a vectored phytobacterial pathogen, Liberibacter asiaticus, and integration with disease control. *Proceedings of the National Academy of Sciences of the United States of America* **117**:3492–3501.

Heather, J. M. and B. Chain. 2016. The sequence of sequencers: The history of sequencing DNA. *Genomics* **107**:1–8.

James, C. 1971. *A Manual of Assessment Keys for Plant Diseases*. American Phytopathological Society, St. Paul, MN.

Jeffers, S. N. and S. B. Martin. 1986. Comparison of 2 media selective for *Phytophthora* and *Pythium* species. *Plant Disease* **70**:1038–1043.

Kõljalg, U., R. H. Nilsson, K. Abarenkov, L. Tedersoo, A. F. Taylor, M. Bahram, S. T. Bates, T. D. Bruns, J. Bengtsson-Palme, and T. M. Callaghan. 2013. Towards a unified paradigm for sequence-based identification of fungi. *Molecular Ecology* **22**:5271–5277.

Lacroix, C., K. Renner, E. Cole, E. W. Seabloom, E. T. Borer, and C. M. Malmstrom. 2016. Methodological guidelines for accurate detection of viruses in wild plant species. *Applied and Environmental Microbiology* **82**:1966–1975.

Lee, H.-Y., L.-C. Chai, S.-Y. Tang, S. Jinap, F. M. Ghazali, Y. Nakaguchi, M. Nishibuchi, and R. Son. 2009. Application of MPN-PCR in biosafety of *Bacillus cereus* sl for ready-to-eat cereals. *Food Control* **20**:1068–1071.

Lindenthal, M., U. Steiner, H.-W. Dehne, and E.-C. Oerke. 2005. Effect of downy mildew development on transpiration of cucumber leaves visualized by digital infrared thermography. *Phytopathology* **95**:233–240.

Luo, Y., S. Gu, D. Felts, R. D. Puckett, D. P. Morgan, and T. J. Michailides. 2017. Development of qPCR systems to quantify shoot infections by canker-causing pathogens in stone fruits and nut crops. *Journal of Applied Microbiology* **122**:416–428.

Madden, L. V., G. Hughes, and F. van den Bosch. 2007. *The Study of Plant Disease Epidemics*. APS Press, St. Paul, MN.

Mahlein, A.-K., E.-C. Oerke, U. Steiner, and H.-W. Dehne. 2012. Recent advances in sensing plant diseases for precision crop protection. *European Journal of Plant Pathology* **133**:197–209.

Nilsson, R. H., K.-H. Larsson, A. F. S. Taylor, J. Bengtsson-Palme, T. S. Jeppesen, D. Schigel, P. Kennedy, K. Picard, F. O. Glöckner, and L. Tedersoo. 2019. The UNITE database for molecular identification of fungi: handling dark taxa and parallel taxonomic classifications. *Nucleic Acids Research* **47**:D259–D264.

Noordzij, M., F. W. Dekker, C. Zoccali, and K. J. Jager. 2010. Measures of disease frequency: prevalence and incidence. *Nephron Clinical Practice* **115**:c17–c20.

Notomi, T., H. Okayama, H. Masubuchi, T. Yonekawa, K. Watanabe, N. Amino, and T. Hase. 2000. Loop-mediated isothermal amplification of DNA. *Nucleic Acids Research* **28**:e63–e63.

Nowakowska, J., T. Malewski, A. Tereba, and T. Oszako. 2017. Rapid diagnosis of pathogenic *Phytophthora* species in soil by real-time PCR. *Forest Pathology* **47**:e12303.

Oerke, E.-C. 2020. Remote sensing of diseases. *Annual Review of Phytopathology* **58**:225–252.

Parnell, S., F. van den Bosch, T. Gottwald, and C. A. Gilligan. 2017. Surveillance to inform control of emerging plant diseases: An epidemiological perspective. *Annual Review of Phytopathology* **55**:591–610.

Peay, K. G. and T. D. Bruns. 2014. Spore dispersal of basidiomycete fungi at the landscape scale is driven by stochastic and deterministic processes and generates variability in plant–fungal interactions. *New Phytologist* **204**:180–191.

Porta, M., editor. 2016. *A Dictionary of Epidemiology*, 6th edition, online version. International Epidemiological Association, Chicago.

Raja, H. A., A. N. Miller, C. J. Pearce, and N. H. Oberlies. 2017. Fungal identification using molecular tools: a primer for the natural products research community. *Journal of Natural Products* **80**:756–770.

Robideau, G. P., A. W. De Cock, M. D. Coffey, H. Voglmayr, H. Brouwer, K. Bala, D. W. Chitty, N. Désaulniers, Q. A. Eggertson, and C. M. Gachon. 2011. DNA barcoding of oomycetes with cytochrome c oxidase subunit I and internal transcribed spacer. *Molecular Ecology Resources* **11**:1002–1011.

Saiki, R. K., S. Scharf, F. Faloona, K. B. Mullis, G. T. Horn, H. A. Erlich, and N. Arnheim. 1985. Enzymatic amplification of beta-globin genomic sequences and restriction site analysis for diagnosis of sickle cell anemia. *Science* **230**:1350–1354.

Scherm, H. A. and P. Ojiambo. 2004. Applications of survival analysis in botanical epidemiology. *Phytopathology* **94**:1022–1026.

Schoch, C. L., K. A. Seifert, S. Huhndorf, V. Robert, J. L. Spouge, C. A. Levesque, W. Chen, and F. B. Consortium. 2012. Nuclear ribosomal internal transcribed spacer (ITS) region as a universal DNA barcode marker for Fungi. *Proceedings of the National Academy of Sciences of the United States of America* **109**:6241–6246.

Seifert, K., G. Morgan-Jones, W. Gams, and B. Kendrick. 2011. *The Genera of Hyphomycetes*. CBS-KNAW Fungal Biodiversity Centre, Utrecht.

Shaner, G. and R. Finney. 1977. The effect of nitrogen fertilization on the expression of slow-mildewing resistance in Knox wheat. *Phytopathology* **67**:1051–1056.

Simko, I. and H.-P. Piepho. 2012. The area under the disease progress stairs: calculation, advantage, and application. *Phytopathology* **102**:381–389.

Soko, T., C. M. Bender, R. Prins, and Z. A. Pretorius. 2018. Yield loss associated with different levels of stem rust resistance in bread wheat. *Plant Disease* **102**:2531–2538.

Stielow, J. B., C. A. Levesque, K. A. Seifert, W. Meyer, L. Iriny, D. Smits, R. Renfurm, G. Verkley, M. Groenewald, and D. Chaduli. 2015. One fungus, which genes? Development and assessment of universal primers for potential secondary fungal DNA barcodes. *Persoonia: Molecular Phylogeny and Evolution of Fungi* **35**:242.

Sun, X., L. Zhang, C. Meng, D. Zhang, N. Xu, and J. Yu. 2020. Establishment and application of a multiplex PCR assay for detection of *Rhizoctonia cerealis*, Bipolaris sorokiniana, and Fusarium spp. in winter wheat. *Journal of Plant Pathology* **102**:19–27.

Tahi, M., I. Kebe, A. B. Eskes, S. Ouattara, A. Sangare, and F. Mondeil. 2000. Rapid screening of cacao genotypes for field resistance to *Phytophthora palmivora* using leaves, twigs and roots. *European Journal of Plant Pathology* **106**:87–94.

Tedersoo, L. and B. Lindahl. 2016. Fungal identification biases in microbiome projects. *Environmental Microbiology Reports* **8**:774–779.

Torfs, S., K. Van Poucke, J. Van Campenhout, A. Ceustermans, S. Croes, D. Bylemans, W. Van Hemelrijck, W. Keulemans, and K. Heungens. 2019. *Venturia inaequalis* trapped: molecular quantification of airborne inoculum using volumetric and rotating arm samplers. *European Journal of Plant Pathology* **155**:1319–1332.

Wakeham, A. J. and T. R. Pettitt. 2017. Diagnostic tests and their application in the management of soil- and water-borne oomycete pathogen species. *Annals of Applied Biology* **170**:45–67.

West, J. S. and R. Kimber. 2015. Innovations in air sampling to detect plant pathogens. *Annals of Applied Biology* **166**:4–17.

Wheeler, D. and D. Johnson. 2016. *Verticillium dahliae* infects, alters plant biomass, and produces inoculum on rotation crops. *Phytopathology* **106**:602–613.

Wu, B., M. Hussain, W. Zhang, M. Stadler, X. Liu, and M. Xiang. 2019. Current insights into fungal species diversity and perspective on naming the environmental DNA sequences of fungi. *Mycology* **10**:127–140.

Yadav, Y., P. K. Maurya, T. Bhattacharjee, S. Banerjee, I. Jamir, A. K. Mandal, S. Dutta, and A. Chattopadhyay. 2018. First evidence on heterotic affinity and combining ability of cultivated okra [*Abelmoschus esculentus* (L.) Moench] inbred lines for tolerance to enation leaf curl virus disease. *Agricultural Research & Technology* **18**:556071.

Zhang, D., X. Zhou, J. Zhang, Y. Lan, C. Xu, and D. Liang. 2018. Detection of rice sheath blight using an unmanned aerial system with high-resolution color and multispectral imaging. *PloS One* **13**:e0187470.

PART 2

Evolutionary ecology of plant–pathogen symbioses

CHAPTER 10

Population ecology of plant disease

Gregory S. Gilbert and Ingrid M. Parker

10.1 Introduction

Individual plants suffer the direct effects of disease (Chapter 8), but aside from the brown blotches on your houseplant, most people notice plant diseases only when they spread dramatically through a population of plants (a field of peas or the pine woods near your house). It is when we consider whole populations that agronomic, ecological, and economic impacts of plant diseases become apparent. This chapter is about how population processes in pathogens result in disease spread, and how population processes in plant hosts respond to and influence pathogens.

Population ecology is concerned with the structure and numerical dynamics of populations. A **population** is made up of individuals of a single species that grow and reproduce in a defined geographic area. Within the traditions of plant pathology, these topics are considered in the context of **epidemiology**. Epidemiologists study how disease intensity changes in a host population to better understand a disease outbreak so decision-makers can determine when and how to intervene. Epidemiology concerns itself with the patterns and processes that drive the distribution, spread, and impact of pathogens in populations of hosts over time and across space. Epidemiology is not just about the rapid, widespread outbreaks of disease that conjure the word "epidemic." It provides conceptual and quantitative frameworks for asking questions about the ecology of plant–pathogen interactions. What proportion of the host population will become diseased, and how quickly? Will disease have a significant impact on the host population's size or dynamics? How does the structure of the host population affect spread of the pathogen population? In much of agronomic epidemiology, the focus is on how spread of a pathogen affects a host population; host dynamics (aside from getting sick and dying) are often not considered. Population ecology expands the view to include population dynamics of both the pathogen and the host.

In this chapter we introduce the quantitative tools that are used to study the numerical dynamics of populations. Here we will consider models that are "non-spatial," in that they do not track the locations of individuals. However, of course in reality disease spread happens across space. In Chapter 11 we consider explicitly spatial aspects of the population ecology of disease.

10.2 Disease progress is a population phenomenon

Although pathogen infection leads to disease on an individual plant, epidemiology considers how a pathogen *population* grows and spreads through a *population* of plants. We talk loosely about the "spread of disease," but it is important to keep in mind that what "spreads" are really living microscopic organisms—like fungal spores—that move from one host to another. Let's construct some models of how disease spreads through host populations.

Models are simplified versions of the world, and simplification requires making assumptions. In this chapter, we will make the assumptions that both plants and pathogens are genetically uniform, that hosts are susceptible to the pathogens of interest,

The Evolutionary Ecology of Plant Disease. Gregory S. Gilbert and Ingrid M. Parker, Oxford University Press. © Gregory S. Gilbert & Ingrid M. Parker (2023).
DOI: 10.1093/oso/9780198797876.003.0010

that environmental conditions are always suitable for disease development, and that neither the host nor the pathogen evolve in important ways over the course of study. These assumptions are seldom met in real life, but making these assumptions helps us get started understanding the fundamental models that form the structural framework for exploring more complex situations. We follow the famous aphorism popularized by the statistician George Box (1979): "All models are wrong but some are useful." We will build on the simple, but useful, models we explore here in Chapter 13 when we look at what happens when we add in population genetics and evolution.

We are often only able to see a snapshot of a population—the current number of individuals and where they are. But populations are dynamic: some individuals reproduce, others die, and sometimes individuals arrive from a distant population while others wander away. Therefore, overall population growth is predicted by *number of births + number of immigrants − number of deaths − number of emigrants*. Using math, we can explore how populations behave and how they respond to density (Box 10.1).

Box 10.1 A primer on population models

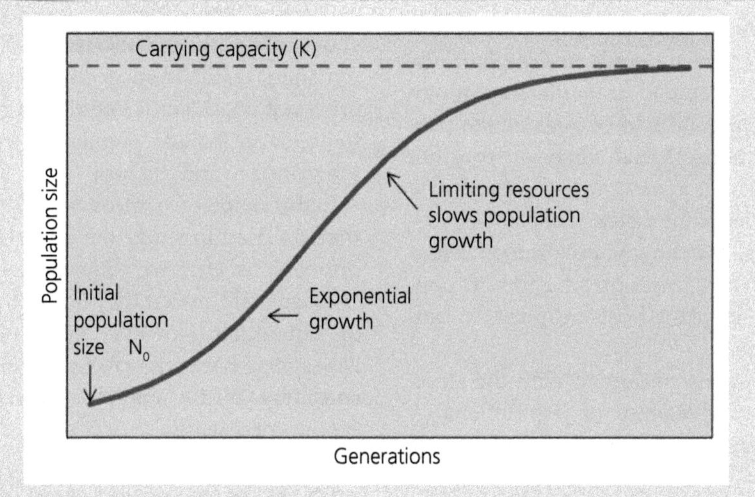

Figure 10.1 General logistic model for population growth.
Figure by Gregory S. Gilbert.

How do populations grow? Models of disease spread are closely analogous to the models used to describe population growth familiar to ecologists. Let's take a moment to visit those models.

Populations grow (or shrink) when there are more (or fewer) individuals entering a population than leaving it. The change in population size, ΔP, is determined by four elements:

$$\Delta P = \#\text{Births} + \#\text{Immigrants} - \#\text{Deaths} - \#\text{Emigrants} \quad \text{[Eq. 10.1]}$$

The increase of a population depends on whether the number of new individuals in the population, #Births + #Immigrants, is greater than the number of individuals leaving the population, #Deaths + #Emigrants, over a given period of time. With N individuals in a population, the population will increase at the rate of $\frac{dN}{dt} = rN$, where r is the **per-capita intrinsic growth rate**. This is the **exponential growth model** and assumes that the per capita rate of increase r is constant no matter the population size. Because r is constant, rN increases as N increases, such that the population grows more and more quickly, producing a classic J-shaped, exponential growth curve, with the population growing to infinity. Of course, real populations cannot grow to infinity—they eventually run out of the resources they need. The **carrying capacity** (K) of a population is the largest number of individuals that can be supported by the available resources (Figure 10.1). When the population size N gets close to K, resources become scarce, causing the

Box 10.1 *Continued*

per-capita birth rate to decline and the per-capita death rate to increase. This is an example of density dependence.

We can model how population size affects population growth by multiplying r by a ratio indicating how far the population is from the carrying capacity, which is $\frac{K-N}{K}$. For example, when N is very small compared to K, $\frac{K-N}{K} \to 1$ ("approaches 1"), so the population would grow at the intrinsic rate r (because $r*1 = r$). As the population nears carrying capacity, $N \to K$ and $\frac{K-N}{K} \simeq 0$, limited resources bring the growth rate to zero ($r*0 = 0$). This is the **logistic growth model**, where

$$\frac{dN}{dt} = rN\left(\frac{K-N}{K}\right) \quad \text{[Eq. 10.2]}$$

The logistic model produces an S-shaped curve of population growth over time, where a population grows exponentially at the start, then growth rate slows to zero as $N \to K$. In logistic growth models, the inflection point (when the population grows the fastest, maximum $\frac{dN}{dt}$) always occurs when the population size N is at one-half the carrying capacity K.

To project how many individuals to expect at a certain time t, we need to integrate both sides of the growth-rate equation. For exponential growth where $\frac{dN}{dt} = rN$, the integrated form of the equation is $N_t = N_0 e^{rt}$, where N_t is the population size at time t, N_0 is the starting population size, and e is the base of the natural logarithm, approximately 2.71828.

For the logistic growth model where $\frac{dN}{dt} = rN\left(\frac{K-N}{K}\right)$, the integrated form of the equation is

$$N_t = \frac{K}{1 + \left(\frac{K}{N_0} - 1\right)e^{-rt}} \quad \text{[Eq. 10.3]}$$

A couple important things to remember:

1. The population increases when the per-capita growth rate $r > 0$ and shrinks when $r < 0$. When $r = 0$, the population size is stable.
2. $\frac{dN}{dt}$ is the absolute rate of increase of the population; when $N \ll K$, resources are not limiting and the population grows exponentially; when the population nears carrying capacity ($N \approx K$), resources become limiting and population growth slows to zero.

Wild plant populations are made up of individuals of different life-history stages—seeds, seedlings, and mature plants. In ecology, **population structure** refers to the relative number of individuals of different ages or stages. Below we discuss how and why population models incorporate plant population structure. As an example, let's consider annual crops. Crops are simplified plant populations where farmers control the distribution, density, and structure of a plant population by deciding when and where to plant seeds or transplant seedlings and when to harvest the crop. From the perspective of a pathogen, a field of crop plants is a host population, but the crop lacks the temporal dynamics and population structure of a wild population. Annual crops provide us with a great starting point to think about the spread of a pathogen population through a host population, primarily because we can initially ignore the dynamics of the host, and because so much of the epidemiological literature comes from agricultural studies.

10.3 Disease cycles

A biological understanding of disease cycles is essential when building mathematical models to characterize plant disease epidemics. Pathogen propagules are produced on one host plant and can potentially infect other host individuals. Such propagules are comprised of whichever units are able to give rise to a new infection: fungal spores, bacterial cells, virus particles, or juvenile nematodes. Plants able to support production of pathogen propagules are called **competent hosts**. The pathogen population grows as new propagules are made and infect more leaves, more branches, or more plants. Pathogen propagules that disperse to neighboring plants during the same growing season are called **secondary inoculum**. Resting spores or fungal mycelium that survive in the soil and initiate infection of host plants in the next generation are called **primary inoculum**, which drives spread across time from season to season (Figure 10.2).

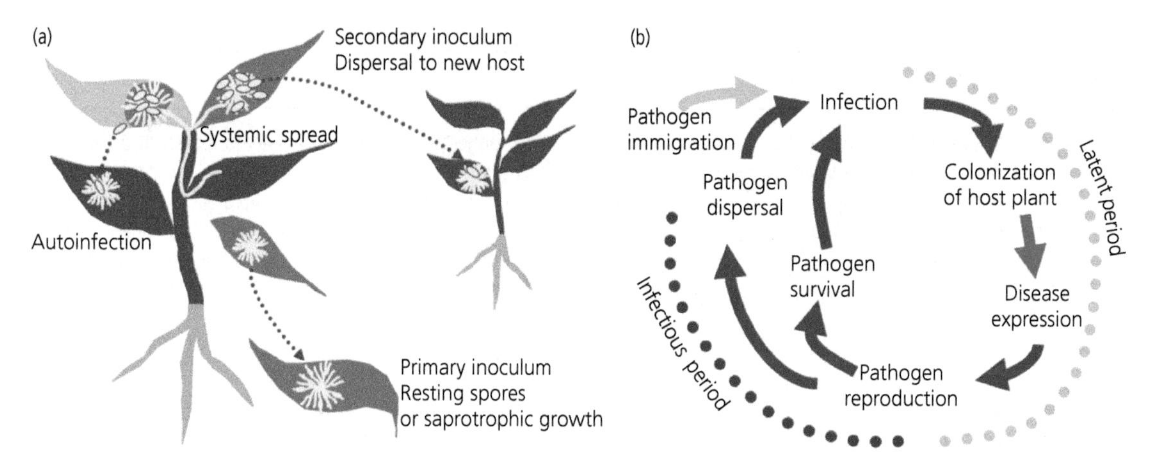

Figure 10.2 Modes of pathogen spread and a generalized plant disease cycle. (a) Pathogens colonize infected plants, leading to disease expression and subsequent pathogen reproduction in competent hosts. Pathogens can spread systemically through the plant, and propagules can lead to autoinfection of distant parts of the same host. Pathogen propagules can disperse to infect susceptible neighbors, leading to spatial spread of the pathogen and increase in disease prevalence in the same growing season. Resting spores that survive through unfavorable conditions as well as saprotrophic growth of facultative pathogens allow pathogen populations to survive in the absence of host plants until the next generation of plants are available. (b) Host infection is initiated from pathogens that survived from previous growing season, that spread from diseased neighbors, or, rarely, from propagules that immigrate from distant sources. The period from infection until the pathogen reproduces is the **latent period**. The period during which pathogen propagules are produced is the **infectious period**. Pathogen propagules may serve for long-term survival or for immediate dispersal to neighboring hosts.
Graphics by Gregory S. Gilbert.

Plant pathologists often categorize disease cycles as one of two broad types: **polycyclic diseases** include both primary and secondary inoculum in the disease cycle, whereas **monocyclic diseases** include only primary inoculum (Figure 10.3). Thus, they differ in how many cycles of infection take place during the lifetime of the host cohort. In polycyclic diseases (Figure 10.3a), each diseased plant produces additional sources of inoculum during a growing season, so disease can grow exponentially through the population. **Disease incidence** (the probability of new infections over time) increases as more diseased plants support production of more pathogen inoculum, until disease progress finally becomes limited at high **disease prevalence** (proportion of individuals diseased), when most of the susceptible plants have already become infected. Polycyclic diseases are those we mostly focus on when considering spatial spread (Chapter 11). In monocyclic diseases (Figure 10.3b), infection comes only from primary inoculum without secondary spread to neighbors in the same generation. For annual crops, this would be one pathogen cycle per plant growing season. As long as pathogen inoculum is available, the host is susceptible, and the environment is favorable, the proportion of diseased plants in the population can increase until there are no more healthy hosts available to infect. That means that disease incidence is limited by the abundance of pathogen inoculum and the availability of non-infected plants. Without generating new sources of secondary inoculum, disease increase is not exponential. Many soil-borne pathogens show this pattern of disease progress.

These two disease cycle models are often called by other names. The monocyclic model is also called the monomolecular model, the Mitscherlich model, or the simple interest model. "Monomolecular" comes from the similarity to physical chemistry models of first-order chemical reactions (described by the German chemist Mitscherlich)—reactions that depend on the concentration of only one reactant. It is called a simple interest model in analogy to one of two ways a bank could pay you interest on a deposit. If the bank pays simple interest, you deposit \$100 with the understanding that the bank will pay you 5% simple interest per year; at the end of the year the bank pays you \$5 interest, because

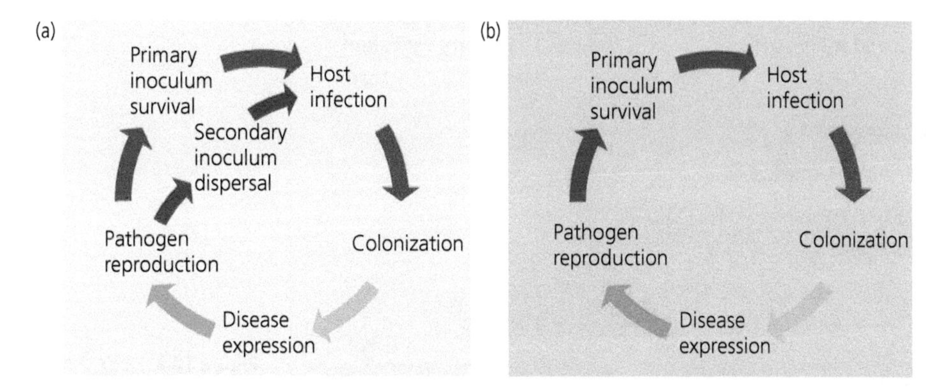

Figure 10.3 Polycyclic and monocyclic models of disease spread. (a) Polycyclic pathogens can reproduce quickly on an infected host, and the resulting secondary inoculum disperses directly to infect and cause disease on neighboring healthy plants. Secondary inoculum can be produced several times during a single growing season. (b) Monocyclic pathogens are able to reproduce only once during the lifetime of their host plant. Resting structures may survive in the soil until the arrival of the next generation of plant hosts, or infection may transmit from generation to generation through infested seed. Plants then become infected from that primary inoculum and develop disease.
Graphic by Gregory S. Gilbert.

the interest is calculated only on the amount you originally deposited—you don't get interest on any interest earnings during the year, so you have \$100 + \$5 = \$105. This is a good analogy to a monocyclic disease model, because the diseased plants do not contribute to additional spread (Paine et al. 2012). The polycyclic model is often called the "compound interest" model, as an analogy to a bank paying compound interest on your deposit. For a \$100 deposit with 5% interest that is compounded daily, after one day the bank calculates the amount of interest due after one day (5%/365 = 0.0137%, or \$0.0137) and adds it to the \$100. On the second day, the bank calculates your 0.0137% daily interest on the total \$100.0137 (rather than on the \$100), and so on each day, adding the accrued interest to the principal. At the end of the year, you will have \$105.13, because of the 13¢ interest gained on the interest. This compound interest analogy is a good fit for polycyclic disease models, where each infected plant pays interest in the form of additional secondary inoculum. The only difference is that the "interest" collected by a spreading plant pathogen can be much higher than what you get from your bank, leading to dramatic differences between monocyclic and polycyclic dynamics.

The distinction between monocyclic and polycyclic disease cycles is easiest to think about for annual crops. Primary inoculum either is already present in the field as a legacy of previous crops or arrived through immigration from outside the local population. If the pathogen is able to produce secondary inoculum that disperses directly to infect neighboring susceptible plants (sometimes taking a brief detour through the body of a vector), a polycyclic cycle develops. If pathogen inoculum is not produced until too late in the growing season to infect neighboring plants in the same cohort, the cycle is monocyclic. Management techniques such as crop rotation, soil amendments, tillage, or disinfestation, litter removal, sanitation, use of resistant cultivars, and use of clean seeds are focused on controlling primary inoculum, whereas management approaches including plant spacing, intercropping, irrigation management, biological control, vector control, and agrochemicals are more focused on secondary inoculum. These management practices are further explored in Chapter 17.

It is a little trickier to think about primary and secondary inoculum, and by extension, mono- and polycyclic diseases, in wild systems. Primary inoculum can persist in a meadow in the decaying remains of annual forbs or as a long-lived bank of oospores that developed in the soil beneath a tree after years of seedlings dying from damping-off. Primary inoculum can initiate new disease outbreaks when resting spores travel on mud-encrusted trucks and machinery along forest roads,

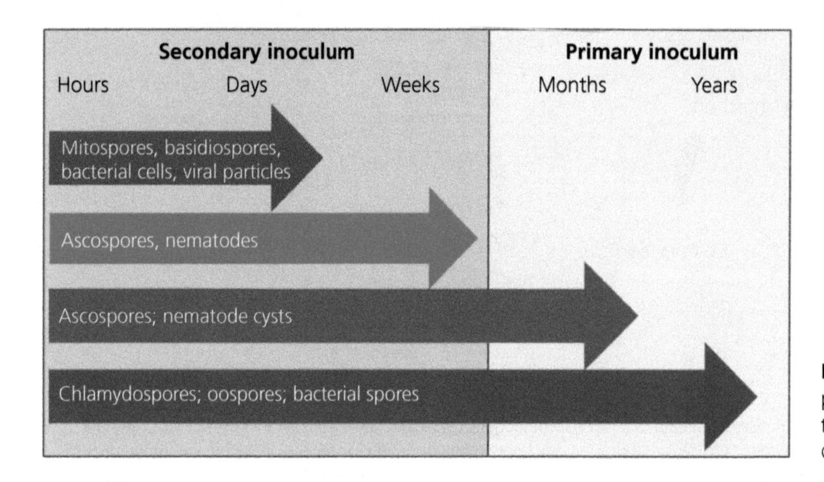

Figure 10.4 Range of times over which plant pathogen propagules can survive in the environment away from a host plant. Graphic by Gregory S. Gilbert.

or when propagules are carried long distances in streams or air currents to create new foci of infection. In wild systems, secondary inoculum need not be applied only to transmission within a single growing season, but can include inoculum produced across the lifetime of the infected plant host. A mature tree with persistent pathogen infection in its evergreen leaves, branches, trunk, or roots may be the source of secondary inoculum that directly infects neighboring plants—even its own offspring—for many years. More generally, we can think of primary inoculum as having an extended period in the environment between host individuals, whereas secondary inoculum transmits quickly from one host to another. Pathogens that only infect hosts after an extended period in the environment follow a monocyclic model. Pathogens that spread directly from one individual to another follow a polycyclic disease model.

10.4 Dispersal over time (surviving in place from one host to next)

Success for a pathogen does not come from killing its host, but rather from successfully reproducing on that host and dispersing (through time or space) to infect another host on which to reproduce (Figure 10.2). Pathogens survive the time between hosts by producing dormant resting spores, desiccation-protected mycelium, and biofilms (Lennon and Jones 2011). Such survival structures constitute primary inoculum. Pathogens that produce long-lived survival structures do not necessarily need to disperse spatially—they can simply wait until the seed of a susceptible host disperses to them, grows, and exudes chemicals that stimulate germination of dormant spores. Propagule types often vary in survival times; some are more suitable as primary inoculum and others as secondary inoculum (Figure 10.4).

Aside from going dormant, there are two ways that some pathogens survive locally in the absence of a particular host: as saprotrophs and on alternative hosts. First, facultative pathogens may grow and reproduce **saprotrophically** on host debris or other organic material in the soil. Then when a susceptible host appears under appropriate environmental conditions, the pathogen infects the plant and completes its life cycle as a pathogen. In some cases, this is a regular part of the pathogen life cycle. For instance, *Venturia inaequalis* causes apple scab on leaves and fruit of apple. Mycelium in leaves that fall in the autumn continues to grow in the leaf as a saprotroph. This mycelium, surviving the winter, then uses the energy from the fallen leaf to produce ascospores to eject in the moist spring to infect the new flush of fresh apple leaves (Stensvand et al. 1998). Other pathogens survive in crop residues (Kerdraon et al. 2019) or in tree stumps (van der Wal et al. 2017).

The second way that some pathogens survive is through reservoirs in **alternative hosts**, which can be essential to pathogen persistence and impacts.

For instance, *Phytophthora ramorum* causes sudden oak death, producing lethal trunk cankers on oak trees (*Quercus* spp.); however, the pathogen is unable to reproduce on oaks. In contrast, *P. ramorum* infects the evergreen leaves of California bay trees (*Umbellularia californica*) where it causes only minor necrosis but is able to grow and produce copious numbers of sporangia (Section 16.4). During moist spring weather, sporangia produced on bay trees infect nearby oaks. **Spillover** (or infection from a reservoir host to other species) is not uncommon. There are many examples of weeds or invasive species serving as reservoirs of fungal pathogens for crops and native plants (Power and Mitchell 2004, Beckstead et al. 2010, Blitzer et al. 2012).

10.5 Disease dynamics: disease progress through a plant population

A successful pathogen population will infect more and more individuals of its host population over time. Let's take a look at how we can model disease progress through a plant population. We start by making two simplifying assumptions. First, only the number of hosts and not their distribution in space is important (we will examine spatial effects in Chapter 11). Second, we will focus on an annual crop, as its simplicity allows us to ignore complicating host dynamics. Disease development on annual crops is limited by the number of susceptible plants in the field and by the window of time between planting and harvest. In other words, within one growing season, a farmer plants seeds in a field, the plants grow, some become diseased, and at the end of the season the farmer harvests the crop and the entire remaining plant population dies. The pathogen population, however, extends temporally beyond the limits of that crop cycle if propagules remain in the soil or crop debris, available to serve as primary inoculum for the following crop.

The rest of this chapter explores models of disease progress through plant populations. As always, models are simplifications of the real system, designed to capture the most important elements. We use an interplay between three representations of the models: verbal models that describe the general features of a system, often through a specific example; graphical models that represent the quantitative patterns of disease spread; and mathematical models that are theoretically grounded and that reproduce observed patterns of disease progress. We recommend that you make the explicit effort to connect the three model representations to each other—we find that is where the most exciting insights arise.

10.5.1 Progress of polycyclic diseases

Polycyclic diseases are those in which the pathogen is able to reproduce on infected hosts and spread to new hosts during the lifetime of the initial host. For annual crops, this would be a single growing season. Let's take as an example *Alternaria solani*, which causes early blight on tomatoes (*Solanum lycopersicum*). A single tomato plant in a well-tended field happens to be growing next to a small bit of surviving *Alternaria*-infested plant debris from a previous crop. Conditions are just right for this primary inoculum to grow and infect the plant, and, following colonization, the plant becomes sick with early blight. After a brief latent period, *A. solani* produces copious amounts of mitospores (secondary inoculum) on the surfaces of diseased areas of the plant. A small puff of air releases spores from the plant and sets them aloft. Some land on neighboring plants and are able to infect them following that night's dew. Several cycles like this of secondary inoculum production and plant infection can happen all in the same growing season. Because newly infected hosts are in turn sources of additional infections, disease prevalence at first increases exponentially (Figure 10.5). However, each diseased plant is no longer "available" for new infection, so eventually the increase in disease incidence slows as most plants in the population become diseased. At the end of the season, the pathogen can overwinter in plant detritus in the soil and serve as primary inoculum to infect plants at the start of the next growing season. Early blight on tomatoes has several features characteristic of a polycyclic disease: (1) an initially infected plant becomes a source of further infection during that growing season, which means (2) disease can spread directly from plant to

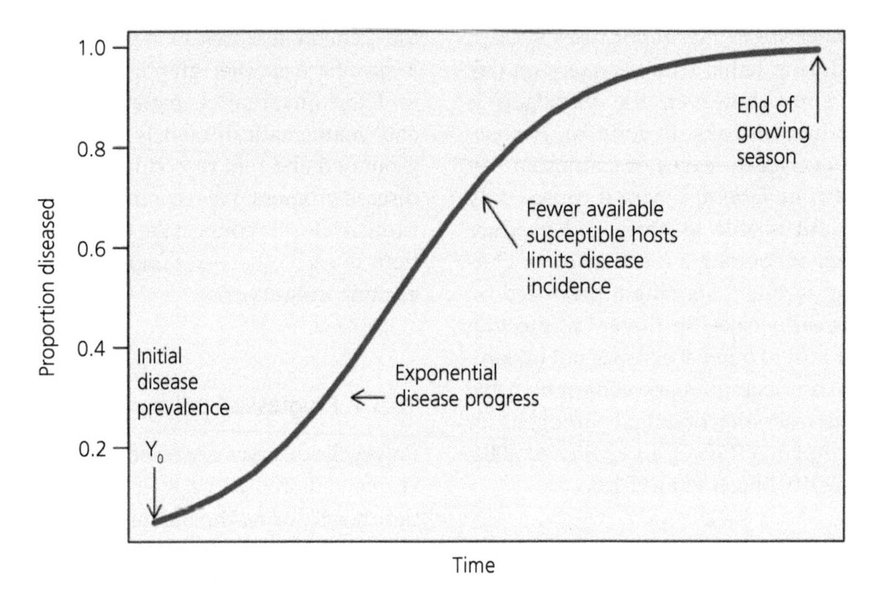

Figure 10.5 Progress of a polycyclic plant disease on an annual crop.
Figure by Gregory S. Gilbert.

plant, so that (3) the rate of increase of the population of diseased plants is dependent on the abundance of both susceptible and infectious hosts in the system.

Our graphical model of the progress of a polycyclic disease through the host population (Figure 10.5) looks a lot like the ecological logistic population-growth curve (Box 10.1), which models the increase in the number of individuals in a population until it reaches carrying capacity. Here, the "carrying capacity" is the total number of hosts in the population, and we follow the progress of disease through this host population as infection by the pathogen converts originally healthy plants into diseased plants. We see exponential increase in **disease prevalence** early on, but **disease incidence** (rate of new infections) slows as the number of diseased hosts approaches the total number of available hosts.

Epidemiologists have adapted the logistic growth model to analyze change in disease prevalence over time. Unlike the logistic population growth model, we are not tracking numbers of individuals but rather the proportion of plants in the population that are diseased, which is bounded by zero and one.

Recall from Box 10.1 that the logistic growth model gives the rate of increase of a population as

$$\frac{dN}{dt} = r\left(\frac{K-N}{K}\right)N \qquad \text{[Eq. 10.4]}$$

where r is the intrinsic rate of increase, N is the number of individuals in the population at time t, and K is the carrying capacity (the maximum N that can be sustained) (Box 10.1).

By analogy, the rate of disease spread for a polycyclic disease can be expressed as:

$$\frac{dY}{dt} = r_L\left(\frac{1-Y}{1}\right)Y = r_L(1-Y)Y \qquad \text{[Eq. 10.5]}$$

where r_L is the rate of increase of disease, Y is the proportion of plants in the population that are diseased at time t, and by extension the proportion of plants still healthy is $1 - Y$. Just as when N approaches the carrying capacity K and the amount of resources available $(K - N)$ approaches zero, as Y approaches 1, the proportion of the population that is still healthy $(1 - Y)$ approaches 0. Because r_L is multiplied by both Y and $(1 - Y)$, when Y is either close to zero or close to one, the rate of disease spread should be very slow, and conversely the disease spread should be fastest when $Y = 0.5$. This is illustrated by an S-shaped curve with an inflection in disease incidence at $Y = 0.5$ (Figure 10.5).

If we take the integral of both sides of $\frac{dY}{dt} = r_L (1 - Y) Y$ we get a predictive equation for disease prevalence in a polycyclic disease:

$$Y_t = \frac{Y_0}{Y_0 + (1 - Y_0)\, e^{-r_L t}} \qquad \text{[Eq. 10.6]}$$

Breaking down Eq. 10.6 into component parts (Figure 10.6) can help make sense of the contributions of both diseased and healthy hosts to disease progress in polycyclic diseases. How does changing Y_0 or r_L affect disease progress?

Here are some take-home messages about polycyclic diseases:

(1) The polycyclic model lets us predict disease prevalence Y at time t (Y_t) if we know two things:

Y_0: the initial prevalence of diseased plants, at time $t = 0$, and

r_L: the rate of increase of disease.

(2) Y_0 is the initial disease prevalence at time $t = 0$. Y_0 is not the amount of primary inoculum, but rather the proportion of plants infected at the beginning of the period of study.

(3) Infection rate parameter r_L depends on the probability of plants becoming infected by secondary inoculum produced on diseased neighbors. The rate of infection r_L is assumed to be constant over time.

(4) When disease prevalence is low, the rate of progress of disease through the population, $\frac{dY}{dt}$, is exponential and depends on r_L.

(5) Because $\frac{dY}{dt} = r_L (1 - Y) Y$, disease progress is limited by both how many susceptible hosts are left in the population $(1 - Y)$ *and* the proportion of diseased hosts producing inoculum (Y). This creates a point of inflection in the disease progress curve when half the plants are diseased, and 1 takes the place of the carrying capacity K in population growth models.

(6) Reducing r_L and Y_0 are critical targets for disease management.

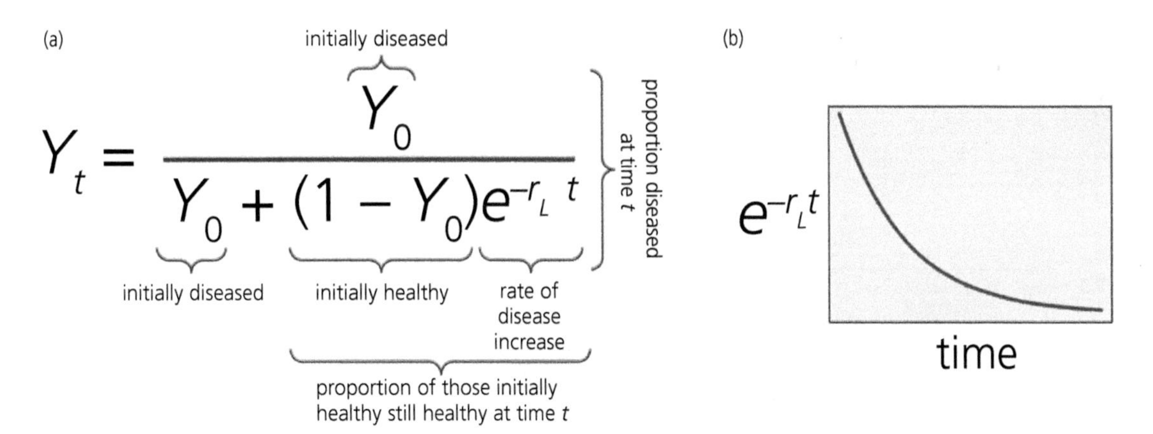

Figure 10.6 Let's break down the polycyclic disease progress model (Equation 10.6) to see what each part of it tells us. (a) Y_t is the proportion of plants that are diseased at time t, and Y_0 is the proportion of plants that were diseased at time 0. In the denominator, the initial starting healthy population $(1 - Y_0)$ is multiplied by the term $e^{-r_L t}$ to calculate how much of the population remains healthy at time t. Because $e^0 = 1$, you can see that at time 0, $e^{-r_L t}$ also equals 1, and the initial ($t = 0$) proportion healthy is $1 - Y_0$. Then the whole denominator at time 0 is just the sum of diseased and non-diseased plants, which is $Y_0 + 1 - Y_0$, or 1. This leaves us with our estimate of Y_t for $t = 0$ as $Y_0/1$, or simply Y_0 (which is our initial proportion of plants diseased, so that checks out). (b) As time t increases, the term $e^{-r_L t}$ gets closer and closer to zero because it is a negative exponential function (for further descriptions of exponential functions, see Section 11.2). So, when multiplied against $(1 - Y_0)$, the proportion that is still healthy gets smaller and smaller with time, and the denominator of the equation also gets smaller and smaller, until Y_t reaches $\frac{Y_0}{Y_0} = 1$ when there are no more plants available to become diseased.
Graphic by Gregory S. Gilbert.

10.5.2 Progress of monocyclic diseases

Now let's construct similar models for monocyclic diseases, starting with a typical disease cycle of a soil-borne, monocyclic pathogen. Imagine a new farmer plants a bare field with strawberry plants, not knowing that the field was **infested** with microsclerotia of the fungal wilt pathogen *Verticillium dahliae*. The drought- and temperature-resistant microsclerotia had been produced during previous crops of lettuce, which is susceptible to the same pathogen. The microsclerotia were dispersed across the field on machinery when preparing the field, acting as primary inoculum. With seasonal rains, the roots of the strawberry plants grow through the soil, sometimes bumping into a microsclerotium. Root exudates stimulate nearby microsclerotia to germinate, producing fungal hyphae that can infect the root. The probability that a plant will get infected depends on the density of primary inoculum and the likelihood that an encountered microsclerotium will germinate and successfully infect the root. Plants generally do not recover from fungal infections, so each newly infected plant increases the overall proportion of the plant population that is diseased. However, because there is a limited number of plants in the field, as more and more plants become diseased (disease prevalence increases) there are fewer and fewer healthy plants available to develop *new* infections, slowing the rate of disease increase (disease incidence declines). Although *V. dahliae* makes conidia that spread through the vascular system of infected plants, it cannot spread to neighboring plants during the same growing season. Instead, it produces microsclerotia in dying host tissue that serve as resting spores in the soil, primary inoculum for the next crop of plants. This monocyclic disease cycle has several important features: (1) infected plants do not become a source of further infection during that growing season, which means (2) disease does not spread from plant to plant, so that (3) the rate of increase of diseased plants is dependent only on the probability that susceptible plants encounter existing primary pathogen inoculum under environmental conditions suitable for disease to develop. From the perspective of disease progress over time, such monocyclic diseases then take the shape of a saturating curve (Figure 10.7).

Unlike polycyclic diseases, monocyclic diseases do not exhibit exponential population growth. The chance of a plant becoming infected and developing disease occurs at a constant rate, r_M. Because infected hosts do not result in new infections, the rate of increase of disease in the population is the same as the infection rate, $\frac{dY}{dt} = r_M$ when $Y \simeq 0$. Because new infections can only take place among

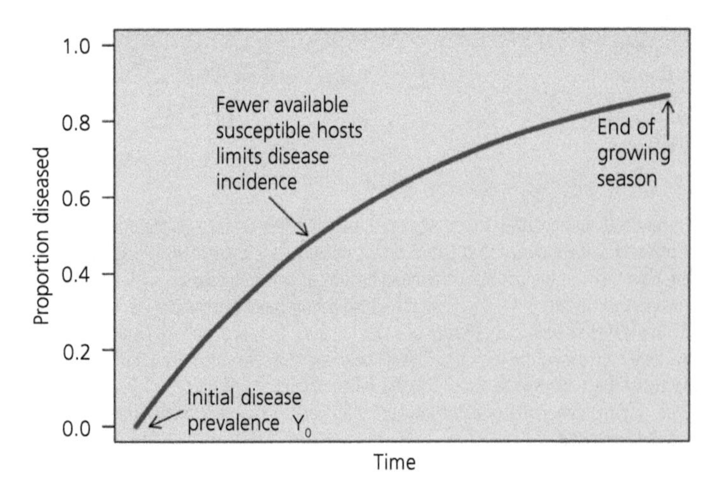

Figure 10.7 Expected disease progress over time for a monocyclic disease.
Figure by Gregory S. Gilbert.

those plants that are still healthy $(1 - Y)$, the rate of increase of disease is

$$\frac{dY}{dt} = r_M (1 - Y) \quad \text{[Eq. 10.7]}$$

Just as in the polycyclic case, as disease prevalence Y gets closer to 1, the proportion of plants still healthy $(1 - Y)$ gets closer to zero, so that the rate of increase of disease $\frac{dY}{dt}$ becomes a smaller and smaller portion of the fundamental rate parameter r_M. That is why the monocyclic disease progress curve becomes flatter and flatter as time goes on (Figure 10.7).

To calculate Y_t, the disease prevalence at time t, from the rate of disease increase $\frac{dY}{dt}$, we take the integral of both sides of Equation 10.7, which gives us a predictive equation for the amount of disease expected for a monocyclic disease:

$$Y_t = 1 - (1 - Y_0)\, e^{-r_M t} \quad \text{[Eq. 10.8]}$$

Just as we did for the polycyclic model, breaking down Equation 10.8 into its component parts (Figure 10.8) can help make sense of what drives disease progress in monocyclic diseases. How does changing Y_0 or r_M affect disease progress?

Here are some take-home messages about monocyclic diseases:

(1) The monocyclic model lets us predict Y_t, disease prevalence at time t, if we know two things:

Y_0: the initial prevalence of diseased plants at time $t = 0$, and

r_M: the rate of increase of disease.

(2) The infection rate parameter r_M is the probability of plants becoming infected by primary pathogen inoculum already present or coming in from outside the population.

(3) There is no period of exponential growth.

(4) Because $\frac{dY}{dt} = r_M (1 - Y)$, disease progress is limited only by how many susceptible hosts are left in the population $(1 - Y)$, not by the proportion of individuals already infected (Y). This also means that there is no point of inflection in the disease progress curve.

10.5.3 Comparing polycyclic and monocyclic disease models

Differences between models for polycyclic and monocyclic disease progress highlight important ecological insights and approaches to disease management. For instance, how would reducing primary inoculum through careful sanitation affect

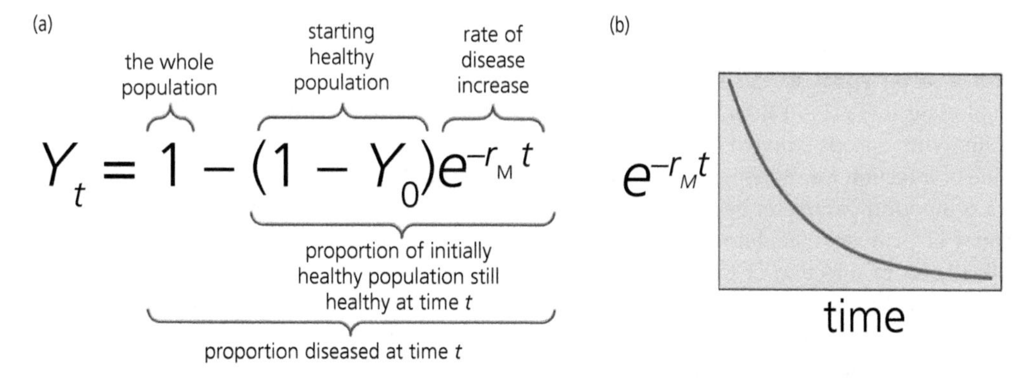

Figure 10.8 Let's break down the model for a monocyclic disease (Equation 10.8) into parts to see what it tells us. (a) Y_t is the proportion of plants that are diseased at time t, and Y_0 is the proportion of plants that were diseased at time 0. A key to understanding how this equation works is to remember that the proportion of plants that are diseased, Y_t, is just 1 – proportion healthy at time t. The whole host population equals 1. The model is just subtracting out how much of the initially healthy population $(1 - Y_0)$ remains healthy over time. To calculate that, the initial starting healthy population $(1 - Y_0)$ is multiplied by $e^{-r_M t}$, which (b) gets closer and closer to zero over time because it is a negative exponential function, just as in the polycyclic model (Figure 10.7). When $t = 0$ the exponent $-r_M t = 0$, and since $e^0 = 1$, the equation simplifies to $Y_t = 1 - 1 + Y_0$. In other words, at time $t = 0$, disease prevalence $= Y_0$. The rate of infection r_M is assumed to be constant over time. Time 0 can be whenever data on disease prevalence are first collected, including before there is any disease at all.
Figure by Gregory S. Gilbert.

polycyclic versus monocyclic diseases for a grower? Because spread is determined entirely by the probability of a plant encountering a primary inoculum propagule and becoming infected, reducing primary inoculum should have a huge and lasting impact on slowing disease progress for monocyclic diseases. On the other hand, for a polycyclic disease, reducing primary inoculum should slow disease progress initially, but because disease progress also includes spread from secondary inoculum production, propagules produced on initially infected plants could rapidly erase any gains in disease reduction from sanitation. For polycyclic diseases, reducing or delaying the production of secondary inoculum (longer latent period or shorter infectious period) will likely have larger impacts on disease progress than would reducing primary inoculum.

So now we know that to predict disease progress, we need to know r_L and r_M. These parameters vary depending on characteristics of the pathogen, the host, and environmental conditions. How do we measure them? Graphs of observed disease prevalence Y over time should produce non-linear shapes similar to Figure 10.5 for polycyclic diseases or Figure 10.7 for monocyclic diseases. Simple transformations of the observed prevalence values allow those relationships to be linearized, which makes them easier to work with. For a monocyclic disease, plotting $\ln(1/(1 - Y))$ vs time produces a straight line, with a slope equal to r_M. For a polycyclic disease, plotting $\ln(Y/(1 - Y))$ vs time produces a straight line, with a slope equal to r_L.

The rate of infection for monocyclic diseases, r_M, is itself a compound parameter because it relies on the amount of primary inoculum present as well as the likelihood of infection of the plant when it encounters the inoculum. That means that r_M can be decomposed into two components where $r_M = QR$, with Q being the initial amount of primary inoculum available and R being a proportional constant that describes the rate of disease progress per unit initial primary inoculum per unit time. If r_M is determined by collecting disease progress data, and there is an independent measure of initial primary inoculum (e.g., the number of microsclerotia per kg of soil in the field), R can then be calculated as $R = r_M / Q$.

In general, the influence of r_L or r_M on disease progress in both types of disease is huge—a much larger effect than that of the Y_0 term. What factors influence r_L and r_M? Agrios (2005) reports observed values of r_M from 0.004 to 0.02 units per day, and r_L from 0.15 to 0.6 units per day, but they are not fixed values for particular plant–pathogen systems. Let's go back to the disease triangle (Figure 1.5). The context of the plant–pathogen interaction, including weather, soil conditions, host density, and host and pathogen genotypes, fundamentally shape the critical parameters of r_L or r_M through their impacts on pathogen infection, growth, reproduction, and survival (De Wolf and Isard 2007). We explore how the ecological impacts of disease may be greater under some conditions than others in Chapter 14, and how disease management can change system conditions to reduce the likelihood of developing disease in Chapter 17.

10.6 Another way to look at disease progress: SEIR compartment models, R, and R_0

The basic disease progress models we have just explored are commonly used in plant disease epidemiology, but there are other very useful frameworks that come out of theoretical epidemiology, which usually draws heavily on diseases of humans and other animals as a frame of reference (Kermack and McKendrick 1927, Brauer et al. 2019). An important framework is that of **compartmental models**, which define a population as a series of non-overlapping, sequential compartments (states) with rates of transition out of one compartment and into the next.

A common compartmental model is known as the **SEIR model**, where individuals are classified as Susceptible, Exposed, Infectious, or Removed (Figure 10.9). Each compartment is linked to the next through a rate of transition. In the SEIR model, susceptible individuals can be exposed to the pathogen (infected), exposed individuals then become infectious, and infectious individuals are removed from contributing to disease spread when they die, recover (rare in plants), or no longer support pathogen reproduction. In plant pathology,

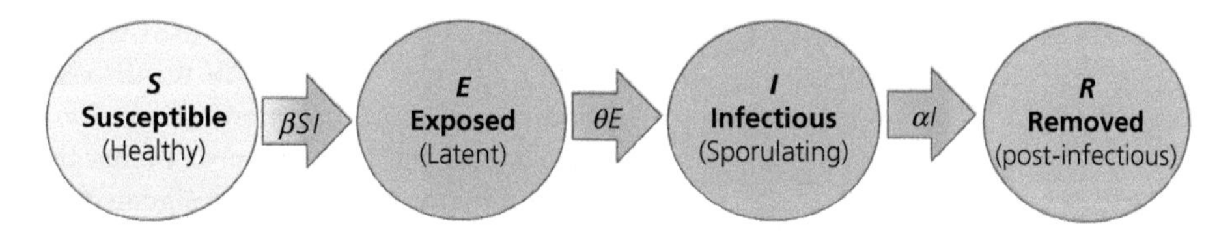

Figure 10.9 Flow chart for the compartmental S-E-I-R (aka H-L-I-R) model. The parameters β, θ, and α are each involved in the transition from one compartment to another, as defined in Eqs 10.11.
Graphic by Gregory S. Gilbert.

this same model is sometimes called the **HLIR** (or **HLSR**) model, following terminology first used by the pioneering South African epidemiologist Vanderplank (1963). The corresponding compartments include (H) healthy (susceptible) hosts, (L) individuals with latent infections, (I) infectious individuals (aka S for sporulating), and (R) removed (post-infectious) individuals.

Such an SEIR model then has the form

$$N = S + E + I + R \quad \text{[Eq. 10.9]}$$

where N is the number of individuals in the population. If $N = 1$, then S, E, I, and R represent the fractions of the whole population in each compartment. In the HLIR version, the equivalent would be:

$$N = H + L + I + R \quad \text{[Eq. 10.10]}$$

For simplicity, and in recognition of the much more widespread use of the SEIR terminology in the scientific literature, we adopt the terminology of SEIR here. But feel free to substitute in HLIR or HLSR if you prefer.

In the SEIR framework, the rates of transition are what govern the dynamics of disease in the system. As represented in the SEIR flow chart (Figure 10.9) we can define a system of differential equations that describe how each compartment changes over time as individuals transition among states:

$$\frac{dS}{dt} = -\beta SI \quad \text{[Eq. 10.11a]}$$

Rate of change in the Susceptible state equals the transition from Susceptible to Exposed (latent infection), which depends on contact between Infected and Susceptible individuals.

$$\frac{dE}{dt} = \beta SI - \theta E \quad \text{[Eq. 10.11b]}$$

Rate of change in the Exposed (latent infection) state, which includes the addition of previously Susceptible individuals, and the loss of individuals that become infectious.

$$\frac{dI}{dt} = \theta E - \alpha I \quad \text{[Eq. 10.11c]}$$

Rate of change in the Infectious state, which includes the addition of previously Exposed (latent) individuals and the loss of individuals that become post-infectious.

$$\frac{dR}{dt} = \alpha I \quad \text{[Eq. 10.11d]}$$

Rate of change in the Removed state, which equals the rate of addition of previously infectious individuals (includes death and recovery with immunity). Note that the rate of transition out of a state (e.g., $-\beta SI$) is the same as the rate of transition into the next state (e.g., $+\beta SI$); all individuals are conserved in the system as a whole. Connecting back to the basic disease cycle (Figure 10.2), θ is the rate of transition from being infected (but latent) to being infectious; $1/\theta$ then gives the average length of the latent period (Figure 10.10). Similarly, $1/\alpha$ is the average length of the infectious period. You can visualize the dynamics of the system as individuals transition through the S, E, I, and R compartments over time (Figure 10.10).

So, how does this SEIR model relate to the monocyclic and polycyclic disease progress models we

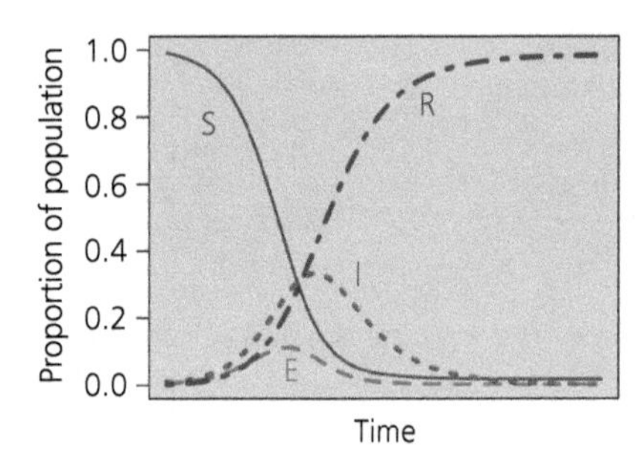

Figure 10.10 Transitions of individuals from Susceptible (S) to Exposed (E), then Infected (I), then Removed (R) compartments in an SEIR model, during an epidemic of a highly infectious pathogen. At any given time, the sum of the values of each compartment is 1. Figure by Gregory S. Gilbert.

have already explored, and how is it used in plant pathology? And what does it have to do with the basic reproductive number R_0 that has become part of everyday language in the wake of the COVID-19 pandemic? Let's start by going back to the logistic model used to describe polycyclic diseases, where Y is the proportion of the plant population diseased, first presented in Equation 10.5 and slightly rearranged as:

$$\frac{dY}{dt} = r_L Y(1-Y) \qquad \text{[Eq. 10.12]}$$

Vanderplank (1963) recognized that this basic logistic model could be made a more realistic descriptor of disease increase by connecting it to the compartmental model. In particular, he noted that not all infected plants are sources of pathogen inoculum at all times: infected plants first pass through a latent (or pre-infectious) period (p) before the pathogen begins reproducing (i.e., Exposed state in the SEIR model). The diseased plant becomes Infectious when pathogen propagules are produced, but this lasts only for a certain time (the infectious period, i) after which plants become post-infectious (Removed state, neither infectious nor susceptible) and no longer contribute to disease spread (Figure 10.2).

Vanderplank (1963) then modified the logistic model (Equation 10.12) to incorporate these compartments, so that the model includes only those plants in the population that were in either the Infectious or Susceptible states. In his formulation (Equation 10.13), the rate of increase of disease at time t, $\frac{dY_t}{dt}$, was a function of the disease transmission rate R, the proportion of infectious plants at time t, and the proportion of plants that are still susceptible at time t. He designated the proportion of plants that were infectious as the proportion of plants that has passed through the latent period (Y_{t-p}) minus the proportion of plants that has passed through both the latent and infectious periods (Y_{t-p-i}) by time t (i.e., total infected–latent–removed; $Y_{t-p} - Y_{t-p-i}$). The proportion of plants that remain healthy (and susceptible) is given as $(1 - Y_t)$, where Y_t is the proportion infected. In short, Vanderplank re-wrote the logistic model as

$$\frac{dY_t}{dt} = R\left(Y_{t-p} - Y_{t-p-i}\right)(1 - Y_t) \qquad \text{[Eq. 10.13]}$$

What is cool, then, is that when $N = 1$ (so that S, E, I, and R represent the proportions of the population in each state; Equation 10.9), then β from Equations 10.11 is equivalent to R from Equation 10.13, and Equations 10.12 and 10.13 fit together as part of the same framework (see Segarra et al. (2001) for details). (Note: R from Equation 10.13 should not be confused with R ("Removed" individuals) from Equation 10.9.). Here, R is the transmission rate of disease, or the number of new infections produced per existing infection per unit time. R_0, then, is a special case of R called the **basic reproductive number**. R is the number of new infections produced per existing infection at the start of an epidemic, when only one individual is infectious and the rest of the population is susceptible. R_0 can be calculated from the parameters in Equations 10.11 as $R_0 = \beta/\alpha$. When R is greater than 1, the amount of disease in the population grows; when R is less than 1, the amount of disease declines. Management of disease outbreaks, once they occur, is all about doing things that bring R down below 1 (Chapter 17).

10.7 Plant populations and the impact of disease

You may have noticed that neither $\frac{dY_t}{dt}$ in the Vanderplank model (Equation 10.13) nor the SEIR model equations (Equations 10.11) include demography of the host plant. Host populations in those models are a fixed size—they cannot grow; individuals only move from one compartment to another. The addition of births and deaths to SEIR-type models is common and, for example, allows for the calculation of the equilibrium proportion of the host population infected for an endemic disease (Brauer et al. 2019). Because plant epidemiology has been almost exclusively focused on managed systems in which populations are fixed, such as agriculture, forestry, and horticulture, these extended models are uncommon for plants.

But outside of managed systems, plant populations are dynamic. As you'll recall from Box 10.1, population numbers fluctuate through reproduction and the death of individuals. (For our purposes, we will ignore immigration and emigration.) The most frequently used plant population models are stage-structured, meaning that the life cycle is divided into stage classes such as seeds, seedlings, saplings, reproductive plants, and senescent plants (Caswell 2006). Individuals may also be characterized by their size (e.g., large versus small adult trees) or age (e.g., one-year-old seeds in the seedbank versus ten-year-old seeds). As in the compartmental models above, individuals transition from one stage to another. But individuals can also add more plants to the population by reproducing, or they can die before transitioning to another stage.

Plant populations tend to follow discrete, annual cycles, so plants are almost always modeled with **discrete-time models**, unlike the continuous-time models that we have seen up to this point. A discrete-time population model predicts numbers of individuals at time $t + 1$ as a function of numbers at time t, with the time step being an annual cycle. In place of the continuous intrinsic rate of increase r, discrete-time models use the finite rate of increase, λ, which is N_{t+1}/N_t, or the proportional change in the population size in one time step. Because $\lambda = e^r$, it is easy to convert between continuous time and discrete time rates. Just as $r > 0$ indicates a growing population, because $e^0 = 1$, $\lambda > 1$ also indicates a growing population.

Box 10.2 walks you briefly through how matrix models are built and analyzed, using as an example Prendeville et al.'s (2014) study on the effect of plant pathogens on wild squash. The researchers collected field data to estimate vital rates and life-history transitions for wild squash from three distinct populations, comparing uninfected plants to plants infected with *Cucumber mosaic virus* (CMV) or *Zucchini yellow mosaic virus* (ZYMV). Even though the average number of gourds produced per plant was indistinguishable across all treatments and populations, the models revealed that virus infection did affect population growth rate, and this population-level response to infection varied among virus species and among host populations.

In Prendeville et al.'s study (2014), they also found that the elasticity for seed dormancy was lower than that for fecundity (the production of new seeds). In such a population, one predicts that a disease that reduces fecundity would have a bigger effect on population dynamics than a disease that reduces long-term persistence of seeds in the soil by the same proportion. All things equal, resistance to the pathogen that affects fecundity would have a bigger impact on the wild gourd population than resistance to the soil-borne seed pathogen.

Godfree et al. (2007) used matrix models to evaluate how breeding resistance to *Clover yellow vein potyvirus* (ClYVV) into white clover (*Trifolium repens*) would affect this species' ability to persist and invade in Australia. White clover is an economically important forage species, but it is also an invasive weed. Godfree et al. planted out virus-free and virus-infected clovers in 21 grassland and woodland sites, and they used these experimental populations to estimate vital rates for their models. Population growth rates were significantly depressed by virus infection overall, and, in four of the 21 sites, protection from virus infection shifted the population from $\lambda < 1$ to $\lambda > 1$ (i.e., from declining to expanding). This implies that ClYVV was responsible for constraining the distribution of white clover, preventing its occurrence in some locations. The original motivation for this work was to assess the risk of genetically engineered (or conventionally bred) white clover with ClYVV resistance; the researchers concluded that ClYVV resistance could lead to adverse ecological impacts associated with weed invasion.

Matrix population models, together with related stage-structured models, are powerful tools that are broadly applied to address questions in both

Box 10.2 Matrix population models

(a)

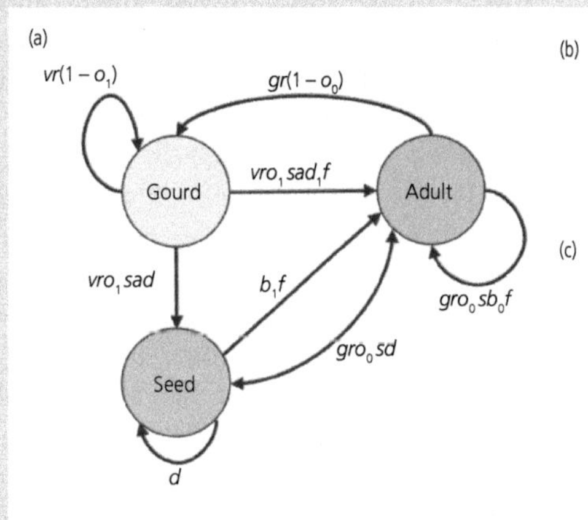

(b)

	Gourd	Seed	Adult
Gourd	$vr(1-o_1)$	–	$gr(1-o_0)$
Seed	vro_1sad	d	gro_0sd
Adult	vro_1sad_1f	b_1f	gro_0sb_0f

(c)

Life-history traits	Symbol
Probability gourds less than 1 yr old open	o_0
Probability gourds more than 1 yr old open	o_1
Probability gourds are not consumed by rodents	r
Probability seeds more than 1 yr old germinate	b_1
Probability gourd more than1 yr old is viable	v
Probability seeds are dormant	d
Proportion of seeds more than1 yr old that are viable	a
Probability of seeds less than 1 yr old germinate	b_0
Probability seedlings survive to flower	f
Average number of gourds per plants	g
Average number of seeds per gourd	s

Figure 10.11 Stage-structured model for wild squash, showing transition between Adult, Gourd, and Seed stages formatted as a (a) life-cycle graph for wild squash, (b) the corresponding matrix of transition probabilities based on (c) key life-history traits.
Graphic by Ingrid M. Parker.

Stage-structured models in discrete time take advantage of the tools of matrix algebra, which represents systems of equations, each of which projects the numbers of individuals in a particular stage at time $t + 1$ as a function of number of individuals in all stages at time t.

Matrix population models are widely used because they offer both simplicity and flexibility. For example, Prendeville et al. (2014) modeled populations of wild squash, *Cucurbita pepo*, by dividing the life cycle into flowering plants, gourds containing viable seeds (the "gourd bank"), and dormant seeds on the soil or buried in it (the "seed bank"). One can represent this model with a life-cycle graph showing the Adult, Gourd, and Seed stages (Figure 10.11a). Arrows represent transitions between life stages on the basis of an annual time step. For example, the arrow from the seed stage to itself represents the proportion of seeds that stay dormant (and viable) in the seedbank for an additional year. The transitions between life stages can be functions of multiple vital rates. For example, the number of new seeds produced is a function of plant survival to produce seeds, as well as the number of seeds produced by a plant that survives. The arrow from Adult to Gourd represents the production of new gourds by adult plants, which is a function of reproduction (g) as well as of the probability of not being eaten by rodents (r) and not breaking open to spill seeds on the ground $(1 - o_0)$.

The life-cycle graph is then translated into a matrix (**A**) that captures all transitions among the three life stages (Figure 10.11b). Transition probabilities are functions of life-history traits of the squash (Figure 10.11c). The matrix is used to project population growth. The population vector $\mathbf{x}$ is the number of individuals of Gourds, Seeds, and Adults. When a population is close to its stable stage distribution (i.e., when the proportion of individuals in each stage remains the same from one time-point to the next), the finite rate of increase of the population can be calculated from the equation $\mathbf{Ax} = \lambda\mathbf{x}$, where λ is the dominant eigenvalue of the matrix **A**. As we have seen, $\lambda > 1$ indicates a growing population and $\lambda < 1$ indicates a population in decline.

Box 10.2 *Continued*

The power of stage-structured population models is that they allow us to study how factors that influence particular stages of the life cycle or particular transitions from one stage to another affect the dynamics of the population overall. If we call a_{ij} the matrix element in the ith row and the jth column, we define the **sensitivity** of λ to changes in a_{ij} as

$$\frac{\delta\lambda}{\delta a_{ij}} = \frac{v_i x_j}{\sum_{k=1}^{N} x_k v_k}. \quad \text{[Eq. 10.14]}$$

The **elasticity** (proportional sensitivity) of λ to changes in a_{ij} is

$$e_{ij} = \frac{a_{ij}}{\lambda} * \frac{\delta\lambda}{\delta a_{ij}}. \quad \text{[Eq. 10.15]}$$

One of the important insights from these models is that having a large effect on one part of the life cycle does not necessarily translate into having a large impact on population trajectories. For example, the effect on population growth (λ) of a large change in seed dormancy may be large or small, depending on how fast the population is growing (or shrinking) and other details of that population's demography.

applied and basic ecology. Perhaps because of its tradition of treating plant populations as fixed, the field of plant pathology has not historically taken advantage of these analytical approaches. Going forward, structured population models could enhance the field of plant pathology by translating the impact of disease on individual aspects of the life cycle into the impact on populations. In this vein, one important application of stage-structured models of plants is in the design and evaluation of biological control of weeds using plant pathogens and other natural enemies (Section 16.5).

10.8 Density-dependent disease in plant populations

The logistic model of population growth (Box 10.1) leads us to expect lots of reproduction and low rates of mortality at low population densities, while at high densities, mortality increases and reproductive rate decreases. In other words, per-capita birth and death rates are generally density dependent. The concept of carrying capacity (Box 10.1) is fundamentally about limiting resources—that resource availability caps the size of a population. Plants compete for resources such as water, light, and nutrients, and their population dynamics can show density dependence driven by that competition.

However, in addition to density dependence caused by limited resources, density dependence can be caused by plant diseases, with fascinating implications for plant populations and communities. When we say plant disease is **density dependent**, we mean that plant disease incidence, prevalence, or severity increase with plant population density. When pathogen prevalence or spread responds to plant host density, it is called **density responsive**. While these concepts are related, and "density dependent" is sometimes used carelessly in place of "density responsive," density dependence implies a response to an organism's own density.

Most plant diseases show density-dependent development (Figure 10.12; Chapter 11). In a classic review, Burdon and Chilvers (1982) examined the overall patterns of density-dependent disease development, and they found widespread empirical evidence for density responsiveness in pathogens. Studies of this phenomenon may compare disease incidence or severity across natural variation in density, or they may experimentally create a range of densities and then track disease development. For instance, Farias et al. (2019) planted soybeans at a range of densities from 12 to 36 plants/m^2 and found that the incidence of stem blight of soybean caused by *Phomopsis longicolla* increased linearly with planting density (Figure 10.13). One mechanism for density responsiveness is that pathogens spread more readily among hosts that are close together. Beyond facilitating the spread of pathogens between plants, high density can produce more shade, creating a moist

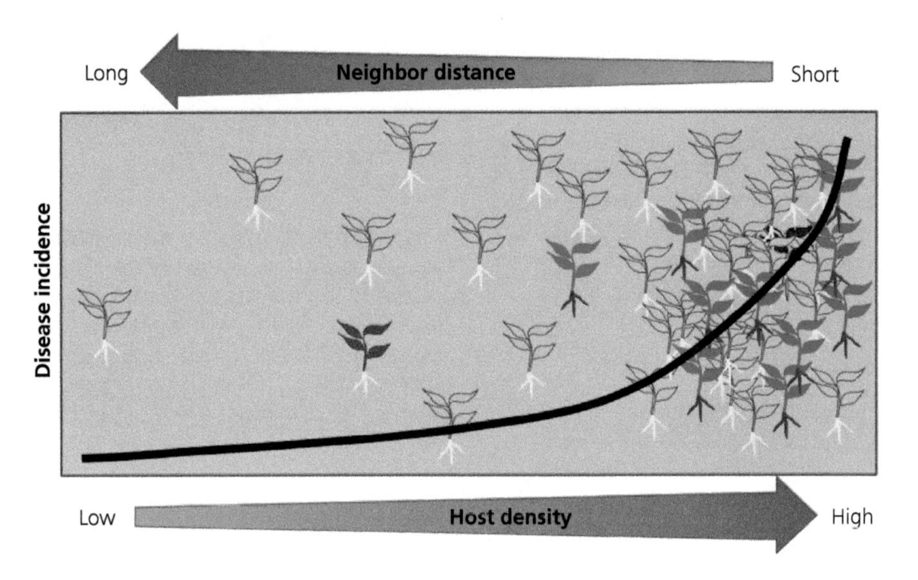

Figure 10.12 Density dependence is frequently observed among plant diseases. Disease incidence (orange plants and black line) is usually greatest at high host density, where the distance between plants is short.
Graphic by Gregory S. Gilbert.

microclimate that is more conducive to infection. High density can also mean more competition, creating stress that can make plants more vulnerable to disease development. In cropping systems, disease management must be considered along with yield when setting planting densities. In their study of soybean, Farias and colleagues also measured crop yield and found no increase in overall yield over their range of plant densities, leading them to recommend the lowest plant density as the one that minimized disease without sacrificing yield.

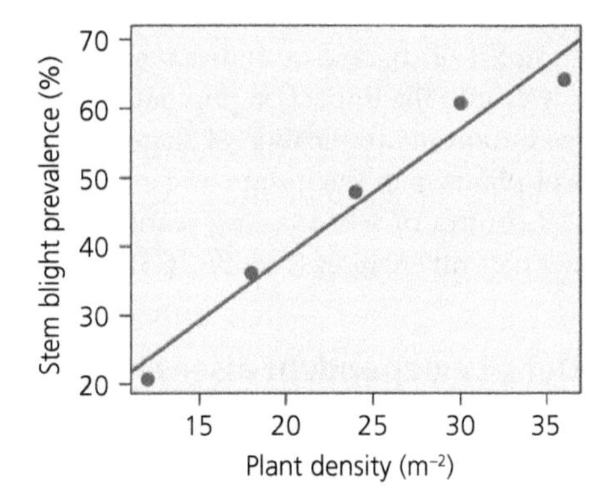

Figure 10.13 Prevalence of stem blight (*Phomopsis longicolla*) increases linearly with plant density in soybean.
Figure by Gregory S. Gilbert, redrawn from data in Farias et al. (2019).

There are two interesting exceptions to density-dependent disease patterns that help us understand the underlying mechanisms of density responsiveness. First, for pathogens such as rusts with obligate alternate hosts, the densities of both hosts are critical and may respond to environmental factors in conflicting ways. For example, with fusiform rust (*Cronartium fusiforme*), disease incidence may be negatively related to density of its pine host, because oaks are obligate alternate hosts that support production of infective aeciospores, and oaks grow poorly in the dark shade of dense stands of pines (Burdon and Chilvers 1982). Second, for some insect-vectored pathogens, the vector is able to disperse long distances and seek out preferred, susceptible hosts. Disease caused by such vector-transmitted pathogens, like many human and other animal pathogens, are often **frequency dependent** rather than density dependent. Here, the probability of developing disease depends on the frequency of diseased plants in the population; spread will be faster when there are close to equal numbers of infected and susceptible hosts.

10.9 Density dependence and the population-level effects of pathogens

Earlier we discussed how pathogens may influence plant population dynamics, and how we use population models to study that (Box 10.2). One limitation of most plant population models is that they ignore density dependence. For example, in order to estimate the response of population growth λ to a reduction in seedling survival, modeling studies often assume the plant population is in an exponential growth phase. This assumption is okay for some scenarios, but density dependence can sometimes have a profound effect on whether (and how) ecological interactions affect plant populations. For example, Sarah Swope and colleagues found that populations of the invasive thistle *Centaurea solstitialis* vary nearly 50-fold in density; they also found evidence of density dependence in population growth of all five of the populations they studied, even the ones with the lowest natural densities (Swope et al. 2017). Simulations showed that in some populations, strong density-dependence in survival meant that attack by an herbivore or pathogen would have hardly any effect. That is, when a pathogen kills individual seedlings, the remaining plants respond to that reduction in density by surviving at higher numbers, resulting in the same number of weedy thistles at the end. As in the study by Prendeville et al. on gourds (Box 10.2), pathogens that attack different parts of the plant life cycle can have very different impacts at the population level, even if they have similar proportional effects on demographic rates.

Density-dependent disease impacts can drive temporal dynamics of host populations, similar to the predator–prey cycles of the lynx and hare often discussed in general ecology. If the prey (hare) reproduces rapidly, the expanding population provides lots of food for the predator (lynx). With abundant food available, the lynx population then grows rapidly, to the point where deaths of hares exceed births, and the hare population shrinks. With fewer prey available to support growth and reproduction, the lynx population in turn declines, until the numbers (and predation rate) are low enough for the hare population to grow again. Hares and lynxes follow a roughly 10-year cycle of predator–prey population dynamics resulting from lagged density-dependent responses (Stenseth et al. 1998).

Now let's imagine a wild population of an annual plant that is susceptible to a moderately virulent pathogen with a narrow host range, so that the success of the pathogen (the predator) depends on the population of the host (the prey), and the growth and reproduction of the host is reduced by the pathogen. In such cases, we might expect to see population dynamics of the host and pathogen coupled in lagged density-dependent cycles similar to predator–prey cycles (Figure 10.14). Very severe infections might kill a host (reducing its fitness to zero), and plants with more severe infections produce fewer seeds. In this way, the intensity of disease is the product of the prevalence in the population and severity on individual plants. Because the plant has an annual life-history strategy (and no long-lasting seed bank), the size of the population in the coming year depends directly on births (seed production) in the present year.

A good example comes from *Hesperolinon californicum*, an annual dwarf flax found on serpentine outcrops in central California. *H. californicum* is attacked by the autoecious rust pathogen *Melampsora lini* (Springer 2009). Both prevalence and severity of rust infection respond to host density, with each rust pustule the product of a separate infection

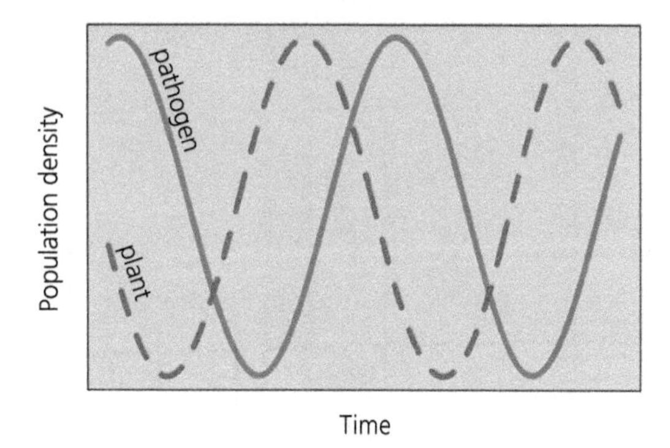

Figure 10.14 Lagged density-dependent cycles of populations of a pathogen (or predator) and a host (or prey). The pathogen population increases in response to increasing host density. After disease impacts reduce host population density, the pathogen population in turn declines.

Figure by Gregory S. Gilbert.

event. The pathogen is polycyclic; an individual plant can be infected numerous times by urediniospores dispersing from diseased neighbors, or even through autoinfection from the plant itself. So being closer to more infected neighbors increases the likelihood of developing more rust pustules (greater severity). At the end of the growing season, the rust produces teliospores that can survive over winter and serve as primary inoculum to initiate the infection cycle the following growing season.

Careful field observation by Yuri Springer (2009) revealed how coupled host–pathogen dynamics play out in this system. We start with an isolated, moderately high-density population of dwarf flax, happily growing on a serpentine outcrop, with very little rust (Figure 10.15). Infected individuals develop pustules, and urediniospores quickly spread the pathogen to neighboring plants. The fecundity of each plant depends on the severity of rust infection. Because the prevalence of rust is low in the population, the number of seeds produced by the population overall is still high, and the next generation begins with an even greater density of plants. This time, overwintering rust infects a larger number of individuals through overwintering teliospores as primary inoculum, followed by a much greater prevalence of rust in the population and greater severity on those plants that are infected. With most plants severely diseased, the overall production of seeds is low, and the production of overwintering inoculum is very high. The following year, the smaller number of seeds available to germinate leads to a low density of dwarf flax. Because primary inoculum is high, there is a relatively high initial prevalence of disease. However, because the plants are at low density, secondary spread is much reduced, so the severity of infection on diseased plants is low. That leads to a milder effect of disease on fecundity, and each plant on average produces more seed than in the previous year, and less overwintering inoculum is produced. The following year, the dwarf-flax population increases, while the primary inoculum available to begin the infection cycle is still low. The coupled dynamics of pathogen and plant populations produce cycles across generations (Figure 10.15).

There are several important criteria that allow for such coupled host–pathogen dynamics. First, the pathogen is host-specialized (at least among locally available host species), so that its population dynamics depend on that host population. Second, the disease has a major effect on the fitness of infected hosts. Third, the life cycles of the host and the pathogen are happening on roughly the same timeframe (i.e., the host doesn't live for centuries while the pathogen has an annual cycle, or there are

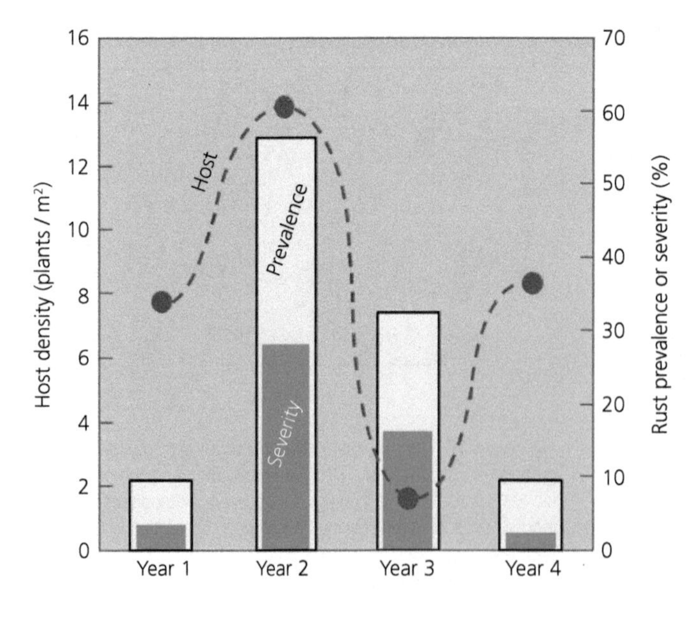

Figure 10.15 A lagged density-dependent cycle in a plant–pathogen system. The rust pathogen *Melampsora lini* on the annual dwarf flax *Hesperolinon californicum* illustrates how severity of disease in one year affects the host population density the following year, and how host population density affects both prevalence and severity of the rust disease.
Figure by Gregory S. Gilbert, adapted from Springer (2009).

no long-lived seed banks or resting spores). Fourth, infection of the host by secondary inoculum is density responsive. Violations of these criteria do not exclude the possibility of density-dependent cycles, but they make the patterns more complex.

10.10 From interacting symbionts to interacting populations

Important economic and ecological impacts of diseases arise when populations of pathogens spread through populations of their host plants. In this chapter we have focused on disease cycles, exploring several kinds of graphical and mathematical models to help us visualize and predict how diseases spread through host populations over time, and how populations of pathogens and plants shape each other's dynamics. Throughout this chapter we have used the simplifying assumption that we can understand disease dynamics without tracking the locations of individuals. In reality, though, individuals in a host population are distributed in space. This makes epidemiology and the population ecology of plant disease fundamentally a spatial phenomenon. As in many other subfields of evolution and ecology, adding the spatial dimension to disease ecology makes things even more exciting—as you will see in the next chapter.

Further reading

Brauer, F., C. Castillo-Chavez, and Z. Feng. 2019. *Mathematical Models in Epidemiology*. Springer, Berlin.

Caswell, H. 2006. *Matrix Population Models: Construction, Analysis, and Interpretation*, 2nd edition. Sinauer, Sunderland, MA.

Madden, L. V., G. Hughes, and F. van den Bosch. 2007. *The Study of Plant Disease Epidemics*. APS Press, St. Paul, MN.

References

Agrios, G. N. 2005. *Plant Pathology*, 5th edition. Elsevier, Academic Press, Cambridge, MA.

Beckstead, J., S. E. Meyer, B. M. Connolly, M. B. Huck, and L. E. Street. 2010. Cheatgrass facilitates spillover of a seed bank pathogen onto native grass species. *Journal of Ecology* **98**:168–177.

Blitzer, E. J., C. F. Dormann, A. Holzschuh, A.-M. Klein, T. A. Rand, and T. Tscharntke. 2012. Spillover of functionally important organisms between managed and natural habitats. *Agriculture, Ecosystems & Environment* **146**:34–43.

Box, G. E. 1979. Robustness in the strategy of scientific model building. In: R. L. Launer and G. N. Wilkinson, editors. *Robustness in Statistics*, pp. 201–236. Academic Press, New York.

Brauer, F., C. Castillo-Chavez, and Z. Feng. 2019. *Mathematical Models in Epidemiology*. Springer, Berlin.

Burdon, J. J. and G. A. Chilvers. 1982. Host density as a factor in plant disease ecology. *Annual Review of Phytopathology* **20**:143–166.

Caswell, H. 2006. *Matrix Population Models: Construction, Analysis, and Interpretation*, 2nd edition. Sinauer, Sunderland, MA.

De Wolf, E. D. and S. A. Isard. 2007. Disease cycle approach to plant disease prediction. *Annual Review of Phytopathology* **45**:203–220.

Farias, M., R. T. Casa, F. Gava, O. A. Fiorentin, M. J. Gonçalves, and F. C. Martins. 2019. Effect of soybean plant density on stem blight incidence. *Summa Phytopathologica* **45**:247–251.

Godfree, R. C., P. H. Thrall, and A. G. Young. 2007. Enemy release after introduction of disease-resistant genotypes into plant–pathogen systems. *Proceedings of the National Academy of Sciences of the United States of America* **104**:2756–2760.

Kerdraon, L., V. Laval, and F. Suffert. 2019. Microbiomes and pathogen survival in crop residues, an ecotone between plant and soil. *Phytobiomes Journal* **3**: 246–255.

Kermack, W. O. and A. G. McKendrick. 1927. A contribution to the mathematical theory of epidemics. *Proceedings of the Royal Society of London, Series A* **115**: 700–721.

Lennon, J. T. and S. E. Jones. 2011. Microbial seed banks: the ecological and evolutionary implications of dormancy. *Nature Reviews Microbiology* **9**:119–130.

Paine, C. T., T. R. Marthews, D. R. Vogt, D. Purves, M. Rees, A. Hector, and L. A. Turnbull. 2012. How to fit nonlinear plant growth models and calculate growth rates: an update for ecologists. *Methods in Ecology and Evolution* **3**:245–256.

Power, A. G. and C. E. Mitchell. 2004. Pathogen spillover in disease epidemics. *American Naturalist* **164**: S79–S89.

Prendeville, H. R., B. Tenhumberg, and D. Pilson. 2014. Effects of virus on plant fecundity and population dynamics. *New Phytologist* **202**:1346–1356.

Segarra, J., M. Jeger, and F. Van den Bosch. 2001. Epidemic dynamics and patterns of plant diseases. *Phytopathology* **91**:1001–1010.

Springer, Y. P. 2009. Edaphic quality and plant–pathogen interactions: effects of soil calcium on fungal infection of a serpentine flax. *Ecology* **90**:1852–1862.

Stenseth, N. C., W. Falck, K.-S. Chan, O. N. Bjørnstad, M. O'Donoghue, H. Tong, R. Boonstra, S. Boutin, C. J. Krebs, and N. G. Yoccoz. 1998. From patterns to processes: phase and density dependencies in the Canadian lynx cycle. *Proceedings of the National Academy of Sciences of the United States of America* **95**:15430–15435.

Stensvand, A., T. Amundsen, L. Semb, D. M. Gadoury, and R. C. Seem. 1998. Discharge and dissemination of ascospores by *Venturia inaequalis* during dew. *Plant Disease* **82**:761–764.

Swope, S. M., W. H. Satterthwaite, and I. M. Parker. 2017. Spatiotemporal variation in the strength of density dependence: implications for biocontrol of *Centaurea solstitialis*. *Biological Invasions* **19**:2675–2691.

van der Wal, A., P. K. Gunnewiek, M. de Hollander, and W. de Boer. 2017. Fungal diversity and potential tree pathogens in decaying logs and stumps. *Forest Ecology and Management* **406**:266–273.

Vanderplank, J. 1963. *Plant Disease: Epidemics and Control*. Academic Press, New York.

CHAPTER 11

Spatial ecology

Gregory S. Gilbert and Ingrid M. Parker

Patterns of plant disease are dynamic across both time and space. In the previous chapter, we focused exclusively on the temporal patterns of disease epidemics, but when diseases spread through populations, inoculum must also move from one host to another, and that requires dispersal across space. In this chapter we consider the great diversity of ways that pathogens spread via their propagules. We then explore how researchers study and model dispersal, introducing the most common experimental and analytical methods. Plant disease epidemiologists combine such quantitative descriptions of pathogen dispersal together with the dynamic models we presented in Chapter 10 to forecast disease spread across space, and we briefly introduce how this is done. We then consider two important topics in spatial ecology. First, we extend our discussion of the role of plant disease in the density-dependent regulation of plant populations and implications for the spatial distributions of plants. Second, we look at how habitat patchiness gives rise to dynamic patterns of colonization, infection, and extinction of both plants and pathogens, a phenomenon called metapopulation dynamics.

11.1 Dispersal (getting from here to there)

All living organisms are limited in how far they can disperse. Even in our highly mobile 21st century, nearly 70% of adults in the United States stay in the state in which they were born, living a median distance of 18 miles (29 km) from their mother (Molloy et al. 2011). Pathogens and their plant hosts are even more dispersal limited than humans because most of their dispersal is passive; propagules are carried on the wind, pulled to the earth by gravity, or splashed by raindrops. Some spores or seeds are carried a bit farther by animals (mammals, birds, insects, nematodes, and even frugivorous fish!). But most individual propagules stay close to home. Limited dispersal has implications for the spread of populations and for coordinated host–pathogen dynamics, as well as for the distribution of species, the diversity of communities, and patterns in the genetic structure of populations, each of which is discussed in subsequent chapters. In this chapter, we focus mostly on the spread of pathogens, for which successful dispersal requires not only being transported over some distance, but also being deposited on a susceptible host. The vectors of some animal-dispersed pathogens transport their cargo directly to another host, while some pathogens are themselves motile and can choose where to go over short distances. Let's take a look at the process and patterns of dispersal of pathogen propagules.

The first step in dispersal is the **release** of the propagules from wherever they are produced, sometimes called **take-off** (Figure 11.1). Release may be caused by outside forces like a gust of wind or the splash of a raindrop. Sometimes propagules are actively released in response to changes in humidity, available free water, or temperature. For example, a combination of soil temperature and convective air movement leads to a burst of fungal spores released around 10 days after forest fires (Camacho et al. 2018). Some pathogens are liberated from the vascular system of their hosts by insect vectors that feed on phloem or xylem contents, or from

The Evolutionary Ecology of Plant Disease. Gregory S. Gilbert and Ingrid M. Parker, Oxford University Press. © Gregory S. Gilbert & Ingrid M. Parker (2023).
DOI: 10.1093/oso/9780198797876.003.0011

Figure 11.1 Fungal spores are released at night and carried away in the breeze from this *Pleurotus* basidioma near Bengaluru, India.
Photo by A. K. Raju, Endemic Greens Pvt. Ltd.

the plant surface when insects simply walk through a colony of the pathogen and get their feet and body covered with spores.

Once released, **transport** processes disperse propagules away from their source. Most pathogen propagules disperse passively on water or air currents and are at the mercy of gravity. Many propagules simply wash off leaves with rain or spray irrigation, falling to the ground (or onto other leaves) with very little horizontal dispersal. Raindrops can transport spores away from the source when they splash on contact with the leaves; large drops carry many spores for short distances, but smaller drops can carry a few spores over long distances. In addition, the raindrop impact creates air vortexes that help surface spores escape the **boundary layer** of the leaf (facilitating take-off), and then the spores are transported through the air.

Airborne propagules disperse through the air on breezes and eddies, until they run into a surface or drop out of the air due to gravity. Particles less than 1 μm diameter (viruses, bacterial spores, some bacteria) can remain suspended in air as **aerosols** for many hours or days, but fungal spores and most bacterial cells are larger and so gravity forces them to settle out of the air within minutes to hours. Some propagules ride upward air currents reaching high above their source plants, where they can be carried extremely long distances, including transcontinental transport.

Pathogen propagules exhibit **adaptive traits** that facilitate transport. Desiccation and damage from UV irradiation limit how long spores, bacterial cells, and virus particles can remain viable in the air. To protect their spores, many fungi with airborne dispersal have dark pigments or thick spore walls. Bacterial spores are predesiccated and packed in protective materials that make them more tolerant to heat and UV. Many fungal spores and bacteria are embedded in sticky polysaccharides (goop) that help them stick to potential vectors or allow them to disperse in small clusters. Some fungal spores have shapes or appendages that facilitate dispersal in water or help them stay airborne longer. Zoospores of oomycetes, along with some bacteria, are flagellated and can swim through water-filled pore spaces in saturated soils or across pools of free water on plant surfaces. These motile cells can swim toward infection courts on a host by swimming up chemical gradients, a process called **chemotaxis**. For instance, zoospores can detect exudates from plant roots and swim several centimeters through saturated soil to the most vulnerable (leaky) part of a plant root, where they then encyst and infect the host.

It's good to remember that many pathogens use multiple mechanisms of dispersal. For example, *Phytophthora lateralis* is an oomycete pathogen that infects Port-Orford cedar (*Chamaecyparis lawsoniana*), causing the lethal Port-Orford-cedar root disease. This impressively large tree is native to the

Pacific Coast of Oregon and northern California. It is also an important landscaping species, but the introduction of *P. lateralis* has resulted in near abandonment of horticultural production in the Pacific Northwest (Betlejewski et al. 2011). *P. lateralis* produces chlamydospores (resting spores) on rootlets that then deteriorate, leaving the chlamydospores free in soil. That soil can be transported long distances by vehicles, heavy equipment, and even the feet of cattle and elk, as well as the boots of hikers. When that soil makes contact with a streambank, or experiences a soaking rain, the chlamydospores of *P. lateralis* germinate and give rise to zoospores that will travel through the water and swim to the roots of a new host. Hyphae of *P. lateralis* can also grow from one cedar tree directly to infect a nearby individual via root grafts. Finally, even aerial dispersal is possible, when detached sporangia are splashed or carried by foggy air onto the foliage of a host tree.

Dispersing propagules **deposit** on a wide range of surfaces, only a small fraction of which are suitable sites on susceptible hosts. Some small fraction of those that land on inhospitable surfaces may be picked up again and continue their journey, but the sad fact for most dispersing pathogen propagules is that they will end their odyssey without arriving at a hospitable port. For those that do deposit on a suitable host, the next challenge is sticking there long enough to mount a successful infection. The polysaccharides (goop) mentioned above can also help anchor propagules to a plant surface. Oomycete zoospores that reach a suitable root secrete a mix of precursor compounds that assemble into a cyst wall that glues to the root surface. Viruses, bacteria, and fungi that are transmitted by insect vectors are often deposited directly inside the host plant through feeding or tunneling by the vector, giving them a huge advantage in efficiency in reaching suitable hosts.

11.2 Describing and modeling pathogen dispersal

Let's explore the conceptual and mathematical framework for how a population of pathogens grows and spreads through a plant population. For illustration, we will place ourselves in two adjacent fields of healthy tomatoes planted at a regular 1-m spacing between plants. In the first field, a bit of mycelium and mitospores of *Alternaria*, an airborne fungal pathogen that causes early blight, overwinters on plant debris and infects a plant in the center of the field. Conditions are just right for rapid colonization, and soon the plant is sick with early blight and covered with mitospores. A small puff of air on an otherwise still day releases the spores from the leaf surface and sets them aloft, and they gently disperse in random directions. Some land on neighboring plants and can germinate and infect after that night's dew. Most travel only a short distance, while a few travel very far. In the center of the second field, a seed contaminated with *Glomerella cingulata* (cause of anthracnose) has grown into an infected plant. Mitospores of *Glomerella* are produced and disperse primarily through raindrop splash. Compared to the wind-dispersed *Alternaria*, dispersal in *Glomerella* is even more concentrated around the source plant.

In both fields, the probability of a spore making it to any particular distance from the source declines rapidly with distance. This distribution of individual distances is called a **dispersal kernel**, and it can be described mathematically by any of a number of **probability density functions** (Turchin 1998). Two functions in particular are widely used and intuitively accessible: the **exponential model** is frequently a good descriptor for splash-dispersed pathogens like *Glomerella*, while the **power-law model** is a good fit for wind-dispersed pathogens like *Alternaria* (Figure 11.2). Let's take a look at how these models can help us understand the observed **dispersion** (pattern of distribution) of spores, and the degree of observed **contagion**, or spatial clumping among diseased plants.

Let's start by looking at the basic mathematical expressions for the exponential model [Eq. 11.1] and power-law model [Eq. 11.2].

$$Y = a_e e^{-b_e s} \quad \text{[Eq. 11.1]}$$

$$Y = a_p s^{-b_p} \quad \text{[Eq. 11.2]}$$

Y is the number of propagules, and s is the distance from the source of inoculum. Both of these equations have exponents, which tells us that the

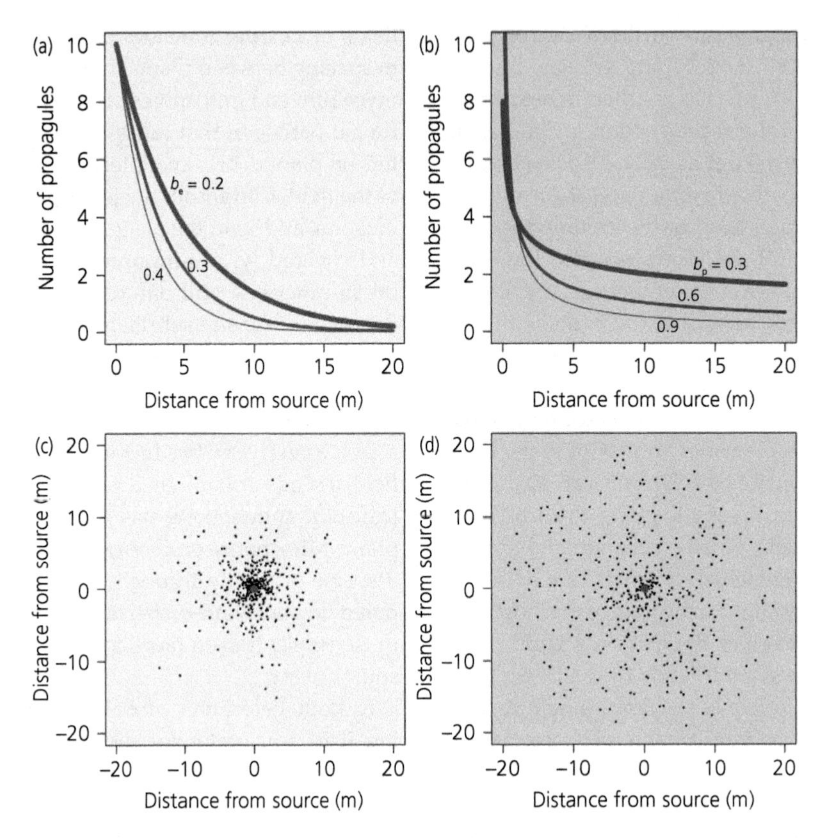

Figure 11.2 Dispersal of spores or other propagules declines rapidly with distance from the source. (a) The exponential model declines gradually and has a thin tail at long distances. This model often does the best job of representing pathogens with splash dispersal. (b) In comparison, the modified power-law model has a much steeper decline at short distances from the source, but then has a fatter tail, with more of the spores dispersing longer distances. The power-law model is often a good descriptor of pathogens with airborne dispersal. For the curves shown in (a) and (b), the exponential models have the parameter $a_e = 20$ and the power-law models have $a_p = 7$; values of b_e and b_p are given as labels on the corresponding curves. Note that the larger the value of b_e or b_p the steeper the decay with distance. Clouds of 500 spores dispersed in random directions from a central source, with travel distance determined by (c) exponential ($b_e = 0.3$) or (d) modified power law ($b_p = 0.6$) models. Whereas the farthest dispersal is similar between the two, the cloud is much more compact for the exponential model, with a much greater proportion of the 500 spores falling within a few meters of the source.
Figure by Gregory S. Gilbert.

responses are curves and not straight lines. Because the exponents are negative, we know that the value of Y declines (rather than increases) with distance. The decay coefficient, $-b_e$ or $-b_p$ (e for exponential, p for power law), determines the steepness of that exponential decay (Figure 11.2a and b). The more negative the value of b_e or b_p the steeper the decay with distance. The last element of these equations is the intercept coefficient, a_e or a_p, which is the number of propagules that are expected at the source.

The base of the exponent differs between the two models. For the exponential model (Equation 11.1) the natural logarithm base e ($e = 2.71828$) is raised to $-b_e s$, in which the decay coefficient b_e is multiplied by the distance. In contrast, in the power-law model (Equation 11.2), distance is the base, raised to the power $-b_p$. This difference matters because it means that the values of b_e and b_p have different units; b_e has units of "per unit distance" (s^{-1}) whereas b_p is unitless. This means that even though the value of b controls the rate of decay in each model, we can't directly compare the numerical values of b_e and b_p.

It is easier to think about and work with linear models than nonlinear models, because straight

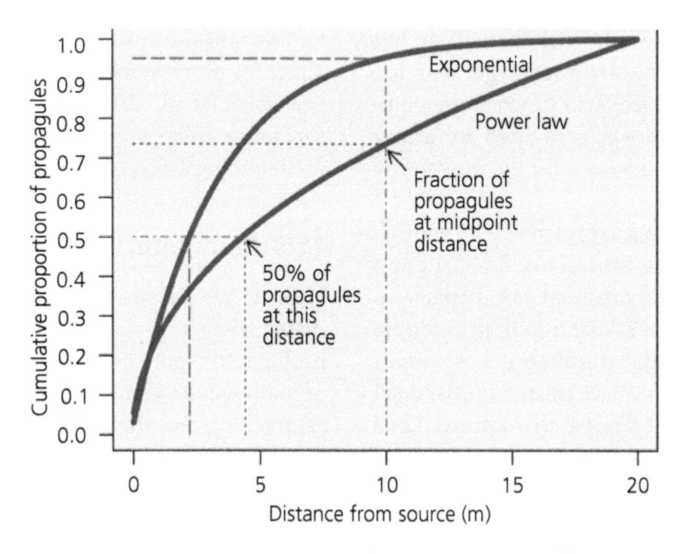

Figure 11.3 The difference between exponential (red) and power-law (purple) dispersal from Figure 11.2 is readily seen if we look at the cumulative number of spores that disperse up to a particular distance from the source. The dashed lines show that 50% of the spores fall within 2.2 m of the source for the exponential model, but those 50% are spread out over 4.4 m for the power-law model ($b_e = 0.3$ and $b_p = 0.6$). Looked at another way, while some spores dispersed to a maximum of 20 m in both models, the midpoint distance of 10 m was exceeded by fewer than 5% of spores under the exponential model but nearly 27% of spores under the power-law model.
Figure by Gregory S. Gilbert.

lines are easier to visualize and to fit to observed data. To linearize an exponential model, we take the logarithm of both sides of the equation. The linearized versions of Equations 11.1 and 11.2 are then:

$$\ln Y = \ln a_e - b_e s \qquad \text{[Eq. 11.3]}$$

$$\ln Y = \ln a_p - b_p \ln s \qquad \text{[Eq. 11.4]}$$

These linearized forms take the familiar form of a line where $y = b + mx$, where b is the intercept (not to be confused with b_e and b_p), and m is the slope. For the power-law model, $y = \ln Y$, $b = \ln a_p$, $m = b_p$, and $x = \ln s$.

There is a small issue with the power-law model that begs for a fix; at distance $s = 0$ (i.e., the source of spores) where there should be a *lot* of spores, the equation is unsolvable because ln 0 is undefined. To take care of this, we usually use the modified power-law model, which adds a small fudge-factor, λ, to all values of s (Equations 11.5 and 11.6). One can think of λ as the radius of the inoculum source.

$$Y = a_p(s + \lambda)^{-b_p} \qquad \text{[Eq. 11.5]}$$

$$\ln Y = \ln(a_p) - b_p \ln(s + \lambda) \qquad \text{[Eq. 11.6]}$$

The curves in Figure 11.2a and b show how many propagules dispersing from a source will reach a certain distance. Another useful way to think about the dispersal of propagules is by looking at the cumulative proportion of propagules that disperse *up to* a particular distance (Figure 11.3). Whereas Figure 11.2 indicates how many propagules should make it a particular distance from the source, Figure 11.3 keeps a running tally of propagules with increasing distance. We see in this case that half of all the spores in the power-law model dispersed farther than 8 m, while in the exponential model half of the spores deposited within just 2.7 m of the source. Only 5% of the spores in the exponential model make it further than 10 m, but some 40% of the power-law spores go that far. The exponential model leads to much more compact, intense spore shadows, compared to farther and more diffuse spore shadows for the power-law model (Figure 11.2).

Some of those dispersing spores are lucky enough to land on a susceptible host plant, infect it, and produce a new crop of spores that can disperse as secondary inoculum. Let's take a look at multiple cycles of spore production through a host population, and how different mechanisms of dispersal influence the patterns of disease that appear. Let's go back to those two tomato fields, infected by either the splash-dispersed *Glomerella* (exponential model) or the wind-dispersed *Alternaria* (power-law model). A single infected plant in the middle of the field is the initial source (Figure 11.4, generation 1). Each source produces 200 spores (small black dots), which then each disperse in random directions to distances determined by the exponential or power-law model. A fraction of those spores happens to hit and infect a host, shown by larger orange or light blue dots. In generation 2, each of those infected plants becomes a new source (dark red or dark blue), with a cloud of 200 spores dispersing from each of those hosts. Each successive generation of spores infects a certain number of new hosts; those new hosts join the already infected hosts to produce yet another cohort of spores, infecting nearly the entire 40×40 m field of plants after six generations of spore production. Notice that the splash-dispersed (exponential) *Glomerella* spreads out as a single, more compact cluster, whereas in the fatter-tailed, wind-dispersed (power-law) *Alternaria*, some satellite foci appear far from the original source, driving a more rapid spread of the pathogen across the whole field.

We examined two important dispersal distributions—the exponential and power-law models—but they are not the only two models of pathogen dispersal. Gaussian (normal) distributions are "thin-tailed" and can be the best option for describing organisms with particularly rare long-distance dispersal. More complex mathematical formulations such as Cauchy models allow for more refined fitting of the shape of decay. See Esker et al. (2007) and Madden et al. (2007) for more details on dispersal models in plant epidemiology.

11.3 Parameterizing models of dispersal

How do we parameterize models that describe the dispersal of a given pathogen? This is a general problem for spatial ecology not unique to the study of pathogens. There are three basic approaches: (1) tracking and trapping of individual propagules to quantify their movement, (2) mechanistic models of propagule movement or biophysical phenomena such as aerodynamics and air currents, and (3) inference of dispersal processes from observed spatial patterns. The same approaches are used to study the dispersal of plants (see an excellent review by Bullock et al. (2006)).

11.3.1 Tracking and trapping

For large organisms like mountain lions or fish, we measure dispersal directly by tracking individuals with GPS collars or pit tags. Too bad we can't put collars on fungal spores! Instead, we trap propagules upon their arrival at different distances after release from a point source, using a variety of active and passive spore traps (Figure 11.5), or using host plants as "traps." Spatial data from traps is sometimes described as a **dispersal gradient,** while data from diseased plants is called a **disease gradient** (Fitt et al. 1987, Rieux et al. 2014). Using experiments with an inoculum source, we can track the locations of diseased plants over time or track the presence of pathogens through their DNA or RNA. For example, Eggeneberger et al. (2016) studied the dispersal of the Gram-positive bacterium *Clavibacter michiganensis* subsp. *nebraskensis*, which is the causal agent of Goss's wilt of corn. They created plots of 400 plants of a susceptible variety, isolated from other plots by rows of a resistant variety. They then created an inoculum point source by either introducing infected corn residue (infected leaves

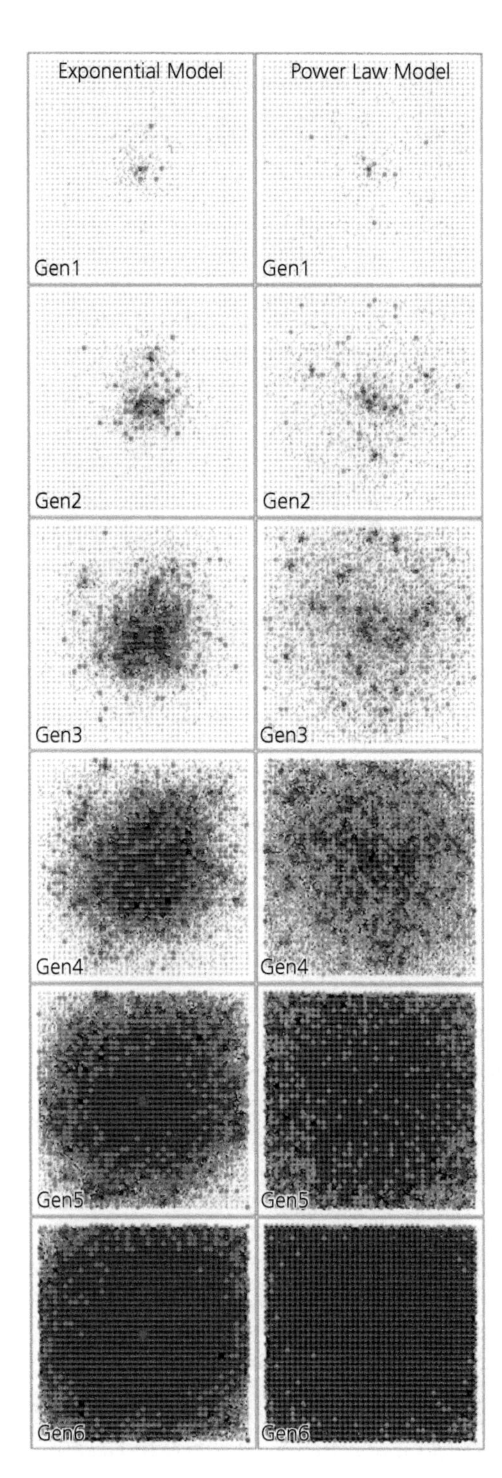

Figure 11.4 Clouds of 200 spores (small black dots) disperse from a single infected, competent host (Generation 1) at the center of a field of tomatoes (green grid points) regularly spaced at 1-m intervals. Splash-dispersed spores disperse from the source with a distribution described by the exponential model (left column) and wind-dispersed spores with a distribution given by the modified power-law model (right column). Those spores that land within 10 cm of the center of a tomato plant "infect" the host (larger orange or light blue dots). In Generation 2, each of those infected plants (dark red or dark blue) in turn supports production of 200 spores that disperse from those hosts, as well as 200 more from the original source. Each generation (down columns) shows the expanding extent of the spore cloud, as well as the increasing extent of diseased plants. Notably, the thin-tailed nature of the exponential model means that disease spreads from a single expanding focus, whereas the fat-tailed power-law model results in satellite foci and a less regular pattern of disease spread. Figures are of simulated data based on Eqs 11.1 and 11.5, with probabilistic dispersal from each spore source, with each spore dispersing in a random direction. For these models, $a_e = 10$, $b_e = 0.3$, $a_p = 4$, $b_p = 0.6$, and $\lambda = 0.1$.
Figure by Gregory S. Gilbert.

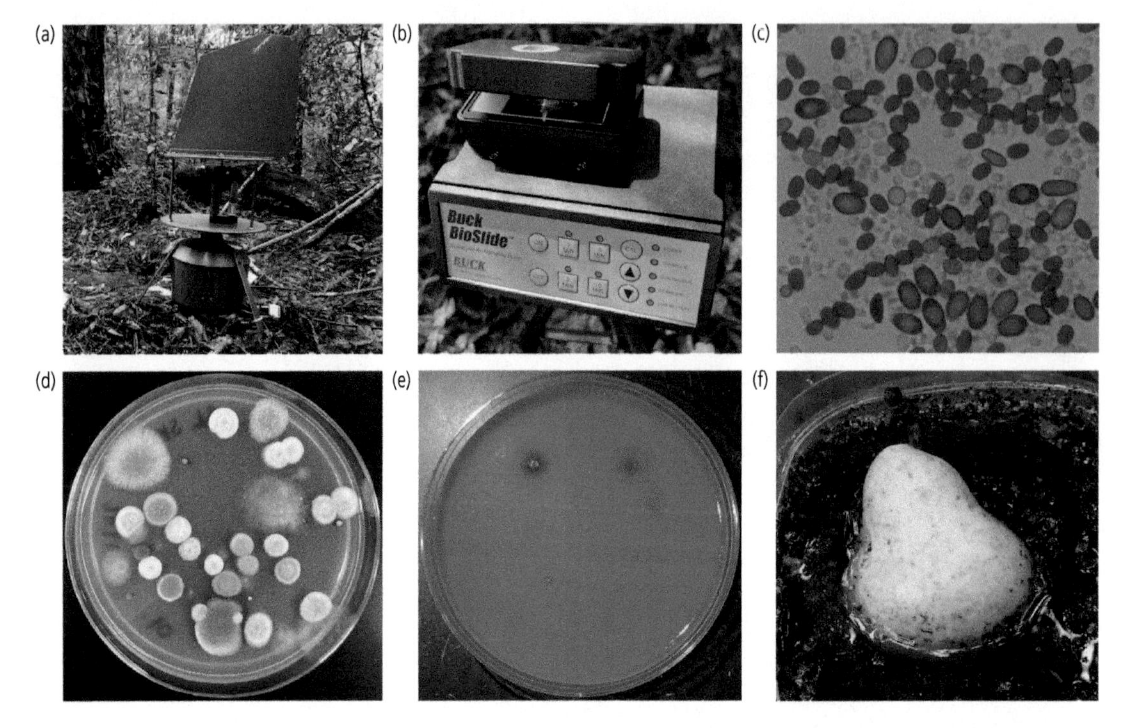

Figure 11.5 Various methods can be used to trap spores from the environment to measure dispersal patterns or seasonal production of fungal spores. Volumetric air samplers like the (a) Burkhard™ volumetric spore trap that collects air particles in a tube, or (b) portable samplers that collect spores on glass slides or filters, allow (c) quantification and identification of fungi using microscopy or molecular methods. Dilution plating or air exposure of (d) general growth media such as malt extract agar or (e) selective growth media such as rose bengal medium for *Botrytis* allow quantitative sampling for fungal communities or targeted pathogens. (f) Presence of *Phytophthora* spp. in soil can be detected by attracting zoospores to green pears placed as bait in flooded soil samples. See Chapter 9 for more methods to measure pathogen abundance.
Photos by Gregory S. Gilbert.

held in place with hardware cloth) or inoculating four plants in the center of the plot. They assessed all the plants in the plot for symptoms of Goss's wilt on multiple dates throughout the growing season; from these data they quantified the distances to newly infected plants. They also collected leaves of asymptomatic plants and estimated the incidence and population density of the pathogen using dilution plating (Figure 9.13). They were able to separate and quantify their particular bacterial pathogen by using a special strain that was resistant to the antibiotic rifampicin, which they added to selective media.

11.3.2 Mechanistic models

Biophysical models of air or water movement guide our predictions of how far different-sized propagules will travel, and how environmental conditions such as wind speed and direction will influence pathogen dispersal. For viruses, bacteria, and fungi that disperse with animal vectors, we use models of animal movement to project the distribution of trajectories of those vectors, combined with the probability of moving from the vector into the host. For self-propelled propagules like nematodes, zoospores, or motile bacteria, we can study movement behavior under a microscope to determine, for example, how much movement of bacteria or spores is directional and how much is random diffusion (i.e., Brownian motion). Movement rules for individuals are then incorporated into models of how populations of individuals will spread over time. Growth and foraging models can be used to estimate the spread of mycelium through soil.

11.3.3 Inference from spatial patterns

Sometimes we cannot track or model the dispersal of propagules themselves; instead, we infer the process of pathogen spread from the spatial pattern of diseased host individuals. This approach requires intensive mapping of individual host plants, where the location of every diseased and healthy plant is recorded, usually over multiple time-points. Figure 11.4 demonstrates how and why this kind of approach works. If you compare the two models at generation 2 or generation 3, you can imagine that a mathematical description of the dispersion of diseased individuals should be able to distinguish between negative exponential dispersal, which results in a single spreading focus and more clumping (called contagion), and power-law dispersal, which results in a broader scatter and multiple new spreading foci. There are many analytical techniques, all under the umbrella of spatial statistics, that are used to describe spatial patterns of contagion in plant disease (Madden et al. 2007). Such analyses can be useful for detecting underlying environmental drivers of disease (e.g., seasonal and directional dispersal by wind), or response to management (e.g., drip versus spray irrigation). They are also useful for developing hypotheses about mechanisms of dispersal. For example, patterns of contagion differ when the pathogen is insect-vectored versus splash-dispersed.

11.4 Spatial ecology and density-dependent disease in plant populations

As we have seen with the logistic model of population growth (Box 10.1), we expect plants growing at low density to have plenty of resources to support survival, growth, and reproduction, while at high densities, mortality increases and reproductive rate decreases. In other words, per-capita vital rates of birth and death are density-dependent. As described in Chapter 10, disease, along with competition for resources such as water and light, can be an agent of density dependence. Let's pull together what we have learned about the spatial aspects of disease to see what it says about density dependence and how it shapes the dynamics of plant populations.

As host density (plants per unit area) increases, the distance between plants must get shorter. Mathematically, the distance between a plant and its nearest neighbor declines with the inverse of the square-root of the density (Figure 11.6), where the average nearest-neighbor distance is a predictable function of density, as long as you know the spacing pattern of the plants. When plants are regularly spaced (like on a grid or lattice), the relationship between nearest-neighbor distance (d_{nn}) and plant density (σ, number per unit area) is:

$$d_{nn} = \frac{1}{\sqrt{\sigma}} \quad \text{[Eq. 11.7]}$$

The expected mean nearest neighbor distance, d_{nn}, similarly declines as a function of the inverse of the square root of density when plants are randomly distributed, but as:

$$d_{nn} = \frac{1}{2\sqrt{\sigma}} \quad \text{[Eq. 11.8]}$$

More complicated formulas exist for other spacing patterns (Cressie 2015). These equations have been used to evaluate the spatial distribution of plants: an observed mean nearest neighbor distance $< \frac{1}{2\sqrt{\sigma}}$ indicates that the plants are more clumped than random, and $d_{nn} > \frac{1}{2\sqrt{\sigma}}$ indicates that the spacing is more regular than random.

Shorter distances among plants translates into greater likelihood of spread of pathogens from

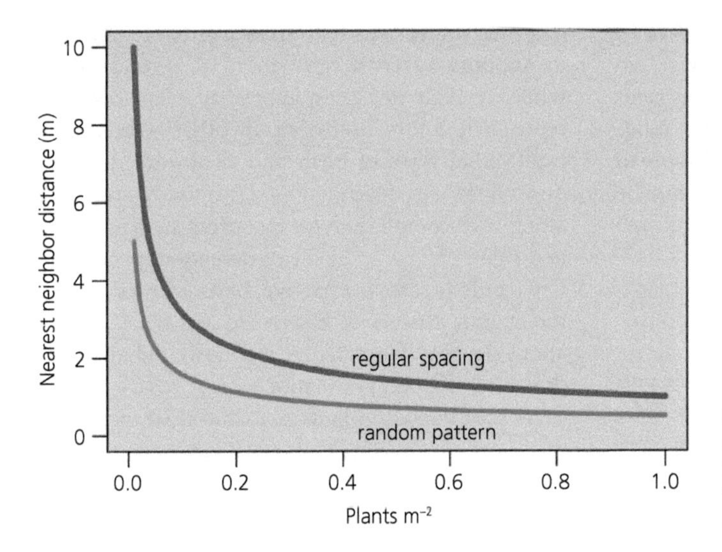

Figure 11.6 Relationship between distance between plants and the overall plant density is non-linear, with nearest neighbor distance declining as the inverse of the square root of the density.
Figure by Gregory S. Gilbert.

infected host plants to neighboring susceptible plants, so the probability of becoming diseased is **density dependent** (Figure 11.7). From the perspective of the pathogen, the growth and spread of the pathogen population is **density responsive** (the pathogen population responds to the density of the host population, not to its own density) (Section 10.8). The effect of host density on the rate of disease spread depends jointly on the distance between hosts and the dispersal ability of the pathogen (Figures 11.7 and 11.8).

11.5 Spatial distribution of plants shapes, and is shaped by, interactions with pathogens

Outside of the regular rows or grids of an orchard or a crop field, most plant populations are spatially clumped. Some clumping is due to the patchiness of suitable environmental conditions (e.g., willows grow in riparian zones), but a lot of it is because the dispersal of plants, like pathogens, is concentrated over short distances. The area where most seeds fall around an adult is called a **seed shadow**, with a gradient of plant density declining with distance. There are two reasons we expect greater disease among plants at high density. First, if one individual becomes infected, subsequent transmission of secondary inoculum (and thus disease incidence) will be greater where plants are more closely spaced. Second, plant patchiness tends to persist over time. A tree or a patch of annual plants will drop large cohorts of seeds in the same location year after year—sometimes for decades. Pathogens that successfully infect the plants that grow from those seeds in turn make yearly deposits to long-lasting propagule banks, which provide a powerful source of inoculum tied to those seed shadows.

This effect of density-dependent disease on plants in wild populations has been most extensively studied for tree seedlings. Trees are easy to study because they produce large crops of seeds over multiple years and provide a well-defined focal point at the center of a seed shadow. A classic study by Carol Augspurger explored the ecological consequences of seed dispersal in tropical trees, including *Platypodium elegans*, a species with wind-dispersed propagules (Augspurger 1983). Augspurger tagged newly germinated seedlings at different distances from their mother trees and followed their fates for a year. She found that damping-off disease incidence declined strongly as a function of distance (Figure 11.9). After a year, survival was much higher for seedlings that dispersed 55–100 m away from the mother tree than for those that dropped 0–20 m away.

Especially when diseases are lethal, like damping-off of seedlings, pathogens can have

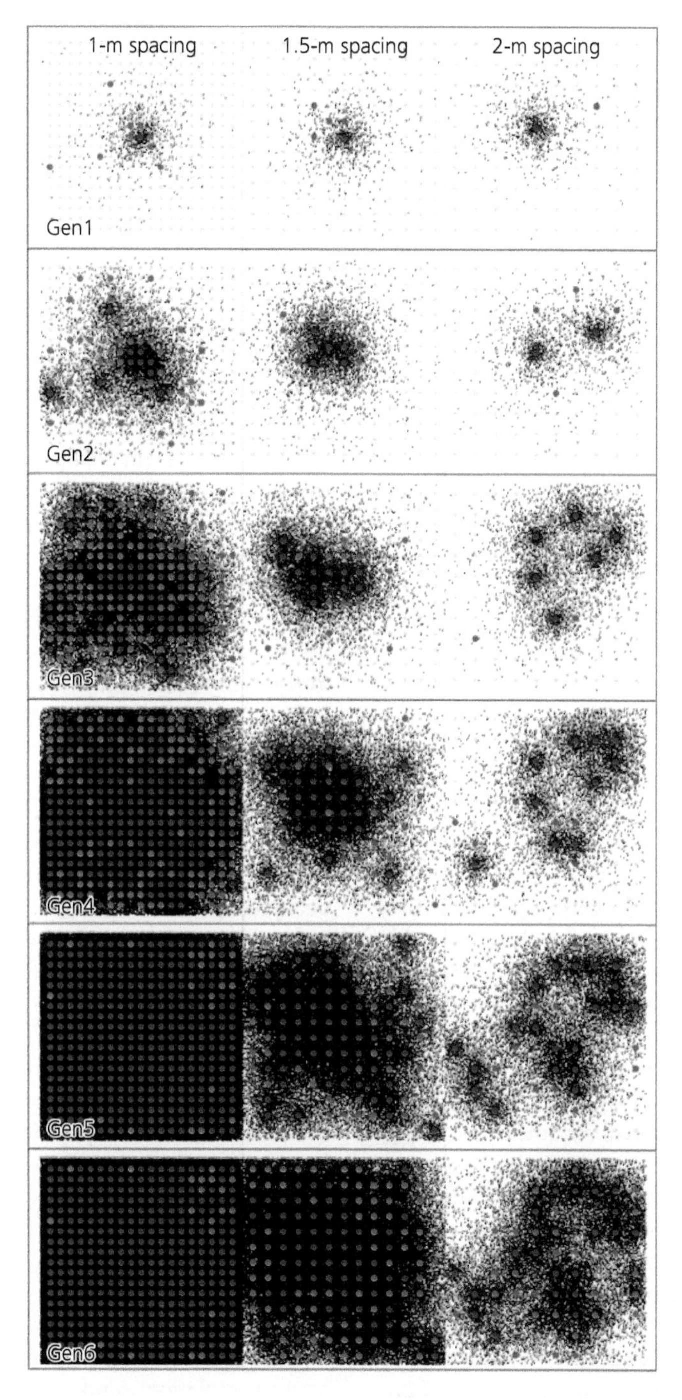

Figure 11.7 Effect of host density (three density scenarios per row) on the spatial spread of pathogens. The size and spread of a pathogen population through a host population responds to the density of the host plants, across six generations (rows, generation numbers in lower left) of secondary spore inoculum. Plants are spaced at 1-, 1.5-, or 2-m distance on a regular grid (density equivalent of 10,000, 4,444, or 2,500 plants ha^{-1}). Each infected plant produces 1000 spores in each successive generation of secondary inoculum, dispersed according to the exponential model ($a_e = 10$, $b_e = 0.4$). In this model, we assume spores do not cross between fields. The rate of spatial spread, as well as the proportion of hosts infected (a measure of the population size of the pathogen), increases more rapidly at higher host density than at lower density, because when plants are more closely spaced, the likelihood is greater that dispersal-limited pathogen propagules successfully disperse to a new host.
Figure by Gregory S. Gilbert.

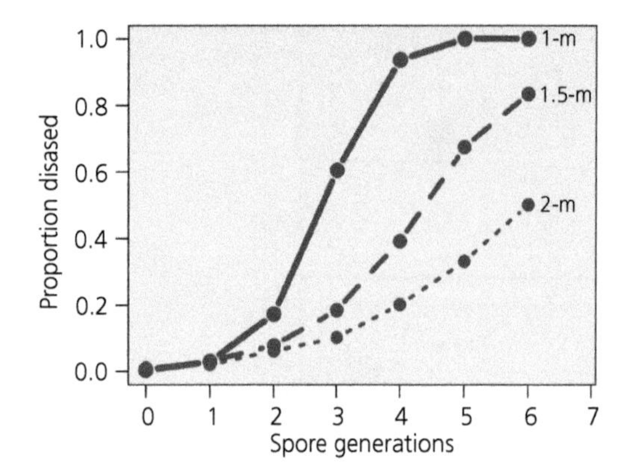

Figure 11.8 Disease progress across spore generations at plant spacing of 1, 1.5, or 2 m, with the simulated spore dispersal shown in Figure 11.7.
Figure by Gregory S. Gilbert.

strong effects on the spatial distribution of their hosts. Here's how it works.

Seeds and germinating seedlings start off highly clumped, but as we have seen, density-dependent disease kills proportionally more plants in high-density areas close to the mother plant. Over time, then, the density of surviving plants close to the seed source will decline disproportionally (Figure 11.9). Importantly, plant populations stay clumped in space (density-dependent mortality does not *over*compensate for dispersal limitation to create a regular spacing among plants), but they are less clumped than they would be in the absence of disease. That is, the mean nearest-neighbor distance d_{nn} gets progressively larger to approach that expected for randomly distributed plants, $\frac{1}{2\sqrt{\sigma}}$ (Equation 11.8).

An example of this shift in spatial clumping comes from the lowland tropical forest on Barro Colorado Island, Panama, where Gilbert et al. (2001) tracked 13,628 seedlings of *Ocotea whitei* over a five-year period. That study found that seedlings were initially extremely clumped around three mother trees, and that over time, the clumping became less pronounced because of the disproportional mortality of seedlings within 6 m of an adult (Figure 11.10).

The impact of density-dependent disease across distance from a maternal tree is a fundamental component of the Janzen–Connell hypothesis for how diversity is maintained in plant communities (Chapter 14 and Janzen 1970, Connell 1971).

11.6 Spatial ecology: Metapopulations

When we adopt a spatial perspective, we quickly realize that populations of plants and pathogens do not occur in a vacuum; they are connected to other populations across the landscape. The phenomena we have discussed up to this point have taken place in isolated populations of hosts, but by considering populations in connected networks (**metapopulations**), we can explain larger-scale phenomena and unintuitive patterns, such as the global persistence of species that are declining in many individual locations.

Metapopulation theory envisions populations as islands or fragmented patches connected across a landscape by dispersal (Levins 1969, Hanski 1999). You might notice that modeling plant populations linked across a metapopulation is analogous to modeling pathogen populations in individual host plants linked across a plant population. Extinction of individual populations is counterbalanced by colonization of empty sites from other populations in the landscape. The change in proportion of patches occupied, p (with $0 \leq p \leq 1$) with time (t), is given as:

$$\frac{dp}{dt} = cp\ (1 - p) - ep \qquad \text{[Eq. 11.9]}$$

where c is the colonization rate and e is the extinction rate. The most fundamental insight from this model is that in a metapopulation there will always be unoccupied (or "empty") sites: if the rates c and e are constant over time, then the expected fraction of occupied sites is

$$1 - \frac{e}{c} \qquad \text{[Eq. 11.10]}$$

There are many interesting implications of thinking about populations, and communities, of plants as linked across space, and this area of research grew into the field of landscape ecology. A key question is the degree to which the dynamics at individual sites in a metapopulation are synchronous or independent of one another. Another is the degree to

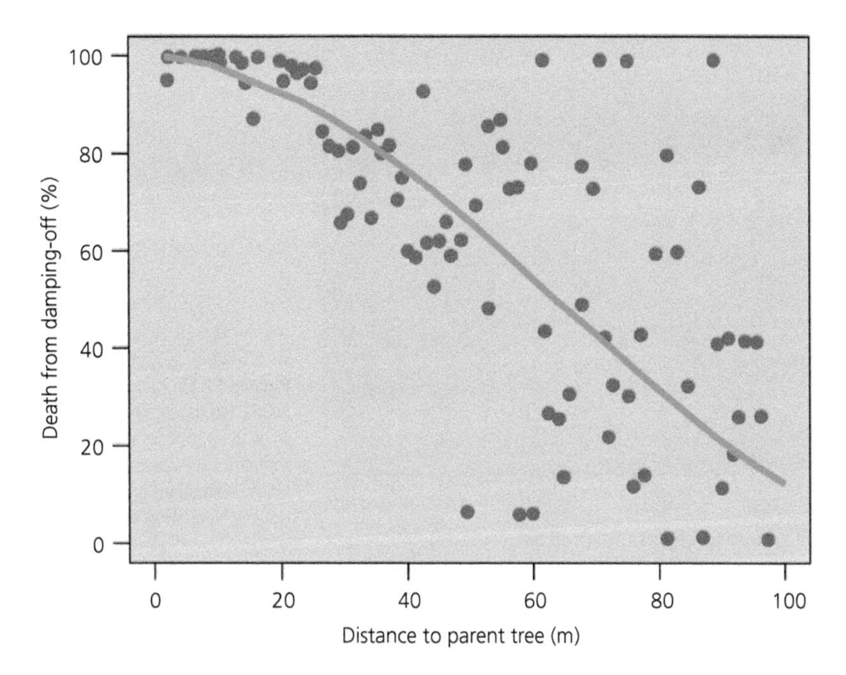

Figure 11.9 In the tropical tree *Platypodium elegans*, seedling mortality is greatest near the parent tree, leading to much higher seedling survival among seedlings from seeds that dispersed far.
Figure recreated from data in Augspurger (1983).

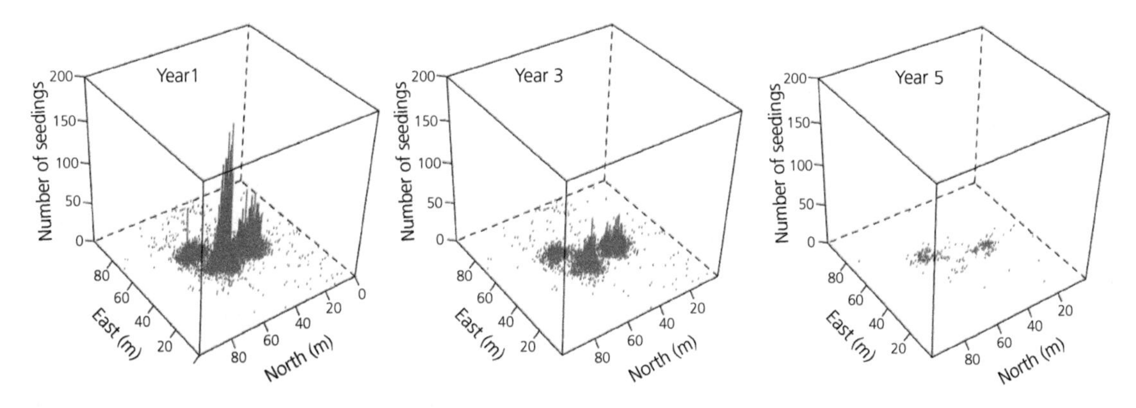

Figure 11.10 Progressive impact of density-dependent disease on spatial clumping of plants. The initial seedling shadow of *Ocotea whitei* in a Panamanian rain forest (Year 1) is highly clumped near each of the three mother trees (the *x* and *y* axes outline a 1-ha area of forest). Density-dependent mortality thins the population at highest density and near the mother plant (Years 3 and 5) so that with each year the plant population becomes less clumped. Notice how the fewest seedlings survived near the mother tree that produced the greatest density of seedlings.
Figure recreated from data in Gilbert et al. (2001).

which the populations all contribute to the colonization of other sites, or alternatively if there are a few populations that are **sources** of propagules and the other populations are **sinks** that are continuously in decline. Metapopulation dynamics are also important in evolution and population genetics, influencing genetic structure and the maintenance of genetic diversity (Section 13.6).

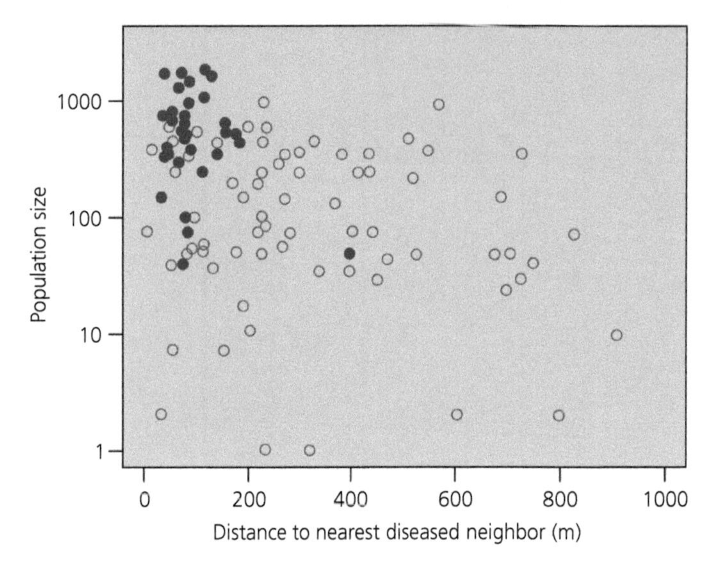

Figure 11.11 Infected (filled purple circles) and uninfected (open green circles) populations of *Filipendula ulmaria* attacked by the rust pathogen *Triphragmium ulmariae*. Disease incidence increases in larger populations and when populations are less isolated from other infected populations. Figure recreated from data in Burdon et al. (1995).

Pathogen colonization of host patches depends both on the presence of a susceptible host and on suitable environmental conditions. Local extinction and colonization are common in natural populations of pathogens. When interactions between hosts and pathogens are set into a metapopulation context, host patch size and connectivity among patches are found to be critical factors influencing disease incidence and persistence. For example, Burdon et al. (1995) studied the herbaceous perennial *Filipendula ulmaria* (Rosaceae) and its highly specialized rust pathogen *Triphragmium ulmariae* across 129 populations in Sweden's Skeppsvik Archipelago. They found higher disease incidence in plant populations that were larger and located closer to the nearest diseased population (Figure 11.11). Larger host populations have lower rates of pathogen extinction, whereas more isolated patches have lower recolonization rates and lower rates of **population rescue**, in which propagules dispersing from neighboring patches save pathogen populations from extinction.

Island archipelagos, as in the previous study, have provided some of our most elegant examples of plant–pathogen metapopulations. For many years, Anna-Liisa Laine and colleagues have studied a metapopulation of *Plantago lanceolata* and its biotrophic, obligate pathogen, *Podosphaera plantaginis* (Erysiphales, powdery mildews), in the Åland Islands of southwest Finland. Laine tracked populations in 3244 meadows over four years and found low disease incidence overall and high rates of extinction and recolonization: of 282 infected host populations, 76% were infected in only one of the four years (Laine and Hanski 2006). Disease incidence increased with host population size and with connectivity, affecting both recolonization and extinction; this suggests a population rescue effect. Other ecological factors were also important; annual variation in environmental conditions strongly affected local extinction of the pathogen, and both roads and wind direction seemed to influence pathogen dispersal.

Most patchy plant populations experience some metapopulation dynamics independently of effects of pathogens, and disease may simply enhance the likelihood of extinctions in some patches. For obligate pathogens, which often have narrow host ranges, host patch extinction necessarily means the pathogen will go extinct in that patch too. Facultative necrotrophs and pathogens with broad host ranges may also go through patch extinction/recolonization events. However, as study organisms, obligate pathogens have the dual attraction of both the undeniable importance of host patch occupancy and the ease of quantifying their prevalence and severity; therefore nearly all the studies of plant–pathogen metapopulation dynamics are on rusts (Burdon et al. 1995, Thrall et al. 2012), smuts (Carlsson and Elmqvist 1992, Alexander et al. 1996, Feurtey et al. 2016), powdery mildews (Laine and Hanski 2006), or downy mildews (Petrželová and Lebeda 2011).

While island archipelagos have provided elegant systems with which to study metapopulation dynamics, the concepts of local extinction, recolonization, and coupled, asynchronous local dynamics can be applied to any wild plant–pathogen system that shows spatial structure across a landscape—which is to say, any wild system!

Further reading

Madden, L. V., G. Hughes, and F. van den Bosch. 2007. *The Study of Plant Disease Epidemics*. APS Press, St. Paul, MN.

References

Alexander, H. M., P. H. Thrall, J. Antonovics, A. M. Jarosz, and P. V. Oudemans. 1996. Population dynamics and genetics of plant disease: A case study of anther-smut disease. *Ecology* **77**:990–996.

Augspurger, C. K. 1983. Seed dispersal of the tropical tree, *Platypodium elegans*, and the escape of its seedlings from fungal pathogens. *Journal of Ecology* **71**:759–771.

Betlejewski, F., D. J. Goheen, P. A. Angwin, and R. A. Sniezko. 2011. *Forest Insect & Disease Leaflet 131*. US Department of Agriculture, Forest Service, Portland, OR.

Bullock, J. M., K. Shea, and O. Skarpaas. 2006. Measuring plant dispersal: an introduction to field methods and experimental design. *Plant Ecology* **186**:217–234.

Burdon, J. J., L. Ericson, and W. J. Muller. 1995. Temporal and spatial changes in a metapopulation of the rust pathogen *Triphragmium ulmariae* and its host, *Filipendula ulmaria*. *Journal of Ecology* **83**:979–989.

Camacho, I., A. Góis, R. Camacho, and V. Nóbrega. 2018. The impact of urban and forest fires on the airborne fungal spore aerobiology. *Aerobiologia* **34**:585–592.

Carlsson, U. and T. Elmqvist. 1992. Epidemiology of anther-smut disease (*Microbotryum violaceum*) and numeric regulation of populations of *Silene dioica*. *Oecologia* **90**:509–517.

Connell, J. H. 1971. On the role of natural enemies in preventing competitive exclusion in some marine animals and in rain forest trees. In: P. J. de Voer and G. R. Gradwell, editors. *Dynamics of Numbers in Populations*, pp. 298–312. Proceedings of the Advanced Study Institute, Osterbeek, 1970, Centre for Agricultural Publication and Documentation, Wageningen.

Cressie, N. 2015. *Statistics for Spatial Data*. John Wiley & Sons, Hoboken, NJ.

Eggenberger, S., M. M. Diaz-Arias, A. V. Gougherty, F. W. Nutter Jr, J. Sernett, and A. E. Robertson. 2016. Dissemination of Goss's wilt of corn and epiphytic *Clavibacter michiganensis* subsp. *nebraskensis* from inoculum point sources. *Plant Disease* **100**:686–695.

Esker, P., A. H. Sparks, M. Bates, W. Dall'Acqua, E. Frank, L. Huebel, V. Segovia, and K. A. Garrett. 2007. Ecology and epidemiology in R: Modeling dispersal gradients. *The Plant Health Instructor*. doi: 10.1094/PHI-A-2008-0129-03.

Feurtey, A., P. Gladieux, M. E. Hood, A. Snirc, A. Cornille, L. Rosenthal, and T. Giraud. 2016. Strong phylogeographic co-structure between the anther-smut fungus and its white campion host. *New Phytologist* **212**: 668–679.

Fitt, B. D., P. Gregory, A. Todd, H. McCartney, and O. Macdonald. 1987. Spore dispersal and plant disease gradients; a comparison between two empirical models. *Journal of Phytopathology* **118**:227–242.

Gilbert, G. S., K. E. Harms, D. N. Hamill, and S. P. Hubbell. 2001. Effects of seedling size, El Niño drought, seedling density, and distance to nearest conspecific adult on 6-year survival of *Ocotea whitei* seedlings in Panama. *Oecologia* **127**:509–516.

Hanski, I. 1999. *Metapopulation Ecology*. Oxford University Press.

Janzen, D. 1970. Herbivores and the number of tree species in tropical forests. *American Naturalist* **104**:501–528.

Laine, A. L. and I. Hanski. 2006. Large-scale spatial dynamics of a specialist plant pathogen in a fragmented landscape. *Journal of Ecology* **94**:217–226.

Levins, R. 1969. Some demographic and genetic consequences of environmental heterogeneity for biological control. *American Entomologist* **15**:237–240.

Madden, L. V., G. Hughes, and F. van den Bosch. 2007. *The Study of Plant Disease Epidemics*. APS Press, St. Paul, MN.

Molloy, R., C. L. Smith, and A. Wozniak. 2011. *Internal Migration in the United States*. Finance and Economics Discussion Series. Divisions of Research & Statistics and Monetary Affairs, Federal Reserve Board, Washington, DC.

Petrželová, I. and A. Lebeda. 2011. Distribution of race-specific resistance against *Bremia lactucae* in natural populations of *Lactuca serriola*. *European Journal of Plant Pathology* **129**:233–253.

Rieux, A., S. Soubeyrand, F. Bonnot, E. K. Klein, J. E. Ngando, A. Mehl, V. Ravigne, J. Carlier, and L. D. L. de Bellaire. 2014. Long-distance wind-dispersal of spores in a fungal plant pathogen: estimation of anisotropic dispersal kernels from an extensive field experiment. *PloS One* **9**:e103225.

Thrall, P. H., A. L. Laine, M. Ravensdale, A. Nemri, P. N. Dodds, L. G. Barrett, and J. J. Burdon. 2012. Rapid genetic change underpins antagonistic coevolution in a natural host–pathogen metapopulation. *Ecology Letters* **15**:425–435.

Turchin, P. 1998. *Quantitative Analysis of Movement: Measuring and modeling population redistribution in animals and plants*. Sinauer, New York.

CHAPTER 12

Physiology and genetics

Gregory S. Gilbert and Ingrid M. Parker

12.1 Introduction

Plants contain all the sugars, nutrients, and water needed by most heterotrophic bacteria, fungi, and oomycetes. However, if all microbes could infect and consume all plants with abandon, our world would be decidedly less green. Instead, while plants have thousands of species in their microbiomes (Chapter 15), most microbes do not cause disease on most plants. Plants defend themselves with a broad array of physical and chemical traits that protect them from enemy attack. Plant pathogens, in turn, have an arsenal of traits to overcome or avoid defenses and manipulate the plant host to benefit themselves.

This chapter outlines the key chemical and physiological traits of plants and pathogens that determine whether a microbe is pathogenic on a particular host (Box 12.1), as well as the genetic basis for those traits and mechanisms of interactions. There is a direct line from the genes encoded in DNA to the transcription of those genes into mRNA, their translation into proteins, to the metabolites and structures that make up the plant or pathogen's phenotype (Figure 6.3, Table 12.1). The evolutionary ecology of plant disease recognizes that traits are the products of evolution and traits themselves shape evolution through ecological interactions, and we will take up the evolution and coevolution of these traits in Chapter 13. After a brief review of constitutive host defenses, we explore the arsenal of traits used by necrotrophic and biotrophic plant pathogens to overcome those defenses. We then look at two types of induced defenses: first, host defenses that act at the local scale and their underlying so-called gene-for-gene structure, and finally, systemic induced defenses, which allow hosts to respond to pathogens far from the point of infection or even in anticipation of attack.

Box 12.1 Some critical terminology in plant–pathogen interactions

Any discussion of the mechanisms and outcomes of plant–pathogen interactions is inevitably laden with much terminology. The specific definitions of the terms are critical to understanding the implications for the evolutionary ecology of the symbiosis, but unfortunately, the terms have historically been defined differently in different research traditions. In this book we use the term **pathogenicity** to mean the ability of a pathogen to cause disease on a particular host. This is a qualitative—yes/no—term. Following the conventions used broadly in human and veterinary disease ecology, **virulence** is a quantitative measure of the degree of damage a pathogen causes to the host plant. Increased virulence usually corresponds to a decrease in fitness of the host. A synonym of virulence is the **aggressiveness** of a pathogen. When a pathogen can reproduce on the host plant, the host is **competent** for that pathogen. When a pathogen can cause disease on a particular host, the host is **susceptible** to the pathogen; otherwise, it is **resistant**. When disease develops on the host but does not affect **host fitness**, the host is **tolerant** of the disease. In the context of gene-for-gene interactions (Section 12.6), the term virulence has traditionally been used differently, referring to the condition that arises when a pathogen does not produce a particular pathogen effector, permitting a **compatible interaction** and producing disease on a host with the corresponding *R*-gene. We are careful to specify that particular use of virulence when it comes up.

The Evolutionary Ecology of Plant Disease. Gregory S. Gilbert and Ingrid M. Parker, Oxford University Press. © Gregory S. Gilbert & Ingrid M. Parker (2023).
DOI: 10.1093/oso/9780198797876.003.0012

Table 12.1 Sequential structure of molecular components that connects the genotype to the phenotype of an organism

Level	Key characteristics
Genome	The entire collection of genes and non-coding DNA found in an organism, (composed of nucleic acids with bases ATCG); usually arranged in chromosomes. Variation in genomic DNA sequence between individuals within a species defines the genotype of an individual.
Transcriptome	A collection of messenger RNAs (mRNA; composed of nucleic acids with bases AUCG) transcribed from DNA as the first step in gene expression. Varies from cell to cell, tissue to tissue, and over time. Changes in response to environmental inputs.
Proteome	A collection of proteins (composed of amino acids) that are translated from the mRNA by ribosomes. Like the transcriptome, it varies from cell to cell, tissue to tissue, over time, and in response to environmental inputs.
Metabolome	Metabolites are all the small-molecule components of an organism, including chemical compounds produced by the actions of proteins. Primary metabolites are related to core functions of growth, including amino acids, sugars, phosphorylated compounds (ATP), and organic acids. Secondary metabolites such as phenolics, terpenoids, alkaloids, and glucosinolates have more specialized functions, often related to defense against enemies. The metabolome represents the organism's chemical phenotype and is critical in determining its morphological phenotype as well.

12.2 Constitutive defenses

Most living plants are protected from attack by most microorganisms. This so-called **non-host resistance** is provided mainly by **constitutive defenses**—defenses always present as part of a plant's normal morphology and chemistry. Physical constitutive defenses include structures like bark, cell walls, trichomes, and the waxy cuticle of leaves (Chapter 2). Plants bear the metabolic costs of producing constitutive defenses whether or not a pathogen is present.

Constitutive defense traits often have multiple critical functions for the plant. For instance, in addition to protecting leaves from fungal infection, the waxy cuticle provides structural support and protects against desiccation, UV irradiation, mechanical injury, insect herbivores, and extreme temperatures (Ziv et al. 2018). Because of their multiple functions, variation in traits like cuticle wax may reflect variation in environmental challenges, herbivory, or a range of other selective forces. Studying the specific influence of pathogens on the trait can be tricky. In addition, defense against pathogens may sometimes be a fortuitous side-effect (Gilbert et al. 2002).

Plant **allelochemicals** may serve as constitutive defenses. An allelochemical is a **metabolite** produced by one species that has detrimental effects on another. Plant allelochemicals can be active against neighboring plants, herbivores, or pathogens. The subset of low-molecular-weight, antimicrobial allelochemicals produced constitutively are sometimes called **phytoanticipins**, because they are produced prior to—in anticipation of—infection by potential pathogens (Pedras and Yaya 2015). Most of the thousands of known antifungal or antibacterial plant allelochemicals fall into one of three chemical classes of organic compounds: phenolics, terpenoids, and the nitrogen-containing alkaloids and cyanogenic glycosides (Wittstock and Gershenzon 2002).

12.2.1 Phenolics

Phenolics (or **phenols**) are a widely distributed group of metabolites that contain 6-carbon aromatic phenyl rings (Figure 12.1). Phenolics may be simple C_6 rings with assorted side groups, or molecules of multiple rings, such as flavonoids (with a C_6–C_3–C_6 structure) or stilbenoids (C_6–C_2–C_6) (Figure 12.1). Flavonoids include the anthocyanins, which provide the blue, red, and purple colors in flowers, fruits, and leaves. Stilbenoids as a group are categorized as defense compounds whose primary function is antimicrobial (Section 12.4). They include pinosylvin, which is produced in the wood of pine trees and is highly toxic to wood-decay fungi. Interestingly, stilbenoids play an important role in human medicine; the most well-studied example is resveratrol, which is used to treat several conditions including diabetes, cancer, high blood pressure, and coronavirus disease.

Long chains of repeating phenolic units form **lignins** $(C_6–C_3)_n$ or condensed tannins $(C_6–C_3–C_6)_n$, highly complex phenol polymers made by

phenol salicylic acid pinosylvin anthocyanin

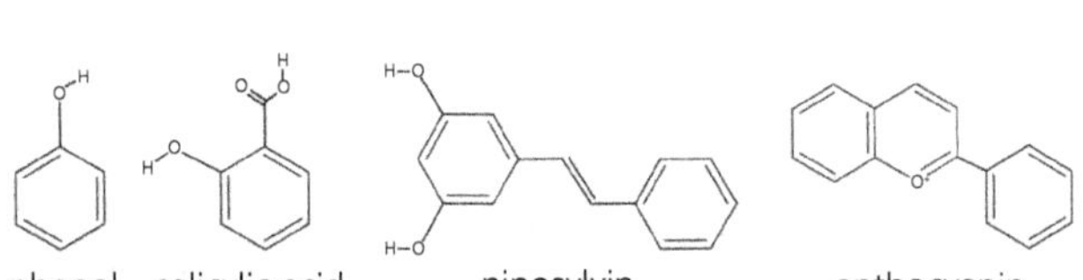

Figure 12.1 Basic structure of phenolics, the largest class of plant secondary metabolites. The basic unit of phenolics is phenol, a 6-carbon aromatic ring with a hydroxyl group (–OH). Salicylic acid, the precursor to aspirin (acetylsalicylic acid), is an important plant hormone and signaling compound. Pinosylvin is a stilbenoid toxin in pines that protects heartwood from fungal decay. Flavonoids such as anthocyanin are polyphenols with many physiological functions as well as roles in defense.
Figure by Gregory S. Gilbert, created with Biorender.com.

isoprene linalool sabinene caryophyllene

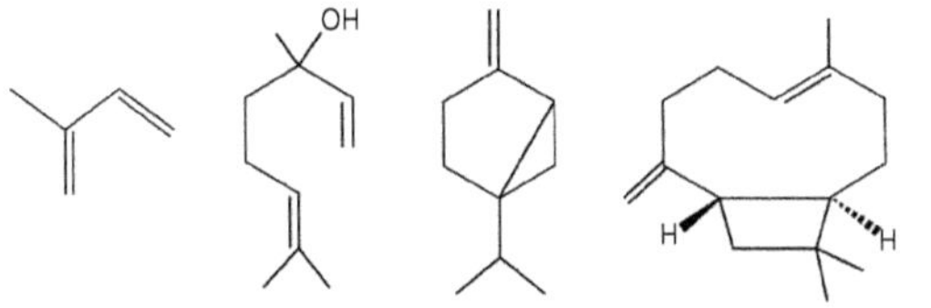

Figure 12.2 Basic structure of terpenoids. The basic unit is isoprene (C_5H_8), an unsaturated, 5-carbon molecule. Linalool is a terpene alcohol found in coriander and lavender. Sabinene is a monoterpene ($C_{10}H_{16}$) found in the essential oils of oaks (*Quercus* spp.), spruce (*Picea* spp.), and bay laurel (*Laurus nobilis*). The sesquiterpene caryophyllene ($C_{15}H_{24}$) is a component of the essential oils of cloves (*Syzygium aromaticum*), *Cannabis*, and hops (*Humulus lupulus*).
Figure by Gregory S. Gilbert, created with Biorender.com.

cross-linking many phenolic precursor molecules. A variety of mechanisms explain the antimicrobial activity of phenolics. Many phenolics bind to components of cell walls (e.g., ergosterol) or membranes of microbes, causing intracellular contents to leak out; others may interfere with the activities of key cellular enzymes.

12.2.2 Terpenoids

Terpenes, or **terpenoids**, are also called **isoprenoids** because they are polymers of the basic unit isoprene (Figure 12.2). Terpenes with two isoprene units are called monoterpenes (C_{10}); with three units, sesquiterpenes (C_{15}), with four units, diterpenes (C_{20}), and so forth. Paradoxically, terpenes are simultaneously uniform and diverse. Every monoterpene has the same chemical formula $C_{10}H_{16}$, and all diterpenes are $C_{20}H_{32}$. But **isomerism** allows these simple components to be arranged in a wide variety of orders and conformations (e.g., chains versus rings), creating molecules with differing properties and activities. Monoterpenes and sesquiterpenes are **volatiles** that contribute the strong scents to essential oils; they are abundant in wood and bark, and they produce the distinctive smells of citrus and cannabis and the sharp flavor of black pepper. Terpenes have many impacts on microbes, including disrupting membranes and cell walls, affecting the activity of fungal mitochondria, interfering with quorum sensing, and inhibiting critical efflux pumps that maintain osmotic balance in bacteria (Nazzaro et al. 2017).

12.2.3 Nitrogen-containing alkaloids and glucosinolates

Two groups of nitrogen-containing metabolites are alkaloids and glucosinolates. **Alkaloids** are a diverse group of heterocyclic organic compounds; "heterocyclic" because they include at least one other type of atom besides carbon in the ring. In alkaloids, the rings include at least one nitrogen atom (Figure 12.3). We are familiar with alkaloids because of their strong effects on the mammalian nervous system; many alkaloids, including caffeine, nicotine, morphine, mescaline, and heroin, have played important roles in human medicinal and cultural practices. Some alkaloids, including sanguinarine, a compound in bloodroot (*Sanguinaria canadensis*) and Mexican prickly poppy (*Argemone mexicana*), show antimicrobial properties. These may be important in plant disease resistance. In the future, alkaloids may contribute novel approaches to chemical disease control (Almadiy and Nenaah 2018).

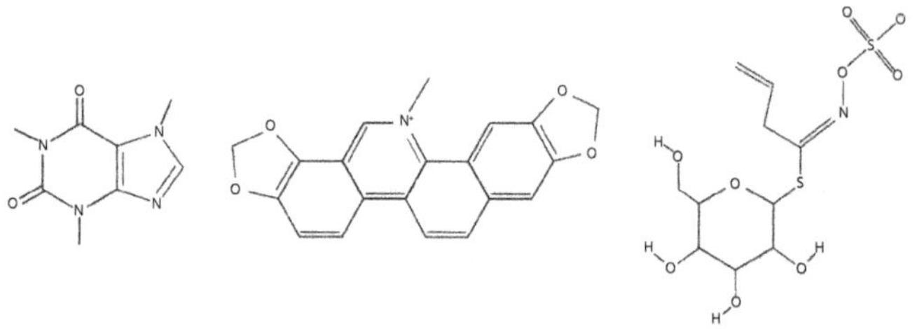

caffeine sanguinarine sinigrin

Figure 12.3 Alkaloids include ring structures that contain at least one nitrogen atom in addition to carbon. Alkaloids can be simple like caffeine or more complex like sanguinarine. Glucosinolates (e.g., sinigrin) include both nitrogen and sulfur in their structures; glucosinolates are what give crops in the Brassicaceae their pungent flavor.
Figure by Gregory S. Gilbert, created with Biorender.com.

Cyanogenic glycosides or **glucosinolates** are organic compounds that include both nitrogen and sulfur (Figure 12.3). They are prominently produced by several families in the order Brassicales (mustard, caper, and papaya families). Glucosinolates are what provide the sharp, distinctive taste to radishes and cabbages. When glucosinolates are hydrolyzed, usually because cell damage leads to mixing with the enzyme myrosinase, they create **isothiocyanates**, also called "mustard oils." In addition to repelling herbivores, mustard oils have potent antimicrobial properties, causing microbial cell membranes to leak cellular contents. Mustard oils have been used to effectively control soil-borne plant pathogens and are being explored as an organic alternative to the soil fumigant methyl bromide (Chapter 17).

12.3 Pathogen arsenal

Pathogens have evolved numerous traits that allow them to avoid or breach constitutive defenses in plants. Some of these are behavioral, such as when fungi, bacteria, and oomycetes take advantage of natural openings in the cuticle—stomata and wounds—to access the interior of the plant (Chapters 2 and 15). Many fungi and oomycetes produce specialized structures such as **appressoria** to penetrate the plant. The appressorium firmly attaches the pathogen to the plant surface and produces a thin hypha called a **penetration peg**, which focuses all the turgor pressure from the appressorium to a small point, perforating the cuticle.

As osmotrophic heterotrophs, plant-associated microbes make a living by secreting chemicals into their living substrate (like leaves or roots) to modify it for their consumption. They secrete extracellular enzymes that break down complex organic molecules into simpler organic and mineral components; these are then taken in through passive or active transport mechanisms (Chapters 3–5). In addition to utilizing extracellular enzymes for feeding their metabolism, microbial plant pathogens use those enzymes to modify their substrate to make it more accessible and hospitable for themselves, likely at the expense of the host plant. Fungi produce a broad array of **cellulases**, **pectinases**, and **proteinases** important in obtaining nutrition. Such enzymes also enable plant pathogens to disrupt the structures of plant tissues and effectively colonize the host. Some fungi produce **laccases**, oxidase enzymes that break down phenolic compounds such as lignin, allowing safe access to more digestible plant components like cellulose and hemicellulose (Rivera-Hoyos et al. 2013). On leaves, fungal appressoria secrete **cutinases** and **lipases** that break down cuticular waxes, facilitating the entrance of the penetration peg into the leaf interior.

Some biotrophic pathogens also push through the cell wall of living host cells using enzymatic softening and physical force, but they need to extract nutrients from the host cell without killing it. They invaginate the plant cell membrane, developing intimate connectivity to the plant cell cytoplasm with a **haustorium**. For a biotroph to be successful, it must be able to avoid host defenses by suppressing them or evading them, and to avoid severely damaging the host cells on which it depends.

A necrotrophic pathogen, on the other hand, has a different job to do. Necrotrophs actively kill plant tissue and then consume the resources released. Necrotrophic pathogens have a wide range of tools to actively kill and then consume host cells. These include **non-host-specific toxins** that interfere with basic cellular functions such as membrane synthesis or mitochondrial function common to all plants. There are also **host-specific toxins** that interact with specific host components; only hosts with the appropriate receptor are susceptible to the disease. Curiously, nearly all the necrotrophic pathogens identified as producing host-specific toxins are in the ascomycete order Pleosporales (Friesen et al. 2008). A host-specific toxin may be translocated through diffusion or through the vascular system, weakening or killing plant cells in advance of colonization by the pathogen. The spread of the toxin produces symptoms of chlorosis and necrosis and releases cellular contents into the apoplast, where they are readily consumed by the pathogen as it spreads through the host tissue.

12.4 Inducible defenses

Once the plant experiences pathogen attack, it responds with a constellation of induced defense mechanisms, including chemical defenses (Figure 12.4). In contrast to the phytoanticipins, which are produced constitutively, induced chemical defenses are produced *de novo* (of the new); that is, host gene expression is turned on in response to induction by the pathogen.

The term **phytoalexin** is used to describe hundreds of low-molecular-weight antimicrobial compounds produced *de novo* by plants in response to an interaction with a potential pathogen (Jeandet et al. 2014). Phytoalexins are chemically diverse, largely distributed across the same chemical classes as the constitutive defenses: phenolics, terpenoids, and alkaloids. Different evolutionary clades of plants tend to produce different kinds of phytoalexins (Jeandet et al. 2014). For instance, the phytoalexins produced by legumes (Fabaceae) are mostly flavonoids or stilbenoids (phenolics); plants in the nightshade family (Solanaceae) produce sesquiterpenoids (terpenoids), glycoalkaloids (alkaloids), phenylpropanoids (phenolics), and polyacetylene chains; grasses (Poaceae) make diterpenoids (terpenes) and flavonoids (phenolics); and mustards (Brassicaceae) produce nitrogen-containing indole alkaloids. Somewhat confusingly, some defensive compounds that are constitutive are also induced, being produced in larger quantities or in specific places after pathogen attack. This makes them both phytoanticipins and phytoalexins.

Phytoalexins kill, deform, or inhibit growth of cells of fungi, oomycetes, and some bacteria. They do this through a variety of mechanisms that include damaging membranes (which makes cells leaky), disrupting mitochondrial function, disorganizing cytoplasm, and blocking cell respiration.

Constitutive defenses	**Local induced defenses**		**Systemic induced defenses**
Bark Cuticle Cell walls Anticipins •Phenolics •Terpenoids •Alkaloids •Glucosinolates	**Pattern-triggered immunity** (RLK-LRR receptors)	**Effector-triggered immunity** (NBS-LRR receptors)	**Signaling pathways:** Salicylic acid (biotrophs) Jasmonic acid (necrotrophs) Ethylene
	Phytoalexins ROS Tyloses Callose Hypersensitive reaction		**Systemic acquired resistance** (initiated by PTI and ETI) **Induced systemic resistance** (initiated by beneficial microbiome)

Figure 12.4 Plants are defended against attack by most microorganisms through a variety of defenses. Constituent defenses include physical and chemical structures (anticipins) formed by plants regardless of attack from pathogens. Constituent defenses are broadly effective against attack by pathogens of all life-history strategies, and so are especially responsible for nonhost resistance (Section 12.2). In contrast, inducible defenses are made in response to the presence of pathogens near host cells. Pattern-triggered immunity (Section 12.5) induces *de novo* production of phytoalexins and other defenses with broad antimicrobial activity. Effector-triggered immunity (Section 12.5) responds with similar defenses, and stimulates a strong and localized cell death process called the hypersensitive response.

In addition to phytoalexin production, plants respond to infection and damage through a variety of other, less-specific defense mechanisms. When a pathogen is detected in the apoplast, there is a rapid flux of ions into and out of the plant cells; hydroxide and potassium ions move out of the cell, while calcium and hydrogen ions move in. The influx of calcium ions triggers cellular responses, including activation of plasma-membrane-localized enzymes that produce superoxide ($^{\bullet}O^{-2}$), which is quickly converted to hydrogen peroxide (H_2O_2), and hydroxyl radicals ($^{\bullet}OH$). Collectively, these molecules are referred to as **reactive oxygen species (ROS)**. These damage pathogen cellular membranes through chemical reactions with lipids. The cells also produce the glucan **callose**, which is a chain of glucose molecules bound by β-1,3 linkages (recall that cellulose is a chain of glucose molecules bound by β-1,4 linkages). Callose is deposited between the cell membrane and the cell wall to reinforce cells and seal off plasmodesmata, slowing pathogen spread. Induced plants lignify affected tissues. They close stomata to prevent the influx of bacteria. They shed infected leaves and parts of leaves. Woody species form wound callus (cankers), create new periderm (corky tissue on the outside of stems), and form tyloses (outgrowths of parenchyma) to plug up damaged and infected xylem vessels.

An important induced defense mechanism is the **hypersensitive response (HR)** (or hypersensitive reaction), which is rapid, localized host cell death around the site of pathogen penetration (Figure 12.5). The cell death is not a result of the pathogen directly killing the host cells, but instead of the host killing its own cells through **programmed cell death**, or **apoptosis** (Balint-Kurti 2019). The basic idea is that when the plant detects an invading pathogen, it quickly kills the host cells at the site of infection, preventing the pathogen from further advance. The HR is similar to other induced defenses in that detection of the pathogen leads to the production of ROS, but in this case the target of the ROS is the plant's own cellular membranes. The chemical reactions kill the plant cells and trigger the production of lignin and callose, which, together with cross-linking of pre-existing cell components, create barriers of reinforced cell walls surrounding the infection, helping to limit further progress

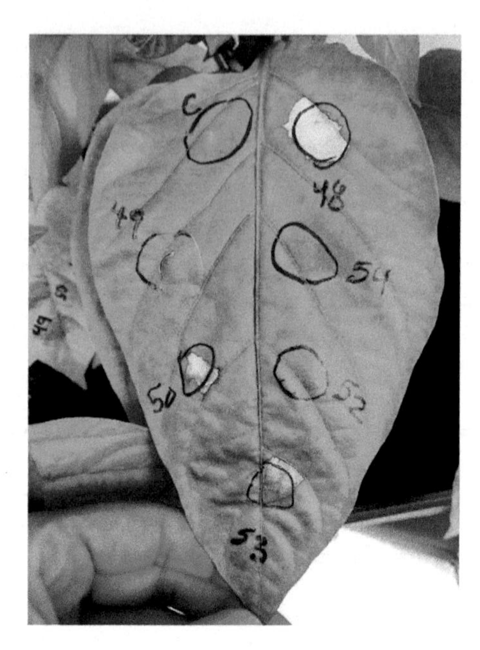

Figure 12.5 Responses to inoculations of different strains of the bacterial pathogen *Xanthomonas campestris* pv. *vesicatoria* on a leaf of bell pepper (*Capsicum annuum*) showing the rapid cell death that characterizes hypersensitive response to strains 48, 50, and 53, but not to other strains or the control.
Photo by Alejandra Huerta.

by the pathogen. This defense is mainly effective against biotrophic pathogens, for obvious reasons! Later we consider how necrotrophs can sometimes take advantage of the hypersensitive response for their own benefit.

For all these inducible defenses to work, the plant must first recognize the pathogen, and then produce a signal to set defenses in action. How does that work?

12.5 How do plants recognize pathogens?

Plants do not have antibody-based immune systems, nor do they have circulating immune cells. They instead rely on the **innate immunity** of each individual cell, coupled with systemic signaling systems that coordinate defense responses (Jones and Dangl 2006). The plant immune system allows plants to recognize potential pathogens and respond with a series of defense mechanisms

Table 12.2 Common acronyms for elements of molecular plant–pathogen interactions

Receptors		
PRR	Pattern recognition receptors	Cell surface receptors that detect ligands listed below
RLK	Receptor-like kinase	Type of transmembrane PRR with a kinase domain
NLR	Nucleotide binding leucine-rich repeat	Intracellular receptors that detect pathogen effectors
LRR	Leucine-rich repeat	Ligand-binding domain of NLRs and many PRRs
NBS	Nucleotide binding site	Oligomerization domain of the NLRs
CC	Coiled-coil domain	Signaling domain of some NLRs
TIR	Toll-Interleukin-1 receptor	Signaling domain of some NLRs
Ligands		
MAMP	Microbe-associated molecular pattern	Microbe cellular product, also used for mutualists
PAMP	Pathogen-associated molecular pattern	Pathogen cellular product, same as MAMP
DAMP	Damage-associated molecular pattern	Plant product associated with infection or damage
HAMP	Herbivore-associated molecular pattern	Cellular product from herbivorous animal
Plant responses		
PTI	Pattern-triggered immunity	Plant detects microbe cellular product in apoplast
ETI	Effector-triggered immunity	Plant detects effector within plant cell
ETS	Effector-triggered susceptibility	Necrotrophic pathogen exploits HR
HR	Hypersensitive response	Programmed death of host cells in response to ETI
ROS	Reactive oxygen species	Oxidizing chemicals that damage cell membranes

that protect against disease development. There are two key layers of pathogen recognition that differ in the specificity of responses. The first layer is called **pattern-triggered immunity (PTI)**, in which **pattern recognition receptors (PRR)** recognize chemical attributes of invading pathogens, called **microbe-associated molecular patterns (MAMPs)**, and induce defense responses that confer broad-spectrum immunity to infection. However, pathogens have evolved many mechanisms for evading receptors and blocking signal transduction to suppress PTI. Suppression of signal transduction is accomplished via **effector** proteins that are translocated from pathogens into host cells, where they bind to or modify specific host proteins. To combat this process, plants have evolved intracellular receptors that can detect specific pathogen effectors, or the modifications induced by these effectors. Activation of these receptors leads to induction of **effector-triggered immunity (ETI)**, which typically includes a hypersensitive response.

We said that plants do not have an adaptive immune response (antibodies and T-cells), but they do have to ward off infection from hordes of genetically distinct potential pathogens. How do they do this? Host genomes include hundreds of genes that code for receptor proteins, each with conserved signaling regions complemented by highly variable regions. The variable regions correspond to and recognize the MAMPs and effector proteins that are products of corresponding genes in the pathogens.

OK, now that you have the overview and a lot of new acronyms (Table 12.2), let's look at each component and how they work together to shape induced responses of plants to pathogen attack. As you read this section, remember that research into the molecular mechanisms and evolution of plant immune responses is proceeding with great speed, and the details are constantly being revised. For example, just a few years ago PTI and ETI were considered quite distinct plant immune responses, but we now know that they are integrated with overlapping elements and linkages (Wang et al. 2022). This has been called "the golden age of PTI and ETI research" (McDowell 2019); exciting new discoveries are on their way!

12.5.1 Pattern-triggered immunity

The plant innate immune system is based on the plant being able to recognize molecules or parts of molecules ("molecular patterns") associated with potential pathogens and respond with PTI in ways that prevent disease development (Saijo et al. 2018).

Groups of microbes have shared chemical characteristics that differentiate them from plants. For example, fungi all produce chitin as a component of their cell walls, and nearly all flagellated bacteria produce flagellin. These chemical components and structures are recognizable **pathogen-associated molecular patterns (PAMPs)** (Boutrot and Zipfel 2017). Some researchers prefer the more general term **microbe-associated molecular pattern (MAMP)** instead of PAMP, so that is what we have adopted but keep in mind that these are simply alternative names for the same thing. **PTI** also (and originally) stands for "**pathogen-triggered immunity**" or "**PAMP-triggered immunity**," but "pattern-triggered immunity" is a convenient, all-inclusive term. PAMPs/MAMPs are **ligands** (molecules that irreversibly bind to receptor proteins), and plants have **pattern recognition receptors (PRRs)** that bind to the ligands. Sometimes, rather than a compound associated with the microbe itself, the PRR binds to a compound generated by the plant as a result of pathogen damage; in this case the ligand is called a **damage-associated molecular pattern**, or **DAMP.**

A critical type of PRR is the **receptor-like kinase (RLK)** (Figure 12.6). RLKs are transmembrane proteins that detect potential pathogens in the apoplast and transmit a signal to the cell cytoplasm, leading to the induction of defense responses (Jamieson et al. 2018). The receptor-like kinase has three domains. First is an outward-facing N-terminal domain that extends into the apoplast and binds to PAMPs, allowing the plant to recognize the presence of the microbe. The most common type of N-terminal domain is a **leucine-rich repeat (LRR)** motif (containing variable numbers of repeats of the amino acid leucine). LRR domains are commonly associated with many cellular functions involving protein–protein or protein–carbohydrate interactions. They are highly variable, and the particular structure of the LRR determines which ligands it will bind to. The second domain in the PRR molecule crosses the plant membrane from the apoplast to the cytosol. Facing inward from the membrane into the cytosol is the third domain, a kinase that triggers a **phosphorylation cascade** when the associated LRR binds to its ligand. The phosphorylation cascade results in immune responses including the production of the phytoalexins, ROS, callose, tyloses, and other induced responses. This is pattern-triggered immunity (PTI).

The PTI response mediated by LRR-RLK receptors is the first layer of innate, induced, immunity. It is effective early in plant-microbe interactions because it detects pathogens even before they have attacked a plant cell. The resulting defense responses are protective against many types of pathogens. Unlike constitutive defenses, these responses are activated only in the presence of

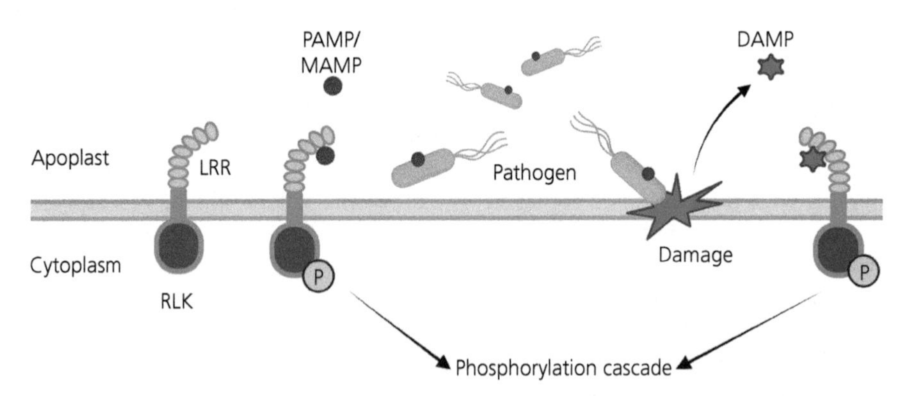

Figure 12.6 Pattern-triggered immunity. LRR-RLK receptors are plant pattern recognition receptors (PRR) with a leucine-rich repeat domain in the apoplast. The LRR recognizes microbe-associated molecular patterns (MAMPs, also called PAMPs, for pathogen-associated molecular patterns), or else products of plant cell damage caused by the pathogen (DAMPs). When an LRR-RLK receptor recognizes one of these ligands, it initiates a phosphorylation cascade in the cell, leading to induced defenses that together are called pattern-triggered immunity, or PTI.
Figure by Ingrid M. Parker, created with BioRender.com.

potential pathogens, and so are potentially less costly to the plant (Section 13.5).

12.5.2 Pathogen effectors and effector-triggered immunity

Effectors are pathogen proteins that manipulate host cellular processes in a way that undermines host immunity and promotes pathogen colonization (Figure 12.7). This is in contrast to the more generic PAMPs/MAMPs and DAMPs detected in the pattern-triggered immunity (PTI) described above. Pathogen effectors can be found in the apoplast but are often active within the cytoplasm of a compromised host cell, where they are busy manipulating the host's physiology. Effectors can have many different functions, including blocking host receptors from recognizing PAMPs, interfering with host gene expression, detoxifying host defenses, and damaging host cells (Wang et al. 2022).

When the plant detects one of these effectors and mounts an immune response, it is called effector-triggered immunity (ETI). Effectors may be detected directly, or indirectly through their modification of a structure in the host cell. Effectors in the apoplast may be detected by the same type of LRR-RLK receptors that are responsible for PTI. But effectors in the cytoplasm are detected by different types of receptors, the **nucleotide-binding leucine-rich repeat receptors (NLR)**, previously called NOD-like receptors or NBS-LRR. They all have a **nucleotide binding site (NBS)** as well as a C-terminal LRR region. Just like the LRR-RLK receptors described above, the LRR domain of the NLR receptors is hypervariable and binds to effector molecules, triggering a cascade of signals and responses. The NBS domain is the regulatory part of the receptor molecule and binds ATP/ADP or GTP/GDP to regulate oligomerization of the NLR protein into a wheel-shaped structure that contains four or five subunits. Finally, NLR receptors fit into two major

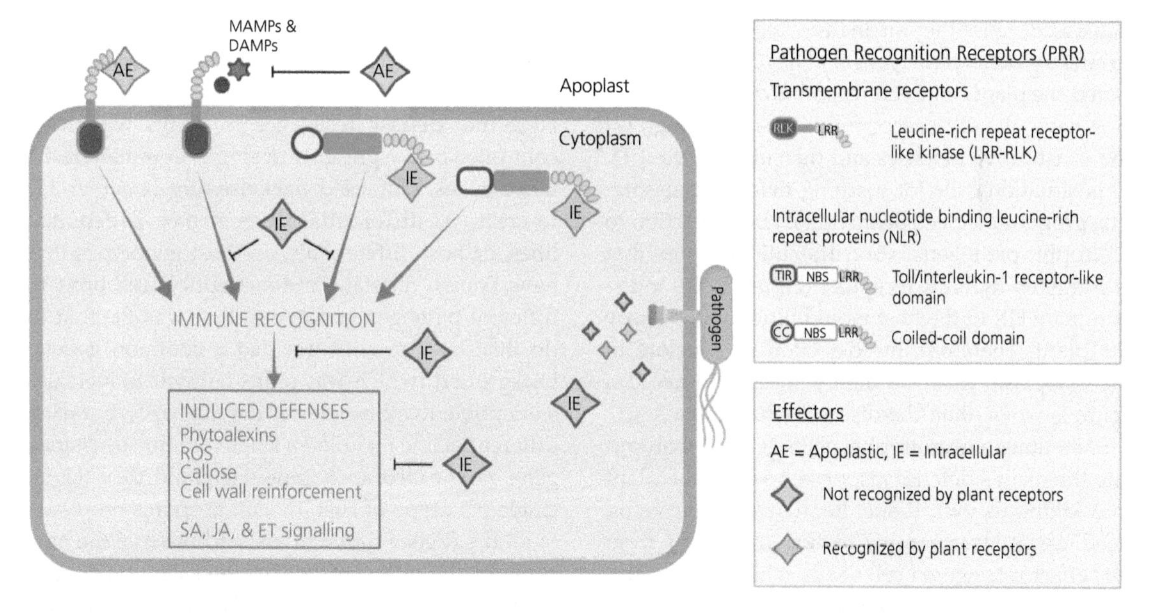

Figure 12.7 Mechanisms of plant immunity and mechanisms by which pathogens evade plant immunity. In the apoplast, transmembrane pathogen recognition receptors (PRRs) recognize microbe-associated molecular patterns (MAMPs) or damage-associated molecular patterns (DAMPs), resulting in a signal transduction cascade leading to induced defenses. Pathogen effectors (proteins directly involved in undermining host immunity) located in the apoplast (AE) may interfere with the recognition function of the receptors. Once a pathogen penetrates the cell and passes effectors into the cytoplasm (IE), those effectors may interfere with steps in the signal transduction cascade, or with any individual component of the induced defense response. Recognition of intracellular effectors occurs via nucleotide-binding leucine-rich repeat receptor (NLR) proteins.
Graphic by Ingrid M. Parker, created with BioRender.com; based on Wang et al. 2022.

classes that differ in a third, N-terminal component, with either a **Toll-interleukin-1 receptor (TIR)** or a **coiled-coil (CC)** domain (Figure 12.7). The TIR and CC domains activate downstream signaling events by distinct mechanisms, but both lead to a rapid influx of calcium ions from the extracellular space that then induce multiple signaling steps, including activation of **transcription factors**.

The domains of the RLK-LRR and NLR receptors have ancient origins and play important roles in signaling across both eukaryotes and prokaryotes. The fusion of NBS and LRR domains into functional NLR receptors first occurred in early land plants. In addition, TIR and CC domains are also broadly conserved across lineages of plants, and the TIR domain shows strong homology with TIR in animals. (Curiously, TIR has been lost from the grass clade *Poales* (Pan et al. 2000), but that's a story for another day.) All these elements are general-purpose structures for signaling based on protein–protein interactions, and they have been put to a wide variety of uses throughout evolution (Dangl and Jones 2001).

The NLR receptors are the plant's third line of defense. Because they are in the cytoplasm and bind to effectors from pathogens that have already penetrated the plant cell, NLR receptors detect microbes that have already successfully passed through all the constitutive defenses and then evaded the PTI. This situation calls for a strong defense response. The programmed cell death of the HR in reaction to biotrophic pathogens is one dramatic response that is triggered by NLR receptors (Figure 12.5). Indiscriminate HR in the host would ultimately destroy the plant. Therefore, the detection of effectors in the cytoplasm must be highly specific to avoid a "cure-is-worse-than-the-disease" situation.

Sometimes a necrotrophic pathogen can manipulate the plant's defense response, tricking the plant into killing its own tissue. Because they thrive on dead tissue, necrotrophic pathogens benefit from this effector-triggered cell death, which in this context is called **effector-triggered susceptibility** (Faris and Friesen 2020). For instance, the host-specific toxin victorin produced by *Cochliobolus victoriae* is recognized by the same CC-NBS-LRR protein that triggers a hypersensitive response in oats (*Avena sativa*) to the biotrophic rust *Puccinia coronata*. In the case of *C. victoriae*, the resulting programmed cell death facilitates infection, rather than blocking it. The necrotrophic pathogens are essentially hijacking plant defense mechanisms.

12.6 Gene-for-gene interactions

Whereas our understanding of the molecular mechanisms of plant immunity is still new and evolving, plant pathologists have been studying the genetic architecture of host resistance for over 80 years. Furthermore, that knowledge has been used in crop breeding for nearly as long. We now dive into a bit of that history and consider the phenomenon of plant resistance from a traditional, gene-for-gene perspective; then we show how a gene-for-gene perspective maps onto the PTI/ETI framework.

The gene-for-gene hypothesis is the idea that for every gene in a pathogen that determines pathogenicity, there is a corresponding gene in plants that determines resistance. It was first articulated by Harold H. Flor, after years of painstaking, groundbreaking work on breeding for plant resistance (Flor 1942, 1955). Flor studied the inheritance of pathogenicity of the rust *Melampsora lini* on flax (*Linum marginale*). Drawing on the knowledge that disease resistance in plants was often controlled by the presence of single, dominant resistance genes, Flor used **backcrossing** (Chapter 17) to create 32 **differential lines** of flax. Differential lines, or host differentials, are host genotypes that have known resistant and susceptible reactions to different pathogen genotypes. Flor was the first to do this; his flax cultivars had a common genetic background (which was a flax cultivar universally susceptible to all isolates of the flax rust), but each differential line included a single, unique resistance gene, now called an *R* gene. He could then take a single genotype of rust, inoculate spores on plants of all the *R* gene differentials, and observe the outcome of the interactions. The rust genotypes that caused disease on a differential were considered **virulent**, and those that did not were considered **avirulent**. The pattern of host differentials on which the rust was virulent or avirulent determined what Flor called the **race** of the pathogen (Figure 12.8a). Flor took advantage of the sexual reproductive cycle of the flax rust to make crosses between virulent

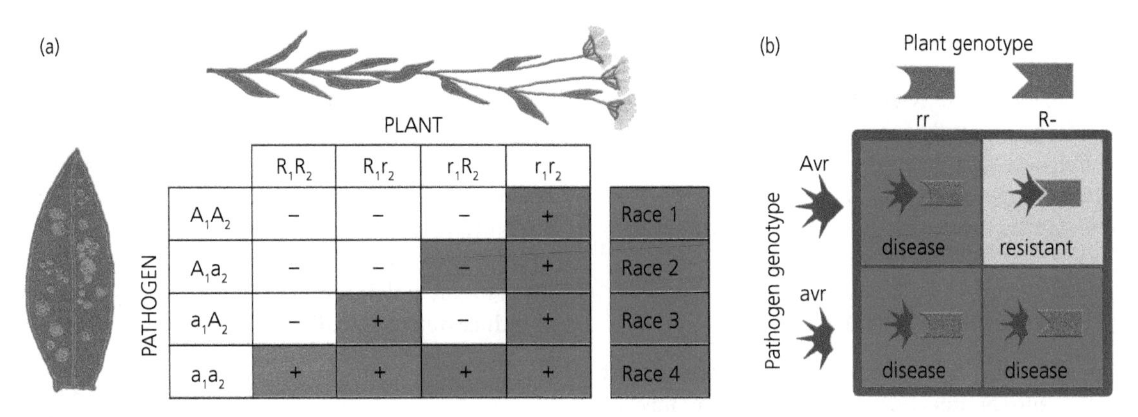

Figure 12.8 Gene-for-gene interactions. (a) As determined by Flor for flax and flax rust, the race of a pathogen is determined by the pattern of gene-for-gene interactions between avirulence alleles in the pathogen and resistance alleles in the host. Here a pathogen population has two avirulence loci (A_1 and A_2), each with two alleles (dominant A_1 and A_2 produce avirulence factors 1 and 2 that correspond to plant resistance genes R_1 and R_2). Pathogens with recessive alleles a_1 and a_2 are not recognized by plants with R_1 or R_2, respectively; plants with recessive alleles r_1 and r_2 do not recognize either pathogen genotype for the corresponding *A* locus. Each combination of alleles at the two avirulence loci produces a distinct pattern of compatibility/incompatibility on different host genotypes and is designated as a distinct pathogen race. (b) Gene-for-gene interactions as understood via the effector-triggered immunity (ETI) system. Pathogen effectors encoded by avirulence (*Avr*) alleles are recognized by NLRs encoded by dominant (*R*) host resistance alleles. Recognition of the pathogen effector by the host receptor represents an incompatible reaction, leading to host ETI and resistance to the pathogen. If the pathogen does not express the avirulence effector, it escapes detection by the host regardless of whether the *R* allele is present, leading to a compatible reaction and susceptible host.
Graphic by Ingrid M. Parker.

and avirulent genotypes, and he analyzed the patterns of virulence/avirulence in the progeny. These consistently showed 3:1 ratios of avirulent:virulent offspring, showing that avirulence was controlled by a dominant allele (*Avr*), with virulence as the recessive allele *avr*.

(At this point, you might be saying: "Wait a second! I thought virulence was a quantitative measure of the degree of damage caused on the host plant" (Box 12.1). Well, this is that special-case use of the term virulence. It comes from Flor's original publications, and we are stuck with it because of historical precedence.)

Flor's work on flax rust showed that each allele at an *R* locus corresponds to an allele for an *Avr* gene in the pathogen. For a plant disease that shows gene-for-gene relationships, different genotypes within the host population have different combinations of resistance alleles, and different genotypes within the pathogen population have different combinations of avirulence alleles. When the host has any *R* allele that matches an *Avr* allele in the pathogen, the host can recognize that pathogen and induce a defense response. If none of them match, the interaction is compatible, disease occurs, and the host is susceptible to the (virulent) pathogen (Figure 12.8a).

So how does Flor's discovery of gene-for-gene relationships between plants and their pathogens map onto what we understand today about plant immune systems? We now know that only some plant diseases show gene-for-gene interactions. These tend to be interactions in which intracellular effectors are recognized by plant NLR receptors, and these effectors have been called "avirulence effectors" (Wang et al. 2022). The avirulence effector is encoded by an avirulence (*Avr*) allele in the pathogen genome. The corresponding NLR protein in the host is encoded by an *R* gene. As Flor discovered, gene expression of *R* genes is dominant, so if a diploid plant has the dominant allele for a particular *R* gene (either homozygous or heterozygous) it will produce that particular variant of the NLR, and only doubly-recessive genotypes will lack the NLR (Figure 12.8b). Avirulence is also dominant, but because most fungi and bacteria are haploid, an allele for a corresponding *Avr* product will be expressed by default, leading to avirulence. Those pathogen genotypes with an allele that produces a slightly different variant of the effector that is not recognized by the *R*-gene product are then virulent. Don't forget that the pathogen must also have all the necessary traits to evade constitutive and pattern-triggered immunity before the *R*-gene/*Avr*-gene pairing determines virulence or avirulence.

12.7 Systemic induction

Plants are not limited to activating defenses right at the site of infection; they also have mechanisms for mounting systemic responses. From the site of attack by a pathogen (or herbivore), signals are carried to distant stems, leaves, and roots and trigger signal transduction pathways and induced defenses. This is called **systemic acquired resistance (SAR)**. The spread of signals throughout the plant body allows the plant to mount defenses in advance of pathogen arrival. Plant hormones (Section 2.7) are these signals. There are three different signaling pathways involving three different plant hormones: salicylic acid, jasmonate, and ethylene.

12.7.1 Three signaling pathways

The primary signaling pathway for SAR in response to biotrophic or hemibiotrophic plant pathogens is the **salicylic acid (SA) pathway** (Figure 12.9). In this pathway, PTI or ETI defenses stimulate the production of the metabolite N-hydroxy pipecolic acid (NHP), which induces the transcription of enzymes that synthesize salicylic acid, which then accumulates in high concentrations. At the same time, induction of the production of glycerol-3-phosphate (G3P) leads to accumulation of that molecule. A set of receptor proteins called **NPR proteins** (for Nonexpressor of Pathogenesis-Related Genes, e.g., NPR1, NPR3, and NPR4) are activated by salicylic acid (and sometimes directly by NHP), and then regulate the transcription of pathogenesis-related genes whose products have antimicrobial activities and serve as local defenses against the pathogens (Hartmann et al. 2018). The high concentration of G3P induces the expression of **small interfering RNAs (siRNAs)** from the *Tas3a* gene (Shine et al. 2022), which is found broadly across the plant kingdom. The siRNAs are readily transported through the plant to organs distant from the attack, where they serve as signals to induce the production of SA and G3P. This again leads to a buildup of salicylic acid in the distant cells, which in turn interacts with the NPR receptor proteins to express SA-dependent defense genes. In addition, a methylated version of SA is highly volatile, and can move through the air to induce defenses in distant parts of the same plant or in neighboring plants (Gao et al. 2021).

SA-dependent defenses are mostly responsive to pathogens that thrive for some time in the plant apoplast before causing tissue damage or cell death, including biotrophic pathogens such as downy mildews, powdery mildews, and the bacterial pathogen *Pseudomonas syringae*. While SA-dependent systemic induction is usually initiated by pathogen infection, some beneficial microbes (like the plant-growth promoting rhizobacteria (PGPR) *Pseudomonas fluorescens*; Sections 15.5 and 17.5) can also induce host defenses against pathogens through the salicylic pathway (Pieterse et al. 2014).

Most systemic responses to necrotrophic pathogens, as well as herbivores, involve the **jasmonic acid (JA) pathway** (Figure 12.9). Jasmonic acid, or jasmonate, builds up in cells in response to DAMPs, **herbivore-associated molecular patterns (HAMPs)**, pathogen effectors, and sometimes metabolites of beneficial organisms in the plant microbiome. Jasmonate activates two receptor proteins (COI1; Coronatine Insensitive 1 and JAZ; Jasmonate Zim) that cause the degradation of JAZ proteins that block transcription factors. The removal of JAZ proteins is a kind of "double negative" that creates a defense response. In fact, many defense responses in cells are normally prevented by the presence of **repressors** such as JAZ proteins. Jasmonate induces the destruction of JAZ proteins, which then allows the expression of defenses that had been repressed (a process called **derepression**). Unlike salicylic acid, jasmonate itself serves as a long-distance signal in plants. Jasmonate transported to distant parts of the plant then derepresses defense genes, leading to SAR against necrotrophs. Some PGPR are also able to induce host defenses through the JA pathway.

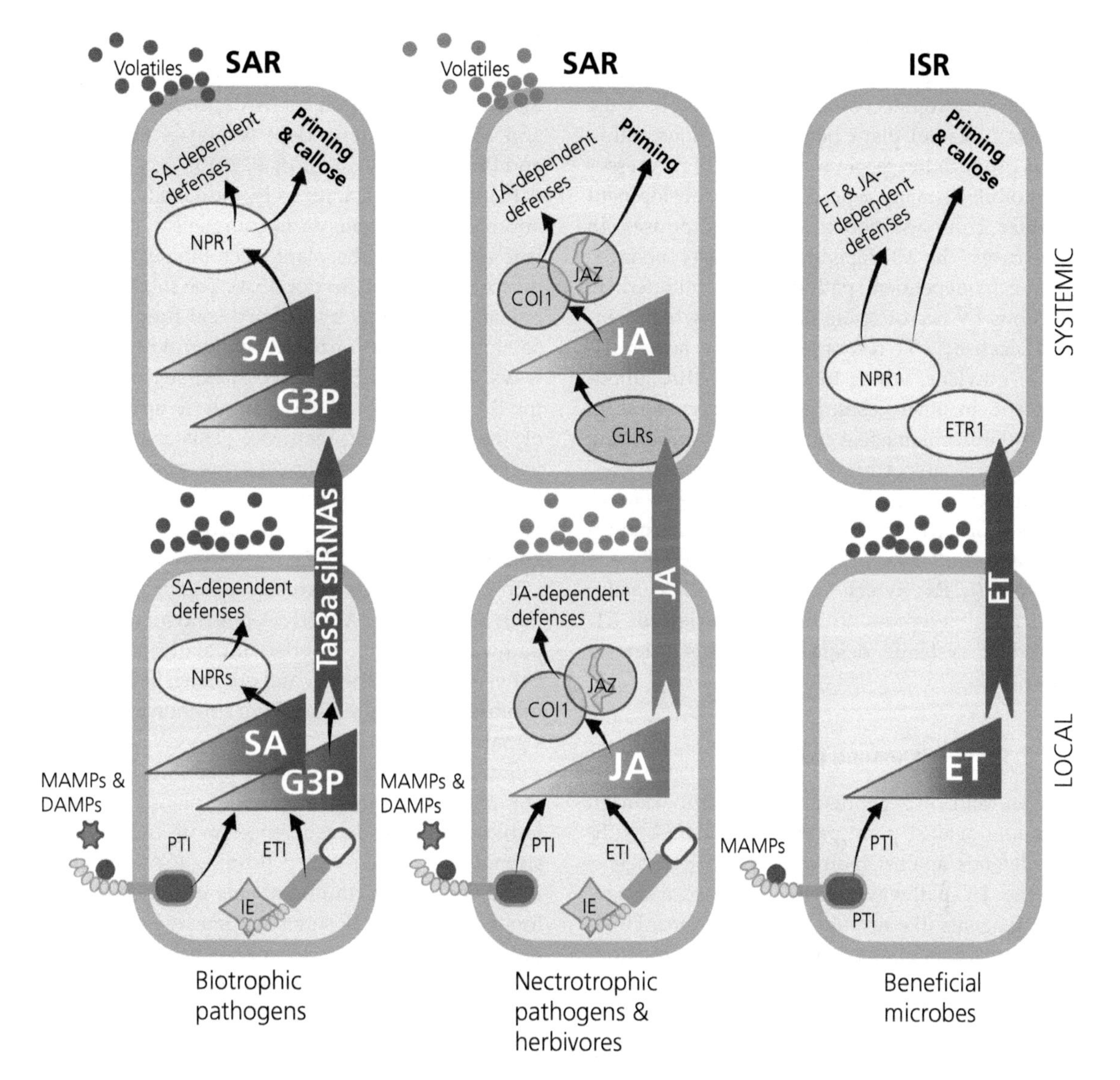

Figure 12.9 Three major pathways of systemic defense responses in plants. Induced defenses act locally (bottom) and signaling pathways allow defenses to be triggered systemically (top). The first level of local induced defenses leads to PTI (pattern-triggered immunity, responding to MAMPS and DAMPS) and ETI (effector-triggered immunity, responding to pathogen effectors—IE). PTI and ETI triggered by biotrophic pathogens induce the production of salicylic acid (SA) and glycerol-3-phosphate (G3P), which accumulate in the cells. Local effects of necrotrophic pathogens and herbivores, in contrast, usually lead to the accumulation of jasmonate (JA), and PTI induced by beneficial microbial symbionts leads to the production of the plant hormone ethylene (although some also induce jasmonate). The SA and JA interact with receptor proteins (NPRs for SA, COI1/JAZ for JA) to induce defense genes locally, where the pathogens were detected. Systemic responses require a signaling molecule to move from the local site of attack to distant parts of the plant. In the SA pathway, G3P induces the expression of small interfering RNAs (siRNAs) from the *Tas3a* gene which then are readily transported to distant plant organs as signaling molecules. In the JA and ET pathways, jasmonate and ethylene themselves are transported throughout the plant and serve the same role. In each case, the signaling molecules induce defense responses in parts of the plant distant from the site of attack by interacting with their corresponding receptors. When the systemic defense responses are induced by pathogens, it is called systemic acquired resistance (SAR); when the systemic defense responses are induced by beneficial organisms in the plant microbiome, it is called induced systemic resistance (ISR). Volatile versions of SA, JA, and ethylene can also move through the air to other parts of the same plant or to neighboring plants, where they can prime plant cells to induce defenses more quickly and strongly in response to pathogen attack.

Graphic by Gregory S. Gilbert, created with BioRender.com; modeled on a schematic in Pieterse et al. (2014).

The third signaling pathway is the **ethylene (ET) pathway** (Figure 12.9). It is also important against necrotrophic pathogens and involves the gaseous (volatile) plant hormone ethylene. Ethylene is produced in response to a variety of stresses and modulates many aspects of plant development (Chapter 2), in addition to defense responses. In some cases, the ET-dependent pathway behaves like the JA-dependent pathway; it is initiated by infections by necrotrophic pathogens, which generate localized PTI responses and the accumulation of ethylene, which then diffuses throughout the plant. In other cases, beneficial organisms in the plant microbiome (e.g., plant growth-promoting rhizobacteria, mycorrhizal fungi, and endophytes; Section 15.5) can stimulate production and accumulation of ethylene. In both situations, accumulated ethylene then interacts with the ethylene receptors (ETRs), which in turn stimulate activity of NPR1 receptors, to induce a variety of ET-dependent systemic defense responses (Pieterse et al. 2014).

12.7.2 Cross-talk among pathways

Both SA- and JA-dependent SAR are important in defending against plant pathogens. The SA pathway defends against biotrophic plant pathogens, and the JA pathway defends against damage-causing agents like necrotrophic pathogens. However, in real life, plants contend with threats from both biotrophic and necrotrophic pathogens, sometimes simultaneously. Unfortunately (for the plant), plants can generally express only one pathway or the other at a time (Kunkel and Brooks 2002). That's because the JA- and SA-signaling pathways are mutually antagonistic; jasmonic acid suppresses the expression of genes in the SA-signaling pathway, and salicylic acid suppresses gene expression in the JA-signaling pathway. For instance, induction of SA-mediated defenses in *Arabidopsis* by the biotrophic bacterial pathogen *Pseudomonas syringae* suppresses JA-mediated defense responses, making tissues of the infected plant more susceptible to the necrotrophic fungal pathogen *Alternaria brassicicola* (Spoel et al. 2007). This cross-talk between signaling pathways appears to be regulated by several proteins including NPR1, which activates SA-dependent defenses and simultaneously blocks JA-dependent responses, and the protein kinase MPK4, which upregulates JA-dependent responses and inhibits SA-dependent responses (Koornneef and Pieterse 2008, Caarls et al. 2015).

Plants are also colonized by a diversity of fungal and bacterial symbionts as commensal or mutualistic elements of the plant microbiome, including mycorrhizal fungi (Section 15.4). The successful colonization of plants by mycorrhizal fungi requires some suppression of SA-dependent defense pathways. Because of SA–JA crosstalk, suppression of the SA-signaling pathway can allow upregulation of the JA-signaling pathway. This may explain why necrotrophic pathogens are suppressed in hosts colonized by mycorrhizae, whereas biotrophic pathogens often thrive in mycorrhizal hosts (Pozo and Azcón-Aguilar 2007).

In contrast to the mutually antagonistic interactions between the JA- and SA-dependent pathways, pathway cross-talk is often mutually reinforcing between both JA and SA and ethylene. In fact, some defense responses require ET in addition to JA or SA signals to function.

Similarly, interactions with microbes or signaling pathways can precondition plant tissues to induce SA-dependent defenses more quickly and strongly—a process called **priming** (Conrath et al. 2015). Our understanding of how priming works is limited (as it is for all signaling pathway cross-talk), but it may involve a variety of mechanisms, including hormonal changes, epigenetic regulation, and transcriptional regulation (Hilker and Schmülling 2019). This is a rapidly growing area of research, as are efforts to understand the ecological importance of priming and the potential to use priming to protect crops from pathogens.

12.7.3 Plant-to-plant signaling

Naturally occurring derivatives of jasmonic acid (methyl jasmonate) and salicylic acid (methyl salicylate), along with ethylene, are highly volatile gases that act as signaling compounds. In addition to diffusing through plant tissues, these volatile compounds can be transported through the air between neighboring plants (Farmer and Ryan 1990). Such inter-plant transmission of jasmonate or ethylene creates an opportunity for "talking trees," where signaling molecules generated from one plant under attack by an enemy induce defense responses in neighboring plants—even plants of a different species (Baldwin et al. 2006). Jasmonic acid released into the soil also allows plants to detect and respond to neighboring plants of the same and different species (Kong et al. 2018). The research on inter-plant communication is fascinating and opens the imagination to all sorts of possibilities; perhaps groups of plant siblings, or even groups of plant neighbors, band together to ward off pathogen attack! However, it remains unclear how often such signaling reaches concentrations great enough to provide protection against plant pathogens in natural situations.

Now that we've explored the mechanics of how these interactions take place at the cellular level, we can turn to patterns of evolution in the traits of plants and pathogens and how they influence the ecology of plant–pathogen symbioses.

Further reading

Buchannan, B. B., W. Gruissem, and R. L. Jones. 2015. Biochemistry and Molecular Biology of Plants, 2nd edition. Wiley–Blackwell, Hoboken, NJ.

Heldt, H-W. and B. Piechulla. 2021. Plant Biochemistry, 5th edition. Elsevier, Amsterdam.

Sessa, G., editor. 2013. *Molecular Plant Immunity*. Wiley–Blackwell, Hoboken, NJ.

References

Almadiy, A. A. and G. E. Nenaah. 2018. Ecofriendly synthesis of silver nanoparticles using potato steroidal alkaloids and their activity against phytopathogenic fungi. *Brazilian Archives of Biology and Technology* **61**: e18180013.

Baldwin, I. T., R. Halitschke, A. Paschold, C. C. Von Dahl, and C. A. Preston. 2006. Volatile signaling in plant–plant interactions: "talking trees" in the genomics era. *Science* **311**:812–815.

Balint-Kurti, P. 2019. The plant hypersensitive response: concepts, control and consequences. *Molecular Plant Pathology* **20**:1163–1178.

Boutrot, F. and C. Zipfel. 2017. Function, discovery, and exploitation of plant pattern recognition receptors for broad-spectrum disease resistance. *Annual Review of Phytopathology* **55**:257–286.

Caarls, L., C. M. Pieterse, and S. Van Wees. 2015. How salicylic acid takes transcriptional control over jasmonic acid signaling. *Frontiers in Plant Science* **6**:170.

Conrath, U., G. J. Beckers, C. J. Langenbach, and M. R. Jaskiewicz. 2015. Priming for enhanced defense. *Annual Review of Phytopathology* **53**:97–119.

Dangl, J. L. and J. D. G. Jones. 2001. Plant pathogens and integrated defence responses to infection. *Nature* **411**:826–833.

Faris, J. D. and T. L. Friesen. 2020. Plant genes hijacked by necrotrophic fungal pathogens. *Current Opinion in Plant Biology* **56**:74–80.

Farmer, E. E. and C. A. Ryan. 1990. Interplant communication: airborne methyl jasmonate induces synthesis of proteinase inhibitors in plant leaves. *Proceedings of the National Academy of Sciences of the United States of America* **87**:7713–7716.

Flor, H. 1942. Inheritance of pathogenicity in *Melampsora lini*. *Phytopathology* **32**:653–669.

Flor, H. 1955. Host–parasite interactions in flax rust—its genetics and other implications. *Phytopathology* **45**: 680–685.

Friesen, T. L., J. D. Faris, P. S. Solomon, and R. P. Oliver. 2008. Host-specific toxins: effectors of necrotrophic pathogenicity. *Cellular Microbiology* **10**:1421–1428.

Gao, H., M. Guo, J. Song, Y. Ma, and Z. Xu. 2021. Signals in systemic acquired resistance of plants against microbial pathogens. *Molecular Biology Reports* **48**:3747–3759.

Gilbert, G., N. Mejia-Chang, and E. Rojas. 2002. Fungal diversity and plant disease in mangrove forests: salt excretion as a possible defense mechanism. *Oecologia* **132**:278–285.

Hartmann, M., T. Zeier, F. Bernsdorff, V. Reichel-Deland, D. Kim, M. Hohmann, N. Scholten, S. Schuck, A. Bräutigam, and T. Hölzel. 2018. Flavin monooxygenase-generated N-hydroxypipecolic acid is a critical element of plant systemic immunity. *Cell* **173**:456–469. e416.

Hilker, M. and T. Schmülling. 2019. *Stress Priming, Memory, and Signalling in Plants*. Wiley Online Library, https://doi.org/10.1111/pce.13526.

Jamieson, P. A., L. Shan, and P. He. 2018. Plant cell surface molecular cypher: Receptor-like proteins and their roles in immunity and development. *Plant Science* **274**: 242–251.

Jeandet, P., C. Hébrard, M.-A. Deville, S. Cordelier, S. Dorey, A. Aziz, and J. Crouzet. 2014. Deciphering the role of phytoalexins in plant–microorganism interactions and human health. *Molecules* **19**:18033–18056.

Jones, J. D. G. and J. L. Dangl. 2006. The plant immune system. *Nature* **444**:323–329.

Kong, C.-H., S.-Z. Zhang, Y.-H. Li, Z.-C. Xia, X.-F. Yang, S. J. Meiners, and P. Wang. 2018. Plant neighbor detection and allelochemical response are driven by root-secreted signaling chemicals. *Nature Communications* **9**:1–9.

Koornneef, A. and C. M. Pieterse. 2008. Cross talk in defense signaling. *Plant Physiology* **146**:839–844.

Kunkel, B. N. and D. M. Brooks. 2002. Cross talk between signaling pathways in pathogen defense. *Current Opinion in Plant Biology* **5**:325–331.

McDowell, J. M. 2019. Focus on activation, regulation, and evolution of MTI and ETI. *Molecular Plant–Microbe Interactions* **32**:5–5.

Nazzaro, F., F. Fratianni, R. Coppola, and V. D. Feo. 2017. Essential oils and antifungal activity. *Pharmaceuticals* **10**:86.

Pan, Q. L., J. Wendel, and R. Fluhr. 2000. Divergent evolution of plant NBS-LRR resistance gene homologues in dicot and cereal genomes. *Journal of Molecular Evolution* **50**:203–213.

Pedras, M. S. C. and E. E. Yaya. 2015. Plant chemical defenses: are all constitutive antimicrobial metabolites phytoanticipins? *Natural Product Communications* **10**:1934578X1501000142.

Pieterse, C. M., C. Zamioudis, R. L. Berendsen, D. M. Weller, S. C. Van Wees, and P. A. Bakker. 2014. Induced systemic resistance by beneficial microbes. *Annual Review of Phytopathology* **52**:347–375.

Pozo, M. J. and C. Azcón-Aguilar. 2007. Unraveling mycorrhiza-induced resistance. *Current Opinion in Plant Biology* **10**:393–398.

Rivera-Hoyos, C. M., E. D. Morales-Álvarez, R. A. Poutou-Pinales, A. M. Pedroza-Rodríguez, R. Rodriguez-Vazquez, and J. M. Delgado-Boada. 2013. Fungal laccases. *Fungal Biology Reviews* **27**:67–82.

Saijo, Y., E. P. I. Loo, and S. Yasuda. 2018. Pattern recognition receptors and signaling in plant–microbe interactions. *The Plant Journal* **93**:592–613.

Shine, M., K. Zhang, H. Liu, G.-h. Lim, F. Xia, K. Yu, A. G. Hunt, A. Kachroo, and P. Kachroo. 2022. Phased small RNA-mediated systemic signaling in plants. *Science Advances* **8**:eabm8791.

Spoel, S. H., J. S. Johnson, and X. Dong. 2007. Regulation of tradeoffs between plant defenses against pathogens with different lifestyles. *Proceedings of the National Academy of Sciences of the United States of America* **104**:18842–18847.

Wang, Y., R. N. Pruitt, T. Nürnberger, and Y. Wang. 2022. Evasion of plant immunity by microbial pathogens. *Nature Reviews Microbiology* **20**:449–464.

Wittstock, U. and J. Gershenzon. 2002. Constitutive plant toxins and their role in defense against herbivores and pathogens. *Current Opinion in Plant Biology* **5**: 300–307.

Ziv, C., Z. Zhao, Y. G. Gao, and Y.- Xia. 2018. Multifunctional roles of plant cuticle during plant–pathogen interactions. *Frontiers in Plant Science* **9**:1088.

CHAPTER 13

Evolution

Gregory S. Gilbert and Ingrid M. Parker

Evolutionary biologists study the processes that drive variation among organisms and why organisms are the way they are. Which traits are key to a species' success, and how did those traits come to be? Evolution shapes traits through a variety of mechanisms that we explore in this chapter; natural selection is one important mechanism that plays out through an organism's interactions with both abiotic and biotic factors in its environment. Plants and microbes have been interacting for as long as there have been plants, shaping each other's evolutionary paths.

This chapter explores the patterns, processes, and implications of evolution in plant–pathogen interactions. It builds on the mechanistic understanding of those interactions developed in Chapter 12, extends those mechanisms into the population biology and genetics of pathogens and their hosts, and develops frameworks for thinking about how evolutionary dynamics shape the ecology and epidemiology of plant disease.

13.1 The evolutionary framework

Evolution is simply a change in the frequency of alleles in a population over time. Evolutionary change can appear quickly from one generation to the next, or accumulate slowly, generating differences among lineages over millions of years. Evolutionary biology incorporates genetics into the study of contemporary heritable change (**microevolution**), while also considering patterns across the history of life (**macroevolution**). As a subset of evolutionary biology, evolutionary ecology emphasizes that the way species interact with each other and with their environment is mediated by the traits (or **phenotypes**) of organisms, which reflect an underlying **genotype**. Phenotypic traits include body plan, modes of dispersal, development, physiology and metabolism, phenology, and chemical composition. When traits affect **fitness**, which is an organism's survival and reproductive success in a particular environment, they will be under selection. As the great biologist Theodosius Dobzhansky noted, "nothing in biology makes sense except in the light of evolution" (Dobzhansky 1950). The outcome of ecological interactions is shaped by traits produced through evolutionary processes, and evolution, in turn, takes place in an ecological arena. For example, when a pathogen evolves to overcome the resistance of a host, it immediately changes the fitness of the pathogen as well as the impact of that pathogen on host fitness. This single evolutionary change drives changes in the ecology of both plant and pathogen by altering the outcome of their interactions.

The ultimate source of variation in phenotypic traits is random **mutation** in the DNA nucleotide sequences, resulting in new **alleles** at a genetic locus. The rate of mutation per base pair per generation is highest for RNA viruses (10^{-4} to 10^{-5}), a bit lower for DNA viruses (10^{-6} to 10^{-8}), and lower still for Bacteria and Eukaryotes (10^{-9} to 10^{-10}) (Sniegowski et al. 2000). However, when considering the relatively rapid generation times of microbes and their astronomically large populations, mutations are not particularly rare. In addition, **horizontal gene transfer** (Section 5.3) can move whole genes and sets of genes between organisms and has played a key role in the evolution of microbes. With slower

The Evolutionary Ecology of Plant Disease. Gregory S. Gilbert and Ingrid M. Parker, Oxford University Press. © Gregory S. Gilbert & Ingrid M. Parker (2023).
DOI: 10.1093/oso/9780198797876.003.0013

generation times and smaller population sizes, host species have an inherent disadvantage when it comes to the emergence of novel genetic variation. But mutation rates vary across the genome within an organism, and resistance loci in plants are mutation hotspots. The leucine-rich repeat (LRR) domains common to most plant receptors involved in pathogen recognition (Section 12.5) occur clustered along the genome in groups; these are subject to slippage during recombination, which leads to new variants (Karasov et al. 2014a), giving plants a little evolutionary boost in terms of the raw material to work with.

Sexual recombination further amplifies the raw genetic diversity generated by mutation, forming new combinations of alleles from parents with different genotypes. However, sexual recombination is not essential for evolution to occur. Asexual (**clonal**) lineages are common in plants as well as most of their pathogens. Because evolution is simply a change in allele frequencies, a new mutation that persists in an asexual lineage also represents evolutionary change.

We can think of evolution as being shaped by both random and non-random forces. In the following sections, we provide a basic introduction to these two types of forces: genetic drift and natural selection.

13.2 Genetic drift

The "random force" of evolutionary change is **genetic drift**, a process by which the frequency of alleles changes from one generation to the next by chance. It is easiest to think about this in small populations (and indeed, genetic drift is a more powerful force in small populations). Imagine a small, isolated patch of plants. If a tree falls and covers most of the plants, only the individuals that happened to escape this fate can go on to reproduce. The alleles carried by those lucky seeds will determine the genetic makeup of the population in the next generation. Genetic change in the population is likely, simply because the small group of plants that survived to reproduce possess a random sample of the alleles at each locus that were present in the original population. (This is why genetic drift is commonly called "sampling error.") The phenotypic traits encoded by the genes of these plants did not provide them with an inherent advantage relative to their neighbors; rather, allele frequencies changed as a simple result of some plants being in the right place at the right time. This is evolution by genetic drift.

Genetic drift is more likely to occur in small populations for the same reason that a small number of coin tosses is unlikely to yield an exact 50:50 ratio. When a plant population shrinks rapidly due to habitat conversion or an extreme event such as a fire or flood, the genotypes of the remaining individuals can represent a skewed subset of the original diversity by chance alone. The rapid reduction in size is called a **population bottleneck**. A bottleneck also occurs when an individual or small group of individuals colonizes a new region—for example during the accidental introduction of a weedy plant to a new continent. This extreme population bottleneck is called a **founder effect**, where the descendants of the "founder" all share combinations of alleles from the same small gene pool. Every pathogen population experiences a founder effect during transmission to a new host. Indeed, founder effects are such a ubiquitous feature of pathogen biology that we should expect the evolutionary dynamics of disease interactions to be strongly influenced by genetic drift.

How do we measure genetic drift? To observe the consequences of genetic drift, we need heritable markers that we can track across populations and across time. As soon as biologists discovered that the structure of molecules can vary across individuals, they began to harness them as genetic markers. There are two classes of molecular genetic markers—those that are gene products like amino acids and proteins, and those that are variants in the DNA sequence itself. The earliest types of markers were proteins with variable electromagnetic properties (**allozymes**). Over time, technology developed for new types of markers, such as **DNA fragment length polymorphisms** (e.g., **microsatellites**, also called SSR or STR); the techniques expanded the number of alleles per locus and the number of loci available for genetic comparisons. Now single nucleotide polymorphisms (**SNPs**) are often used because of the immense number of loci provided by whole-genome sequencing.

To quantify random changes in allele frequencies (i.e., evolution by genetic drift, not selection), it is helpful to have genetic markers that are neutral (not under selection). Because enzymes are important for metabolism, even subtle structural variation in enzymes may be the target of natural selection, which is why allozymes are generally no longer used to study genetic drift. A DNA nucleotide change may be in a noncoding (intergenic) region, or it may be in a coding region (a gene, which codes for a protein). Microsatellite loci in noncoding regions of DNA or synonymous nucleotide substitutions that do not change any amino acids in the resulting gene product (Section 13.4) are particularly useful for quantifying patterns of genetic drift.

13.3 Population genetics and genetic structure

One of the effects of genetic drift is to cause separate populations or separated groups of individuals to become genetically different from each other over time, shaping the distribution of genetic variation among and within populations, which is called **genetic structure**. As the distance between two individuals increases, their genetic similarity decreases—a pattern called **isolation by distance** (Wright 1938). This can happen even over short spatial scales when mating occurs primarily with close neighbors.

The force that counteracts genetic drift is **gene flow**, also known in population genetics as **gene migration**, which is the movement of alleles via dispersal. In pathogens, gene flow occurs via the dispersing stage of the life cycle. In plants, gene flow occurs through the movement of both seeds and pollen. Thus pollination ecology strongly influences plant genetic structure; for example, a plant with highly mobile pollinators will experience gene flow over a larger distance and show less genetic differentiation among local patches. Genetic structure is also influenced by the **mating system** (Section 2.5); selfing increases homozygosity as well as genetic differentiation among lineages and populations.

Plant pathologists use genetic structure to gain important information about the behavior of pathogens, the epidemiology of particular diseases, and host–pathogen interactions. One application is to infer the mode of reproduction of a pathogen. For many fungi we don't know if they reproduce entirely asexually or also have a sexual phase. Sexually reproducing organisms evolve faster because recombination generates a pool of genetic variation for selection to act on. Thus, sexual reproduction has a big effect on a pathogen's expected rate of evolution in response to resistance genes, pesticides, and other elements of disease control (Chapter 17). To determine the frequency of sexual reproduction in a pathogen, we can inspect the degree of heterozygosity in a pathogen strain, the amount of genetic differentiation among strains, and the distribution of alleles (or multilocus genotypes) among strains or populations (Taylor et al. 1999).

The way genetic structure has been traditionally quantified is with Wright's F statistics and related measures (F for "fixation index," Wright 1949), which characterize the distribution of genetic variation across different scales: between populations, among individuals within populations, and within individuals (i.e., heterozygosity). F_{ST}, the differentiation between populations or groups, is used for many applications in population genetics (Holsinger and Weir 2009), including the study of natural selection across the genome (described below). More recently, new approaches provide a more nuanced picture of population structure, such as Bayesian clustering methods, which assign individuals to groups in a probabilistic framework (e.g., STRUCTURE, Pritchard et al. 2000), and estimate genetic **admixture** between groups. Relationships among populations are also explored with phylogenetic trees (Box 13.3), reconstructing historical patterns of divergence and gene flow (Avise 1989).

An important application of the study of genetic structure is to test for host specialization. For example, *Exserohilum turcicum* causes northern leaf blight, a devastating foliar disease of maize and sorghum in sub-Saharan Africa, India, Asia, Latin America, and the northeastern United States. Nieuwoudt et al. (2018) collected *Exserohilum turcicum* isolates from neighboring fields of maize and sorghum. They screened the isolates with 12 microsatellite markers and calculated a version of F_{ST} between neighboring pairs of maize and sorghum fields, and across fields in different geographic regions. They found that *Exserohilum* isolates were more similar to isolates collected from the same host in a far-away field than to isolates from the opposite host in a neighboring field. This suggests that isolates infecting maize and those infecting sorghum were not mixing, essentially behaving like separate populations. Nieuwoudt and colleagues also studied the reproductive strategy of the pathogen by testing for departures from Hardy–Weinberg equilibrium (Box 13.1), finding evidence for a mix of both sexual and clonal reproduction.

Population genetics can also be used to reconstruct the origins and spread of an introduced pathogen. The chestnut blight fungus, *Cryphonectria parasitica*, was introduced into North America and Europe in the early 1900s and rapidly caused the devastation of hardwood forests on the east coast of the United States (Section 16.3). Milgroom and colleagues used molecular markers (first with restriction fragment length polymorphisms (Milgroom et al. 1996), later with microsatellites (Dutech et al. 2012)) to reconstruct the introduction history of the pathogen. Patterns of genetic structure clearly showed more similarity between introduced populations in North America and populations in Japan (relative to those in China); this was also consistent with historical records of plant imports from Japan (Milgroom et al. 1996). The authors used Approximate Bayesian Computation (Estoup and Guillemaud 2010) to reveal support for a scenario of multiple independent introductions of the pathogen, with population admixture across North America (Dutech et al. 2012).

13.4 Natural selection

As we said in the introduction, evolution is shaped by both random and nonrandom forces; the "nonrandom" force of evolution is **natural selection**. Evolution in plant populations can proceed rapidly in the face of an aggressive pathogen, as shown by the case of *Puccinia chondrillina* (Figure 13.2),

Box 13.1 Hardy–Weinberg Equilibrium

To detect evolutionary processes like genetic drift, selection, inbreeding, and population mixing, we compare patterns of genetic structure to the expectations of **Hardy–Weinberg equilibrium**. The principle of Hardy–Weinberg equilibrium is that in the absence of outside forces, allele frequencies in the population will remain the same from one generation to the next, and genotype frequencies will follow simple predictions.

For a locus with two alleles,

A (with frequency p)
a (with frequency q)

three genotypes are possible: AA, Aa, and aa, with expected frequencies p^2, $2pq$, and q^2 (Figure 13.1)

Because these three genotypes together make up the whole population, by definition

$$p^2 + 2pq + q^2 = 1$$

These genotype frequencies represent a "null model." Deviations from these expected frequencies can be used to test for gene flow, genetic drift, and selection, as well as patterns of non-random mating such as inbreeding.

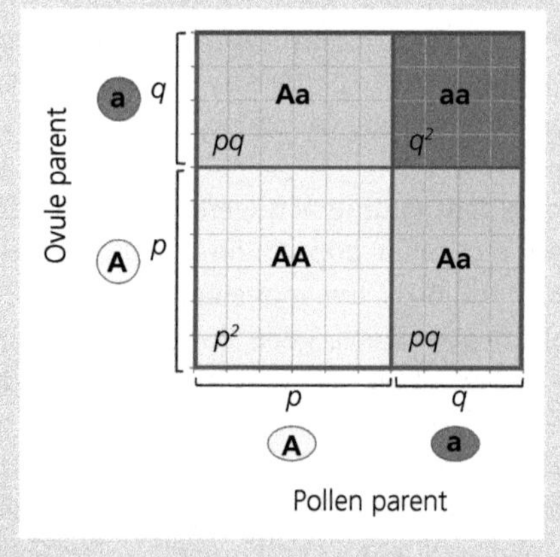

Figure 13.1 Hardy–Weinberg expectations for a plant population at a single locus with two alleles, A (frequency $p = 0.6$) and a (frequency $q = 0.4$). The resulting expected frequencies of the genotypes are $p^2 = 0.36$, $2pq = 0.48$, and $q^2 = 0.16$.
Graphic by Ingrid M. Parker based on design by Tebeuszek, CC BY-SA 3.0 <https://commons.wikimedia.org/wiki/File:Schemat_punneta2.svg> via Wikimedia Commons.

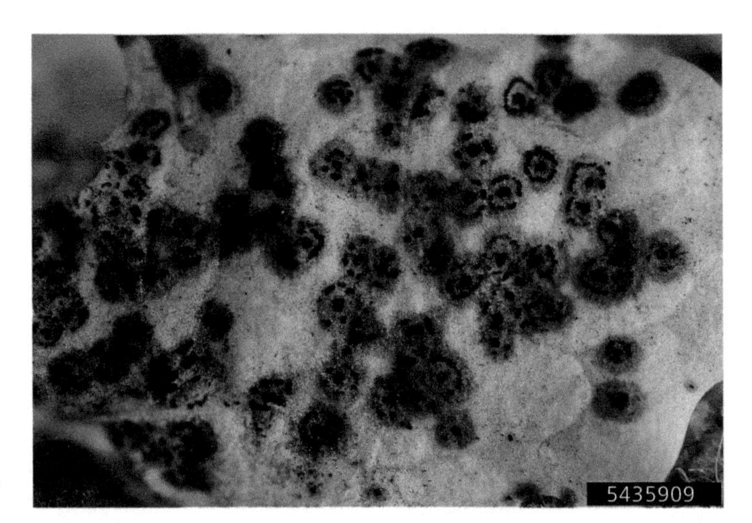

Figure 13.2 *Puccinia chondrillina* on *Chondrilla juncea*.
Photo by Eric Coombs, Oregon Department of Agriculture, Bugwood.org.

a highly host-specific rust fungus introduced into Australia and the United States to control populations of the invasive host plant rush skeletonweed (*Chondrilla juncea*) (Supkof et al. 1988). After introduction, the rust quickly reduced skeletonweed populations to low levels, a great success in the biological control of an invasive weed! But in Australia, the skeletonweed population was actually made up of three genetically different clones, which could be distinguished by their leaf width. The narrow-leaf type was originally much more common than the two other clone types, but it was also the only type that was susceptible to the rust. After the introduced rust decimated the narrow-leaf clone population, the resistant clones began to expand. In the presence of rust, the fitness of the resistant plants was much greater than the susceptible, narrow-leaved plants. The rust fungus caused strong natural selection on skeletonweed, leading to a rapid evolutionary change in the skeletonweed population from mostly susceptible to mostly resistant (Hanley and Groves 2002). There is now a concerted effort to introduce new strains of *P. chondrillina* able to attack the other clones of skeletonweed.

Pathogens can cause evolution in their hosts even when they do not have lethal effects. For instance, an epidemic of the rust *Melampsora epitea* swept through a population of the willow tree *Salix viminalis* (Salicaceae) that showed variation in resistance (Verwijst 1993). Although the rust did not kill its host directly, the disease reduced the height of susceptible genotypes, providing resistant genotypes a competitive advantage. Within three years, many of the shorter trees had died, causing a shift toward resistant genotypes.

Charles Darwin's theory of evolution by natural selection can be thought of as having four key points. First, organisms show variation in traits. Darwin didn't know about genetic mutations, but he was a keen observer of nature who wrote books on everything from barnacles to carnivorous plants, and he noted that variation is a universal truth among living things. Second, traits can be inherited from parents to offspring. Third, drawing from the earlier work of Thomas Robert Malthus (1826), Darwin pointed out that organisms have many more offspring than can survive to maturity, leading to intense competition among them. Fourth, because of this competition, any trait that provides an advantage, even a small advantage, will increase the probability that its owner will survive and leave offspring, passing on the trait.

Natural selection can be thought of simply as the relationship between fitness and a trait. Alleles that reduce fitness are selected against and may be lost from the population, while those that enhance fitness will increase in frequency. First let's think about a continuous (also called "quantitative") trait, like the thickness of a spore wall. When fitness always increases (or decreases) with the value of the trait, that represents **directional selection** in favor of (or against) the trait (Figure 13.3a). Because spore

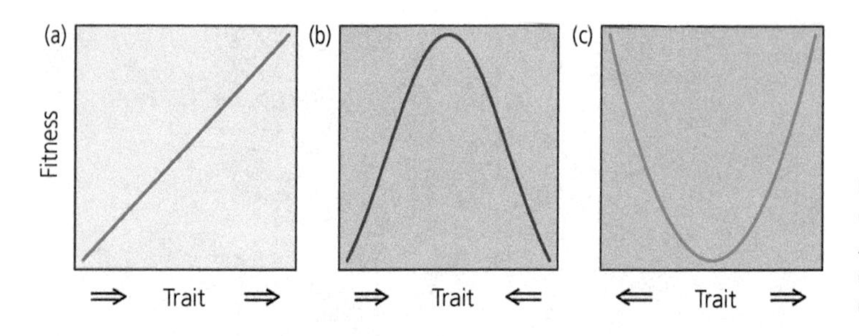

Figure 13.3 Three types of selection: (a) directional, (b) stabilizing, (c) disruptive. Arrows along the *x*-axis indicate the predicted direction of evolutionary change in the trait. Graphic by Gregory S. Gilbert.

wall thickness protects spores against desiccation, we imagine that in a dry year, thicker spore walls will enhance survival, and fungal genotypes with thicker spores will have more successful transmission and higher fitness. That's directional selection. However, thicker spore walls also make the spore heavier, which could reduce dispersal distance and the probability of reaching a new host individual. Therefore, under certain conditions, pathogen fitness might be highest at an optimum spore wall thickness that is not too thin or too thick—that is an example of **stabilizing selection** (Figure 13.3b). Under these two cases, natural selection will act to reduce genetic diversity in the population over time, because it culls out individuals with the "wrong" value of the trait (in directional selection) or with extreme values of the trait (in stabilizing selection). On the other hand, sometimes natural selection will act to maintain genetic diversity. The spore with the intermediate phenotype, a medium-thick wall, might be a "jack-of-all-trades-master-of-none" that doesn't do anything well. In that case, there may be **disruptive selection** on individuals (Figure 13.3c) to produce either thin-walled spores and achieve high fitness through very effective transmission, or to produce thick-walled spores and achieve high fitness with high spore survival rates. When multiple selective forces (in this case via both transmission and survival) favor different values of a trait, they will maintain genetic diversity in a population; we call that **balancing selection. Fluctuating selection**, in which fitness varies over space and/or time, is also a mechanism for balancing selection and helps to maintain genetic diversity, counteracting the effects of genetic drift and directional selection.

Now let's also think about discrete ("qualitative") traits, such as variants for a plant *R* gene. Natural selection in host–pathogen interactions often involves **frequency-dependent selection**, where the fitness of a particular genotype depends on its frequency in the population (the proportion of individuals with that genotype) (Barrett 1988). For example, if there is one dominant host genotype to which the local pathogens are well adapted, a novel host genotype with a different resistance mechanism will have a fitness advantage while it is rare. This is negative frequency-dependent selection. Negative frequency-dependent selection is another type of balancing selection, often fluctuating over time. Later in the chapter (Section 13.6), we discuss coevolutionary dynamics that involve fluctuating, frequency-dependent selection between pathogens and hosts. We can think of the pathogen population as a critical component of the host's fluctuating environment, and vice versa.

Evolutionary biologists have a variety of tools to study evolution by natural selection (Linnen and Hoekstra 2009). We can measure selection directly by quantifying fitness as a function of the value of a trait of interest. For a categorical (yes/no) trait, such as pigmented versus non-pigmented spores, we measure the strength of selection as the difference in mean fitness between the categories. For a continuous trait, like spore survival time or leaf area, the strength of selection is the slope of the linear relationship between fitness and the trait (as in Figure 13.3a). Because many traits are correlated with each other, comprehensive studies of natural selection measure indirect as well as direct selection on traits, and consider directional, stabilizing, and disruptive selection.

Matthew Parker measured pathogen-mediated selection on the annual legume *Amphicarpaea bracteata* in a wild population near Chicago, IL

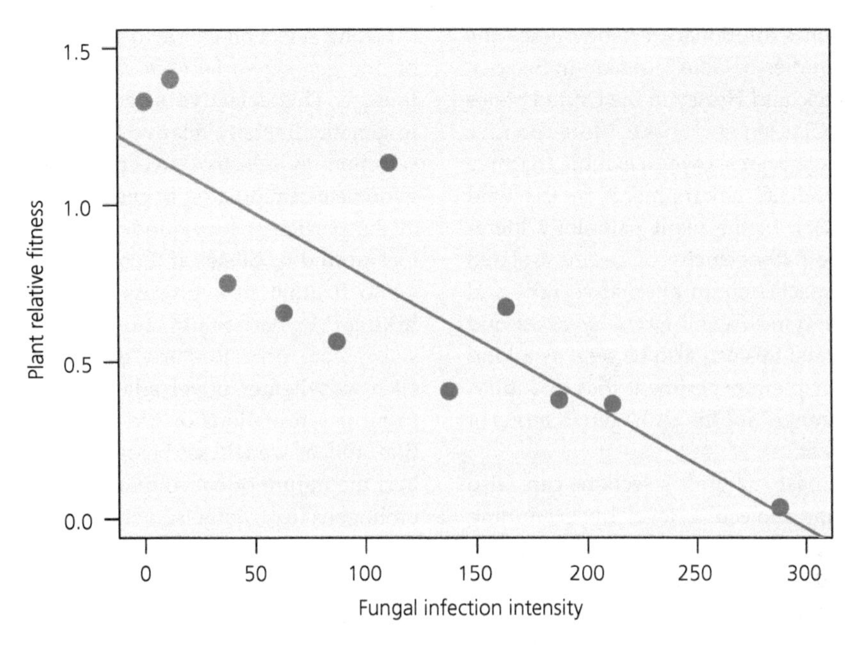

Figure 13.4 Natural selection on the annual legume *Amphicarpaea bracteata* caused by infection by the host-specific fungus *Synchytrium decipiens*. The strength of selection was quantified from the regression of relative fitness on susceptibility to infection as a phenotypic trait. Relative fitness was an individual's total seed biomass produced divided by the population mean, and fungal infection intensity was the number of sori on stems and leaves of young plants. The selection differential in this example was −0.335 in phenotypic standard deviation units (with the trait standardized to mean = 0 and SD = 1).

Figure by Gregory S. Gilbert based on data in Parker (1986).

(Parker 1986). *A. bracteata* is attacked by *Synchytrium decipiens*, a host-specific chytrid fungus that reduces plant growth and increases mortality. Parker used intensity of infection, quantified as the number of reproductive structures (sori) on 2-week-old plants, as an indicator of plant susceptibility. Susceptibility was treated as a complex phenotypic trait representing a suite of underlying molecular, morphological and physiological traits of the plant. Parker calculated the directional selection differential as the regression of relative fitness (total seed biomass produced per individual divided by the population mean) on susceptibility (infection intensity) (Figure 13.4).

13.5 Response to selection: Adaptive evolution

If there is heritable variation for a trait under selection, adaptive evolution will occur and can be measured as the response to selection. Sometimes adaptive evolution transforms a plant disease from a small nuisance to a major problem. Stripe rust of wheat (*Puccinia striiformis* f.sp. *triciti*) was only a minor concern in the south-central United States until its sudden emergence as a severe agricultural problem in 2000. Milus et al. (2006) compared pathogen isolates collected after this outbreak to isolates collected before 2000. They found that new variants of the pathogen were better adapted to tolerate the higher temperatures of the late season, with greater spore germination and a shorter latent period than ancestral pathovars.

A classic tool in testing for evidence of past natural selection is the **reciprocal transplant experiment**, in which organisms from contrasting environments are transplanted into their home environment and into the other's home environment. When the organisms each perform best in their own environment (a "home-site advantage") it is interpreted as evidence for **local adaptation.** Reciprocal transplant experiments have been a mainstay of plant evolutionary ecology since the 1920s with the studies of Göte Turesson in Sweden and Clausen, Keck, and Heisey in the United States (Turesson 1922, Clausen et al. 1940). More recently, microbes also have been shown to exhibit adaptive differentiation to local environments in the wild (Chase et al. 2021). In the plant pathology literature, we test the pathogenicity of strains isolated from one host species on an alternative host and vice versa, identifying fungal *formae speciales* and bacterial pathovars; this can also be seen as a kind of reciprocal transplant experiment that quantifies "home site advantage" for the pathogen in terms of its fitness on a host.

Evidence of past natural selection can also be inferred at the molecular level by examining the nucleotide bases (DNA sequence) of the genome (Aguileta et al. 2009). Within a protein-coding gene, a nucleotide change that does not change the resulting amino acid is a **synonymous substitution** (e.g., both TGT and TGC code for cysteine), whereas a nucleotide change that creates a codon for a different amino acid is a **nonsynonymous substitution**, also called a missense substitution (e.g., TGT codes for cysteine but TGG codes for tryptophan). We can compare the number of nonsynonymous substitutions (dN) to the number of (presumed neutral) synonymous substitutions (dS) to measure selection on the gene. The expectation under neutrality is $dN/dS = 1$, with $dN/dS < 1$ indicating negative selection, also called **purifying selection** because it removes novel variants. In contrast, $dN/dS > 1$ indicates that the gene was under positive selection. As an example of this approach, Schweizer et al. (2018) used dN/dS to investigate selection in lineages of biotrophic smuts evolving on different host species. In addition to purifying selection, they found widespread evidence for positive selection, especially on presumptive effector genes that code for secreted proteins. The authors followed up with gene deletion experiments to test whether the genes under positive selection coded for virulence phenotypes. Studies like this one, combining genomics tools with experimental tests of gene function, provide powerful insights into the evolution of plant–pathogen interactions and into mechanisms of virulence and resistance.

Strong selection can lead to a rapid substitution of one genetic variant for another across a whole lineage. This **selective sweep** causes a reduction in genetic diversity relative to other loci not under selection. A selective sweep is observable in the genome as a reduction in genetic diversity not only at the particular locus under selection but also at loci around it. Alleles at those surrounding loci can go to fixation in a process called **genetic hitchhiking** (Maynard Smith and Haigh 1974). Selective sweeps are used to study evolutionary questions such as whether novel adaptations tend to arise from new mutations or pre-existing genetic variation, and, as we will see below, provide key insights into the nature of coevolution between plants and pathogens (e.g., Bergelson et al. 2001).

13.5.1 Pathogen fitness and the evolution of virulence

In order to understand adaptive evolution in plant pathogens, we need to understand what fitness means for a microbe. Importantly, causing disease on a plant host only benefits the pathogen if it enhances pathogen fitness. The pathogen has no particular stake in whether the host flourishes or dies, beyond the effect this has on the pathogen's own survival, growth, reproduction, and dispersal. For that reason, it is not possible to even guess at the direction or strength of selection on any particular trait without considering it in the context of the life-history strategy of the pathogen.

In Chapter 10, we discussed how population growth can be expressed as the intrinsic rate of increase, r (or the discrete-time finite rate of increase of the population, λ), which integrates all demographic rates into one combined measure. Evolutionary biologists think of fitness as the "population growth rate" of a genotype—where the genotype will increase in frequency if its fitness is higher than the population average. Fitness is not just survival (which is why modern evolutionary biologists don't like to use the phrase "survival of the fittest"): it is the integration of performance at all life stages. Traits that increase the fitness of a plant pathogen are those traits that allow the pathogen to better infect, grow, reproduce, disperse to new host individuals, and survive adverse conditions.

Nowhere is this perspective more clearly articulated than when we consider the evolution of virulence. (For a quick recap of what is meant by virulence, pathogenicity, susceptibility, tolerance, and related terms, refer to Box 12.1.) Pathogens face an evolutionary dilemma because greater virulence (more aggressive use of the host) can translate into greater reproductive potential, but if a pathogen kills its host too quickly it risks missing the opportunity to be transmitted to other host individuals. In the conventional wisdom of medicine and veterinary science, this was used to argue that pathogens should evolve toward lower virulence. However, disease ecologists showed with mathematical models that lower virulence is not always the predicted outcome (Anderson and May 1982). Rather, the evolution of pathogen virulence depends on biological details related to transmission and pathogen fitness. Pathogens for which transmission does not depend on a live host (e.g., pathogens able to grow saprotrophically and reproduce on plant debris), and pathogens with very efficient transmission (e.g., those that produce massive numbers of propagules or durable resting structures), are more likely to evolve high virulence. Conversely, biotrophic pathogens that rely on a healthy host to grow and reproduce, and pathogens with poor between-host transmission, may evolve towards a more benign association with their host (Jarosz and Davelos 1995).

13.5.2 Evolutionary tradeoffs and the cost of resistance

Evolution nearly always involves trade-offs. **Trade-offs** place constraints on adaptation; they can also result in balancing selection and maintain genetic diversity. As described above, trade-offs between reproduction and transmission influence the evolution of pathogen virulence. In our example of spore wall thickness, maximizing fitness involves a trade-off between two aspects of the pathogen's life history: dispersal and spore survival. Some trade-offs are clearly ecological; for example, plant phenology could be under conflicting selection pressures if summer drought favors early-flowering individuals but pathogen pressure favors later-flowering individuals. Other trade-offs are structural. Seeds with tough seed coats are better protected from soil-borne pathogens but cannot imbibe water as quickly—so they germinate more slowly. In roots, cells in the zone of elongation must be flexible to grow and so cannot have fully hardened cell walls, but that flexibility makes them leaky and susceptible to infection. Other trade-offs occur at the biochemical level, where production of a particular defense chemical depends on precursors that are also involved in other biochemical pathways. Limited availability of the precursor creates an unavoidable trade-off between production of different compounds with different physiological functions.

Trade-offs are ubiquitous where limited metabolic resources (energy and materials) are invested. If a plant invests heavily in production of a chemical or structural defense, those resources are unavailable for investment in growth and reproduction. In the absence of pathogen attack, this cost of defense should select against the production of the defense. Plants that employ inducible defenses (Section 12.4) can reduce the metabolic cost of defense by producing that defense only when actively threatened by pathogen attack. In fact, the existence of inducible defenses is itself an argument that defense must carry a fitness cost. Plant biologists tend to use the term **cost of resistance** rather than cost of defense, and here resistance includes both the production of constitutive and inducible defenses as well as the cost of maintaining resistance alleles (Brown and Rant 2013). Cost of resistance in plants was originally studied in the context of crop breeding; Vanderplank (1963) asked whether breeding for resistance carries with it a trade-off of lower yield, and why resistant genotypes sometimes decline in places where the corresponding pathogen is absent.

Models of the maintenance of genetic polymorphisms at resistance loci often include a key assumption that there is a cost of resistance (Antonovics and Thrall 1994). Yet measuring the cost of resistance accurately is challenging; studies must compare the fitness of plants that differ only in their resistance genotype, and must do so in the absence of the pathogen. In a review of 88 such studies done on crops and weeds, Bergelson and Purrington (1996) found significant evidence for a cost of resistance in about half of the studies. Crops were more likely to show a cost of resistance than non-crops, and plants were more likely to demonstrate a cost of resistance to pathogens than to herbivores. Even within a species, the cost of resistance is variable. In the model annual plant *Arabidopsis thaliana*, the cost of *R*-gene resistance varies strongly across loci and depends on the context of multiple pathogens and competing plants (Karasov et al. 2014b). We have surprisingly few studies of the cost of disease resistance in perennial plants or in wild plant populations outside of *Arabidopsis*. In a 3-year common garden study on the herbaceous perennial *Plantago lanceolata*, Susi and Laine (2015) measured costs of both qualitative and quantitative resistance against the powdery mildew fungus *Podosphaera plantaginis*, and found that the cost of resistance for *P. lanceolata* varied greatly over time and with plant age. This kind of context dependency is probably common, but more studies are needed! The paucity of studies in wild plant populations is surprising given the fundamental importance of cost of resistance in plant evolutionary ecology. Future studies will need to explore not just whether a cost of resistance exists, but how big the costs are and the nature of the trade-offs underlying them, both of which are fundamental for predicting evolutionary outcomes.

Cost of resistance can reflect complex biochemical interactions as well as trade-offs in responses to different antagonists. The plant hormones involved in coordinating induced defenses (salicylic acid, jasmonate, and ethylene) experience regulatory cross-talk resulting in trade-offs in defense across biotrophic and necrotrophic pathogens as well as herbivores (Section 12.7). These hormones also have roles in numerous plant functions besides defense, and regulatory cross-talk in these hormonal signals can add other important costs to the host plant (Karasov et al. 2017). For instance, salicylic acid can inhibit the production of auxin, which plays a critical role in cell elongation and apical dominance. In addition, "misfired" induction of defenses in response to microorganisms that are not pathogenic can lead to significant costs.

How a plant allocates resources to defenses in general is a response to fundamental evolutionary trade-offs involving resource availability. For instance, nearly all plants grow better when given more light. Yet some species absolutely require high light conditions to survive (light-demanding species) and others are able to tolerate lower light conditions (shade-tolerant species). Shade-tolerant species dominate low-light conditions such as the forest understory not because they require low-light conditions, but because the light-demanding species are not able to compete under those conditions. Light-demanding species often have high growth rates and a life-history strategy that takes advantage of abundant light resources to grow and reproduce quickly. According to the Resource Availability Hypothesis of Lissy Coley and colleagues (1985), such plants tend to invest little in defenses against pathogens and herbivores because they can afford to lose leaves, which they quickly replace. In contrast, shade-tolerant species are limited by the light resources available to produce new leaves. They are slow growing and tend to invest heavily in defense. Similarly, plants with long-lived evergreen foliage tend to invest more in defense than deciduous plants. As a consequence, fungal infections and disease are more common in light-demanding than shade-tolerant species (García-Guzmán and Heil 2014). Finally, across all organisms (not just plants), theory predicts that longer-lived organisms should invest more in defense than shorter-lived species (Bruns et al. 2015). This result is driven by an increased risk of disease and greater pathogen prevalence in longer-lived hosts, and it contradicts the expectation that evolution will necessarily be slower in longer-lived hosts because of their slower generation times.

13.5.3 The evolution of host range

A few pathogens are extreme generalists, able to cause disease and complete their life cycles on a very wide diversity of host species. They tend to be aggressive necrotrophs that rapidly kill host tissue and reproduce on the plant residue. For instance, the soil-borne fungal pathogen *Thanatephorus cucumeris* causes disease on at least 419 host species from 86 plant families. At the other extreme, a few pathogens are highly specialized on only a single host species or even on particular genotypes of a plant species. Pathovars of the bacterium *Pseudomonas syringae* show extreme host specialization associated with hundreds of virulence genes found in different combinations (Sarkar et al. 2006). While examples of extreme generalist and specialist pathogens are striking, the host range of plant pathogens falls on a continuum between these two extremes, with most pathogens being moderately polyphagous.

Figuring out exactly what should count as a host is a remarkably complicated question (Morris and Moury 2019). Often, we count as hosts those plants on which a pathogen causes disease symptoms after infection (Gilbert and Webb 2007). But we know that many microbes are active parts of the plant microbiome, including many latent pathogens that infect the host but only cause disease symptoms under particular conditions (Chapter 15). Should the host range include those plants on which the microbe can grow, even if it does not cause disease symptoms? What about hosts that exhibit disease symptoms but do not support pathogen reproduction? Should the host range include species that are susceptible if inoculated, but that fall outside the geographic range of the pathogen, or should it only include ecologically available host species? The definition of the host range of a pathogen depends on whether we are concerned about impacts on plants, epidemiology of the pathogen, structure of plant–pathogen networks, or evolution of interactions.

From an evolutionary perspective, the only hosts that matter to a pathogen are those on which it can reproduce (**competent hosts**). The two strongest targets of selection will be the ability to infect a host and the ability to produce propagules. Whether a pathogen produces disease symptoms that are devastating to the host (high virulence) or are insignificant (low virulence) is of little consequence to the pathogen except in how those symptoms affect pathogen fitness. For instance, *Phytophthora ramorum*, the cause of sudden oak death, kills oak trees (*Quercus* spp.), but it cannot reproduce on them. There may be little reason to evolve to be less aggressive, however, because *P. ramorum* infects many dozens of other host species, and some of them suffer minor symptoms (like California bay *Umbellularia californica*) but allow for abundant production of sporangia. That some of those sporangia happened to kill nearby oak trees is of little evolutionary consequence to the pathogen; the host range of the pathogen is thus an important modulator of evolution of virulence.

Why do some pathogens associate with many hosts while others attack only a few? It seems like it would benefit a pathogen to have a lot of alternative host options, so why don't all pathogens evolve to infect more and more kinds of hosts? As with so much in evolution, the answer is in a trade-off, this one described as "a jack-of-all-trades is master of none." To illustrate, Bruns et al. (2014) looked at 30 strains of *Puccinia coronata* f.sp. *avenae* (cause of oat crown rust) that varied widely in the number of *Avena sativa* genotypes they could infect; strains carried from 2 to 23 *avr* alleles. The strains that carried more *avr* alleles (i.e., broader ranges of host genotypes) had longer latent periods and made smaller pustules (thus fewer spores) than did strains with narrower host ranges. This cost of pathogenicity mirrors the cost of resistance in plants, described above, and is an important constraint to the evolution of broader host ranges. Basically, pathogens cannot be pathogenic on all hosts and hosts cannot be resistant to all pathogens because of trade-offs. We often end up somewhere in the middle.

Cost of pathogenicity can shape host ranges in another way. A pathogen can expand its current host range by adding a new host, or it can shift from one host to another. There is evidence from viral plant pathogens that a mutation that favors infectivity or virulence in one host may have the opposite effect in another host, a phenomenon called **antagonistic pleiotropy** (García-Arenal and Fraile 2013). These trade-offs have the effect of limiting host range expansion, and instead favor host switching or specialization. However, cross-host costs of pathogenicity appear to be more common in viruses than other pathogens (Sacristán and García-Arenal 2008).

Rapid changes in host range can result from mutations in **MAMPs** or **effectors** (Section 12.5) that allow a pathogen to evade detection on a new host. Hybridization and horizontal gene transfer can facilitate changes in host range by providing entirely new effectors (e.g., toxins or detoxifying enzymes) that enable pathogenicity on a new host. For example, horizontal transfer of the gene that encodes the host-selective toxin ToxA from *Stagonospora nodorum* to *Pyrenophora tritici-repentis* led to the emergence of a novel disease, tan spot of wheat (Mehrabi et al. 2011).

Within a set of plants acting as local hosts of a pathogen, adaptation can change the pathogen's interactions with each of its hosts. For instance, the foliar necrotroph *Stemphylium solani* is the dominant pathogen of both native and introduced "clovers" (species in the genera *Trifolium* and *Medicago*) in coastal California grasslands. In a serial passage experiment, we inoculated, re-isolated, and re-inoculated *S. solani* isolates on each of 11 different clover species (Gilbert and Parker 2010). The fungus rapidly evolved the ability to infect its particular serial-passage host more effectively over four generations. Then focusing on the introduced hosts, we went back to their native range (Mediterranean Europe) and collected plant genotypes for the same species found in California. We compared infectivity and aggressiveness on the introduced genotypes familiar with *S. solani* and European genotypes for which *S. solani* was a novel pathogen. Interestingly, California *S. solani* was much more effective at infecting familiar genotypes than European genotypes, suggesting an evolutionary increase in the ability to infect over the generations since host introduction.

Most of the above changes in host range are based on evolutionary changes in the pathogen arsenal. If a host evolves resistance to the pathogen, local host range would contract. Once again, evolution plays out in an ecological arena. For a polyphagous pathogen, the consequences of the loss of one host depend on how strongly that host influences pathogen population dynamics. For a host-specific, obligate pathogen, the "loss of one host" would mean local extinction. Therefore, the emergence of resistance creates strong selection pressure on a host-specific pathogen, potentially leading to reciprocal, coevolutionary responses. This is what we turn our attention to next.

13.6 Coevolutionary dynamics

Reciprocal selection between interacting species leads to a dynamic evolutionary process of **coevolution** (Thompson 1994). For example, when a trait of a pathogen causes selection on a corresponding trait in a plant, and that plant trait in turn drives selection that further shapes the trait in the pathogen, the plant and pathogen are coevolving. While there are many examples of how selection for particular traits in a host plant protect it against disease and increase host fitness, and similarly numerous examples of how selection for particular virulence traits in a pathogen increase its fitness, the reciprocal "co-" part of coevolution is remarkably difficult to demonstrate in plant–pathogen interactions (Burdon and Thrall 2009). Let's look at what is needed for coevolutionary dynamics.

13.6.1 Coevolution in a gene-for-gene system

The gene-for-gene system (Section 12.6) theoretically sets up an ideal situation for coevolutionary dynamics. Imagine a host plant with a biotrophic pathogen that blocks seed production. A compatible reaction (pathogen *avr* with host *rr*) prevents host reproduction (host fitness zero), whereas an incompatible reaction (pathogen *Avr* genotype with host *RR* or *Rr*) is fatal to the pathogen (pathogen fitness zero). Start with a population of all susceptible plants (*rr*). A mutation (*r* to *R*) in the LRR domain of the host NLR protein that recognizes the pathogen effector protein switches the interaction from compatible to incompatible. That mutation provides a strong selective advantage for the new host phenotype, and its frequency will increase with each host generation. Eventually the plant population is dominated by resistant individuals, and the pathogen has a low probability of encountering a susceptible host. This creates a strong selective advantage in the pathogen population for mutations (*Avr* to *avr*) that change the effector protein in a way that evades detection by the host NLR receptor. Because most of the host population is suddenly susceptible, the new *avr* genotype comes to dominate the pathogen population through a selective sweep. This in turn creates strong selective pressure on the host population, favoring new mutations in the NLR genes. Such cycles can continue indefinitely with reciprocal selection between the pathogen and host for molecular traits. Other heritable traits related to pathogenicity and resistance (outside the *R–Avr* gene-for-gene framework) can also be subject to coevolutionary dynamics.

It is important to note that simply documenting that a pathogen trait evolved to overcome a host defense does not demonstrate coevolution. As illustrated by the above scenario, the key characteristics of *co*-evolution are that (1) heritable resistance traits in the host provide the selective pressure for an increase in frequency of a trait in the pathogen, and (2) the heritable pathogenicity trait in turn creates the selective pressure for an increase in frequency of a resistance trait in the host.

13.6.2 Arms race versus trench warfare, and the Red Queen Hypothesis

There are two primary models of the genetic basis of coevolution in plant–pathogen systems. They are generally described using military terminology as the arms-race model and the trench-warfare model.

The classic model of antagonistic coevolution is like an **arms race**: a novel weapon evolves in one partner, the other partner evolves a counter-weapon, and so on, leading to a runaway escalation (Dawkins and Krebs 1979). Under the arms-race-model, pathogens accumulate more and more powerful tools in their arsenal, and plants accumulate more and more potent resistance mechanisms. Novel variants arise and go quickly to fixation in selective sweeps (Figure 13.5a). Consequently, genetic polymorphisms are transient rather than maintained over a long period, and genes for pathogenicity and resistance are expected to show low allelic diversity in pathogen and host populations at any particular time.

The second model is like **trench warfare**, drawing on an image of back-and-forth advances and losses between the pathogen and host (Stahl et al. 1999). Under the trench-warfare model, multiple alleles for effectors coexist in the pathogen population and multiple *R*-gene alleles coexist in the host population, and instead of selective sweeps, frequency-dependent processes lead to balancing selection. Individual alleles increase and decrease in a dynamic polymorphism, in contrast to the rapid substitution of alleles under the arms-race model (Figure 13.5b).

So, is plant–pathogen evolution more like an arms race or more like trench warfare? The earliest test of this question using genomic data was in the plant model system *Arabidopsis thaliana* (Stahl et al. 1999). Analyzing DNA sequences around *Rpm1*, a gene involved in the recognition of *Pseudomonas*

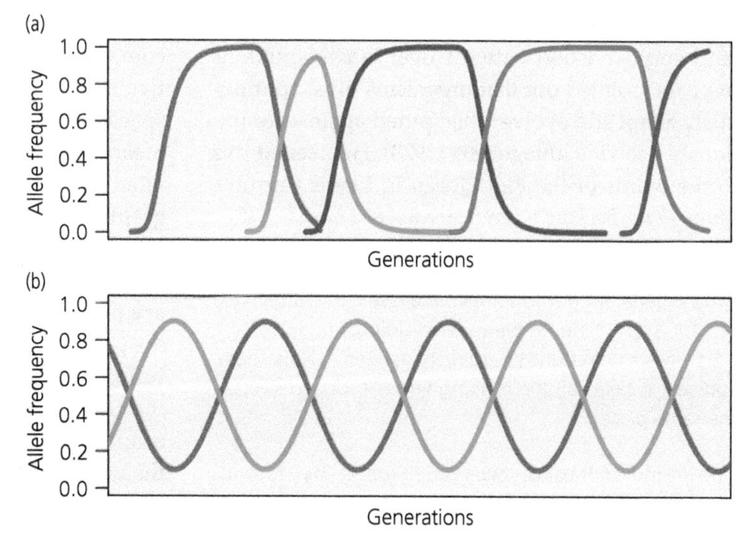

Figure 13.5 Contrasting hypotheses for how alternate alleles at the same locus change over time under the arms-race model versus the trench-warfare model of antagonistic coevolution. (a) In the arms-race model, each new allele for host resistance or for pathogen pathogenicity (aka "virulence," *sensu* Section 12.6) increases quickly to fixation, only to be replaced by the next new allele, in recurring selective sweeps. Genetic polymorphisms are fleeting. (b) In the trench-warfare model, multiple alleles (here two alleles) coexist in the population in a dynamic polymorphism maintained by frequency-dependent, balancing selection. Genetic polymorphisms are long-lasting, sometimes even predating taxonomic lineages. Figure by Gregory S. Gilbert.

pathogens, Stahl and coworkers discovered that alleles for resistance and susceptibility (caused by the deletion of *Rpm1*) have coexisted for millions of years. This long-lived polymorphism maintained by balancing selection was clear evidence for the trench-warfare model. In contrast, a whole-genome sequencing study of the anther-smut fungi *Microbotrium lychnidis-dioicae* and *M. silenes-dioicae*, parasites of *Silene latifolia* and *S. dioica*, found evidence for many genes that experienced selective sweeps (Badouin et al. 2017). These genes under positive selection included candidate genes for effectors, which were identified from a subset of genes that were upregulated during infection. Thus empirical studies support both arms-race and trench-warfare models for the genetic basis of coevolution. The recent explosion of genomics tools and data for nonmodel organisms has transformed research on this topic (Märkle et al. 2021, Amandine et al. 2022). Future advances will come by combining the study of both pathogen and host genomes, considering quantitative traits as well as Mendelian (e.g., gene-for-gene) traits, and putting genome coevolution into an ecological context, incorporating spatial structure, environmental variation, population dynamics, and pathogen life history, among other factors.

Antagonistic interactions like those between plants and pathogens, as well as predators and prey, parasites and their animal hosts, and herbivores and plants, are an ever-present threat to the survival of a population. Leigh Van Valen, a paleontologist studying extinction patterns over tens of millions of years, pointed out that organisms must continuously adapt and evolve when pitted against continuously evolving antagonists (1973). He likened this to the words of the Red Queen in Lewis Carroll's *Through the Looking Glass* (Carroll 1871):

> "Well, in our country," said Alice, still panting a little, "you'd generally get to somewhere else—if you ran very fast for a long time, as we've been doing."
> "A slow sort of country!" said the Queen. "Now, here, you see, it takes all the running you can do, to keep in the same place."

The same metaphor was later used by Hamilton (1980), Bell (2019), and others to argue that pathogen–host interactions drive the maintenance of sexual reproduction, commonly known as the **Red Queen Hypothesis** (Clay and Kover 1996). (For a history of the Red Queen Hypothesis see Lively (2010).) Testing the Red Queen Hypothesis using a comparative approach, Busch et al. (2004) found evidence that outcrossing rates increased with the number of fungal pathogens associated with a plant species (from fungi reported in Farr et al. 2003; see Box 15.1). Sexual reproduction enables rapid evolution of resistance in host populations (and pathogenicity for pathogens) through the dramatic effect of recombination on the generation of novel genotypes, essential under the arms-race model. Recombination also allows for the recovery of genotypes that were previously selected against and are now favored (through frequency-dependent, fluctuating selection), as in the trench-warfare model. In fact, some use the term Red Queen dynamics interchangeably with trench warfare to refer to that dynamic polymorphism (Amandine et al. 2022). The importance of pathogens in maintaining sexual reproduction in plants is still an open area of research.

13.6.3 Evolution in metapopulations and the geographic mosaic of coevolution

All evolution, including the evolution of plant–pathogen interactions, varies across space. Spatial variation applies to everything in the disease triangle: plant, pathogen, and environmental conditions, including microclimate, plant community composition (including the presence of alternative hosts for the pathogen), and other interacting species (including other pathogens). This variability means that selection and resulting patterns of adaptation differ among populations, creating a **geographic mosaic of selection** (Thompson 2005). In some locations, reciprocal selection between plant and pathogen are strongly coupled (**coevolutionary hotspots**) while in other locations, selection by the partner will be weak or lacking entirely (**coevolutionary coldspots**). The remixing of trait values through gene flow results in geographic patterns of coadaptation—and lack of coadaptation—between the partners (Thompson 2005).

Some variation happens along predictable gradients; for instance, fog is more common along the coast than further inland. Other aspects of spatial variation involve patchiness: riparian forests occur along moist river courses but are absent in the drier landscape between; alpine plant communities are found on cool mountain tops separated by warmer valleys; and islands are patches of terrestrial plants surrounded by an inhospitably wet sea. Earlier, we described how a network of connected populations, or **metapopulation**, is dynamic because both plants and pathogens can colonize, or go locally extinct from, individual patches of habitat (Section 11.6). Extinction is most common in small patches, and colonization is most common when patches are close together; disease incidence is expected to be greatest when plant populations are large and close together (Figure 11.11). Pathogens may enhance patch extinction rates in a plant metapopulation.

Metapopulation structure strongly influences evolution. For one thing, plant populations that are more highly connected to other populations may evolve higher levels of disease resistance (Jousimo et al. 2014). Second, the metapopulation as a whole creates numerous, independent genetic bottlenecks for both the host and the pathogen, with associated founder effects (Section 13.2). Thus, each colonized patch is likely to have a different combination of host and pathogen genotypes, each a subset of the genetic diversity of the overall metapopulation. When coupled with limited gene flow among patches, these patchy patterns create independent evolutionary trajectories in the plant, the pathogen, and the interaction between them (Thrall and Burdon 1997). Because coevolution is cyclical in a tightly coupled host–pathogen pair, the pathogen may have the upper hand with most plants susceptible in some individual patches, while in other patches, the plant may have the upper hand and most plants are resistant.

For example, in Australia, Thrall et al. (2012) studied the evolutionary dynamics of six patches of a metapopulation of *Linum marginale* (flax) infected by its host-specific rust *Melampsora lini*, collecting both seeds and fungal spores every other year for a 6-year period. They measured changes in allele frequencies by sequence analysis of two *Avr* loci in the pathogen. In a clever greenhouse experiment, they then recreated the temporal dynamics of evolution of pathogenicity in the pathogen and resistance in the host by inoculating plants with spores from pathogens from two years before seed collection, the same year as seed collection, and two years after seed collection. They found that the evolutionary dynamics varied across different patches, with some patches showing adaptive changes in pathogen genotypes leading to greater pathogenicity and/or adaptive changes in host genotypes leading to greater host resistance. In other patches, either plant or pathogen showed changes that were maladaptive—reducing pathogenicity or reducing resistance. The changes were not synchronized across patches. Because severe infections by the rust make hosts susceptible to dying in the subsequent winter, changes that led to greater disease incidence also increased the chances of host extinction in a patch.

13.7 Macroevolution and plant disease

Macroevolution refers to the evolution of species and higher taxonomic groups. The same evolutionary processes—mutation and recombination, gene flow, genetic drift, and selection—are at work, but over immense periods of time. Then add the extinction of lineages and the appearance of new lineages through speciation. You can think of macroevolution as the study of patterns of biodiversity over geologic time. To reconstruct patterns of speciation, extinction, and changes in traits, we use fossil evidence and comparisons among extant taxa, relying heavily on DNA sequence information. We name and classify living organisms through the practices of **systematics**, which is the scientific field of discerning evolutionary relationships among taxa.

13.7.1 Phylogenetic trees

Evolutionary relationships among lineages are represented as **phylogenetic trees** (Box 13.2). The branches of a phylogenetic tree are hypotheses for how and when particular lineages began to evolve independently of each other. Creating and interpreting phylogenetic trees is generally much harder for microbes (and even plants) than it is for animals, because microbes can easily exchange genes across independent lineages, leading to converging branches in what is called a **reticulate phylogeny**.

(For a highly entertaining summary of this topic, read David Quammen's (2018) *New York Times* best-seller *The Tangled Tree: A Radical New History of Life*). **Horizontal gene transfer** occurs commonly in bacteria through conjugation, transformation, and transduction (Section 5.3). Horizontal gene transfer also occurs in fungi and oomycetes through viral transfer, extra-chromosomal DNA, and hybridization through anastomosis and mitotic recombination. Gene transfers can sometimes be across entire kingdoms; for example, horizontal gene transfer from bacteria to nematodes allowed the acquisition of novel enzymatic activities (Quist et al. 2015). In plants, hybridization across divergent lineages is also common (Soltis and Soltis 2009). Plants are much more likely to produce viable offspring following hybridization than are animals, sometimes by doubling chromosome number in a process called **polyploidization**. An **allopolyploid** is produced from the hybridization of two plant species and retains both chromosomes from each of the parents. (**Autopolyploids** are created from chromosome doubling with two parents of the same species; this does not result in reticulate evolution). Because plants (unlike most animals) are modular, can reproduce clonally, and/or can self-fertilize, the allopolyploid offspring may give rise to a new lineage that is reproductively isolated from either of the parental lineages. This is essentially instant speciation, as well as reticulate evolution. We now know that allopolyploidy in plants and horizontal gene transfer in microbes have contributed in many important ways to evolution. We are only beginning to unpack the implications of this revelation for plant–pathogen interactions (Soanes and Richards 2014, Lacroix and Citovsky 2018).

Box 13.2 How to read and interpret a phylogenetic tree

Phylogenetic trees are graphical representations of hypotheses about evolutionary descent and relationships among taxa (Figure 13.6). Trees are used because evolution produces patterns that are tree-like with many branching paths, rather than a ladder-like progression from bottom to top. The tips of the branches are individual taxa (terminal taxa), such as extant species. The branches connect together at **nodes**. A node represents a hypothesized ancestral taxon common to all the descendent lineages. Those ancestral taxa likely no longer exist, but much of their genomes still do, in various versions, among their descendants. The common ancestor and all of the branches that are descended from it form a **clade**. Phylogenetic trees are hierarchical, with clades nested within each other; small clades with just a few taxa are parts of larger clades with many taxa (Figure 13.6). Clades are sometimes given taxonomic designations, like genus, family, or order.

The **topology** of the tree is the pattern of branching of the tree, using shared characters like genetic sequences and phenotypic traits to arrange taxa hierarchically and represent the evolutionary hypothesis of patterns of common descent. The length of the branches on the tree are usually drawn with lengths proportional to the time since that lineage first evolved, usually measured in millions of years (**Myr**). While the topology reflects the overall pattern of which taxa are more closely or more distantly related and which taxa descended from which ancestors, the branch lengths provide a quantitative estimate of how long the lineages have been evolving. The time from the present to the origin of most recent common ancestor (**MRCA**) of a clade—the age of the node—is usually determined by the earliest appearances of those taxa in the fossil record (dated using radiometric methods and stratigraphic associations of where the fossils were found) together with calculations from the number of differences in base pairs in DNA sequences, assuming a particular rate of accumulation of mutations called a **molecular clock**. The molecular clock uses assumptions of how quickly mutations are expected to arise in a lineage (including things like background mutation rates and generation times) to extrapolate backwards to estimate how long it would take to reach the level of genetic divergence observed today. The age of a node representing the MRCA is often given in units of **Ma (mega-annum)**—millions of years ago. Each of the descendent lineages from that node have been evolving independently of each other since they became reproductively isolated; each branch represents many generations and accumulated neutral and non-neutral mutations without substantial gene flow between lineages. The pairwise **phylogenetic distance** between two taxa is given as twice the age of the node for the MRCA, measured in units of Myr (Figure 13.6).

Box 13.2 *Continued*

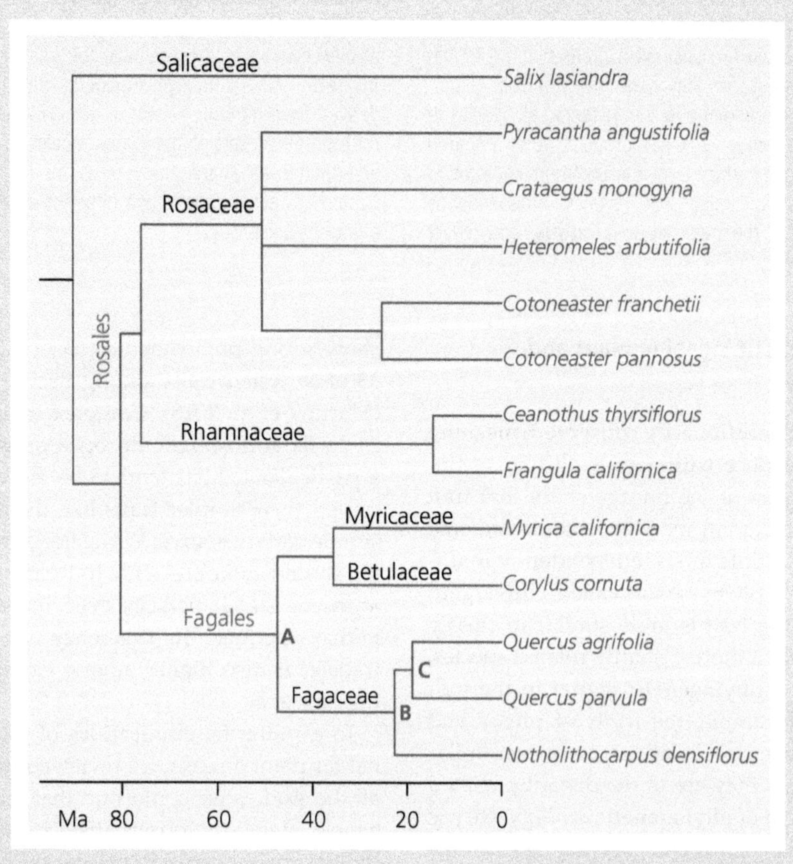

Figure 13.6 Phylogenetic tree of the "fabids" clade of Rosids among the woody plants on the UCSC Forest Ecology Research Plot showing the hierarchical relationship among clades. Time goes from the right at present (0 Ma) back to 100 Ma at left. Ancestors are on the left, and descendant lineages to the right. Terminal taxa are at the branch tips as extant species. Node A shows the most recent common ancestor for all species in the clade that includes families Fagaceae, Betulaceae, and Myricaceae: these three families are included in the order Fagales. Within the family Fagaceae (clade of taxa distal to node B), *Notholithocarpus* is distinct from the two species in genus *Quercus* (node C). The age of the nodes can be read from the scale at the bottom. The pairwise phylogenetic distance is twice the age of the most recent common ancestor for a pair of species. For instance, the MRCA for *Quercus agrifolia* and *Notholithocarpus densiflorus* is at node B, with an age of 23 Ma; the phylogenetic distance between those two species is then 46 Myr. The phylogenetic distance from *Q. agrifolia* to each of the species in the *Rosales* clade is 160 Myr. Note that there is not a consistent age at which clades are designated as orders, families, or genera; *Ceanothus* and *Frangula* have a shorter phylogenetic distance than do species within *Quercus* or *Cotoneaster*; three families in the order Fagales have a common ancestor that is more recent than the node for the family Rosaceae. Criteria for these clade designations vary among systematists and the families being studied.

Figure by Gregory S. Gilbert.

Box 13.2 *Continued*

Clades are often characterized by their **shared derived characters**; when a trait (character) appears in the descendants of a common ancestor that had that trait, it suggests that the trait is **homologous**, and that the trait has been **phylogenetically conserved**. Traits that are similar across organisms because they were inherited from a common ancestor are homologous. For instance, all species in the clade of the family Fagaceae produce nuts, a shared derived character inherited from the common ancestor of all Fagaceae. When similar traits evolved independently in different clades, the traits are **analogous**, and reflect **convergent evolution**. For instance, toothed margins of leaves, C_4 photosynthesis, and fleshy fruits to encourage animal dispersal of seeds have each evolved independently numerous times among flowering plants. A set of taxa that all evolved from a common ancestor are considered **monophyletic**—of a single phylogenetic origin. A set of taxa derived from multiple ancestral groups are **polyphyletic**; for example, desert plants in the families of Euphorbiaceae and Cactaceae share many traits for life in arid conditions, but because they have different ancestors, they are a polyphyletic group.

13.7.2 Phylogenetic signal in plant and pathogen traits

Many traits are **evolutionarily conserved**, meaning the trait evolved once within an evolutionary lineage, and members of the lineage share that trait because of their common ancestor. In contrast, some traits evolved multiple times independently in different lineages, a phenomenon called **convergent evolution.** When a trait is more similar in closely related species than more distantly related species, we say there is a **phylogenetic signal** in the trait (Box 13.3). For example, the fruits of plums and peaches, which are close relatives, are more similar to each other than they are to the distantly related walnut. In the field of **phylogenetic ecology**, we use patterns of physiological and morphological traits across evolutionary lineages to understand the role of those traits in the distribution and abundance of species. Phylogenetic ecology provides tools to generalize what we know about some species to their close relatives, and it helps create a unified understanding from case studies in ecology.

Many traits of both plants and microbes show a phylogenetic signal (Figure 13.8, Box 13.3, and Gilbert and Parker 2016). Phylogenetically conserved traits in plants include plant architecture, growth, and metabolism, while some traits related to water use efficiency and photosynthesis are less strongly conserved (i.e., they are more labile). Similarly, many functional traits of microorganisms show strong phylogenetic conservatism, including 83 of 89 tested traits among Bacteria and Archaea (Martiny et al. 2013). Complex traits like the ability to fix atmospheric nitrogen or association with a particular habitat tend to be more strongly conserved than simpler traits like the ability to use a simple carbon source. While horizontal gene transfer should interfere with trait conservatism, phylogenetic signal persists even in Bacteria. This is perhaps because the frequency of horizontal gene transfer is also higher among closely related taxa (Friesen et al. 2008).

To explore the implications of phylogenetic signal for plant disease, we reviewed the literature for all the studies we could find that measured phylogenetic signal or conservatism for features related to interactions between plants and microbes. Traits included those associated with the pathogen arsenal and host defenses, traits specific to the interaction outcomes themselves, and traits related to relevant aspects of morphology and physiology of both pathogen and host (Gilbert and Parker 2016). Nearly all the traits showed significant phylogenetic signal (Figure 13.8).

The minority of traits that do not show a phylogenetic signal are particularly informative. We see a phylogenetic signal when the members of some clades have a particular trait but those of other clades do not. There are two very different ways to *not* show a phylogenetic signal. First, the trait may be so ancient that it is widespread across

Box 13.3 Phylogenetic signal

Pairs of taxa with shorter phylogenetic distances are more closely related than those at longer phylogenetic distances; across their genomes, close relatives should have fewer accumulated differences in DNA sequences and overall should have more similar phenotypes (a greater proportion of shared traits). The pattern in which trait similarity between taxa declines with the phylogenetic distance between them is called a **phylogenetic signal** (Figure 13.7) (Gilbert and Parker 2016). Phylogenetic signal is thus a measure of the extentof phylogenetic conservatism. Traits can be binary (they have it or they do not) or continuous (amount or degree). Evolutionary biologists have developed several metrics for the strength of phylogenetic signal of traits across a phylogeny: Fritz and Purvis' D is used for binary traits, and Blomberg's K, Pagel's λ, and Moran's I are used for continuous traits (Gilbert and Parker 2016). The slope of the negative relationship between trait similarity and phylogenetic distance is an indicator of phylogenetic signal (Figure 13.7b).

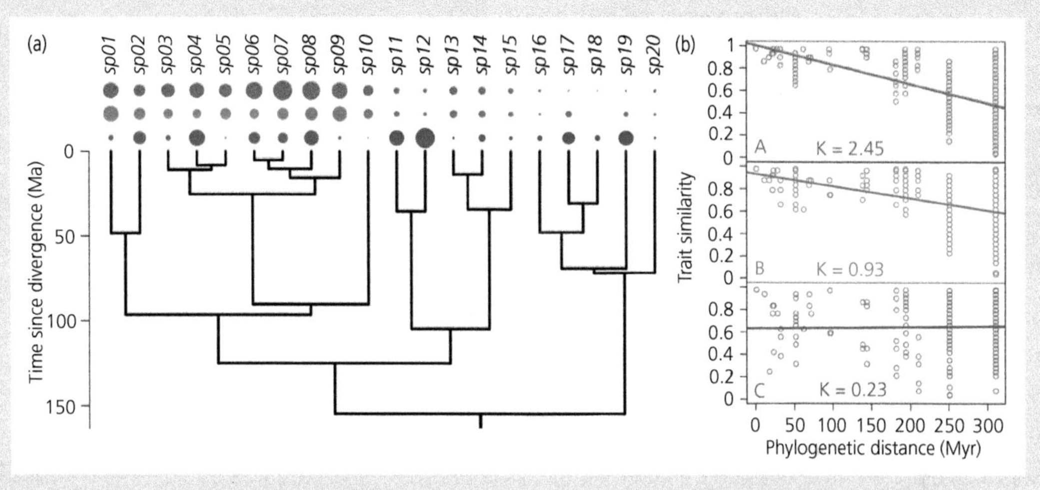

Figure 13.7 Phylogenetic signal. (a) The topology of the phylogenetic tree illustrates the hypothesized evolutionary history of 20 species. The timescale ranges from the present, at the branch tips, to approximately 150 Ma (million years ago) at the root. Species are shown in italics at branch tips. The distribution of three traits (A, B, C) across the branch tips is shown at the top, with the size of the circle representing the measured value of the trait. Trait A (in green) shows a pattern expected with strong phylogenetic conservatism: Species 1–10 form a clade, and all species in that clade show a large value for trait A, while species in other clades do not. In contrast, values for trait C (in blue) appear seemingly random across the clades, suggesting it is a highly labile trait or that the trait has evolved independently several times. Finally, trait B (in orange) shows an intermediate degree of conservatism.
(b) Phylogenetic signal reflects the relationship between trait similarity and phylogenetic distance. Each point in the scatterplot is one species pair, with a measure of trait similarity between them. Phylogenetic distance between any two species is twice the time since divergence; e.g., the most recent common ancestor between Species 1 and 15 is 102 Ma, so their phylogenetic distance is 102×2 = 204 Myr of independent evolution. The slope for trait A is steep, revealing a strong phylogenetic signal, while the slope for trait C is flat, showing the lack of phylogenetic signal we expect for labile traits. Blomberg's *K* is calculated for each of the three cases.
Reproduced (with modification) with permission from the Annual Review of Phytopathology, Volume 54 © 2016 by Annual Reviews, http://www.annualreviews.org

all the clades. For example, the three plant signaling pathways (ethylene, jasmonic acid, and salicylic acid) do not show phylogenetic signal because they are present in nearly all plants—these very ancient signaling mechanisms work well and are nearly universally retained. The other way to *not* show a phylogenetic signal is when a trait is evolving very rapidly within lineages, even within populations. For example, we found that pathogen effectors and the LRR domain of receptor proteins involved in plant resistance did not show a phylogenetic signal. These molecular traits are evolving rapidly in the

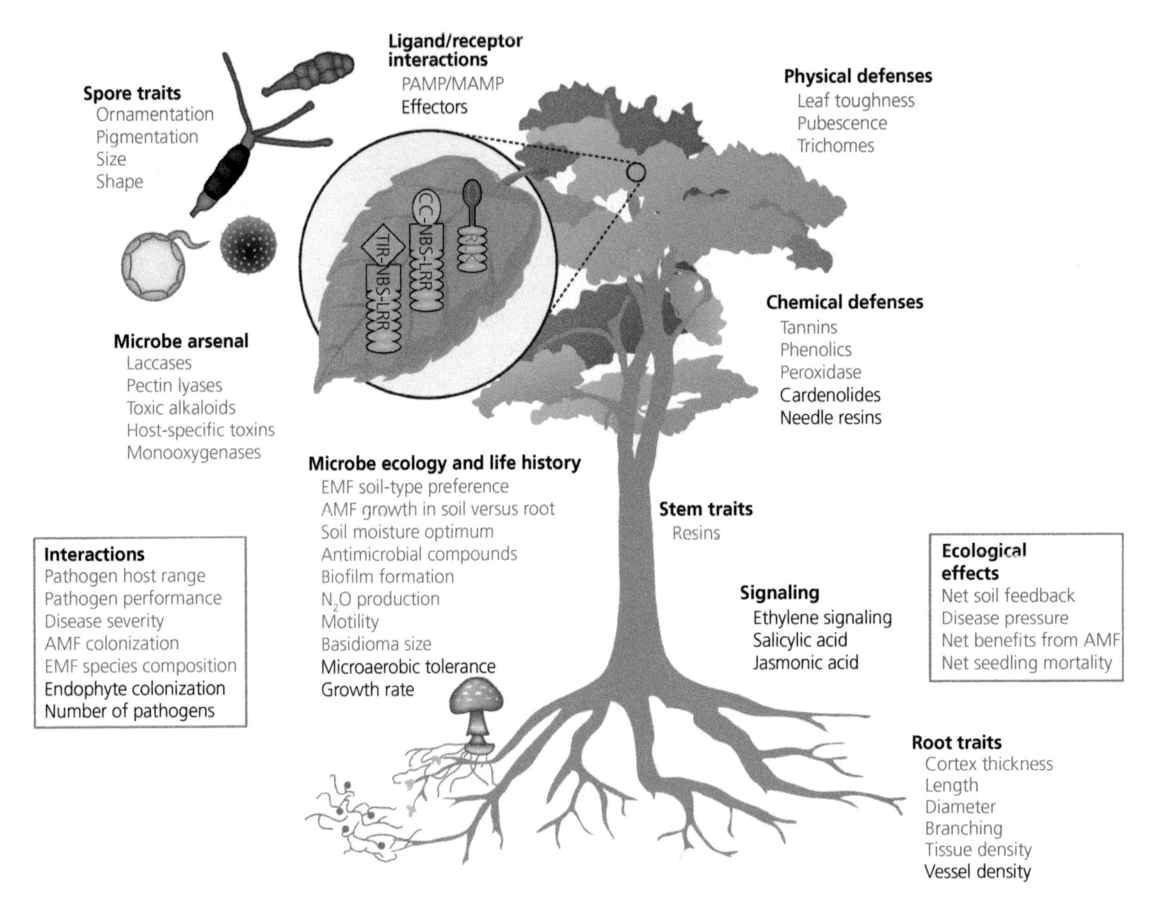

Figure 13.8 Schematic of important features of plant–microbe interactions for which phylogenetic signal or phylogenetic conservatism has been studied, including plant traits, microbial traits, and features of the interaction itself. Blue labels indicate features for which a phylogenetic signal has been demonstrated; black labels indicate features for which studies did not find a significant phylogenetic signal, or features expected to be highly labile, such as rapidly coevolving receptors. Abbreviations: AMF, arbuscular mycorrhizal fungi; CC, coiled-coil; EMF, ectomycorrhizal fungi; LRR, leucine-rich repeat; MAMP, microbe-associated molecular pattern; NBS, nucleotide binding site; PAMP, pathogen-associated molecular pattern; RLK, receptor-like kinase; TIR, Toll–interleukin-1 receptor.

Reproduced with permission from the Annual Review of Phytopathology, Volume 54 © 2016 by Annual Reviews, http://www.annualreviews.org

local context of a host–pathogen arms race (Gilbert and Parker 2016).

13.7.3 Phylogenetic signal in host ranges

As described in Section 13.5, most pathogens are not narrowly limited to a single host species, nor are they able to infect all plant species; often, they attack a limited set of related host species. For example, Barrett et al. (2009) found that about half of fungal plant pathogens are restricted to a single plant family. Because a fungus that causes disease on one plant species is more likely to also cause disease on close relatives, this suggests that taking a phylogenetic ecology approach to the study of plant disease might teach us something new about pathogen host ranges.

One way to measure the phylogenetic signal in host range is to think of the ability of a pathogen to cause disease on a particular host as a trait in itself. Gilbert and Webb (2007) calculated the phylogenetic distance between the source host (from which a fungal pathogen was isolated) to the target host (host that received the fungal inoculum) for 578 cross-inoculated source-target pairs in rainforest plots in Panama. They found a very strong phylogenetic signal in host range. These necrotrophic pathogens were highly likely to cause disease on closely related species, and the probability declined continuously with phylogenetic distance between the source and its target. Similar patterns were found when examining global host-range databases for a diversity of types of plant pathogens (Gilbert et al. 2012).

Patterns of evolutionary relatedness turn out to be useful for interpreting and predicting many phenomena in plant–pathogen interactions. Phylogenetic signal in host range influences the dynamics of pathogens in plant communities (Section 14.5), how pathogens can help maintain plant diversity (Section 14.7), what kinds of impacts introduced pathogens are likely to have (Section 16.4), how pathogens help prevent introduced plants from becoming weeds (Section 16.5), and which plant species combinations might be most effective in intercropping or rotation sequences to manage disease in crops (Section 17.4).

Further reading

Burdon, J. J. and Laine, A. L. 2019. *Evolutionary Dynamics of Plant Pathogen Interactions*. Cambridge University Press.

Herron, J. C. and S. Freeman. 2014. *Evolutionary Analysis*, 5th edition. Pearson, London.

Milgroom, M. G. 2015. *Population Biology of Plant Pathogens: Genetics, Ecology, and Evolution*. APS Press, St. Paul, MN.

Thompson, J. N. 1994. *The Coevolutionary Process*. University of Chicago Press.

References

Aguileta, G., G. Refregier, R. Yockteng, E. Fournier, and T. Giraud. 2009. Rapidly evolving genes in pathogens: methods for detecting positive selection and examples among fungi, bacteria, viruses and protists. *Infection, Genetics and Evolution* **9**:656–670.

Amandine, C., D. Ebert, E. Stukenbrock, R. C. R. de la Vega, P. Tiffin, D. Croll, and A. Tellier. 2022. Unraveling coevolutionary dynamics using ecological genomics. *Trends in Genetics* **38**:1003–1012.

Anderson, R. M. and R. M. May. 1982. Coevolution of hosts and parasites. *Parasitology* **85**:411–426.

Antonovics, J. and P. H. Thrall. 1994. The cost of resistance and the maintenance of genetic polymorphism in host–pathogen systems. *Proceedings of the Royal Society of London, Series B, Biological Sciences* **257**:105–110.

Avise, J. C. 1989. Gene trees and organismal histories: a phylogenetic approach to population biology. *Evolution* **43**:1192–1208.

Badouin, H., P. Gladieux, J. Gouzy, S. Siguenza, G. Aguileta, A. Snirc, S. Le Prieur, C. Jeziorski, A. Branca, and T. Giraud. 2017. Widespread selective sweeps throughout the genome of model plant pathogenic fungi and identification of effector candidates. *Molecular Ecology* **26**:2041–2062.

Barrett, J. 1988. Frequency-dependent selection in plant–fungal interactions. *Philosophical Transactions of the Royal Society of London, Series B, Biological Sciences* **319**: 473–483.

Barrett, L. G., J. M. Kniskern, N. Bodenhausen, W. Zhang, and J. Bergelson. 2009. Continua of specificity and virulence in plant host–pathogen interactions: causes and consequences. *New Phytologist* **183**:513–529.

Bell, G. 2019. *The Masterpiece of Nature: The Evolution and Genetics of Sexuality*. Routledge, Abingdon, Oxfordshire.

Bergelson, J., M. Kreitman, E. A. Stahl, and D. Tian. 2001. Evolutionary dynamics of plant *R*-genes. *Science* **292**:2281–2285.

Bergelson, J. and C. B. Purrington. 1996. Surveying patterns in the cost of resistance in plants. *American Naturalist* **148**:536–558.

Brown, J. and J. Rant. 2013. Fitness costs and trade-offs of disease resistance and their consequences for breeding arable crops. *Plant Pathology* **62**:83–95.

Bruns, E., M. L. Carson, and G. May. 2014. The jack of all trades is master of none: a pathogen's ability to infect a greater number of host genotypes comes at a cost of delayed reproduction. *Evolution* **68**: 2453–2466.

Bruns, E., M. E. Hood, and J. Antonovics. 2015. Rate of resistance evolution and polymorphism in long-and short-lived hosts. *Evolution* **69**:551–560.

Burdon, J. J. and P. H. Thrall. 2009. Coevolution of plants and their pathogens in natural habitats. *Science* **324**: 755–756.

Busch, J. W., M. Neiman, and J. M. Koslow. 2004. Evidence for maintenance of sex by pathogens in plants. *Evolution* **58**:2584–2590.

Carroll, L. 1871. *Through the Looking Glass*. Ch. II, The garden of live flowers. Macmillan. Project Gutenberg, https://www.gutenberg.org/files/12/12-h/12-h.htm#link2HCH0002.

Chase, A. B., C. Weihe, and J. B. Martiny. 2021. Adaptive differentiation and rapid evolution of a soil bacterium along a climate gradient. *Proceedings of the National Academy of Sciences of the United States of America* **118**:e2101254118.

Clausen, J., D. D. Keck, and W. M. Hiesey. 1940. *Experimental Studies on the Nature of Species. I. Effect of Varied Environments on Western North American Plants.* Carnegie Institution of Washington, Washington DC.

Clay, K. and P. X. Kover. 1996. The Red Queen Hypothesis and plant/pathogen interactions. *Annual Review of Phytopathology* **34**:29–50.

Coley, P. D., J. P. Bryant, and F. S. Chapin, III. 1985. Resource availability and plant antiherbivore defense. *Science* **230**:895–899.

Dawkins, R. and J. R. Krebs. 1979. Arms races between and within species. *Proceedings of the Royal Society of London, Series B, Biological Sciences* **205**:489–511.

Dobzhansky, T. 1950. Evolution in the tropics. *American Scientist* **38**:209–221.

Dutech, C., B. Barrès, J. Bridier, C. Robin, M. Milgroom, and V. Ravigné. 2012. The chestnut blight fungus world tour: successive introduction events from diverse origins in an invasive plant fungal pathogen. *Molecular Ecology* **21**:3931–3946.

Estoup, A. and T. Guillemaud. 2010. Reconstructing routes of invasion using genetic data: why, how and so what? *Molecular Ecology* **19**:4113–4130.

Farr, D. F., A. Y. Rossman, M. E. Palm, and E.B McCray. 2003. Fungus–Host Distributions, Fungal Databases, Systematic Botany and Mycology Lab. ARS/USDA. http://nt.ars-grin.gov/fungaldatabases/.

Friesen, T. L., J. D. Faris, P. S. Solomon, and R. P. Oliver. 2008. Host-specific toxins: effectors of necrotrophic pathogenicity. *Cellular Microbiology* **10**:1421–1428.

García-Arenal, F. and A. Fraile. 2013. Trade-offs in host range evolution of plant viruses. *Plant Pathology* **62**:2–9.

García-Guzmán, G. and M. Heil. 2014. Life histories of hosts and pathogens predict patterns in tropical fungal plant diseases. *New Phytologist* **201**:1106–1120.

Gilbert, G. S., R. Magarey, K. Suiter, and C. O. Webb. 2012. Evolutionary tools for phytosanitary risk analysis: phylogenetic signal as a predictor of host range of plant pests and pathogens. *Evolutionary Applications* **5**:869–878.

Gilbert, G. S. and I. M. Parker. 2010. Rapid evolution in a plant–pathogen interaction and the consequences for introduced host species. *Evolutionary Applications* **3**: 144–156.

Gilbert, G. S. and I. M. Parker. 2016. The evolutionary ecology of plant disease: A phylogenetic perspective. *Annual Review of Phytopathology* **54**:549–578.

Gilbert, G. S. and C. O. Webb. 2007. Phylogenetic signal in plant pathogen-host range. *Proceedings of the National Academy of Sciences of the United States of America* **104**:4979–4983.

Hamilton, W. D. 1980. Sex versus non-sex versus parasite. *Oikos* **35**:282–290.

Hanley, M. E. and R. H. Groves. 2002. Effect of the rust fungus *Puccinia chondrillina* TU 788 on plant size and plant size variability in *Chondrilla juncea*. *Weed Research* **42**:370–376.

Holsinger, K. E. and B. S. Weir. 2009. Genetics in geographically structured populations: defining, estimating and interpreting F_{ST}. *Nature Reviews Genetics* **10**:639–650.

Jarosz, A. M. and A. L. Davelos. 1995. Tansley Review No. 81: Effects of disease in wild plant populations and the evolution of pathogen aggressiveness. *New Phytologist* **129**:371–387.

Jousimo, J., A. J. Tack, O. Ovaskainen, T. Mononen, H. Susi, C. Tollenaere, and A.-L. Laine. 2014. Ecological and evolutionary effects of fragmentation on infectious disease dynamics. *Science* **344**:1289–1293.

Karasov, T. L., E. Chae, J. J. Herman, and J. Bergelson. 2017. Mechanisms to mitigate the trade-off between growth and defense. *The Plant Cell* **29**:666–680.

Karasov, T. L., M. W. Horton, and J. Bergelson. 2014a. Genomic variability as a driver of plant–pathogen coevolution? *Current Opinion in Plant Biology* **18**:24–30.

Karasov, T. L., J. M. Kniskern, L. Gao, B. J. DeYoung, J. Ding, U. Dubiella, R. O. Lastra, S. Nallu, F. Roux, and R. W. Innes. 2014b. The long-term maintenance of a resistance polymorphism through diffuse interactions. *Nature* **512**:436–440.

Lacroix, B. and V. Citovsky. 2018. Beyond Agrobacterium-mediated transformation: horizontal gene transfer from bacteria to eukaryotes. *Agrobacterium Biology* **418**: 443–462.

Linnen, C. R. and H. E. Hoekstra. 2009. Measuring natural selection on genotypes and phenotypes in the wild. *Cold Spring Harbor Symposia on Quantitative Biology* **74**: 155–168.

Lively, C. M. 2010. A review of Red Queen models for the persistence of obligate sexual reproduction. *Journal of Heredity* **101**:S13–S20.

Malthus, T. R. 1826. An essay on the principle of population or a view of its past and present effects on human happiness, an inquiry into our prospects respecting the future removal or mitigation of the evils which it occasions by Rev. TR Malthus. Reeves & Turner, London.

Märkle, H., S. John, A. Cornille, P. D. Fields, and A. Tellier. 2021. Novel genomic approaches to study antagonistic coevolution between hosts and parasites. *Molecular Ecology* **30**:3660–3676.

Martiny, A. C., K. Treseder, and G. Pusch. 2013. Phylogenetic conservatism of functional traits in microorganisms. *ISME Journal* **7**:830–838.

Maynard Smith, J. and J. Haigh. 1974. The hitch-hiking effect of a favourable gene. *Genetics Research* **23**:23–35.

Mehrabi, R., A. H. Bahkali, K. A. Abd-Elsalam, M. Moslem, S. Ben M'Barek, A. M. Gohari, M. K. Jashni,

I. Stergiopoulos, G. H. Kema, and P. J. de Wit. 2011. Horizontal gene and chromosome transfer in plant pathogenic fungi affecting host range. *FEMS Microbiology Reviews* **35**:542–554.

Milgroom, M. G., K. Wang, Y. Zhou, S. E. Lipari, and S. Kaneko. 1996. Intercontinental population structure of the chestnut blight fungus, *Cryphonectria parasitica*. *Mycologia* **88**:179–190.

Milus, E., E. Seyran, and R. McNew. 2006. Aggressiveness of *Puccinia striiformis* f. sp. *tritici* isolates in the south-central United States. *Plant Disease* **90**:847–852.

Morris, C. E. and B. Moury. 2019. Revisiting the concept of host range of plant pathogens. *Annual Review of Phytopathology* **57**:63–90.

Nieuwoudt, A., M. Human, M. Craven, and B. Crampton. 2018. Genetic differentiation in populations of *Exserohilum turcicum* from maize and sorghum in South Africa. *Plant Pathology* **67**:1483–1491.

Parker, M. A. 1986. Individual variation in pathogen attack and differential reproductive success in the annual legume, *Amphicarpaea bracteata*. *Oecologia* **69**:253–259.

Pritchard, J. K., M. Stephens, and P. Donnelly. 2000. Inference of population structure using multilocus genotype data. *Genetics* **155**:945–959.

Quammen, D. 2018. The Tangled Tree: A Radical New History of Life. Simon & Schuster, New York.

Quist, C. W., G. Smant, and J. Helder. 2015. Evolution of plant parasitism in the phylum Nematoda. *Annual Review of Phytopathology* **53**:289–310.

Sacristán, S. and F. García-Arenal. 2008. The evolution of virulence and pathogenicity in plant pathogen populations. *Molecular Plant Pathology* **9**:369–384.

Sarkar, S. F., J. S. Gordon, G. B. Martin, and D. S. Guttman. 2006. Comparative genomics of host-specific virulence in *Pseudomonas syringae*. *Genetics* **174**:1041–1056.

Schweizer, G., K. Münch, G. Mannhaupt, J. Schirawski, R. Kahmann, and J. Y. Dutheil. 2018. Positively selected effector genes and their contribution to virulence in the smut fungus *Sporisorium reilianum*. *Genome Biology and Evolution* **10**:629–645.

Sniegowski, P. D., P. J. Gerrish, T. Johnson, and A. Shaver. 2000. The evolution of mutation rates: separating causes from consequences. *Bioessays* **22**:1057–1066.

Soanes, D. and T. A. Richards. 2014. Horizontal gene transfer in eukaryotic plant pathogens. *Annual Review of Phytopathology* **52**:583–614.

Soltis, P. S. and D. E. Soltis. 2009. The role of hybridization in plant speciation. *Annual Review of Plant Biology* **60**:561–588.

Stahl, E. A., G. Dwyer, R. Mauricio, M. Kreitman, and J. Bergelson. 1999. Dynamics of disease resistance polymorphism at the *Rpm1* locus of *Arabidopsis*. *Nature* **400**:667–671.

Supkof, D. M., D. B. Joley, and J. J. Marois. 1988. Effect of introduced biological control organisms on the density of *Chondrilla juncea* in California USA. *Journal of Applied Ecology* **25**:1089–1096.

Susi, H. and A. L. Laine. 2015. The effectiveness and costs of pathogen resistance strategies in a perennial plant. *Journal of Ecology* **103**:303–315.

Taylor, J., D. Jacobson, and M. Fisher. 1999. The evolution of asexual fungi: reproduction, speciation and classification. *Annual Review of Phytopathology* **37**: 197–246.

Thompson, J. N. 1994. *The Coevolutionary Process*. University of Chicago Press.

Thompson, J. N. 2005. *The Geographic Mosaic of Coevolution*. University of Chicago Press.

Thrall, P. H. and J. J. Burdon. 1997. Host–pathogen dynamics in a metapopulation context: The ecological and evolutionary consequences of being spatial. *Journal of Ecology* **85**:743–753.

Thrall, P. H., A. L. Laine, M. Ravensdale, A. Nemri, P. N. Dodds, L. G. Barrett, and J. J. Burdon. 2012. Rapid genetic change underpins antagonistic coevolution in a natural host–pathogen metapopulation. *Ecology Letters* **15**:425–435.

Turesson, G. 1922. The plant species to the habitat. *Hereditas* **3**:211–350.

Van Valen, L. 1973. A new evolutionary law. *Evolutionary Theory* **1**:1–30.

Vanderplank, J. 1963. *Plant Disease: Epidemics and Control*. Academic Press, New York.

Verwijst, T. 1993. Influence of the pathogen *Melampsora epitea* on intraspecific competition in a mixture of *Salix viminalis* clones. *Journal of Vegetation Science* **4**: 717–722.

Wright, S. 1938. Size of population and breeding structure in relation to evolution. *Science* **87**:430–431.

Wright, S. 1949. The genetical structure of populations. *Annals of Eugenics* **15**:323–354.

CHAPTER 14

Community ecology

Gregory S. Gilbert and Ingrid M. Parker

14.1 Introduction

Plants exist in communities of interacting species, competing for nutrients, water, and light. They are also interacting with a diverse community of pathogens and other symbionts. Even crop plants in monoculture production interact with other plants: weeds, plants on the field margins, patches of wild habitat on the landscape, cover crops, or other harvestable species in rotation. Each plant interacts with a number of pathogen species, and many of those pathogens may be shared with other plants in the community. In this chapter we explore the ecology of plant disease in the context of plant communities.

We use the word **community** to refer to the collection of taxa of a particular trophic level, growth form, or taxonomic group present together in a particular location, like a tree community or a fungal community. We view communities as assemblages of species that interact in space and time. Communities may be ancient and composed of species with long shared evolutionary histories, or they may be assembled *de novo* like a garden or a weedy abandoned lot. Communities are characterized by their species composition and by their structure, as measured by the diversity and relative abundances of those species.

Many factors influence plant communities, including humans. Wild plant communities are shaped primarily by their abiotic environment, and by species interactions between plants and other plants, animals, and microbes. Wild plant communities are dynamic, changing over time, and varying across space. **Ecological succession** is the process of change in community composition and structure over time after a disturbance, such as fires, landslides, glaciers, or the abandonment of an agricultural field. On the other hand, agricultural plant communities are shaped primarily by choices made by people, informed by management constraints and economics. However, those choices are also contingent on the abiotic environment, plant–plant interactions, animals, and microbes. Human and non-human drivers shape characteristics and dynamics of all communities to varying degrees.

Plant pathology has traditionally focused on the impacts of a single pathogen species on a single host species. For instance, in the journal *Plant Disease*, some 54% of research articles involve a single pathogen species on a single host species, and only 7% of the studies include both multiple pathogens and multiple hosts (Table 14.1). This prevalence of one-to-one studies reflects our focus on reducing the impact of diseases in conventional agricultural production. However, understanding the ecology and evolution of plant–pathogen interactions requires recognizing and appreciating the implications of complex networks of plants and pathogens, including how those pathogens affect plant–plant interactions in a community setting. This chapter places plant disease into the context of communities of plants interacting with communities of pathogens.

In thinking about plant disease in a community context, we have three broad goals. First, understand the influences of diseases on the diversity, structure, composition, and dynamics of communities. Second, understand how plant community composition and structure influence disease

The Evolutionary Ecology of Plant Disease. Gregory S. Gilbert and Ingrid M. Parker, Oxford University Press. © Gregory S. Gilbert & Ingrid M. Parker (2023).
DOI: 10.1093/oso/9780198797876.003.0014

Table 14.1 Tally of the number of pathogen and host species in all the research articles published in 2015 in the journal *Plant Disease*

	Number of host species					
Number of pathogen species	**1**	**2**	**3**	**4**	**≥5**	**Total**
1	107	16	4	3	11	141
2	16	2			2	20
3	11		2	1	2	16
4	5	1				6
5+	11				3	14
Total	150	19	6	4	18	197

Just over half (54%) of all publications focused on a single pathogen on a single host species (violet cell). The green area highlights the largely empty research space in the realm of community ecology (less than 7% of the studies)—where multiple pathogens interact simultaneously with multiple plant species.

processes. And third, use that knowledge to better design and manage communities to reduce unwanted impacts of disease. In this chapter we primarily explore the impacts of disease on plant communities. In Chapter 15, we turn to how pathogens and other microbes interact with each other in a community context, and in Chapter 17, we look at applying what we know about disease in plant communities to agriculture, forestry, and conservation.

14.2 Some definitions and general truths about biological communities

The makeup of a community, in terms of the species that are in it, is called **community composition**. The number of species is the **species richness**. Some species are common and some are rare; the **relative abundance** of a species is calculated as its proportion of the total number of individuals, area covered, or biomass of all the species. The degree to which there is a similar relative abundance among species is called **evenness**. Ecologists usually use the term **diversity**, to mean a combination of richness and evenness, with several different ways to quantify it. The spatial scale used to measure diversity depends on the questions, systems, and organisms of interest. We can describe the diversity of plant species in a sample quadrat, in a garden, in a county, or in a national park, or we can measure the number of fungal species in a handful of soil, a single plant leaf, a tree, a plant population, or a forest plot.

Ecologists have long observed that larger areas (such as large islands) support a greater number of species than smaller areas (such as small islands) (MacArthur and Wison 1967). The larger the area examined, the more species will be encountered. In part, this is because all organisms are patchy in space—a species is never present in all the places it *could* be. As a result, greater sampling effort uncovers a larger total number of species, until all the possible species are eventually found (Box 14.1, Figure 14.1a). Because a community sample seldom uncovers all the species in an area, we use a **species accumulation curve** to show the cumulative number of species encountered as the sampling effort increases, until the curve saturates near the true number of species present. Sampling effort can be measured as area, individuals, time, or DNA sequences examined. Species accumulation curves for plant species in a study plot generally saturate with reasonable time and effort. In contrast, saturated curves are seldom the case for hyperdiverse microbial communities, which are most often still on the steep part of the accumulation curve even after intensive sampling (Figure 14.1b).

Biological communities tend to have a few common species and many rare ones (Figure 14.2). In the case of microbial communities, including pathogens, this means that datasets will often be dominated by species that only appear a single time ("singletons"). For example, in the dataset on fungi present on cover crops in Figure 14.1b, of the 1125 fungal taxa we encountered, 401 (35%) were detected only once. It is logical to assume that the most common microbial species will have the largest ecological impacts on plants, and researchers will often exclude singletons from analyses for

Box 14.1 Measuring diversity

Quantifying and comparing species richness can be a little bit complicated. Species accumulation curves highlight the importance of scale and sampling effort (Figure 14.1). When sampling effort differs among study areas, **rarefaction** can be used to estimate the number of species you would expect to encounter for a random subset of samples of a given size, and in some cases, to predict how many species to expect with greater sampling effort (Chao et al. 2014). Rarefied estimates of richness are the "industry standard" for comparison of species richness across samples of different intensity, although alternative approaches are being explored (Willis 2019).

Evenness is maximized when all species are equally represented in a community, and it is low if particular species are very dominant. Biological communities tend to have a few common species and many rare ones. The distribution of relative abundances across species is flatter for a community with greater evenness, and shows a strong right skew when evenness is low (Figure 14.2).

Diversity metrics combine richness and evenness. There are entire books written about the many (too many) metrics that have been proposed as measures of diversity (Magurran and McGill 2011), and we will not wade into the decades of unresolved debates over their relative merits. Regardless of this debate, there are a few metrics that you can feel comfortable using because they are commonly encountered in the literature and simple to calculate. First is the Simpson Index (D; Equation 14.1), which gives the probability (from 0 to 1) that two individuals taken from the community at random are of the same species:

$$D = \sum \left(\frac{n_i [n_i - 1]}{N [N - 1]} \right) \qquad \text{[Eq. 14.1]}$$

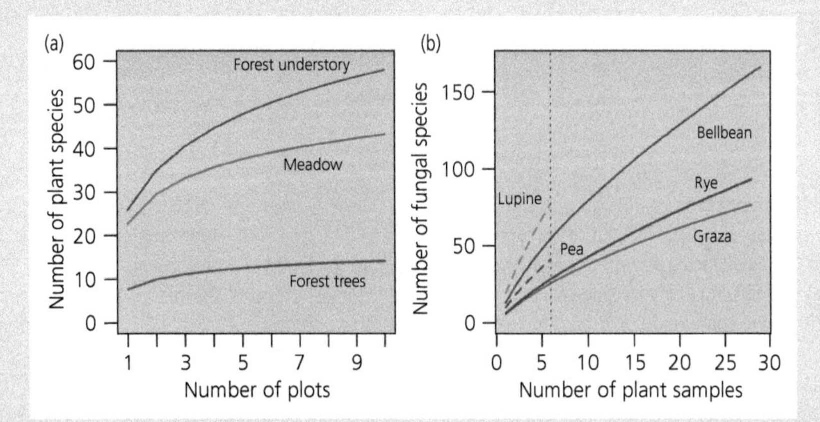

Figure 14.1 Species accumulation curves showing the cumulative number of species observed with increasing sampling effort. (a) The number of herbaceous plant species encountered in ten 20-m radius sample plots in a meadow, and similarly for tree and understory plant species in a nearby forest on the UC Santa Cruz campus. The number of tree species found was much lower than for the herbaceous plants, with much greater diversity in the forest understory than in the meadow. That all the species accumulation curves tend to flatten out (saturate) as more plots are sampled suggests that most of the common plant species in those habitats have been encountered. This suggests it is reasonable to compare species richness between the habitats, using data from 10 sample plots. (b) The leaves of five crops growing on the UCSC campus farm were colonized by a tremendous diversity of endophytic fungi (measured using high-throughput sequencing of fungal internal transcribed spacer (ITS) barcodes; Section 9.8). Bell bean, rye, and graza were sampled with equal intensity, and we can conclude that fungal species richness on bell bean is much greater than on rye or graza. However, none of the species accumulation curves saturate (many new fungal species are still encountered with each additional sample), so we must be cautious about comparing fungal richness between crops with different sampling efforts. For instance, lupine and pea were sampled less intensively. To compare across different sampling efforts, we use rarefaction to determine how many fungi would be expected if sampling were equivalent. In this case, there were 166 fungal species found on bell bean and only 77 on lupine; however, rarefaction suggests we would expect to find only about 55 species of fungi on bell bean if we had collected the same number of samples as for lupine (vertical dotted line). Foliar fungal diversity is actually greater on lupine than on bell bean! Randomization procedures in rarefaction allow estimations of error and statistical comparisons. Comparing species richness across habitats (species, locations, treatments) requires comparing the number of species observed for an equivalent sampling effort.
Figure by Gregory S. Gilbert.

Box 14.1 *Continued*

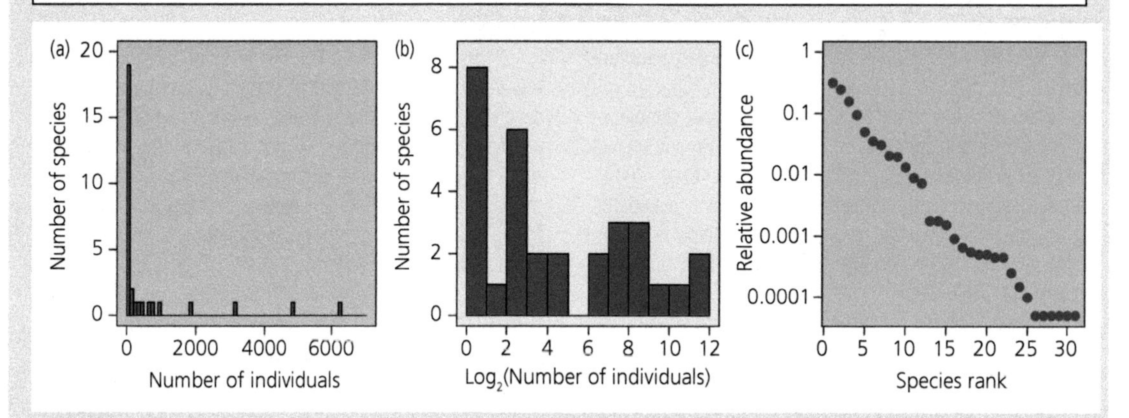

Figure 14.2 In biological communities, there are often many rare species and just a few species that are orders of magnitude more abundant. Therefore, relative abundance is best visualized on a logarithmic scale. Each graph represents the number of individuals of woody species on the UCSC Forest Ecology Research Plot. (a) A right-skewed histogram on an arithmetic scale shows that most species are rare and only a few are common. (b) The histogram bins are for $\log_2$ of the abundance (sometimes called Preston octaves), giving a clearer representation of relative abundances. (C) A Whittaker rank-abundance plot, showing the relative abundance of each species (on a $\log_{10}$ scale) from the most to least abundant.
Figure by Gregory S. Gilbert.

where n_i = number of individuals of the ith species and N = the total number of individuals. The index is often expressed as 1–D or 1/D, so the index increases with greater diversity.

The Simpson Index is particularly sensitive to dominance (D for dominance) in the community, and has the unfortunate property of being more sensitive to species richness in communities of low diversity and more sensitive to relative abundance in communities of high diversity. A measure of evenness ($E_{1/D}$) that is independent of effects of richness can be derived from the inverse of the Simpson Index (Equation 14.2):

$$E_{1/D} = \frac{(1/D)}{S} \quad \text{[Eq. 14.2]}$$

where D is the Simpson Index and S is the total number of species.

Next comes the Shannon Index (H'; Equation 14.3), a statistic based on information theory, where the value of H' increases either with greater species richness or with greater evenness in the community:

$$H' = -\sum p_i \ln p_i \quad \text{[Eq. 14.3]}$$

where p_i is the relative abundance (e.g., proportion of individuals) of the ith species.

This integration of the two components of diversity into a single metric (similar to the issue with D) can complicate both the behavior and interpretation of the Shannon Index, and many ecologists have fruitlessly argued against its use. However, whereas the Simpson Index is most appropriate for count data, the Shannon Index can be used with any measures of abundance, as long as comparisons are made among comparably sampled groups or treatments. Finally, Pielou developed a useful index that extracts the evenness component from H' (J'; Equation 14.4); J' ranges from 0 (a near-monoculture) to 1 (all species equally represented):

$$J' = \frac{H'}{H_{max}'} \quad \text{[Eq. 14.4]}$$

where H' is the Shannon Index and $H'_{max} = \ln S$, where S is the total number of species.

simplicity. However, the roles and relative importance of rare and common species is a compelling and unresolved question in plant disease ecology.

We have special terminology for describing diversity over different spatial scales. **Alpha** (α) **diversity** describes the variety of organisms that share a relatively homogeneous and contiguous place. Alpha diversity may refer to the combination of richness and evenness, although originally this term was used for richness alone. The differences in species composition from one place to another is **beta** (β) **diversity**. The term **gamma** (γ) **diversity** describes the total diversity across a landscape of heterogeneous habitats, a combination of α and β diversity. Beta diversity is particularly important in community ecology, because it allows us to examine how community composition and structure vary along environmental gradients, or how they respond to disturbances or management practices. Beta diversity can be expressed through a variety of indices, and is visualized through one of several dimension-reducing analyses called **ordination** (Box 14.2).

Box 14.2 Community composition and beta diversity

The number of species is only part of what characterizes a community. Species composition—the identity of which species are there and their attributes—is of course what gives communities their unique character. Ecologists use a range of tools to describe and compare the species composition of communities: how similar are they? How are they different?

Differences in composition can be quantified as β (beta) diversity, the turnover of species found in different communities. Change in species composition along an environmental gradient (e.g., as we go up in elevation into the mountains, or travel from the equator toward the poles) is an example of β diversity. So is variation among samples or over time at a single location. Two simple and common measures of similarity between samples using only presence or absence of species are the Jaccard Similarity Index (J; Equation 14.5) and a slight modification called the Sørensen Similarity Index (C_s; Equation 14.6).

$$J = \frac{a}{a+b+c} \quad \text{[Eq. 14.5]}$$

where a = number of species found in both samples b and c, b = number of species found only in sample b, and c = number of species found only in sample c. $J = 0$ if there are no species in common and 1 if all the species are found in both samples. The complement of J, $1 - J$, is then a measure of dissimilarity (β *diversity*), and can be used as a distance measure in ordination (Figure 14.3b).

$$C_s = \frac{2a}{2a+b+c} \quad \text{[Eq. 14.6]}$$

uses the same designations as Equation 14.5. The complement 1-C_s measures dissimilarity.

When relative abundances of species are available, the Bray–Curtis Quantitative Dissimilarity Index (Equation 14.7) is used as an abundance-weighted measure of similarity.

$$BC = 1 - \frac{2C_{ab}}{S_a + S_b} \quad \text{[Eq. 14.7]}$$

where S_a and S_b are the total number of individuals of all species found in sites a and b, respectively, and C_{ab} is the sum of the number of individuals of those species found in both samples a and b, but including only the lower of the two counts for each species.

One use of dissimilarity measures is to see how differences among samples within a habitat type compare to the differences between habitat types. For example, we sampled plant community composition in forest understory and meadow habitats on the UC Santa Cruz campus, estimating the relative abundance of the 65 plant species across 20 sample plots (Figure 14.3). The community composition is much more similar (lower β diversity) among plots within a habitat (mean Jaccard distance within forest = 0.44, within meadow = 0.41) than between pairs of forest and meadow plots (mean Jaccard distance = 0.97). Overall there were 32 species found only in the forest, 26 species found only in the meadow, and 7 species found in both, giving an overall Jaccard Dissimilarity of $1 - J = 1 - 7/(7 + 32 + 26) = 0.892$.

Differences in community composition can be visualized using multivariate **ordination** techniques. We can think of differences in composition as the distance between

Box 14.2 *Continued*

Figure 14.3 Differences in plant communities across habitat types on the UC Santa Cruz campus. (a) Satellite image (from Google Earth) of the UCSC campus, showing the location of 6-ha study areas on the Forest Ecology Research Plot (teal) and Great Meadow (orange). Presence or absence of 65 plant species was recorded in each of ten 20-m radius sample plots in forest understory and ten in the meadow. (b) A principal coordinates analysis (PCoA) ordination-based plant species composition in forest understory and meadow plots. Points closer together have more similar species composition. The plant communities of the two habitats are strongly differentiated along PCo Axis 1.
Figure by Gregory S. Gilbert.

samples in an imaginary space, where species themselves are the dimensions (we call this "species space"). For instance, imagine a community with only two species, A and B. Samples may have relatively equal amounts of A and B, or they may have lots of A and only a little of B, or more B than A. If each sample is a point in our imaginary "species space," the distance between two points (samples) will be short for two samples with similar composition. In contrast, the distance would be large between a sample dominated by A and a sample dominated by B. Most communities have many more than two species, but because our physical experiences are rooted in three dimensions, it is hard to visualize hyperdimensional species-space, of say, 65 dimensions. Ordination techniques reduce the number of dimensions, combining data from all species into composite scores, and visualizing differences between sites or samples as distances that we represent graphically. (For details on the use and relative merits of different approaches, see Minchin 1987, ter Braak 1995, Anderson and Willis 2003). While the mathematical details vary across methods (e.g. Non-Metric Multidimensional Scaling, Principal Components Analysis, Principal Coordinates Analysis, Discriminant Function Analysis), all ordination graphs show samples with similar species composition closer together (Figure 14.3b).

14.3 Specialists and generalists and everything in between

Just how many different plant pathogens must a particular plant deal with? We can think of a plant's **potential suite of pathogens** as all the pathogen species, anywhere in the world, that could cause disease on a host species if they were brought together under favorable environmental conditions. A pathogen species somewhere in the world that has never had the opportunity to encounter a plant species, but has all the necessary traits to be able to cause disease on that plant, can be thought of as **preadapted** to that host and would be included in its potential suite of pathogens. Many of these preadapted pathogens may have evolved as pathogens of related plant species with non-overlapping biogeographic ranges. Why do we care about this theoretical suite of potential pathogens? One reason is because the human-driven global movement of plants and pathogens dramatically increases the chances of bringing such potential plant–pathogen partners together (Section 16.4).

In the meantime, only those pathogens that coincide geographically with a host plant can cause disease; we call this a **biogeographic filter**. It is possible for a pathogen species to have a broad host range at a global scale, yet locally act as a specialist, if only one of its susceptible hosts is present (Fox and Morrow 1981). Finally, the basic tenets of the disease triangle (Figure 1.5) further restrict this list of pathogens to only those for which the local conditions are favorable for infection and disease development. This **environmental filter** then delimits the local **realized suite** of pathogens. Changes in environmental conditions, for example because of global climate change (Section 16.6), will change which pathogens successfully pass through local environmental filters.

Like plants and their suites of pathogens, each pathogen associates with only a subset of possible plant hosts. Because they require live host cells, biotrophic pathogens tend to show high host specificity, and the relatively simple underlying genetic architecture of gene-for-gene interactions (Section 12.6) made these systems ideal for studies of population biology and evolution. For instance, 80% of the pathogen species mentioned in Jeremy Burdon's (1987) transformative book *Diseases and Plant Population Biology* were highly specialized to just a few host species or genera. The extra attention received by highly specialist plant–pathogen interactions has led to a common perception that most disease interactions are specialized.

However, there are broad host-generalists in addition to the narrow specialists, and there are a lot of pathogens in between that are moderately **polyphagous** (Section 13.5). Barrett et al. (2009) found that of the 1252 fungal pathogens with recorded host ranges, 14% were known from just a single host species, another 47% were known from multiple species within a single plant family, and 39% were found on hosts from multiple families. Gilbert and Webb (2007) found that a typical fungal pathogen in the rainforest of Panamá was able to infect the leaves of about a quarter of the tree species in the surrounding area.

Interacting communities of plants and pathogens form an ecological network. Because individual hosts are linked to each of their pathogens, and each pathogen is linked to its hosts, the interactions can be visualized as a **bipartite network**. Using the tools of network analysis, features such as the extent of specialization can be calculated as the degree of **connectivity**, and variation among hosts in the size of their suite of pathogens can be easily seen in the network diagram (Figure 14.4). When a subset of hosts and pathogens interact with each other more commonly than they do with other species in their community, they create a module; the degree to which the network is bundled into modules is called the **modularity** of the network. Network modularity governs the spread of polyphagous pathogens through a plant community.

What are the relative impacts of specialist and more generalist pathogens on plant communities? This is an open question: to our knowledge, no one has yet attempted to quantify the relative impacts of all components of its pathogen community on even a single host species. The relative impact of specialists and generalists varies strongly across types of disease (Chapter 8). Root-rot and damping-off diseases are caused by necrotrophic pathogens that tend to be more host generalist. Biotrophic pathogens such as rusts tend to be more host specialist. The diseases that have the largest impacts on plants and harvestable plant products in agricultural systems are not necessarily the same diseases that are most impactful on plants in wild ecosystems. In addition, the translation from impact on individual host plants to impact on populations passes through the filter of plant demography (Section 10.7). For example, a pathogen that kills seedlings may have a 100% impact on survival of individual plants, but it may have a more subtle effect at the population level than a pathogen that reduces adult tree survival by only 10%. To answer the question of "What disease has greatest impact on this species?" requires understanding the details of biology at multiple scales.

A pathogen's degree of specialization is the key to the way it shapes plant distribution, abundance,

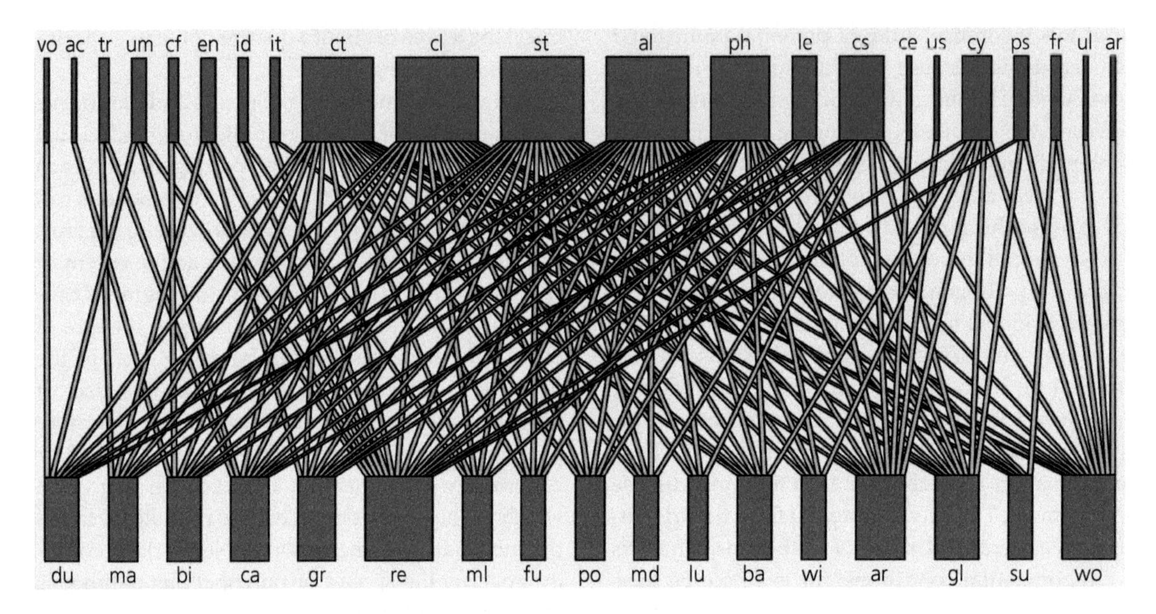

Figure 14.4 A bipartite network of fungal pathogens found on leaves of clover species (*Trifolium* and *Medicago*) in a California coastal prairie (as described in Table 15.1) illustrates the structure of interactions between pathogens and plants in a wild community. Plant host species are listed across the bottom in green and are connected to each of the fungal pathogens found on that species (blue, across the top). While a few pathogens were limited to only one host, most were found on several different host species. Similarly, some host species were associated with many different fungi (e.g., re: *T. repens*), whereas others with just a few (e.g., du: *T. dubium*).
Figure by Gregory S. Gilbert.

and diversity. A specialist pathogen may help maintain plant diversity by causing negative density dependence, whereas a generalist pathogen may reduce plant diversity by spreading from one host species onto a more susceptible one. Specialization is a recurring theme in the evolutionary ecology of plant disease. Keeping that in mind, let's take a closer look at the role of diseases in shaping plant communities.

14.4 What maintains plant diversity?

Why are there so many kinds of plants? For all their wondrous diversity, plants are all pretty ecologically similar, requiring access to the same small set of resources: light, water, and nutrients. The **competitive exclusion principle** (sometimes called Gause's Law) posits that two species with similar resource requirements cannot coexist for long, because whichever species is the better competitor for those resources will crowd out the other one (Box 14.3). It then follows that, all else equal, the diversity of an assemblage of plant species will decline over generations, because the better competitors will win out over the inferior competitors. However, since many plant communities are indeed quite diverse, it appears that all else must not actually be equal. There must be some mechanisms to counteract competitive exclusion and allow species coexistence.

Chesson (2000, 2018) divided possible mechanisms that promote species coexistence into two major categories: equalizing and stabilizing.

14.4.1 Equalizing mechanisms

All organisms face trade-offs in allocating resources towards survival, growth, or reproduction; these life-history trade-offs prevent a species from being best at everything, and serve as **equalizing mechanisms**. An equalizing mechanism reduces differences in average fitness between species. For example, if a plant species invests in traits that allow it to grow very quickly in response to abundant resources (e.g., large, thin leaves with a

Box 14.3 Intraspecific competition, interspecific competition, and competitive exclusion

Individuals of a species compete for limiting resources with other individuals of the same species through intraspecific competition. Intraspecific competition leads to the self-regulating population dynamics captured in the logistic growth model (Equation 10.2), and repeated here for species 1:

$$\frac{dN_1}{dt} = r_1N_1\left(\frac{K_1 - N_1}{N_1}\right) \qquad \text{[Eq. 14.1]}$$

where r_1 is the per-capita intrinsic growth rate for species 1, N_1 is the number of individuals, and K_1 is the carrying capacity of species 1. When another species competes with species 1 for limited resources, the individuals of species 2 effectively occupy a portion of the carrying capacity available to species 1; the relative competitive coefficient for the interspecific competition effect of species 2 on species 1 is given as α_{12}, so that:

$$\frac{dN_1}{dt} = r_1N_1\left(\frac{K_1 - N_1 - \alpha_{12}N_2}{N_1}\right)$$

$$\frac{dN_2}{dt} = r_2N_2\left(\frac{K_2 - N_2 - \alpha_{21}N_1}{N_2}\right) \qquad \text{[Eqs 14.2]}$$

This pair of equations is known as the Lotka–Volterra competition equations. The Lotka–Volterra competition equations provide a theoretical framework for the competitive exclusion principle, that two species competing for the same resource cannot stably coexist, leading to the competitive exclusion of the subordinate competitor. We can imagine that there are also intraspecific competition coefficients α_{ii} and α_{jj} in front of N_1 and N_2, respectively, but because the values α_{12} and α_{21} are relative competition coefficients, the value of the intraspecific coefficients is always 1.

Peter Chesson (2000) noted that these classic Lotka–Volterra equations could be made more intuitively accessible by envisioning absolute, rather than relative, intraspecific and interspecific competition coefficients (i.e., not scaling the coefficients to the carrying capacity) such that:

$$\frac{dN_1}{dt} = r_1N_1\left(1 - \alpha_{11}N_1 - \alpha_{12}N_2\right)$$

$$\frac{dN_2}{dt} = r_2N_2\left(1 - \alpha_{22}N_2 - \alpha_{21}N_1\right) \qquad \text{[Eqs 14.3]}$$

Species 1 and 2 can then coexist if intraspecific competition is greater than interspecific competition, such that $\alpha_{11} > \alpha_{21}$ and $\alpha_{22} > \alpha_{12}$. Stabilizing mechanisms (Chesson 2000) including niche partitioning, the storage effect, density-dependent disease development, and plant–soil feedbacks help maintain diversity by increasing the strength of intraspecific competition (α_{11} and α_{22}) compared to interspecific competition (α_{21} and α_{12}).

high concentration of chlorophyll), the plant cannot simultaneously invest resources in defenses like toxic chemicals and tough leaves. Thus there is an investment tradeoff between rapid growth and resistance to attack from natural enemies (Coley et al. 1985; Section 13.5).

Under certain conditions, plant pathogens can be an equalizing mechanism. If a pathogen causes more damage on a competitively dominant host species, it will "equalize" the average fitnesses of that dominant species and slower-growing plants. This can happen (and is expected) when the dominant species attains its competitive advantage by growing rapidly and appropriating resources like light. The **growth-resistance tradeoff** means that slower-growing, more resistant species will gain a relative advantage in the presence of pathogens. This equalizing effect can be caused by generalist as well as specialist pathogens. We found just such a case with the generalist foliar pathogens (*Stemphylium* spp., *Alternaria* spp., and *Cladosporium* spp.) in our study of coexistence of clover (*Trifolium*) and burr-clover (*Medicago*) species, where larger, fast-growing host species suffered more infection and disease than the slowest growing species (Parker and Gilbert 2018). We might expect common plant pathogens with broad host ranges (such as *Pythium* spp. or other damping-off pathogens (Augspurger and Wilkinson 2007)) to act as equalizing mechanisms to sustain plant diversity where the local plant species vary in their allocation to growth and disease resistance.

14.4.2 Stabilizing mechanisms

A **stabilizing mechanism** is a factor that makes intraspecific interactions more negative than interactions between species. Stabilizing mechanisms include any process that reduces the strength of interspecific competition, such as **niche partitioning**, which is specialization through different use of habitats or resources. Another type of stabilizing mechanism is temporal variation, when different species do best in different time-periods, depending on their response to environmental conditions. As long as populations are able to persist (for example in a seed bank) through their own "bad years," then having divergent responses to annual variation maintains diversity in a phenomenon known as the **storage effect**. When variation in performance is based on variation in conditions across space, it is called a spatial storage effect. Finally, an important set of stabilizing mechanisms provide a fitness benefit to locally rare species through negative density- or frequency-dependent processes, called a **rare-species advantage**. For example, because the prevalence and severity of plant disease so often increases with the density of the host (Chapters 10 and 11), the **conspecific negative density dependence** driven by disease can be a critical stabilizing mechanism contributing to the maintenance of plant diversity. Let's look at several ways that plays out.

At first, we might expect pathogens to reduce the diversity of plant communities, since they can kill their host plants and suppress host populations. There are indeed examples of pathogens that have had strong negative effects on individual host species in wild ecosystems, and when these hosts are dominant or otherwise ecologically important, the pathogen can have a large impact on ecosystem properties and functions. Interestingly, most such transformative diseases are from epidemics of pathogens introduced into a new biogeographic range (Section 16.4). In most such cases, the dramatic impacts are to the populations of one or a few host species. A notable exception is the introduction of the oomycete *Phytophthora cinnamomi* into Western Australia; its broad host range and high virulence greatly reduced species diversity in areas it invaded (Section 16.4).

However, most plant disease in wild systems is not associated with epidemics of novel, invasive pathogens, but instead is the result of routine ecological interactions between plants and pathogens that have long evolutionary histories. Perhaps counterintuitively, the impact of normal, everyday, local disease in wild plant communities may play a critical role in maintaining diversity. The key is through stabilizing mechanisms—let's look at how that works.

14.4.3 Gillett's Theory of Pest Pressure

Ecologists in the mid-20th century focused a great deal on exploring how species diversity is maintained through niche partitioning. In a symposium address in 1959, botanist J. B. Gillett broke from that tradition and put forward a new framework for how plant diversity could be maintained by pests and pathogens (Gillett 1962). Gillett mused about the tremendous diversity of trees in tropical forests, and wondered how, if all the tree species basically have the same needs of water, light, and nutrients, "can it be seriously suggested that a rather uniform area of Amazonian rain forest provides, in 3.5 hectares of land, anything like 179 separate ecological niches for trees?" (p. 39). Instead, he postulated that natural enemies like herbivores and pathogens, which can evolve to have host-specific impacts and respond to host density, provide a powerful mechanism for the maintenance of species diversity. His beautiful essay (well worth reading in the original) outlines many ways in which pathogens can shape plant communities, calling pest pressure "the inevitable, ubiquitous factor in evolution which makes for an apparently pointless multiplicity of species in all areas in which it has time to operate" (p. 40).

Perhaps Gillett's most important contribution was the recognition that density-dependent disease dynamics provide an elegant mechanism for host-specific pathogens to maintain plant species diversity. Through Virgil's maxim *parcere subiectis, et debellare superbos* ("spare the vanquished, and beat down the dominant") (p. 40), he introduces the idea that "the reproductive capacity of common species will be worn down by an ever growing burden of pests, while scarcer species ... will, relatively speaking, escape" (Gillett 1962). Any plant species that becomes numerically abundant in a community would foster the spread of its specialist pathogens, leading to a subsequent decrease in abundance. Locally uncommon species have the advantage of escaping pest pressure from specialists, leading to increased population growth. All the species in an assemblage are subject to the same dynamic; whichever species becomes highly abundant is likely to stay dominant only until its enemies catch up; they then reduce its average fitness difference compared to neighboring species. This rare-species advantage prevents competitive exclusion and maintains species diversity in the system.

14.4.4 Janzen–Connell Hypothesis

Tropical ecologists Dan Janzen (1970) and Joseph Connell (1971) independently extended Gillett's theory into a spatial context. Like Gillett, they wondered how so many tropical tree species are able to coexist in a pretty small area. In their quest to understand why competitively dominant tree species did not drive inferior competitors to local extinction, Janzen and Connell made a critical connection between density-dependent impacts of natural enemies and observed spatial patterns of trees. They noted that trees make a lot of seeds, but most of the seeds do not disperse very far from the mother tree (Section 11.5). Those seeds give rise to a **seedling shadow** around the mother tree, with a very high density of offspring beneath the canopy of the mother tree and declining density at greater distances (Figure 11.10). They reasoned that there were two ways that natural enemies (including pathogens) of those seeds and seedlings could have greater impacts on offspring near the mother trees. First, host-specific natural enemies could disperse directly from the mother tree to the offspring (e.g., herbivores or pathogens in foliage in the tree crown could fall onto seedlings below), creating a gradient of natural enemy pressure that declines with distance from the source tree. Second, the high density of seeds and seedlings near the mother tree could promote the reproduction and spread of pathogens from one seedling to the next, as well as attracting seed predators and herbivores. The result is a gradient of pest pressure that depends on offspring density. Taken together, pest pressure on the offspring of a tree would be greatest close to the mother tree where seedling density is highest, and so the probability of seedling survival should increase with dispersal distance. The two spatial gradients—decreasing dispersal and increasing survival—combine to create a spatial–temporal process where cohorts of offspring start off highly clumped and become less and less clumped over time (Figure 11.10).

What does the change in clumping described by Janzen and Connell have to do with species diversity? The massive number of seedlings in a seedling shadow gives that species numerical dominance compared to all other tree species. Each year that the mother tree makes more seeds reinforces this monospecific dominance. All else equal, seeds from any other species that disperse into that seedling shadow would have a tiny probability of prevailing in the intense competition for sun and space. However, the pathogens and other natural enemies make all else *un*equal by specifically reducing the growth and survival of the locally dominant species. Therefore, a seed that dispersed from a distant tree of a different species would have a greater probability of successful recruitment than a seed dispersed near its mother in a monospecific seedling shadow. The Janzen–Connell Hypothesis thus adds a spatial component to Gillett's temporal dynamics as a mechanism for maintaining plant diversity in wild ecosystems.

There is now abundant evidence that pathogens can drive this kind of conspecific negative density-dependence and the spatial patterns predicted by the Janzen–Connell Hypothesis. The pioneering work of Carol Augspurger (Chapter 11 and Augspurger 1983, 1984) provided the first empirical evidence from studies of damping-off disease of seedlings of rainforest trees; Gilbert et al. (1994) showed that those disease patterns were also apparent for canker disease in decades-old saplings in the same rain forest. Packer and Clay (2000) showed that this phenomenon is not limited to tropical rainforests, finding the same patterns for cherry tree seedlings in Indiana, USA. In the half-century since Augspurger's original research, scores of observational, experimental, theoretical, and review studies examined which kinds of natural enemies have the greatest effects in which kinds of plant communities, and whether the scale and strength of impacts are enough to maintain observed levels of plant diversity (e.g., Bagchi et al. 2010, Mangan et al. 2010, Comita et al. 2014, Chisholm and Fung 2020, Jia et al. 2020). A meta-analysis of 154 tests found strong support for both density-dependent and distance-dependent mortality in both tropical and temperate systems, with effects on seedlings stronger than on seeds (Comita et al. 2014). The weight of the evidence suggests that Janzen–Connell effects are widespread at the population level. However, for Janzen–Connell effects to maintain species diversity at the community level, locally uncommon species should also experience less disease and greater average performance than abundant species. Removing the effects of disease should reduce diversity. This prediction has had considerably fewer tests. In a rainforest in Belize, Bagchi and colleagues (2014) found that there is a strong increase in plant species richness and a shift in plant species composition from the seed stage to the seedling stage, consistent with the Janzen–Connell hypothesis. They showed this was caused by disease by excluding seed and seedling pathogens by applying fungicides for 17 months; this resulted in lower plant species richness, as predicted.

Open questions remain on the importance of Janzen–Connell effects relative to other equalizing and stabilizing mechanisms, and whether some plant species are more susceptible to Janzen–Connell effects than others (and why). Finally, what about specialization—does the Janzen–Connell hypothesis still apply when pathogens are not host specific? Let's now turn to the effects on plant communities of pathogens that grow on multiple hosts.

14.5 Apparent competition and pathogen spillover

When one plant species (A) has a negative effect on the growth of a second plant species (B), we often assume that competition for resources between the species is the cause of the negative effect. However, the two species may share a pathogen (or other enemy) that affects B more than A. Acting as a source of a shared pathogen is an indirect mechanism by which the presence of species A can suppress species B, an example of **apparent competition** (Holt 1977).

When a pathogen is more virulent on one host than another, apparent competition can have dramatic effects on the more vulnerable host. A striking example of this is the sudden oak death pathogen *Phytophthora ramorum*, which causes disease on both California bay (*Umbellularia californica*) and coast live oak (*Quercus agrifolia*) (Section 16.4). On California bay, the pathogen causes necrosis on the tips of leaves, but has little impact on host growth, survival, or reproduction. However, it serves as an exceptionally competent host, producing massive amounts of sporangia that disperse readily in spring rains. When the sporangia land on a nearby oak, infection can occur on branches and the trunk, not only on leaves. Infections in the trunk lead to death of cambial tissue, developing spreading sunken, oozing cankers that girdle and kill the tree. The pathogen does not reproduce on the "dead-end" oak host. This means that oaks living near a California bay are more likely to die of sudden oak death than those distant from bays, whereas bays are indifferent to having oak neighbors. If we did not know that a shared pathogen drives this lopsided relationship, we might conclude that bay is a superior competitor, outcompeting the inferior oak.

The dynamic between sudden oak death on bay and oak is a classic example of **pathogen spillover**. When a polyphagous pathogen spreads from a **reservoir host** on which the pathogen is abundant to another host, the pathogen spillover can drive disease dynamics. We follow the definition of Power and Mitchell (2004), that "Pathogen spillover occurs when epidemics in a host population are driven not by transmission within that population but by transmission from a reservoir population," although some use the term in a more restricted sense of transmission from domesticated (reservoir) species to wild species (Daszak et al. 2000).

The impact of pathogen spillover can be difficult to see at first glance, when disease causes slower growth, rather than necrosis or death. The aphid-transmitted *Barley yellow dwarf virus* (BYDV) infects a wide range of grass species, causing reduced growth and fecundity. In California grasslands,

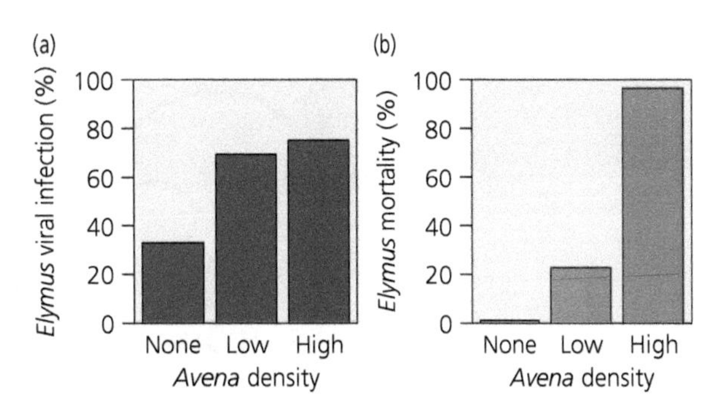

Figure 14.5 Greater density of the invasive annual *Avena fatua* (wild oat), a reservoir of *Barley* and *Cereal yellow dwarf viruses* (BYDV and CYDV), (a) increases viral infection of the native perennial grass *Elymus glaucus*, and (b) increases mortality of the native host, a result of both direct and apparent competition.
Figure by Gregory S. Gilbert using data from Malmstrom et al. (2005).

the invasion of annual grasses from Mediterranean Europe has displaced the native perennial bunchgrasses in most places. BYDV is found in both native and non-native species, but aphids are attracted to and reproduce more on the non-native annual *Avena fatua* than on native perennials—*Avena* serves as an important reservoir species for the virus and the vector. In experimental plots, natural BYDV infection of the native *Elymus glaucus* doubled from 34% in the absence of *Avena* to 75% when growing with high *Avena* density. Annual mortality of the perennial *Elymus* following severe stunting increased from 1% without *Avena* to nearly complete mortality at high densities of *Avena* (Malmstrom et al. 2005) (Figure 14.5). Although direct resource competition also contributes to the high mortality of *Elymus* in the presence of *Avena*, apparent competition likely plays a role. A modeling study suggests that virus spillover from the non-native reservoir hosts such as *Avena* may even reverse competitive outcomes, contributing to the dominance of non-native annuals in California grasslands (Borer et al. 2007).

The consequences of pathogen spillover for host populations and communities depend on the symmetry of disease impacts on infected hosts, as well as the competence of the hosts to support pathogen reproduction. The case of sudden oak death falls at one end of the gradient, where asymmetries in impact and competence are extreme. The brunt of the negative impacts fall on the oak (lethal to oaks, negligible on bay), whereas bay is an exceptionally competent host and oak supports almost no pathogen reproduction (Cobb et al. 2010). In this case the pathogen promotes a competitive advantage of bay over oak, which could lead to the expansion of bay populations at the direct expense of oaks. In many ways, the outcome of the symbiosis between *Phytophthora ramorum* and *Umbellularia californica* functions indirectly as a mutualism, despite presenting as an antagonistic disease. At the other end of the symmetry gradient is a situation in which a pathogen that is shared between two hosts has similar impacts and reproduction on each host. In this case the pathogen dynamics will respond to the combined population density of the hosts, and strong disease impacts could lead to synchronized host population dynamics. However, spillover would have little impact on competitive interactions between the two hosts.

The probability that two plant species share a pathogen, enabling pathogen spillover and apparent competition, is not random. The **phylogenetic signal** in the host range of pathogens (Section 13.7) means that spillover is more likely between closely related plant species. Within plant communities, we expect that polyphagous pathogens will move among individuals of different species, and any given focal plant will be vulnerable to pathogen spillover, particularly from closely related species that are abundant (Figure 14.6). Parker et al. (2015) showed how this phylogenetic pathogen spillover can influence disease pressure across species in a plant community; in a California grassland, disease increased not only with a focal host's conspecific abundance (density-dependent disease development), but also with the abundance of closely related species. Similarly, in a tropical forest, Liu et al. (2012) demonstrated how pathogen spillover onto close relatives can create a "phylogenetic Janzen–Connell effect." Among seedlings that dispersed into the seedling shadow of a mature tree, pathogen spillover from the tree and its seedlings

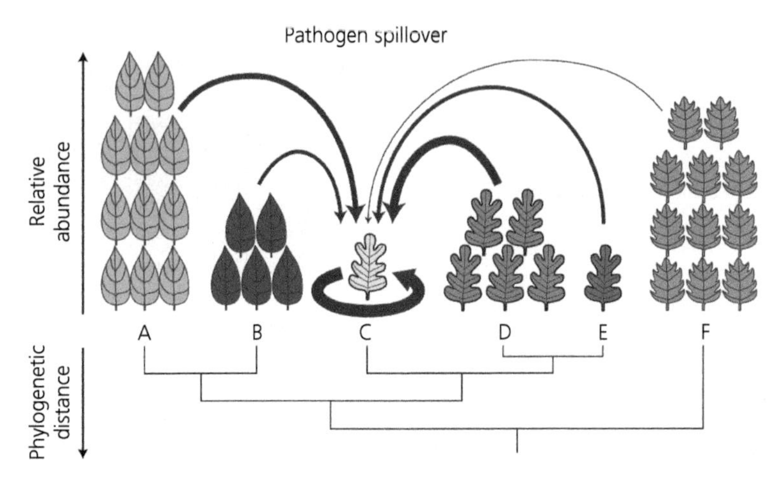

Figure 14.6 Pathogens can spread among individuals of a host species (thick arrow for focal host C), but can also spill over from reservoirs in neighboring species that are competent hosts for the shared pathogens. A phylogenetic signal in host range of pathogens means that close relatives are more likely than distant relatives to share pathogens (e.g., D & E are more closely related to C than is F, and thus more likely to share pathogens with C). Among neighboring competent hosts, those that are more abundant are likely to contribute more inoculum pressure (e.g., D and E are equally likely to share pathogens with C, but D is more abundant and would have a greater influence on disease dynamics in C).
Figure reproduced from Parker et al. (2015); drawn by Karen Tanner.

will lead to disease on close relatives of that tree species. In contrast, seedlings of distantly related species will reap the benefits of the rare-species advantage (Section 14.4). Within the framework of plant–pathogen networks, spillover is most likely to occur within modules (Section 14.3); we expect modularity to reflect the phylogenetic structure of the community. Untangling how the structure of host ranges of polyphagous pathogens affects disease dynamics in complex plant communities is currently a major challenge to plant disease ecologists.

14.6 Mutualistic plant–microbe symbioses and plant communities

In this book we focus on antagonistic symbioses, but mutualistic plant–microbe symbioses may also affect interactions among plants and the composition and diversity of plant communities. Just as specialist pathogens provide a mechanism for conspecific negative density dependence, by the same logic microbial mutualists are expected to create positive feedback loops that should lead to increased growth, reproduction, and possibly dominance of their hosts (Thrall et al. 2007). For instance, the seed-transmitted fungal endophyte *Epichloë coenophialum* (former anamorph name *Neotyphodium coenophialum*) colonizes above-ground tissue of the grass *Festuca arundinacea*, but not other species of co-occurring grasses. *Neotyphodium* increases the growth of the host plant and produces toxins inside plant tissue that protect the host from herbivores (Section 15.8). Clay and Holah (1999) established experimental plots with about 80% *Festuca* and the rest a mix of other grass species that recruited from the soil seed bank. Half the plots had *Festuca* colonized by *E. coenophialum*, and the other half free of the endophyte. After four years, in the endophyte(+) plots the proportion of plot biomass comprising *Festuca* increased to 90%, while in the absence of the endophyte, *Festuca* declined to 60% of the biomass; total biomass did not change. In four years, plots with *E. coenophialum* had nearly become a *Festuca* monoculture ($H' = 0.34$), while communities where the endophyte was absent were a diverse mix of other grass species and forbs ($H' = 0.79$).

In addition to foliar fungal endophytes, mutualistic symbioses include a number of below-ground plant–microbe interactions such as mycorrhizal fungi, nitrogen-fixing rhizobia, and plant-growth-promoting rhizobacteria (PGPR) that promote the growth and reproduction of their host plants, especially under conditions of nutrient or water limitation (Sections 15.3–15.5). Like pathogens, microbial mutualists have host ranges that vary from specialist to generalist, and their host ranges also show a phylogenetic signal. Specialist microbial mutualists can favor their host at the expense of other plants in the community, while generalists may spill over from one host species to another or sometimes even literally connect plants to each other (i.e., mycorrhizal networks) (Marler et al. 1999). These indirect positive effects can mediate competitive interactions among plants in the presence of microbes. In some cases, microbial mutualisms can help maintain plant diversity, if subordinate competitors benefit disproportionately from mutualistic associations, upending competitive hierarchies (Grime et al. 1987). The combination of these below-ground microbial mutualists and below-ground pathogens contributes to the phenomenon of plant–soil feedbacks.

14.7 Plant–soil feedbacks

When a plant causes changes in the soil that in turn affect the subsequent growth or fitness of plants in that soil, we call it a **plant–soil feedback** (Figure 14.7) (Bever et al. 2010). Those changes can either benefit or harm the plant, and they may be caused by factors that are either biotic or abiotic—but we will focus on the biotic, microbial mechanisms. A positive feedback can occur when a growing plant cultivates, or "cultures," certain mycorrhizae or PGPR in the soil microbiome, which then benefit the next plant of the same species to grow there. On the other hand, a negative feedback occurs when the plant promotes soil-borne pathogens. The overall effect of plant–soil feedback is the net sum of negative and positive effects.

Plant–soil feedbacks are generally studied by comparing the growth of a plant species in its "home" soil, which is soil that was cultured by the same species, to its growth in "away" soil, which was cultured by other species.

Some ecologists restrict the concept of plant–soil feedback to reciprocal, relative effects of feedback on pairwise comparisons between two plant species (Crawford et al. 2019). In an early such pairwise experiment, James Bever (1994) studied plant–soil feedbacks across four species in an old-field plant community. Strong negative feedback harmed the performance of plants when grown in home compared to away soil. Plant–soil feedbacks can be used to predict changes in plant communities over time;

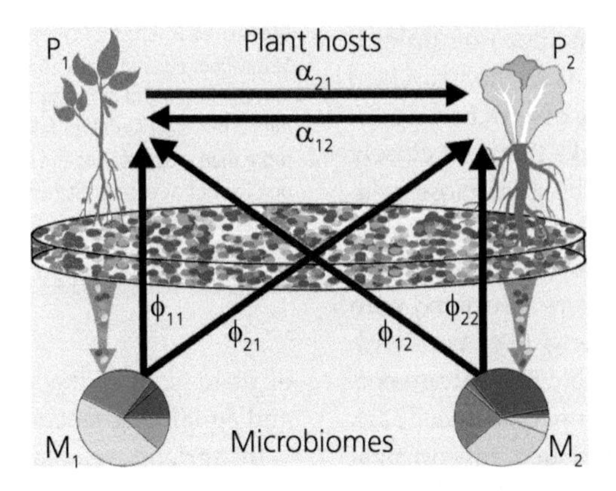

Figure 14.7 Plant–soil feedback can have impacts within host species and across species. As roots of each of plants P_1 and P_2 grow through the soil, they selectively enhance the abundance of separate subsets of pathogenic (warm colors) and beneficial (cool colors) microbes from the soil (microbiomes M_1 and M_2, respectively). The net effect of those microbes on the next generation of the same plant species is designated as ϕ_{11} or ϕ_{22}; feedback is negative if the pathogens outweigh the beneficials, and positive if the balance is beneficial. Other species of plants may respond differently to those microbes; the effect of M1 on P2 is given as ϕ_{21}, and of M_2 on P_1 as ϕ_{12}. The overall intensity of the pairwise feedback interaction would then be thought of as $I_S = \phi_{11} + \phi_{22} - \phi_{21} - \phi_{12}$ (Equation 14.4). In addition, plants growing as neighbors are in competition for resources; following the classic Lotka–Volterra competition equations (Box 14.3, Equations 14.2 and 14.3) the competitive effect of P_2 on P_1 is given as α_{12} and the inverse as α_{21}. In addition to the direct microbial effects of feedback on performance of subsequent generations of plants, the feedback can alter the strength of α_{12} or α_{21}, creating a kind of apparent competition (Section 14.5).
Graphic by Gregory S. Gilbert and Ingrid M. Parker, inspired by Abbott et al. (2021).

a species with a strong negative feedback might be replaced in that location by a neighboring species in the next generation. Thus, negative plant–soil feedbacks could result in temporal shifts of the locations or abundances of common species in the old-field community (e.g., stabilizing mechanisms).

From paired feedback experiments, it is possible to define a pairwise feedback interaction metric, I_s, where the sum of the intraspecific feedback effects (ϕ_{11} is the average performance of plant 1 growing in soil conditioned by plant 1, ϕ_{22} the same for plant 2) minus the interspecific effects (ϕ_{12} is the performance of plant 1 growing in soil conditioned by plant 2, and ϕ_{21} for plant 2 in soil 1) (Figure 14.7). That is:

$$I_s = \phi_{11} + \phi_{22} - \phi_{21} - \phi_{12} \quad \text{[Eq. 14.4]}$$

If plant performance is best when species are grown in their own soil, I_s is positive; if the relative performance of a species is worse in its own soil, I_s is negative. In a meta-analysis of over 1000 pairwise comparisons, the average feedback was negative, which is required for plant–soil feedbacks to contribute to plant diversity as a stabilizing mechanism (Crawford et al. 2019). Because the feedback is always measured relative to the paired species, the interpretation of I_s can sometimes be unintuitive. For example, pairs of plants in the Crawford study that were phylogenetically distant had significantly greater negative feedbacks than did closely related pairs of plants; this is because close relatives are more likely to share the same pathogens (Section 13.7), so the away soil would be similar to the home soil. Similarly, plant pairs associated with the same guild of mycorrhizal fungi (e.g., both with arbuscular mycorrhizae or both with ectomycorrhizae, Section 15.4) had stronger negative feedback effects. That is because mismatched mycorrhizal guild pairs have the potential for positive interactions with microbes that benefit their own species without crossover benefits to the other host. Negative plant–soil feedbacks are clearly common and have the potential to help maintain plant diversity, but since more than three-quarters of all the studies in the Crawford et al. (2019) study were conducted in lab or greenhouse settings rather than in the field, we are still developing a picture of the importance

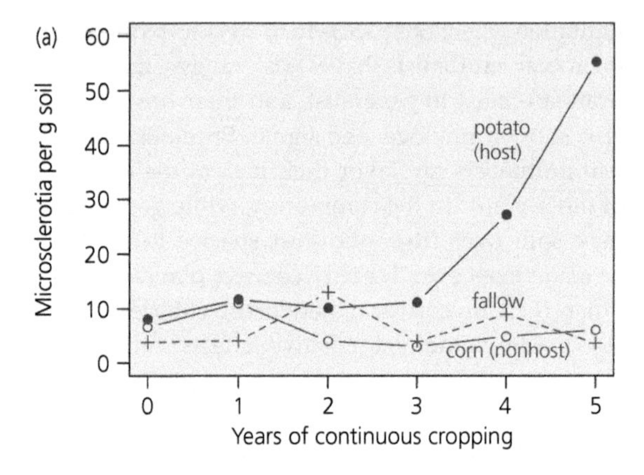

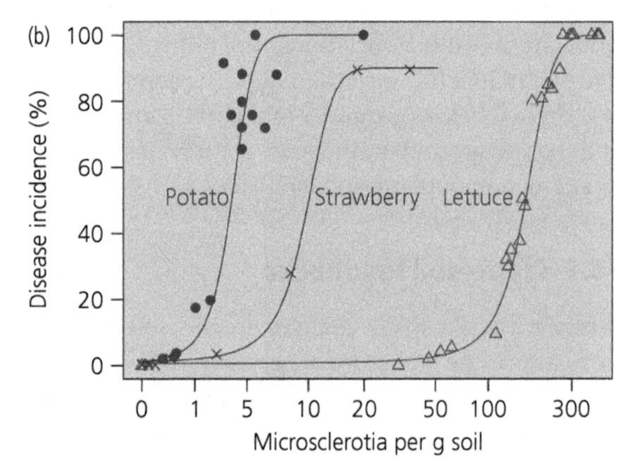

Figure 14.8 Effect of cropping on density of microsclerotia of *Verticillium dahliae* in the soil. (a) Microsclerotia increase in successive years of cropping to a susceptible host (potato), but not with a non-host (corn), or fallow. Microsclerotia can survive in soil for more than a decade. (b) The density of microsclerotia in soil is a predictor of disease incidence in susceptible crops.

Figures created by Gregory S. Gilbert using data from (a) Davis et al. (1994); (b) Nicot and Rouse (1987) (potato), Wu and Subbarao (2014) (lettuce), and Harris and Yang (1996) (strawberry).

of plant–soil feedbacks relative to other stabilizing and equalizing mechanisms in wild communities.

In agriculture, plant–soil feedback is a fundamental reason behind the practice of **crop rotation** (Section 17.4). Continuous cropping of the same plant species can lead to a buildup of pathogen inoculum in the soil (Figure 14.8). The shift in the soil microbial community to include more pathogens can create a negative feedback on performance of the next generation of that crop. Since many soil-borne pathogens produce long-lived resting structures, pathogen abundance can build with each successive generation of a host species and leave a long legacy in the soil. Rotation through a sequence of crops that do not share pathogens can help prevent such a buildup.

Plant–soil feedback via soil microbial communities can drive the dynamics of **succession** in plant communities. European coastal dune vegetation follows a quite predictable temporal succession of plant species, with marram grass (*Ammophila arenaria*) first colonizing and stabilizing shifting dune sands, and then gradually being replaced by a sequence of species (*Carex arenaria, Elymus athericus,* and *Hippophae rhamnoides*). Wim van der Putten and colleagues (1993) conducted a series of elegant experiments that showed that a buildup of soil-borne pathogens with differential impacts on each of the hosts causes successive species declines; species that are able to thrive in earlier successional stages are unable to thrive in soils collected from later in the successional sequence. The accumulated succession of different pathogens through a plant–soil feedback process creates a one-way succession path for coastal dune vegetation. Early successional species like *Ammophila* can colonize fresh sand dune areas, but once its pathogens have accumulated and its population has declined in that area, later successional species replace it.

In western North America, the soil-borne fungus *Phellinus weirii* causes severe laminated root rot on dominant Douglas fir (*Pseudotsuga menziesii*) and mountain hemlock (*Tsuga mertensiana*), spreading along roots through soil to create foci or mortality centers that measure 10–100 m radius. These mortality centers create gaps in the forest canopy, where other tree species that are less susceptible to the pathogen—such as rot-resistant pines, western hemlock, and western redcedar—can colonize and thrive (Hansen and Goheen 2000). This creates a patchwork of species composition in the forest that has flavors of both plant–soil feedback and Janzen–Connell processes. Interestingly, while the impacts on species composition are large, the effects of the mortality centers on plant diversity can be either positive or negative, depending on the local availability of species able to resist the pathogen.

So far, we have focused on the effects of pathogens on diversity, structure, and dynamics of plant communities. Let's now turn to how the structure of the plant community affects pathogen communities and the diseases they cause.

14.8 Community structure modulates the impacts of plant disease

Large, dense monocultures of a single plant species, as often practiced in industrial agriculture, provide ideal conditions for disease development; individual plants are surrounded by conspecifics at short nearest-neighbor distances that allow even dispersal-limited polycyclic pathogens to readily spread through a susceptible host population (Section 11.4). In contrast, there is a long-standing observation that plants growing in communities with high diversity suffer much less disease (Elton 1958).

Rottstock and colleagues (2014) looked at whether plant species diversity affected the diversity, prevalence, and severity of plant diseases. They took advantage of the Jena Experiment, a biodiversity manipulation that includes 82 large experimental plots that vary in species richness (1–60) as well as functional diversity. Rottstock et al. examined the number of guilds of aboveground plant pathogens observed (including rusts, powdery mildews, downy mildews, smuts, and leaf-spot diseases) among all the plant species. They found that the diversity of pathogens (measured as the cumulative number of pathogen guilds observed across each of the plant species in a plot) increased with the diversity of plant species. This might suggest that greater plant diversity leads to greater disease. But they also found that the incidence of disease among plants in a plot, as well as the severity of disease on individuals, decreased with increasing diversity of plant species. They also noted that co-infections—multiple fungal guilds on the same individual plant—were less common in more diverse plots. Taken together, more diverse plant communities may support a greater richness of plant pathogens, but plant diversity reduces the impact of those pathogens. Of course, interactions among pathogens and other plant-associated microorganisms are also important; later we will take up communities of microbes and how those interactions affect disease pressure (Chapters 15 and 17).

The phenomenon of increasing host diversity reducing the intensity of disease in the community is often called the **dilution effect** (Ostfeld and Keesing 2012, Civitello et al. 2015, Young et al. 2016). The dilution effect can incorporate several different mechanisms that reduce the spread of pathogens and the impacts of plant disease in more diverse communities.

First, because the spread and prevalence of disease is very often density dependent, increased plant diversity can reduce disease pressure by reducing host density. This is because total plant abundance is often a zero-sum game: for a given amount of resources (water, light, nutrients, space), there is a maximum density of individuals or amount of plant biomass that can occupy an area. Adding species nearly always results in reduced density of each component species. For instance, in a 1-ha plot where a regular grid of 50-cm spacing (40,401 plants) is planted with either one or two crop species, the nearest-neighbor distance between individuals of the same species would be 50 cm in the monoculture, but 70 cm with two species (Equation 11.7). If the same number of individual plants were distributed randomly across a meadow, a monospecific stand would have a mean nearest-neighbor distance between conspecifics of 25 cm, while in an equal mix of two species there would be 35 cm between conspecifics (Equation 11.8). This reduced host density can limit pathogen spread and disease intensity (Figure 11.6).

Second, most plant pathogens are passively dispersed and have no control over where they land. Interspersed non-host plants among individuals of host plants create physical barriers that pathogens stick to as they disseminate from source hosts. Increasing plant diversity means that a bigger fraction of plants are dead-end, non-host species for a pathogen; this reduces disease pressure on each component host species (Section 17.4).

Finally, simply increasing the species richness in a community would not reduce disease pressure if all the component species were susceptible to the same pathogens ... species composition matters, too! Increasing the proportion of the community comprising non-hosts or non-competent hosts should **reduce** disease spread; increasing the proportion of competent hosts can instead **amplify** disease spread.

Plant species richness may help moderate disease impacts in wild plant communities, and it can be manipulated in agroecosystems as a mechanism for disease control (Section 17.4). However, plant species diversity will only suppress disease when the component species differ in the pathogens to which they are susceptible and for which they are competent. All these interactions between plant communities and pathogens also depend on the development of microbial communities on plant hosts. In Chapter 15 we look at the plant microbiome and how it influences disease processes.

Further reading

Gurevitch, J., S. M. Scheiner, and G. A. Fox. 2021. *The Ecology of Plants*, 3rd edition. Oxford University Press.

Magurran, A. E. and B. J. McGill. 2011. *Biological Diversity: Frontiers in Measurement and Assessment*. Oxford University Press.

Mittelbach, G. G. and B. J. McGill. 2019. *Community Ecology*, 2nd edition. Oxford University Press.

References

Abbott, K. C., M. B. Eppinga, J. Umbanhowar, M. Baudena, and J. D. Bever. 2021. Microbiome influence on host community dynamics: Conceptual integration of microbiome feedback with classical host–microbe theory. *Ecology Letters* **24**:2796–2811.

Anderson, M. J. and T. J. Willis. 2003. Canonical analysis of principal coordinates: a useful method of constrained ordination for ecology. *Ecology* **84**:511–525.

Augspurger, C. K. 1983. Seed dispersal of the tropical tree, *Platypodium elegans*, and the escape of its seedlings from fungal pathogens. *Journal of Ecology* **71**:759–771.

Augspurger, C. K. 1984. Seedling survival of tropical tree species—interactions of dispersal distance, light-gaps, and pathogens. *Ecology* **65**:1705–1712.

Augspurger, C. K. and H. T. Wilkinson. 2007. Host specificity of pathogenic *Pythium* species: implications for tree species diversity. *Biotropica* **39**:702–708.

Bagchi, R., R. E. Gallery, S. Gripenberg, S. J. Gurr, L. Narayan, C. E. Addis, R. P. Freckleton, and O. T. Lewis. 2014. Pathogens and insect herbivores drive rainforest plant diversity and composition. *Nature* **506**: 85–88.

Bagchi, R., T. Swinfield, R. E. Gallery, O. T. Lewis, S. Gripenberg, L. Narayan, and R. P. Freckleton. 2010. Testing the Janzen–Connell mechanism: pathogens cause overcompensating density dependence in a tropical tree. *Ecology Letters* **13**:1262–1269.

Barrett, L. G., J. M. Kniskern, N. Bodenhausen, W. Zhang, and J. Bergelson. 2009. Continua of specificity and virulence in plant host–pathogen interactions: causes and consequences. *New Phytologist* **183**:513–529.

Bever, J. D. 1994. Feedback between plants and their soil communities in an old field community. *Ecology* **75**:1965–1977.

Bever, J. D., I. A. Dickie, E. Facelli, J. M. Facelli, J. Klironomos, M. Moora, M. C. Rillig, W. D. Stock, M. Tibbett, and M. Zobel. 2010. Rooting theories of plant community ecology in microbial interactions. *Trends in Ecology & Evolution* **25**:468–478.

Borer, E. T., P. R. Hosseini, E. W. Seabloom, and A. P. Dobson. 2007. Pathogen-induced reversal of native dominance in a grassland community. *Proceedings of the National Academy of Sciences of the United States of America* **104**:5473–5478.

Burdon, J. J. 1987. *Diseases and Plant Population Biology*. Cambridge University Press, Cambridge.

Chao, A., N. J. Gotelli, T. Hsieh, E. L. Sander, K. Ma, R. K. Colwell, and A. M. Ellison. 2014. Rarefaction and extrapolation with Hill numbers: a framework for sampling and estimation in species diversity studies. *Ecological Monographs* **84**:45–67.

Chesson, P. 2000. Mechanisms of maintenance of species diversity. *Annual Review of Ecology and Systematics* **31**:343–366.

Chesson, P. 2018. Updates on mechanisms of maintenance of species diversity. *Journal of Ecology* **106**:1773–1794.

Chisholm, R. A. and T. Fung. 2020. Janzen–Connell effects are a weak impediment to competitive exclusion. *American Naturalist* **196**:649–661.

Civitello, D. J., J. Cohen, H. Fatima, N. T. Halstead, J. Liriano, T. A. McMahon, C. N. Ortega, E. L. Sauer, T. Sehgal, and S. Young. 2015. Biodiversity inhibits parasites: broad evidence for the dilution effect. *Proceedings of the National Academy of Sciences of the United States of America* **112**:8667–8671.

Clay, K. and J. Holah. 1999. Fungal endophyte symbiosis and plant diversity in successional fields. *Science* **285**:1742–1744.

Cobb, R. C., R. K. Meentemeyer, and D. M. Rizzo. 2010. Apparent competition in canopy trees determined by pathogen transmission rather than susceptibility. *Ecology* **91**:327–333.

Coley, P. D., J. P. Bryant, and F. S. Chapin. 1985. Resource availability and plant antiherbivore defense. *Science* **230**:895–899.

Comita, L. S., S. A. Queenborough, S. J. Murphy, J. L. Eck, K. Y. Xu, M. Krishnadas, N. Beckman, and Y. Zhu. 2014. Testing predictions of the Janzen-Connell hypothesis: a meta-analysis of experimental evidence for distance- and density-dependent seed and seedling survival. *Journal of Ecology* **102**:845–856.

Connell, J. H. 1971. On the role of natural enemies in preventing competitive exclusion in some marine animals and in rain forest trees. In: P. J. de Voer and G. R. Gradwell, editors. *Dynamics of Numbers in Populations*. Proceedings of the Advanced Study Institute, Osterbeek, 1970, Centre for Agricultural Publication and Documentation, Wageningen, pp. 298–312.

Crawford, K. M., J. T. Bauer, L. S. Comita, M. B. Eppinga, D. J. Johnson, S. A. Mangan, S. A. Queenborough, A. E. Strand, K. N. Suding, and J. Umbanhowar. 2019. When and where plant-soil feedback may promote plant coexistence: a meta-analysis. *Ecology Letters* **22**:1274–1284.

Daszak, P., A. A. Cunningham, and A. D. Hyatt. 2000. Emerging infectious diseases of wildlife – threats to biodiversity and human health. *Science* **287**:443–449.

Davis, J., J. Pavek, D. Corsini, L. Sorensen, A. Schneider, D. Everson, D. Westermann, and O. Huisman. 1994. Influence of continuous cropping of several potato clones on the epidemiology of Verticillium wilt of potato. *Phytopathology* **84**:207–214.

Elton, C. S. 1958. *The Ecology of Invasions by Animals and Plants*. University of Chicago Press.

Fox, L. R. and P. A. Morrow. 1981. Specialization: species property or local phenomenon? *Science* **211**:887–893.

Gilbert, G. S., S. P. Hubbell, and R. B. Foster. 1994. Density and distance-to-adult effects of a canker disease of trees in a moist tropical forest. *Oecologia* **98**:100–108.

Gilbert, G. S. and C. O. Webb. 2007. Phylogenetic signal in plant pathogen-host range. *Proceedings of the National Academy of Sciences of the United States of America* **104**:4979–4983.

Gillett, J. B. 1962. Pest pressure, an underestimated factor in evolution. *Systematics Association Publication Number* 4:37–46.

Grime, J., J. Mackey, S. Hillier, and D. Read. 1987. Floristic diversity in a model system using experimental microcosms. *Nature* **328**:420–422.

Hansen, E. M. and E. M. Goheen. 2000. *Phellinus weirii* and other native root pathogens as determinants of forest structure and process in western North America. *Annual Review of Phytopathology* **38**:515–539.

Harris, D. and J. Yang. 1996. The relationship between the amount of *Verticillium dahliae* in soil and the incidence of strawberry wilt as a basis for disease risk prediction. *Plant Pathology* **45**:106–114.

Holt, R. D. 1977. Predation, apparent competition, and the structure of prey communities. *Theoretical Population Biology* **12**:197–229.

Janzen, D. 1970. Herbivores and the number of tree species in tropical forests. *American Naturalist* **104**:501–528.

Jia, S., X. Wang, Z. Yuan, F. Lin, J. Ye, G. Lin, Z. Hao, and R. Bagchi. 2020. Tree species traits affect which natural enemies drive the Janzen–Connell effect in a temperate forest. *Nature Communications* **11**:1–9.

Liu, X. B., M. X. Liang, R. S. Etienne, Y. F. Wang, C. Staehelin, and S. X. Yu. 2012. Experimental evidence for a phylogenetic Janzen–Connell effect in a subtropical forest. *Ecology Letters* **15**:111–118.

MacArthur, R. A. and E. O. Wison. 1967. *The Theory of Island Biogeography*. Princeton University Press.

Magurran, A. E. and B. J. McGill. 2011. *Biological Diversity: Frontiers in Measurement and Assessment*. Oxford University Press, New York.

Malmstrom, C. M., A. J. McCullough, H. A. Johnson, L. A. Newton, and E. T. Borer. 2005. Invasive annual grasses indirectly increase virus incidence in California native perennial bunchgrasses. *Oecologia* **145**:153–164.

Mangan, S. A., S. A. Schnitzer, E. A. Herre, K. M. L. Mack, M. C. Valencia, E. I. Sanchez, and J. D. Bever. 2010. Negative plant–soil feedback predicts tree-species relative abundance in a tropical forest. *Nature* **466**:752–755.

Marler, M. J., C. A. Zabinski, and R. M. Callaway. 1999. Mycorrhizae indirectly enhance competitive effects of an invasive forb on a native bunchgrass. *Ecology* **80**:1180–1186.

Minchin, P. R. 1987. An evaluation of the relative robustness of techniques for ecological ordination. In: I. C. Prentice and E. Maarel. *Theory and Models in Vegetation Science*, Proceedings of Symposium, Uppsala, July 8–13, 1985, pp. 89–107. Springer, Berlin.

Nicot, P. and D. Rouse. 1987. Relationship between soil inoculum density of *Verticillium dahliae* and systemic colonization of potato stems in commercial fields over time. *Phytopathology* **77**:1346–1355.

Ostfeld, R. S. and F. Keesing. 2012. Effects of host diversity on infectious disease. *Annual Review of Ecology, Evolution, and Systematics* **43**:157–182.

Packer, A. and K. Clay. 2000. Soil pathogens and spatial patterns of seedling mortality in a temperate tree. *Nature* **404**:278–281.

Parker, I. M. and G. S. Gilbert. 2018. Density-dependent disease, life-history trade-offs, and the effect of leaf pathogens on a suite of co-occurring close relatives. *Journal of Ecology* **106**:1829–1838.

Parker, I. M., M. Saunders, M. Bontrager, A. P. Weitz, R. Hendricks, R. Magarey, K. Suiter, and G. S. Gilbert. 2015. Phylogenetic structure and host abundance drive disease pressure in communities. *Nature* **520**:542–544.

Power, A. G. and C. E. Mitchell. 2004. Pathogen spillover in disease epidemics. *American Naturalist* **164**:S79–S89.

Rottstock, T., J. Joshi, V. Kummer, and M. Fischer. 2014. Higher plant diversity promotes higher diversity of fungal pathogens, while it decreases pathogen infection per plant. *Ecology* **95**:1907–1917.

ter Braak, C. J. F. 1995. Ordination. In: R. H. G. Jongman, C. J. F. Ter Braak, and O. F. R. van Tongeren, editors. *Data Analysis in Community and Landscape Ecology*, pp. 91–274. Cambridge University Press.

Thrall, P. H., M. E. Hochberg, J. J. Burdon, and J. D. Bever. 2007. Coevolution of symbiotic mutualists and parasites in a community context. *Trends in Ecology & Evolution* **22**:120–126.

Vanderputten, W. H., C. Vandijk, and B. A. M. Peters. 1993. Plant-specific soil-borne diseases contribute to succession in foredune vegetation. *Nature* **362**:53–56.

Willis, A. D. 2019. Rarefaction, alpha diversity, and statistics. *Frontiers in Microbiology* **10**:2407.

Wu, B. and K. Subbarao. 2014. A model for multiseasonal spread of Verticillium wilt of lettuce. *Phytopathology* **104**:908–917.

Young, H. S., I. M. Parker, G. S. Gilbert, A. S. Guerra, and C. L. Nunn. 2016. Introduced species, disease ecology and biodiversity–disease relationships. *Trends in Ecology & Evolution* **32**:41–54.

CHAPTER 15

The plant microbiome

Gregory S. Gilbert and Ingrid M. Parker

15.1 Plants as habitats for microbial communities

Plants are food. They are full of carbohydrates, proteins, fats, mineral nutrients, and water—that is why we eat them. Plants are also food for a tremendous diversity of microbes. Fungi, oomycetes, and bacteria are all **osmotrophic heterotrophs** (they digest what they live in), and all parts of every plant are full of microbes of different kinds. That assemblage of microbes is the plant **microbiome**.

In this chapter we first look at plants as a complex habitat for communities of microbes, and then explore how interactions among microbes shape the health of plants, including the development of disease. Those microbe–microbe interactions include the same range of context-dependent outcomes described for plant–microbe symbioses (Figure 1.3): mutualism, parasitism, competition, amensalism, and commensalism. The collective balance of microbe–microbe interactions across the microbiome, in conjunction with the kind of symbiotic relationship each microbe has with the host plant, determine the net effect of the microbiome on plant health. When plant-parasitic microbes dominate the microbiome, the plant becomes diseased. One could consider the microbiome to be included in the "environment" vertex of the disease triangle (Figure 1.7), but because the dynamics and impacts of the microbiome are largely products of microbe–microbe and microbe–plant interactions, it is qualitatively different from, say, humidity or nutrients.

There are three main messages from this chapter. First, the plant is a rich habitat for a broad diversity of microbes with a range of ecological effects. Second, treating plant disease as the simple outcome of an interaction between a single pathogen and a single host is risky, because the microbiome can have a major modulating effect on plant–pathogen interactions. Third, by actively manipulating the microbiome and the ways microbes interact with each other, we can improve plant health. Such manipulations are the basis for biological control of plant disease (Section 17.6).

We start off by looking at characteristics and key players in the microbiomes associated with different parts of plants: we focus on roots and leaves where the plant microbiome has been best studied, and then touch on wood and vascular tissue, and finally fruits and seeds. We take a small detour to introduce three common kinds of plant–microbe mutualisms: rhizobia, mycorrhizae, and plant growth-promoting rhizobacteria. Then we go a little deeper into the range of microbe–microbe interactions that shape microbiomes, and how those interactions affect the influence of microbes on their hosts.

15.2 The rhizosphere effect

Around the dawn of the 20th century, a number of researchers noticed that bacteria were more numerous around plant roots than in the rest of the soil (the "bulk soil"). In 1904, Lorenz Hiltner called the volume of soil affected by roots the **rhizosphere** (*ρίζα* or *rhiza* = root; this chapter is packed with funky words derived from Greek) (Hartmann et al. 2008). The changes in the density and composition of microbial communities around roots is known

The Evolutionary Ecology of Plant Disease. Gregory S. Gilbert and Ingrid M. Parker, Oxford University Press. © Gregory S. Gilbert & Ingrid M. Parker (2023).
DOI: 10.1093/oso/9780198797876.003.0015

as the **rhizosphere effect**. Microbial growth in soil is primarily carbon-limited, and growing roots leak or actively excrete carbon-rich compounds (up to 40% of the plant's photosynthate)! Plants can also increase soil moisture, and these changes, along with other physical and chemical alterations to the soil around roots, drives a ten-fold increase in density of bacteria (Lynch and Whipps 1990). This nutrient-rich environment not only increases bacterial abundance but also selects for and amplifies a particular subset of bacteria that thrive in association with the plant. This results in qualitatively different composition (and lower diversity) of bacterial communities in the rhizosphere than in bulk soil (reviewed in Gilbert et al. 1994, Müller et al. 2016). At the root surface, called the **rhizoplane**, a subset of the rhizosphere bacteria thrives by creating a biofilm embedded in self-produced extracellular polysaccharides on the root surface. An even more restricted (and less diverse) subset is able to colonize the root system internally as endophytes, in the **endorhizosphere** (*ενδο* or *endo* = within) (Figure 15.1). The rhizosphere region influenced by the root but not in direct contact with the root is called the **ectorhizosphere** (*ἐκτός* or *ektos* = outside). Each of these three zones of the rhizosphere are habitats for dynamic microbial communities that are shaped by a combination of plant–microbe and microbe–microbe interactions, with large impacts on plant nutrition and plant disease development. Rhizosphere microbes may benefit, harm, or have no effect on the associated plant host.

Just like plants, microbes can alter the chemical composition of the rhizosphere habitat. **Quorum sensing** (Section 5.4) among the high densities of rhizosphere bacteria initiates their production of extracellular chitinase and protease enzymes that mineralize inaccessible, complex forms of nitrogen and make the nutrients available for uptake by the plant and other microbes (DeAngelis et al. 2008). Fluorescent species of the bacterium *Pseudomonas* produce potent iron-binding molecules

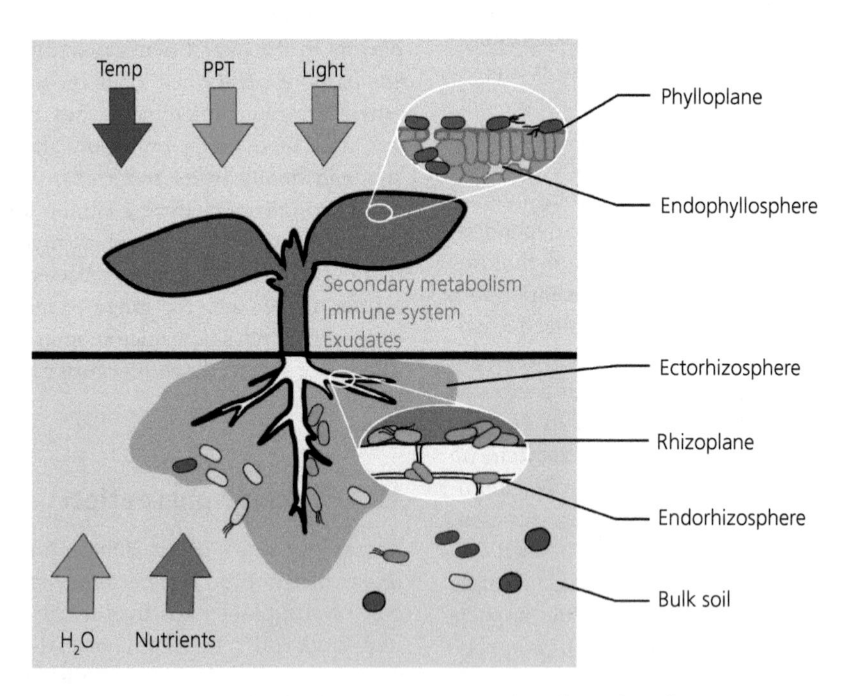

Figure 15.1 Microbiomes of the rhizosphere and phyllosphere, and the abiotic and biotic factors that influence them. The rhizosphere includes the endorhizosphere (the environment inside the root), the rhizoplane (where microbes are living directly on the surface of the root), as well as the ectorhizosphere (the region out beyond the root surface). Similarly, the phyllosphere includes elements of the microbiome on the surface of leaves (phylloplane) and inside of leaves (endophyllosphere).

Graphic by Ingrid M. Parker.

(**siderophores**) that allow them to efficiently scavenge iron; they also produce phosphatases that solubilize and make soil-fixed phosphorus available (Rai et al. 2017). In the highly competitive environment of the microbiome, many bacteria produce **antibiotics** (chemicals that kill or inhibit growth of other microorganisms) that suppress neighboring microbes to gain a competitive advantage.

Communities of rhizosphere bacteria vary across plant hosts and across different soils, but some subsets of bacteria are consistently favored through the rhizosphere effect—especially the Actinobacteria, Bacteroides, Firmicutes, and Proteobacteria—while other groups fail to thrive, compared to the bulk soil (Müller et al. 2016). The functional traits of bacteria in the rhizosphere include the ability to use a wider diversity of simple carbon sources, the production of fewer extracellular enzymes associated with breaking down complex carbon sources, and resistance to a wider range of antibiotics, compared to bacteria from the bulk soil (Gilbert et al. 1993).

Let's now take a brief detour from plant diseases to look at several groups of rhizosphere symbioses that generally have a direct, positive effect on the host plant: nitrogen-fixing rhizobia, mycorrhizae, and plant growth-promoting rhizobacteria. Later in the chapter (Section 15.8) we examine a range of direct and indirect microbe–microbe interactions that shape the plant microbiome, including enhancing or mitigating the effects of plant pathogens.

15.3 Nitrogen-fixing Rhizobia

The most well-known mutualism between plant roots and bacteria occurs between plants in the Fabaceae family (commonly called legumes) and a group of bacterial partners collectively called rhizobia. **Rhizobia** can fix atmospheric nitrogen—converting nonreactive dinitrogen gas (N_2) into biologically available ammonia (NH_3). **Nitrogen fixation** is important because all terrestrial organisms are bathed in air that is 78% N_2 gas, yet this atmospheric nitrogen is not available to most organisms because they are unable to break the strong triple bonds that bind the two N atoms together. Rhizobia and some other **diazotroph** bacteria can fix nitrogen with the help of **nitrogenase** enzymes through the following chemical transformation:

$$N_2 + 16ATP + 8e^- + 8H^+ \rightarrow 2NH_3 + H_2 + 16ADP + 16P_i \quad \text{[Eq. 15.1]}$$

Nitrogen fixation demands a lot of energy; note that 16 ATP are used to process one molecule of N_2 into NH_3 (Equation 15.1). The resulting ammonia (NH_3^0) is actually toxic, but it is quickly reduced to the less toxic ammonium form (NH_4^+). If released into the soil, ammonium is transformed by a suite of soil-borne nitrifying bacteria into nitrite (NO_2^-) and then nitrate (NO_3^-), which is the preferred form of nitrogen for most plants.

While some diazotrophs fix nitrogen as free-living organisms in soil, rhizobial bacteria form an endorhizosphere symbiosis with plants. Complex signaling between the bacteria and the plant is required for the rhizobia to colonize the root. **Signaling** requires more than just sensing the environment; it requires both a sender and a receiver, with a response from the receiver. The plant releases flavonoid compounds that attract rhizobia to root hairs, where the rhizobia then produce signaling compounds called **nod factors**, chemicals that induce root hairs on the plant to curl around the rhizobial cells, forming an **infection thread** through which bacteria enter root cells. Rapid root cell division then encapsulates a microcolony of the rhizobial cells inside a specialized organ called a **nodule** (Figure 15.2), within which the rhizobial cells shed their walls to transform into **bacteroids** and begin the process of nitrogen fixation. In the nodule, the plant provides a supply of leghemoglobin, a red-colored, complex chemical very similar to hemoglobin in human blood (that's why nodules are pink inside). Leghemoglobin has a high affinity for oxygen and binds up most of the oxygen in the nodule. The resulting low-oxygen environment is critical, because rhizobial nitrogenase is inhibited by oxygen. Hence the plant regulates the oxygen concentration in nodules so that it is low enough for nitrogen fixation to take place yet high enough to support aerobic respiration in the bacteria. The fixed NH_4^+ is provided to the plant, which incorporates it into amino acids. The added nitrogen from fixation explains why pulses (the seeds of legumes,

Figure 15.2 Nitrogen-fixing rhizobia nodules on roots of a legume.
Photo by Gregory S. Gilbert.

like pinto beans, lentils, and garbanzos) are high in protein, and why growing leguminous crops and incorporating their remains into the soil is a good way to enhance soil fertility.

Rhizobial bacteria are mostly in the α-Proteobacteria clade (genera *Rhizobium*, *Mesorhizobium*, *Bradyrhizobium*, *Sinorhizobium*, and others), with a few in the β-Proteobacteria clade (*Paraburkholderia*). In addition, the actinobacterium *Frankia* is able to create similar nitrogen-fixing nodules with some plants in the Fagales and Rosales orders, notably in the woody genera *Alnus*, *Casuarina*, *Morella*, and *Ceanothus*.

15.4 Mycorrhizae

On the fungal side, the most well-known mutualisms with plant roots are the **mycorrhizae** (*μύκης* or *mycos* = fungus)—a diverse assemblage of fungus–plant root symbioses that are nearly ubiquitous in the plant world. In general, mycorrhizal fungi colonize plant roots and extend their mycelium into the surrounding soil, where their high surface area and extracellular enzymes make them extremely effective foragers for soil water and nutrients. The fungi receive sugars from the host plant, which allows them to grow (remember carbon is a limiting nutrient for most soil microbes); in exchange, the fungi provide the host with an assortment of nutrients, especially nitrogen and phosphorus, that are limiting for the plant. Note that unlike rhizobia, mycorrhizal fungi do not fix nitrogen, they are just efficient foragers of nitrogen in the soil. Nearly all plant species form mycorrhizae, with exceptions in just four plant families: Proteaceae (southern hemisphere woody plants), Caryophyllaceae (carnation family), Brassicaceae (mustard family), and some Amaranthaceae (spinach family).

For plants, the most **limiting nutrient** is usually nitrogen in temperate soils and phosphorus in tropical soils (although there is variation across soil types everywhere). Since fungi are exceptionally good at accessing nitrogen and phosphorus from soils but are in need of reliable carbon sources, there is a basis for a good market exchange between carbon-wealthy plants and nutrient-wealthy fungi. Under nutrient-limited conditions, the mycorrhizal symbiosis is a classic example of a mutualism, with both the plant and fungus benefitting. As plant disease ecologists, however, we always look at plant–microbe symbioses through the lens of the disease triangle (Figure 1.6), and it provides an important perspective here. In an environment with abundant mineral nutrients and water, the benefits of the symbiosis to the plant are diminished, but the cost in sugars provided to the mycorrhizal fungus remains the same. Under such conditions, which are common in agriculture, the mycorrhizal fungus is parasitic on the host plant and can significantly reduce the growth of the plant (Ryan and Graham 2002, Jacott et al. 2017).

This general overview of mycorrhizal symbioses obscures important variation in both the biology of the symbiosis and the outcomes of the symbiosis for the plant. There are four major types of mycorrhizae that differ in both the kinds of fungi involved and the physical arrangement of the root–fungus interactions: arbuscular mycorrhizae, ectomycorrhizae, orchid mycorrhizae, and ericoid mycorrhizae. Let's look at each of those four types, and then explore the impacts of mycorrhizal networks a bit more.

15.4.1 Arbuscular mycorrhizae

Arbuscular mycorrhizae (often abbreviated as AM) are formed by some 80% of all plants. Arbuscular mycorrhizae are ancient: the very earliest land plants, 450 million years ago, had fungal associates that look very similar to modern arbuscular mycorrhizae—even before plants evolved root systems! Evidence suggests that the symbiosis evolved once and has been maintained throughout the history of land plants. They are formed by fungi in the Glomeromycota clade in association with herbs, grasses, bryophytes, most tropical trees, and some temperate tree species. Glomeromycota make no easily visible reproductive structures, and are most readily seen by clearing (removing the pigments from) fine roots, staining the fungi within them, and examining them with a microscope (Figure 15.3). Within the root, AM produce **arbuscules**, which are highly branched **haustoria** invaginating the root cells (Figure 1.4) and separated from the plant cytoplasm by a plasma membrane. Arbuscules serve for metabolite exchange between the fungus and the plant. AM fungi also produce **vesicles**, oblong lipid-storage structures made inside host cells; these can serve as **chlamydospores**. The extensive and intimate colonization of root tissue, along with their production of vesicles, gives arbuscular mycorrhizae two older common names: the **endomycorrhizae** and the **vesicular–arbuscular mycorrhizae** (VAM). Outside the root, the mycelium produces large **glomerospores** (80–500 μm) that can serve as

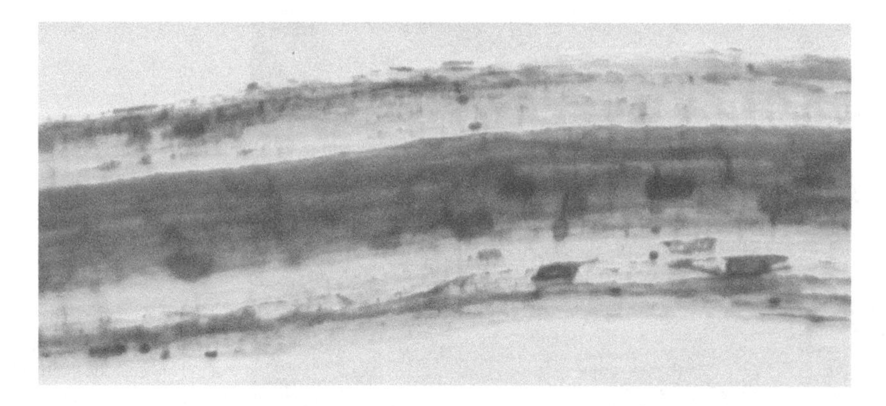

Figure 15.3 Cleared and stained root of arbuscular mycorrhizae on strawberry root. Dark blue structures are fungal vesicles. Photo by Gregory S. Gilbert.

resting structures. Arbuscular mycorrhizae tend to be able to colonize a wide range of host species, but preferences among locally available hosts in plant-fungus pairings are common (Sepp et al. 2019). In addition, the outcome of the symbiosis for the host plant (beneficial, neutral, or harmful) is extremely context dependent and varies tremendously among host species colonized by the same fungal strain (Bennett and Groten 2022).

Just as for legume–rhizobial nodule symbioses, establishing and maintaining such an intimate relationship between plant roots and arbuscular mycorrhizal fungi requires complex and extended signaling between the partners (MacLean et al. 2017). In fact, rhizobia and arbuscular mycorrhizae share an important signaling system called the **common symbiosis pathway** (CSP). The CSP includes at least ten plant proteins involved in a signal transduction pathway that begins after the plant perceives a signal from the fungus (or bacterium). The CSP then leads to the induction of additional pathways and responses specific to each kind of symbiont, in a back-and-forth conversation (Genre and Russo 2016). The CSP may also be involved in infection processes from other kinds of mycorrhizae (Cope et al. 2019) and some plant pathogens (Rey et al. 2014).

15.4.2 Ectomycorrhizae

Most temperate-zone trees and a few tropical tree species form **ectomycorrhizae.** Here the fungi (mainly basidiomycetes, plus a few ascomycetes fungi) colonize an emerging lateral root, forming layers of hyphae surrounding the root to create a kind of pseudoparenchymatous tissue at the rhizoplane, called the **sheathing mantle** (Figure 15.4). The hyphae also push into the root, growing and repeatedly branching in the apoplast to surround cells of both epidermis and cortex with what is called the **Hartig net** (after Robert Hartig, the founder of forest pathology). Ectomycorrhizae are readily visible to the naked eye on the tips of tree roots because the sheathing mantle is apparent as a thickened, club-like, or densely branched structure on the root tips. The close contact between the Hartig net and the root cells allows for efficient metabolite exchange. The fungal mycelium extends into the soil where it absorbs mineral nutrients, but ectomycorrhizal fungi generally have a reduced ability to decompose complex organic matter compared to saprotrophic fungi. That means they get most of their carbon nutrition from the host plant.

Ectomycorrhizal associations with hosts have evolved independently numerous times within

Figure 15.4 Ectomycorrhizal fungi form a red mycelial sheathing mantle around the plant root, with light-colored mycelium extending into the soil.
Photo by Thomas Bruns.

different clades of fungi. Unlike their microscopic counterparts in the arbuscular mycorrhizae, most ectomycorrhizae make large, macroscopic **basidiomata** or **ascomata** for sexual reproduction (Figure 15.4). Many of the wild mushrooms (Basidiomycetes) and truffles (Ascomycetes) prized by chefs are ectomycorrhizal fungi. The host ranges of ectomycorrhizal fungi tend to be more restricted than those of arbuscular mycorrhizal fungi, which is why foragers, when hunting particular kinds of mushrooms, will look for stands of specific tree species that are likely to support their favorites.

15.4.3 Orchid mycorrhizae

The Orchidaceae is a remarkably diverse family of flowering plants. Orchids form mycorrhizae with a variety of Basidiomycetes, especially from the orders Cantharellales and Sebacinales, including the plant pathogen *Ceratobasidium* (anamorph *Rhizoctonia*). An unusual feature of orchids is that they make massive numbers of very tiny, semi-microscopic seeds that lack endosperm. Endosperm in most plants provides the energy needed for a seed to germinate and begin growing. The way orchid seedlings survive without endosperm is through their obligate dependence on a mycorrhizal symbiosis! The fungi penetrate the seeds and stimulate seed germination; then the embryo grows into a **protocorm**, which is a mass of mostly totipotent cells. The fungi penetrate parenchyma cell walls in the protocorm and form a knotted, intracellular mass of hyphae called a **peloton**. The fungi provide nutrients and water to the protocorm as it grows, until the plant can begin photosynthesis. During the protocorm stage (which lasts days to weeks), the orchid is entirely and obligately **mycoheterotrophic** (Section 7.2), obtaining carbon from the fungi, which acquire their carbon through decomposition of organic matter. In contrast, the fungi in the partnership have rich lives independent of the orchid hosts, as plant pathogens, saprotrophs, or ectomycorrhizal fungi. An individual protocorm may be colonized by multiple different species of fungi, and as the plant grows it will often change mycorrhizal associates. Many orchids are mycorrhizal throughout their lives, while some become mycorrhiza-free as mature plants. Strains of mycorrhizal fungi isolated from mature orchids are often not as effective at supporting germination and early growth as strains found in protocorms.

15.4.4 Ericoid mycorrhizae

Plants in the Ericaceae (blueberry family) have taken a different approach—actually three approaches—to mycorrhizal symbioses. Most genera in the family associate with Ascomycete fungi (*Pezoloma* and some other fungi in the order Helotiales) to create **ericoid mycorrhizae**. Fungi penetrate the cell walls of epidermal cells of very fine roots and pack them with hyphae; the fungus then creates a loose mantle of mycelium around the outside of the root. The associations may be more transient than arbuscular or ectomycorrhizal associations, lasting a few weeks on any individual fine root before degrading. This is the most common type of mycorrhizal symbiosis in the Ericaceae family, but it first evolved about 100 million years ago, while the three most basal clades in the family are older than this symbiosis and have different approaches. First, the rare, Asian genus of *Enkianthus* forms arbuscular mycorrhizae. Second, the arbutoid clade, which includes Mediterranean-climate trees and shrubs in *Arbutus* and *Arctostaphylos*, form **arbutoid mycorrhizae** with ectomycorrhizal fungi. Arbutoid mycorrhizae differ from normal ectomycorrhizal associations because, in addition to a Hartig net, fungal hyphae penetrate extensively through plant cell walls, creating what is sometimes called an **ectendomycorrhiza**. Third, plants in the Monotropoidea clade form **monotropoid mycorrhizae**, also with ectomycorrhizal fungi. What's fascinating is that the host plants in the Monotropoidea all lack chlorophyll, and are thus totally dependent on their mycorrhizae to provide them not only with mineral nutrients but also with a source of carbon (Section 7.2). Where do the fungi get the carbon? By forming ectomycorrhizal associations with nearby trees, from which they capture photosynthetic sugars and share it with their mycoparasitic hosts.

15.4.5 Common mycorrhizal networks

Monotropoid mycorrhizae are a very particular example of a **common mycorrhizal network**: multiple plants connected together by a common fungal mycelium through which nutrients (or other materials) are able to move between plants. Such networks can be established with any type of mycorrhizae, and there is substantial evidence that carbon, nitrogen, phosphorus, micronutrients, and even signaling compounds can be transported among plants through the mycorrhizal fungal mycelium (Wipf et al. 2019). For example, Song et al. (2014) planted pairs of tomato plants in rectangular pots divided by a stainless steel 25-μm filter separating the two plants; this filter allowed fungal hyphae, but not plant roots, to pass between the two sides of the pot. In half of the pots, the divider included an additional membrane that blocked hyphae as well, preventing a mycorrhizal network from forming between the plants. Both sides were inoculated with the AM fungus *Funneliformis mosseae*. When one of the two plants (the "donor" plant) had caterpillars chewing on it, defensive enzymes in the jasmonate (JA) pathway increased in *both* plants, but only when they were connected by a common mycorrhizal network. When Song et al. did the experiment using a plant with a mutation defective in JA biosynthesis as the "donor" (attacked) plant, there was no increase in defensive enzymes in the "recipient" plant, even in the presence of the mycorrhizal network. This showed that the origin of the defensive compounds in the "recipient" tomato plant was from the donor tomato plant via the common mycorrhizal network. Such networks can connect well-established mature trees with their seedlings, supplementing resources at a critical time for these young plants living in a dark environment and with a small root system. In this "source-sink" scenario, resource-limited individuals (sinks) benefit from the transfer of resources from wealthier source plants.

The nature and role of common mycorrhizal networks in forests and other ecosystems has become a subject of spirited debate. Some have suggested that these networks form the basis of an intentional super-organism of sorts, with trees engaging in altruistic sharing of food and information with kin and non-kin alike (Wohlleben 2016, Simard 2021). While there is clear evidence that multiple plants can tap into a common mycorrhizal network, and that resources and signals move between plants through the fungal mycelium, it is still far from clear how ecologically important or widespread such networks are (Henriksson et al. 2021). There are reasons why many scientists are skeptical about the idea that mycorrhizal networks create cooperative forest communities of the kind that have become popular in the public imagination. First, the necessary conditions rarely occur at the required scale. Although an exceptional fungal clone may extend over a large area (Section 3.3), most of the time the root system of any individual plant is interacting with a large number of independent, fairly localized fungal individuals, often of various species, and those fungi compete with each other for access to the host's carbon. Second, we can't forget that fungi act in their own self-interest. What does that mean for the fungus? First and foremost, it needs to ensure access to a stable food supply for the mycelium to grow and reproduce so it can pass its genes on to continuing generations. In studies of mycorrhizal networks, the outcomes for different plant hosts associating with a single strain of mycorrhizal fungus can range from positive to negative, and similarly, fungi may receive carbon at a higher rate from some plants than from others. The presence of multiple partners creates the opportunity for "cheaters," which benefit from the system while skimping on their own contributions. Sometimes the interests of the two partners align perfectly, and sometimes they do not. Because very few studies in biology have found support for the idea that non-human organisms act in ways that systematically counteract their self-interest in the long term, it is hard to imagine a mechanism by which plants would promote fungal strains that benefit unrelated neighboring plants in the network (Flinn 2021).

Some scientists have proposed thinking about common mycorrhizal networks as like a market economy, where all participants have commodities on offer as well as needs they must fulfill (Kiers et al. 2011). In this view, every participant, both plant and fungal, may exploit the law of supply and demand to get the most in exchange for what they provide. A key requirement of a functional mycorrhizal market economy is the ability to detect the reception of nutrients ("commodities") and distinguish among partner individuals; a second requirement is the ability to "sanction" (i.e., withhold resources from) partners that provide few resources and to promote the most generous partners. Researchers have developed wonderfully creative approaches to test these ideas in experimental studies. For example, Toby Kiers and colleagues (2011) categorized three strains of *Glomus* by their degree of cooperativeness by quantifying the number of carbon atoms sequestered by the fungus per atom of phosphorus (P) provided. Then they grew *Medicago truncatula* plants with the three strains simultaneously in the presence of $^{13}CO_2$; the stable isotope of carbon can be tracked as it makes its way through photosynthesis into plant sugars and then on to the fungal symbiont. By extracting total fungal RNA after 24 hours and then partitioning it among the three fungal strains using strain-specific primers and quantitative polymerase chain reaction (qPCR, Chapter 9), the researchers found that the plants provided more carbon to more cooperative fungi. Moreover, they set up trials in Petri plates of plant roots with pairs of fungi of the same strain; when they provided additional P to one of the two fungi, the plant rewarded that individual with more carbon (Figure 15.5). This field is evolving rapidly through the development of new technologies, such as tagging individual molecules with fluorescent quantum-dot nanoparticles and watching them move through fungal hyphae with confocal microscopy (van't Padje et al. 2021). A greater understanding of what drives the behavior of fungal and plant partners will help inform our understanding of big-picture questions like the nature of common mycorrhizal networks and the nature of cooperation in biology.

15.5 Plant growth-promoting rhizobacteria and fungi

The rhizosphere is colonized by a tremendous phylogenetic diversity of bacteria with equally diverse kinds of relationships with the plant. The overall composition of rhizosphere bacterial communities has been recognized for a long time for its potential to protect plants against pathogens (Gilbert et al. 1994). A small subset of free-living rhizosphere bacteria have particularly beneficial effects on the host plant, and are collectively known as **plant growth-promoting rhizobacteria (PGPR)** (Beneduzi et al. 2012). This loose designation arose from observations in the 1970s that a subset of rhizosphere bacteria improved plant performance through a variety of mechanisms beyond the symbiotic nitrogen fixation associated with rhizobia. PGPR include many kinds of bacteria, but strains from the genera *Pseudomonas* and *Azotobacter* (γ-proteobacteria), *Berkholderia* (β-proteobacteria), the *Cytophaga–Flavobacterium–Bacteroides* cluster (CFB), and *Bacillus* (Firmicutes) are especially common. PGPR benefit plants by mediating plant interactions with both nutrients and other organisms, sometimes manipulating hormone signaling of the plant; understanding the effects of PGPR is a very active

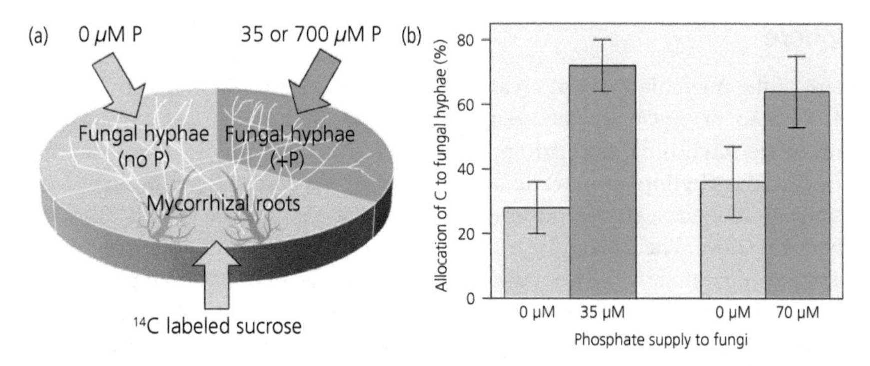

Figure 15.5 An *in vitro* experiment to test how plants control the contribution of carbon to different fungal partners. (a) The mycorrhizal fungus *Glomus intraradices* grew from the roots into each of the upper compartments of the plate; one compartment was provided with supplemental phosphorus at either 35 or 700 μM. (b) Providing the root with sugar containing labeled carbon revealed that the plant root allocated more carbon to the fungal hyphae that provided phosphorus to the root.

Graphics by Gregory S. Gilbert, redrawn from data in Kiers et al. (2011).

area of research. In fact, the list of known benefits for plants from PGPR reads like a crazy quilt of possibilities. PGPR can improve host nutrition by solubilizing phosphorus or by enhancing the availability of iron and other micronutrients. The free-living diazotroph *Azotobacter* fixes N_2 in the rhizosphere and produces auxin and gibberellins that stimulate plant shoot and root development (Aasfar et al. 2021). Some PGPR can act as helper organisms that enhance the establishment of mutualistic mycorrhizae. Molecules produced by PGPR can induce host resistance or tolerance to pests or adverse abiotic conditions, including excessive toxic metals. Rhizosphere *Flavobacterium* produces volatiles that stimulate fruit ripening in the host plant and inhibit oomycete pathogens (Sang et al. 2011). Many PGPR are directly antagonistic to plant pathogens (through antibiosis, parasitism, or competition, below), helping protect plant hosts from infection (Beneduzi et al. 2012). Managing populations of particular strains of PGPR, as well as managing the soil to enrich rhizosphere communities with PGPR, shows great potential for enhancing crop productivity and biological control of plant pathogens (Section 17.5).

Beyond bacteria, some fungi and oomycetes in the rhizosphere (in addition to mycorrhizae) are also beneficial to their host plants. These may produce antibiotics or parasitize pathogenic fungi as they approach roots, reducing infection and disease development in the hosts (Section 15.10).

15.6 Phyllosphere

The **phyllosphere** is the microbial habitat created by plant leaves (φύλλο or *phyllo* = leaf). Some researchers think of the phyllosphere as primarily the surface of the leaf (the **phylloplane**), while others include the tissues within the leaf (the **endophyllosphere**) (Vacher et al. 2016). The phylloplane and the endophyllosphere, separated by the leaf cuticle, differ considerably as habitats for microorganisms.

The endophyllosphere is an environment rich in nutrients and water where microbes are in close contact with living plant cells that produce sugar. Leaves regulate moisture and temperature within tolerable limits of cellular growth, and protect their cells from the damaging effects of UV radiation. This is why **endophytic** (εντός or *endos* = within and φυτό or *phyte* = plant) fungi or bacteria can be found inside pretty much every part of every leaf on every plant. Microbes grow in the apoplast, within parenchyma cells, and even within sieve-tubes and xylem vessels.

The phylloplane is a more challenging place to make a living as a microbe. For most endophytes, including many plant pathogens, the phylloplane is a gateway to the riches of the endophyllosphere; it is a transient habitat where fungal spores germinate and bacteria grow just long enough to penetrate into the inner leaf through stomata, wounds, or by breaching the cuticle. But for resident **epifoliar** microbes (επί or *epi* = on, the foliage), also known as **epiphylls**, the phylloplane is a thriving and dynamic habitat where microbes interact with each other and with the plant in ways that affect host health and the development of plant disease.

The structures and chemical traits of leaf surfaces define microhabitats that critically influence the assemblages of microbes that thrive on and in leaves. Further contributing to the habitat of the leaf surface is the **boundary layer**, which is a zone of still air up to a few millimeters thick, formed by air friction at the leaf surface. It slows the exchange of water vapor, CO_2, and heat between the leaf and the surrounding air, creating a phylloplane microclimate that is more humid, lower in CO_2, and warmer than the surrounding air. Larger leaves and those with trichomes tend to have thicker boundary layers, because these features create more friction for the air passing by the leaf. **Stomata** (Section 2.2) are critical leaf features for phyllosphere microbes, both because of their role in maintaining a moist environment at the leaf surface and as an entryway from the phylloplane into the endophyllosphere. When stomata are open for gas exchange, they are also open conduits allowing bacteria and fungi to pass through the cuticular barrier (Figure 15.1). Some foliar bacterial pathogens, such as *Xanthomonas campestris*, produce signals that induce stomatal opening, facilitating their entry into the leaf, where they thrive in the apoplast and cause disease (Beattie 2011). Some rust fungi can orient their growth based on ridge spacings of epidermal cells on leaves, and they differentiate to produce **appressoria** necessary for infection through the stomata

(Section 12.3), responding to the particular size and shape of guard cells on leaves of their host plant (Hoch et al. 1987).

Moisture critically influences microbial growth and presents a particular challenge to phylloplane microbes; conditions on a leaf surface can change from abundant free water (visible droplets) to desiccating dryness and back in the span of minutes. Extended periods with free water can be critical for fungal spore germination and leaf infection (Bradley et al. 2003), and support the rapid growth of epifoliar bacterial populations (Beattie 2011). Free water forms on leaf surfaces following precipitation, dew formation, and condensation of fog on leaves (Figure 15.6). Because films of water on leaves inhibit gas exchange and thus photosynthesis, most plants have evolved **hydrophobic** (water-repelling), waxy cuticles and assorted leaf morphologies (e.g., dissected leaves, elongated drip-tips on leaves) that help shed puddles of water from the leaf surface. However, microscopic water films persist on leaves, and several mechanisms bring water from inside the leaf to the surface, including pathways through the cuticle around trichomes and veins, and recondensation of water vapor that is released from stomata and trapped within the boundary layer. These localized water sources directly support microbial growth and survival, and are critical in leaching, solubilization, and chemical transformations of nutrients in the phylloplane. Epifoliar microbes are densest in areas where water is most available; bacteria or fungal spores can survive for days on moist leaves but perish quickly at low relative humidity. Epifoliar microorganisms have a variety of adaptations to survive water stress, including cell aggregation and the production of extracellular polysaccharides and surfactants that can actually increase the wettability of the leaves (Beattie 2011).

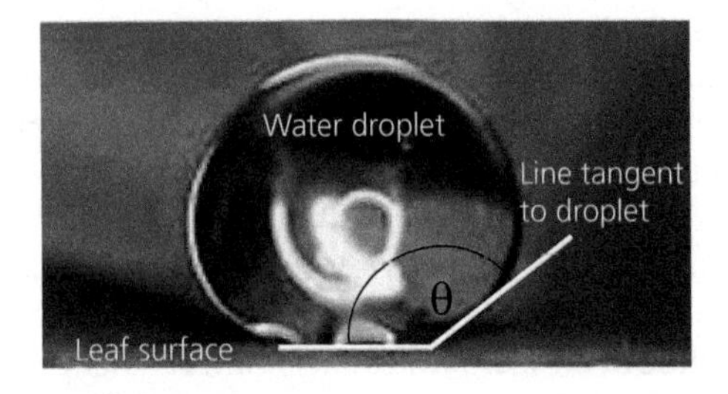

Figure 15.6 Free water forms on leaf surfaces following rain, dew, and fog deposition, and is critical to microbial growth. The angle θ is large for drops on a hydrophobic leaf like this, and small on a hydrophilic leaf.
Photo used with permission from Bradley et al. (2003).

Along with moisture, acquiring nutritional resources also poses a challenge for microbes in the phylloplane. A source of nutrition for epifoliar microbes comes from the plant's own leaky transport system (Vacher et al. 2016). During sugar transport in the leaf, some of the sucrose in the apoplast is broken down into simpler glucose and fructose. A fraction of these simple sugars is then leached by diffusion, along with other nutrients, through the cuticle and onto the phylloplane. Plants vary in the permeability of their cuticles, and enzymes produced by some epifoliar microbes increase the permeability of the cuticle to enhance the rate of leaching. Whereas leakage on the leaf surface pales in comparison to the quantity of leakage in the rhizosphere, it does provide a critical resource base for phylloplane fungi and bacteria. Additional nutrients accumulate on the phylloplane from particles in the air, and sap-sucking insects in the Hemiptera order (like aphids and scale insects) can coat the leaves of their host plants with their digestive waste product, nutrient-rich "honeydew."

As in rhizosphere communities, microbe–microbe interactions including competition, parasitism, and antibiosis also shape the composition and dynamics of microbe communities that thrive on the phylloplane, microbes that colonize the interior of the leaves as endophytes, and some that span both worlds. Let's look at a few charismatic ways in which phyllosphere microbes make a living and interact.

15.7 Epifoliar communities

On the upper (adaxial) surface of leaves, epifoliar bacteria are at their greatest density along leaf veins and trichomes, while on the lower (abaxial) surface they are particularly abundant around stomata, where you find high moisture and nutrients (Vacher et al. 2016). Bacterial assemblages on the phylloplane are dominated by α- and γ-Proteobacteria, and include free-living diazotrophs, methanol-consuming bacteria, and a diversity of plant pathogens.

Much of what we know about phyllosphere communities comes from decades of work on the behavior and dynamics of a model organism, *Pseudomonas syringae*, bacteria that are not only plant pathogens but ubiquitous elements of the biosphere that regulate the formation of rain and snow (Xin et al. 2018). This **species complex** in the γ-Proteobacteria is related to the fluorescent pseudomonads common in the rhizosphere. It includes important plant pathogens with at least 50 different **pathovars** (host-specific genotypes) comprising 13 distinct phylogenetic clades. *P. syringae* is a common phylloplane inhabitant of many leaf surfaces, where populations at low background levels can grow exponentially in response to hard rain events; at high abundance, the bacteria can enter the endophyllosphere through wounds and stomata and cause disease.

In addition to directly causing disease, *P. syringae* has another important impact for which it is more famous—driving frost damage on plants. *P. syringae* produce proteins on their outer membranes that serve as important catalysts in the formation of ice crystals; these are called **ice-nucleation proteins** (coded by the *ina* gene). *P. syringae* ice nuclei are found in rain, clouds, and rivers, and are thought to be important in formation of clouds and rain patterns (Morris et al. 2013b). They also raise the temperature at which frost damage occurs on plant leaves. When *P. syringae* is abundant in the phyllosphere, frost forms on leaf surfaces at temperatures several degrees warmer than if they were not present (Arny et al. 1976). In one of the first applications of genetic engineering in agriculture, Steven Lindow[1] and colleagues discovered that they could selectively deactivate the *ina* gene in a strain of *P. syringae*. This created an "ice-minus" strain that was identical to the wild-type strain in all ways except in the production of the ice-nucleating protein (Wilson and Lindow 1994). When they established phylloplane populations of the ice-minus bacteria on leaf surfaces, the ice-minus strain preemptively outcompeted the wild-type strains and protected crops from frost damage. The use of the ice-minus strain of *P. syringae* in 1987 was the first authorized environmental release of a genetically modified organism (Wilson and Lindow 1994).

In addition to bacteria, the phyllosphere microbiome is rich in yeasts (e.g., *Aureobasidium*) (Fonseca and Inácio 2006) and a polyphyletic guild of filamentous Ascomycetes (mostly Dothidiomycetes) that have converged to share traits that allow them to thrive on the phylloplane. Phylloplane fungi (also called **epifoliar fungi**) have hyphae and spores with dark pigment that protects them from UV irradiation; they are readily spread by water splash and germinate very quickly when moist (Flessa et al. 2021). Many of these fungi are found only associated with leaf surfaces. Some epifoliar fungi are broad host generalists, while others seem to show strong host specificity (Gilbert et al. 2007). They are most common in warm, moist habitats such as forest understories, and so are especially common in the wet tropics. Among the epifoliar fungi are sooty molds (Capnodiales), which subsist on nutrient-rich honeydew excreted by piercing and sucking insects, along with nutrients leaked to the leaf surface by the plant itself (Abdollahzadeh et al. 2020). Some plant species are particularly good hosts for sooty molds (e.g., citrus, California bay), supporting a dense growth of black mycelium (Figure 15.7). Surprisingly, while such growth can reduce the light reaching the leaves, this seldom harms the host (Insausti et al. 2015). Ants that tend aphids actively gather the honeydew from the aphids for their own consumption; this can reduce the food source available for sooty molds. As a result, excluding ants can lead to an explosion of epifoliar fungi (Queiroz and Oliveira 2001).

Epifoliar fungi alter the leaf surface as a habitat for other microbes. They add complexity to the structure of the leaf surface and so affect other phyllosphere microbes, including other epiphytes and plant pathogens. The surface tension enables water to persist where phylloplane fungal hyphae join the leaf surface, slowing evaporation compared to the open cuticle; this zone along the fungal hyphae is

[1] In addition to the many distinguished contributions to microbial ecology and plant pathology for which Dr. Lindow was elected to the National Academy of Sciences, we would like to recognize his singularly important role in introducing the authors of this book. Thanks, Steve!

Figure 15.7 Shielded scale insects feeding on the phloem of California bay produce honeydew that coats the leaf surface and supplies nutrients needed to support a rich growth of epifoliar sooty mold fungi.
Photo by Gregory S. Gilbert.

readily colonized by phylloplane bacteria (Vacher et al. 2016). Some epifoliar fungi, in addition to growing across the leaf surface, will penetrate the cuticle and produce haustoria that extract nutrients from epidermal cells in the leaf interior, blurring the distinction between the phylloplane and endophyllosphere (Zeng et al. 2020). By secreting cutinase or otherwise penetrating the cuticle, epifoliar fungi can enhance movement of water and nutrients from the leaf endosphere to the phylloplane, facilitating the growth of other microbes. On the other hand, some epifoliar fungi are parasites on other fungi (Section 15.10).

15.8 Foliar endophytes

The insides of leaves are rich places for fungi to grow as **endophytes**. Here we use the term endophyte to mean microbes living inside plant tissue. While bacteria can also grow as endophytes, the literature is far richer on the ecology of endophytic fungi, and here we focus primarily on them. Sometimes the term endophyte is used more narrowly to refer exclusively to fungal species that never produce symptoms and are mutualistic (Clay 1990). However, the symbioses between host plants and endophytic fungi defined more broadly take on many forms: obligate mutualists that protect the plant host against pests, latent pathogens that could cause disease later in response to changes in the environment or host development, or commensals that colonize the leaf tissue and obtain nutrients from the apoplast without causing either measurable damage or benefit to the host. Endophytes may also be "sit-and-wait decomposers," colonizing the living leaf tissue without damaging the host, well positioned to capture the nutrition embedded in the leaf once it senesces and falls from the host.

Endophytes are diverse. As difficult as it is to know how many kinds of pathogens are associated with a plant species (Box 15.1), pathogens comprise only a fraction of the endophytic community. Betsy Arnold and colleagues (2000, 2001) found a striking hyperdiversity of fungi in the leaves of understory shrubs in Panamanian rainforest, with over 100 fungal taxa among cultures grown from a few individuals of a single plant species (and they even ignored the rarest taxa)! When they sampled one host species at two sites 500 m apart, only half of the fungal taxa were shared between the two sites, indicating striking beta diversity (high species turnover) in tropical endophyte communities.

Outside of tropical settings, the diversity of fungal endophytes may be much lower. For instance, during four years of sampling we identified only 23 fungal species in 17 species of co-occurring clovers

Box 15.1 How many pathogens are on a plant species?

The most recent (10th) edition of the *Dictionary of the Fungi* lists close to 100,000 described species of fungi, which is a clear underestimate of the global fungal diversity. Mycologists have long tried to estimate how many species of fungi there are in the world (likely a few million) based on the number of known plant species multiplied by the number of fungal species per plant species, and adjusting for specificity of fungi for different plant hosts (e.g. Hawksworth 2001, Blackwell 2011). Unfortunately, the values used for the last two components of the calculation have been little more than informed guesses. Why don't we have better estimates for how many kinds of fungi (or other pathogens) are associated with different species of plants? Let's look at two approaches to getting those data: analysis of published literature and direct analysis of the mycobiome.

There is a massive amount of scientific literature reporting which fungi are associated with which plants, and the USDA Fungus–Host Distributions database (https://nt.ars-grin.gov/fungaldatabases/fungushost/fungushost.cfm; (Farr et al. 2003)) has gathered together much of that literature; as of April 2021, it includes 417,820 unique fungus–host combinations, including 99,062 fungus names and 64,966 host plant names (L. Castlebury, pers. com.). Those records also include geographic information, identifying the country, and sometimes state, from which it was reported. From this database you can readily look up which fungi are known from a particular host, or which hosts have been reported for a given fungus. It would seem that calculating the number of fungal species per host would be a simple matter. For instance, we can see that crops such as maize, rice, soybean, lemons, and tomato have some 988, 619, 597, 262, and 203 known fungal associates, respectively. There are parallel databases for other kinds of plant symbionts. From Plant Viruses Online (http://bio-mirror.im.ac.cn/mirrors/pvo/vide/), we can add to that list some 57, 23, 109, 6, and 135 viruses, respectively. Nemabase, compiled by UC Davis researchers (http://nemaplex.ucdavis.edu) lists 149, 105, 80, 29, and 144 species of nematodes on those crops. Unfortunately, systematic tallies of other kinds of pathogens are much spottier. But it is clear that globally each plant species faces a lot of potential pathogens, provided the plant and pathogen find themselves in the same place under conditions favorable to disease development.

However, there are at least four significant limitations to using such databases for estimating pathogen richness on a host. We'll focus on the USDA database to illustrate. First, the list includes not only pathogens, but fungi associated with plants that could be harmful, benign, or even beneficial. Whether this is a problem depends on the particular question asked.

Second, the database does not adjust for synonymous fungal names, which would produce an overestimate of the number of fungal species on any given host. Synonymous listings result when anamorph (e.g., *Colletotrichum gloeosporioides*) and teleomorph (e.g., *Glomerella cingulata*) names of the same fungus (Box 3.3) are listed separately, as well as from simple changes in the taxonomic nomenclature (e.g., *Asterina anomala* is now *Limacinula anomala*). Fortunately, synonomies are now being noted in the list, and the USDA maintains a Fungal Nomenclature database where one can manually check for such duplications and adjust appropriately (https://nt.ars-grin.gov/fungaldatabases/nomen/nomenclature.cfm). From a list of 331 taxa of fungi and oomycetes reported on six host species, we found that eliminating synonyms reduced the number of fungal taxa to 266; across the six host species, after correcting for synonyms the number of pathogen taxa was 88.1% (± 7.5) of the original number of taxa reported on the host.

Third, economically important crops and forest trees attract much more research than less important crops or wild plant species. There is a close relationship between the number of fungal species reported from a plant host and how much research of any kind is done on that host (Figure 14.5). One reason for this relationship is biologically trivial: more researchers, with more funding, will uncover a greater depth of fungal associates on a valuable crop species (crop species on average had 12-fold more citations than wild species). But there is also a biogeographical reason: important crops or forest trees planted widely around the world will encounter broader sets of pathogens and interact with them under a greater range of environmental conditions. Indeed, the broadly distributed host species can have up to an order of magnitude more known fungal pathogens than those with small geographic distributions (Miller 2012), and geographic range size is one of three universal predictors of parasite richness (along with host body size and population density) across all kinds of host taxa (Kamiya et al. 2014).

Finally, there is a general bias of much greater reporting of plant pathogens in geographic regions with strong traditions and support for plant disease research. This means that the diversity of plant pathogens is underrepresented for crops and wild plants from many regions of the world.

Table 15.1 Fungal species associated with leaves in 17 species of clovers (*Trifolium* and *Medicago*) over four years in a coastal grassland in California

	Host species*																
Fungus species	**ba**	**bi**	**fu**	**gr**	**ma**	**md**	**ml**	**wi**	**wo**	**ar**	**ca**	**du**	**gl**	**lu**	**po**	**re**	**su**
Alternaria sp.	■	■	■	■	■	■	■	■	■	■	■	■	■	■	■	■	■
Cladosporium sp.	■	■	■	■	■	■	■	■	■	■	■	■	■	■	■	■	■
Stemphylium spp.	■	■	■	■	■	■	■	■	■	■	■	■	■	■	■	■	■
Colletotrichum trifolii	■	■	■	■	■	■	■		■	■	■	■	■		■	■	■
Phoma sp.	■	■	■	■				■	■	■	■	■	■		■	■	
Acrodictys sp.		■															
Arthrinium sp.									■								
Cercospora sp.										■							
Chaetopsina fulva		■														■	
Cochliobolus spicifer				■		■	■		■	■	■		■	■		■	
Cylindrocarpon sp.	■							■	■	■			■			■	
Epicoccum nigrum				■	■											■	
Fusarium roseum									■								■
Idriella sp.							■									■	
Itersonilia sp.	■															■	
Leptosphaerulina trifolii				■						■			■		■	■	
Pseudopeziza trifolii									■				■			■	
Trichurus sp.					■						■						
Ulocladium atrum									■								
Uromyces minor											■	■				■	
Uromyces striatus										■							
Volutella sp.												■					

Identifications represent 1178 identified fungi (rusts and fungal cultures) from 1005 leaves. The first five species represent 95% of all fungi encountered (Parker and Gilbert 2007).

*Host species codes are, in order: California natives (yellow): *T. barbigerum*, *T. bifidum*, *T. fucatum*, *T. gracilentum*, *T. macrei*, *T. microdon*, *T. microcephalum*, *T. willdenovii*, *T. wormskjoldii*. Species introduced from southern Europe (blue): *M. arabica*, *T. campestre*, *T. dubium*, *T. glomeratum*, *M. lupulina*, *M. polymorpha*, *T. repens*, *T. subterraneum*.

(*Trifolium* and *Medicago* spp.) in a coastal California grassland; nearly all fungi were found on multiple host species (Table 15.1) (Parker and Gilbert 2007). Jean Langenheim and her students found only 23 species of foliar endophytes in leaves of coastal redwoods (*Sequoia sempervirens*) across California (Rollinger and Langenheim 1993). Focusing on beta diversity, they found very little variation in the dominant fungal species across the entire 700 km range. However, within sites there was habitat differentiation in fungal communities: leaves from basal sprouts generally hosted a particular subset of fungal species found in canopy foliage of the same tree (Espinosa-Garcia and Langenheim 1990). As suggested by these examples, endophyte diversity shows a latitudinal gradient, with greater prevalence and species diversity in tropical regions. Interestingly, more species diversity in the tropics does not imply more diversity at higher taxonomic levels (Arnold and Lutzoni 2007).

One challenge to this kind of work is that not all fungi are readily culturable. Recent technological advances allow DNA sequencing of entire communities at once, a process known as metabarcoding (Section 9.8). Metabarcoding of plant tissue with primers for fungal ITS barcodes has revealed even greater species richness of fungi associated with plant leaves than older cultural techniques (Peay et al. 2016). For example, more than 4000 taxa of endophytic fungi were associated with the leaves of a single tree species in Hawaii (Zimmerman and Vitousek 2012)! Barge et al. (2019) used DNA metabarcoding of endophytic fungi of leaves of the tree *Populus trichocarpa* across its entire geographic range, which encompasses areas with both rainy and dry climates. They found more than 1200 fungal taxa from 50 orders, 92% of which were Ascomycetes. The composition of the fungal communities varied with distance between samples and across years, but the strongest driver of fungal community composition was climate.

15.8.1 Horizontal and vertical transmission in fungal endophytes

A major axis of life-history variation in the evolutionary ecology of fungal endophytes is horizontal versus vertical transmission. **Horizontal transmission** involves movement through the environment from the propagule source (e.g., a pathogen reproducing on an infected, compatible host plant) to the new host. Microbes that reproduce only on living hosts, those that reproduce on senescent host tissue, and those that reproduce as saprotrophs on other organic substrates in the environment pass between hosts via horizontal transmission. **Vertical transmission** involves direct passage to the progeny of infected individuals, either through infected seeds or clonal propagules. The distinction between these two life-history strategies has enormous evolutionary implications for the endophytic symbiosis. Vertically transmitted endophytes depend on the successful survival, growth, and reproduction of the host plant for their own propagation, whereas horizontally transmitted endophytes are relatively untethered from the fitness of their host. What does this imply for the evolution of virulence in endophytes? Horizontally transmitted endophytes can have impacts that range from beneficial to harmful for the host, while vertically transmitted endophytes should be constrained to the beneficial side of the symbiotic outcome spectrum (Figure 1.3). This is what we predict based on evolutionary theory, and it is also what we see in nature.

One of the best-studied examples of vertical transmission involves Ascomycetes in the genus *Epichloë* that are endophytes in plants of the Poaceae (Clay and Schardl 2002). The ancestral, sexually reproducing species of *Epichloë* colonize meristematic tissue of their host plants and produce perithecia in a stroma on the plant culm; this blocks development of the host inflorescence, prevents production of seed, and limits the host to clonal reproduction (Chapter 14). Asexual species of *Epichloë* (anamorph name *Neotyphodium*) have repeatedly evolved from these sexual ancestors. Instead of preventing sexual reproduction, these fungi colonize the developing seeds of the host plant and are vertically transmitted to the seedlings that grow from those seeds. These vertically transmitted endophytes provide a variety of benefits to their host grasses, including enhanced tolerance to drought (Decunta et al. 2021) and resistance to pathogens and herbivores through the production of potent alkaloid toxins (Pérez et al. 2020).

15.9 Endophytes of other plant parts: fruits, seeds, and wood

Other parts of plants are also commonly colonized by endophytic microorganisms but are generally much less well studied. We will take a brief look at endophytes of fruits, seeds, and wood, including the traits that make them particularly suited for those habitats. Similar to the designation of phyllosphere and rhizosphere habitats, the fruit as habitat for microbes is called the **carposphere**, the seed habitat is often called the **spermosphere**, and woody habitat in the stems and branches is called the **caulosphere**.

Fruits of course are rich in sugars and are attractive foods for everything from bacteria to bears. Although bacterial growth in fruit is often limited by low pH, fleshy fruits are frequently colonized by yeasts and filamentous fungi as they develop, and those microbes can cause fruit rot while still on the plant or soon after ripening. In the process, the microbes produce a variety of alcohols, volatile organic compounds, and toxins that make the fruits unpalatable for other organisms (Gonçalves et al. 2019). Many of the chemicals produced by fruit-inhabiting microbes likely mitigate microbe–microbe interactions as the microscopic **frugivores** (fruit-feeders) compete with each other for the fruit substrate. However, fruit-rot toxins may also be important in a different scramble for the fruits—competition with frugivorous mammals and birds (Section 8.6).

The seed microbiome is critical for plant health and food safety, and it includes important plant pathogens as well as influential mutualists. **Seed-borne** fungi and bacteria colonize the seed coat, endosperm, or the embryo through various pathways, from the early stages of ovule development in the flower through seed dispersal and beyond. A subset of seed-borne microbes is **seed-transmitted**, vertically passed down across generations as permanent features of the microbiome for that host species (Section 15.8). Some seed-associated fungi are **xerotolerant**, that is, particularly tolerant of the low water activity in seeds (Box 3.4). Other seed-borne microbes have a more transient association with seeds and primarily colonize the growing seedling. Seeds are low-moisture environments, so, inside the seed itself, microbes are generally dormant and low in abundance. When seeds imbibe water and begin to germinate, seed-borne microbes take off. Seed-borne pathogens can greatly reduce plant fitness, so plants invest in chemical and structural defenses to protect against them (Dalling et al. 2020).

Dried seeds are important food and feed sources for humans and livestock (e.g., maize, wheat, rice, sorghum, quinoa, and peanuts). Unfortunately, a number of the seed-associated fungi (especially *Aspergillus*, *Fusarium*, and *Penicillium* spp.) produce highly toxic and carcinogenic **mycotoxins**, including aflatoxins, fumonisins, deoxynivalenon, ochratoxin, zearalenone, ergot alkaloids, patulin, and citrinin. While mycotoxin-producing fungi may colonize seeds pre or post harvest, primary growth of the fungi and production of mycotoxins occurs during grain storage, with mycotoxin production favored when grains are stored at high moisture content, under warm conditions, or at high relative humidity (Neme and Mohammed 2017). Dangerous levels of mycotoxin are present in at least a quarter of the global food crop (Eskola et al. 2020). Hundreds of millions of people, especially in sub-Saharan Africa, tropical Americas, and tropical Asia, consume mycotoxin-containing cereals at levels that substantially increase mortality and morbidity.

Perhaps the least-studied plant microbiome is that which is found in wood. Healthy woody tissue in tree trunks and branches is colonized by a diversity of fungi and bacteria (Rodríguez et al. 2011, Martín et al. 2013). Some may be actively involved in initiating wood decay, while others may persist indefinitely with minimal growth or impact on the host, but then quickly activate when the tree becomes weakened or dies. Fungi that colonize a living tree are in a prime position for rapid growth as saprotrophs once the host loses its ability to contain their spread and reproduction. That is why dying trees can sprout dramatic displays of basidiomata or ascomata on their trunks seemingly overnight.

15.10 Mechanisms of microbe–microbe interactions

In the microbiome, intimate microbe–microbe interactions directly shape the composition, diversity, structure, and functioning of the microbial community. Those interactions include both direct and indirect mechanisms, with antagonistic and beneficial outcomes. Let's look at some of the main mechanisms of microbe–microbe interactions.

15.10.1 Parasitism and predation

Just as all plants have microbial symbionts, plant-associated microbes have their own microbial symbionts, including fungi, oomycetes, bacteria, and viruses. Microbes that are parasites of plant parasites are called **hyperparasites**, and can be important in reducing plant disease. Who can parasitize whom is scale dependent; viruses can be parasites on everything (since they are the smallest); bacteria can parasitize fungi, oomycetes, and nematodes; fungi and oomycetes can parasitize each other (skinnier ones parasitize fatter ones) as well as macroparasites. At the microbial scale, the line between parasitism and **predation** is often fuzzy; a great example is that fungi are predators of nematodes and nematodes consume fungi (Figure 15.8) (Thakur and Geisen 2019). These tiny organisms have important trophic impacts on the plant microbiome and consequently on their plant hosts.

Viruses

All kinds of plant pathogens are themselves hosts to viruses. Viruses of bacteria are called **phages**, and are important both in population dynamics of bacteria and in the horizontal transfer of plasmids containing important functional genes among bacteria species (Chapter 5). Phages infect bacterial cells by injecting their genetic material into the cell. After infection, a phage uses one of two strategies for reproduction: **lytic phages** replicate to high numbers, packaging the viral DNA or RNA into coat proteins until the bacterial cell bursts, while **temperate phages** integrate their genes into the bacterial chromosome, where viral DNA replicates with each host cell division. (Consider the parallels to the vertical and horizontal transmission of seed-borne pathogens described above.) Phages sometimes also passively capture plasmids from the bacterial cytoplasm within their coat protein and transmit those to a new bacterial cell during infection (Figure 5.3c). On lysis of the host cell, the phage virus particles spread into the environment, where they can be taken up by other bacteria, beginning the cycle again. The expression of viral genes in temperate phages, or of the bacterial genes on phage-transferred plasmids, can confer important functional traits such as resistance to antibiotics. Phages are ubiquitous and diverse, vary in their host ranges, and can affect both the size and phenotypes of their bacterial host populations (Koskella and Taylor 2018). Because phage viruses necessarily kill their host cell to spread, their activities can limit bacterial population growth. Phage activity may act as a stabilizing mechanism (Section 14.4), helping to maintain diversity in the bacterial communities by preventing competitive exclusion.

Fungi (as well as oomycetes) are also susceptible to viruses; viruses that infect fungi are often called **mycoviruses**, and can have a range of impacts from harming the fungi to providing them with heat tolerance or enhanced virulence (Hillman et al. 2018). In plant pathology, some important mycoviruses are those that cause reduced virulence, or **hypovirulence**, in their host fungi. For instance, the chestnut blight fungus, *Cryphonectria parasitica* (Section 16.4), is susceptible to a number of double-stranded DNA viruses including CHV1, which causes stunted growth patterns and reduced (hypo-)virulence of the fungus on its chestnut host. Introducing the virus into a developing chestnut blight canker can transform a virulent fungal mycelium to a relatively benign hypovirulent one, allowing the host to respond with a variety of defenses, and halting development of disease. The virus spreads to new fungal mycelia through hyphal anastomosis. *C. parasitica* (like many ascomycetes) has multiple genes that control vegetative compatibility—generally, the alleles at each of the **vegetative compatibility** (vc) genes must be matching to allow successful **anastomosis** and **plasmogamy** between two fungal hyphae (Section 3.5). That means that the virus can spread easily among individual mycelia of the same vc type, but not between mycelia of incompatible vc types. In Europe, where *C. parasitica* has few vc types, the CHV1 virus has spread widely, slowing the impact of chestnut blight on chestnut populations. In contrast, North American populations of *C. parasitica* have a high diversity of vc genes, so that any two hyphae that happen to run into each other on a host are unlikely to be vegetatively compatible. That has limited the spread of CHV1, even with intensive efforts to use it to restore chestnut populations in North America (Feau et al. 2014). This partly explains why the once mighty North American chestnut has been decimated throughout its range (Section 16.4) whereas European Chestnut populations are doing relatively well.

Parasitic fungi and oomycetes

Some filamentous fungi and oomycetes are hyperparasites on other fungi and oomycetes. For instance, *Pythium oligandrum* coils around the hyphae of other species of plant parasitic *Pythium* as well as some true fungi, then penetrates into the hyphal cytoplasm as an intracellular parasite, disrupting the integrity of the hypha and reducing growth rates of the host organism (Berry et al. 1993). Basidiomata are frequently hyperparasitized by **fungicolous** fungi, and a number of epifoliar fungi are hyperparasites on other epifoliar fungi. Hyphal parasites can greatly reduce the growth of the host fungi, and even kill them. This makes them good candidates for biological control of fungal pathogens (Section 17.6). Curiously, the same *Pythium oligandrum*, which is used in the biological control of plant pathogens, can be used to treat athlete's foot, a skin disease caused by fungi (Gabrielová et al. 2018).

Predation by, and of, nematodes

Plant parasitic nematodes evolved from fungus-feeding ancestors (Quist et al. 2015), and fungus-feeding nematodes remain voracious predators of plant- and soil-associated fungi (Thakur and Geisen 2019). Fungus-feeding nematodes are ectoparasites on fungi, using stylets to suck out hyphal contents. With turn-about being fair play, a fascinating guild of fungi consume nematodes, either as parasites (*Purpureocillium lilacinum* is a common parasite of nematode eggs) or as predators. Since fungi obtain their nutrition as osmotrophic heterotrophs, it is startling to imagine them as predators that catch their food. However, a number of fungi, mostly Ascomycetes in the order Orbiliales, have evolved clever structures that trap nematodes (Figure 15.8). Such traits include hyphae that grow as inflatable lassos that catch nematodes passing through them, and sticky "lollipops" that entangle nematodes long enough for the fungi to attack with enzymes and penetrating hyphae. Once caught, the nematodes are slowly consumed with extracellular enzymes or by penetrating their bodies and degrading them from within (Jiang et al. 2017).

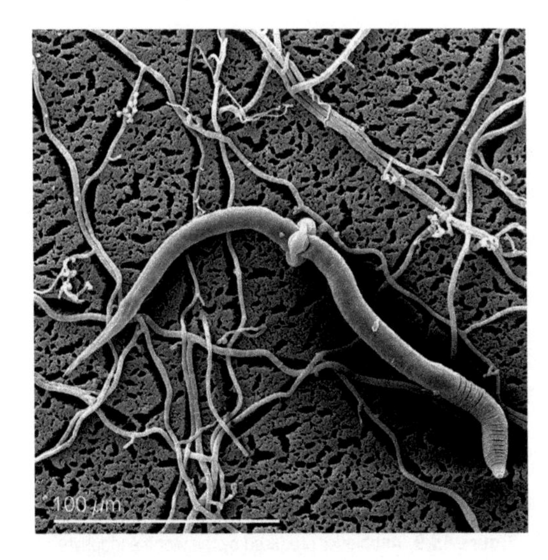

Figure 15.8 Predatory *Drechslerella dactyloides* fungus uses an inflatable lasso to trap, and then consume, a nematode.
Photo used with permission, originally published in Liu et al. (2012).

15.10.2 Competition

As described above, microbial activity is concentrated in those regions of the plant with abundant carbon and water resources, and that creates an important arena for intense competition among microbes for other nutrients such as nitrogen, phosphorus, and micronutrients. There are three general types of competitive interactions among microbes: preemptive competition, exploitative competition, and interference competition.

Preemptive competition is where an early-arriving competitor dominates a habitat and monopolizes a resource. Preemptive competition is an important process in structuring the plant microbiome; the rapid population growth of microbes means that the first to establish often has a strong numerical advantage and controls the resources at that location. Later arrivals, even if they would be stronger competitors in an equal match-up, are at a disadvantage because they have to actively displace those that were there first. We've already noted the importance of such preemptive competition in phylloplane bacteria, fruit- and seed-inhabiting fungi, and foliar endophytes that are sit-and-wait decomposers, prepared for leaf senescence. The benefit of being "first at the table" has strongly shaped the evolution of many microbe life histories.

Exploitative competition is where competitors differ in their abilities to acquire a resource, and the more effective competitor uses up the resource. Ecologists think of exploitative competition in two different ways, and each is associated with microbial traits. In the first view, exploitative competition is conceived in terms of the rate at which nutrients diffuse and are taken up. Associated traits include those that simply promote fast growth, allowing a population to capture more of a limited resource more quickly than its more sluggish neighbors. This type of competitive ability is also enhanced by traits such as the secretion of enzymes or other extracellular chemicals that enhance efficiency at capturing limiting nutrients quickly (Ghoul and Mitri 2016). However, competition can also be thought of as the extent of resource depletion at equilibrium (MacArthur 1972). In this view, a population will consume and reduce a resource until there is not enough left for the population to grow; the resource concentration at that level is known as R^* ("R-star"), and the species with the lowest R^* will outcompete the others (Tilman 2007). In this case, important competitive traits are those that scavenge nutrients to low levels and reduce R^*. For instance, iron is a limiting nutrient for microbes in the rhizosphere, and many organisms produce iron-chelating siderophores that efficiently scavenge iron for uptake. Bacteria that produce potent siderophores increase their own access to scarce iron resources while simultaneously limiting the ability of neighboring fungi and bacteria to grow. Such bacteria essentially manipulate nutrient availability in the substrate to enhance their own competitive advantage.

Interference competition is where competitors physically prevent other organisms from accessing a resource. Both preemptive and exploitative competition are the inadvertent consequences of each microbe doing its best to obtain the resources it needs, and in the process it deprives others of those resources. However, microbes have also evolved myriad chemical tools to enhance their own access to limiting resources by actively interfering with their competitors. The primary tools of such interference competition are antimicrobial compounds, commonly called antibiotics. Let's take a deeper look at antibiosis in microbe–microbe interactions.

15.10.3 Antibiosis

Bacteria and fungi common in the rhizosphere and phyllosphere produce a wide diversity of antibiotics that are detrimental to other microorganisms, including plant pathogens (Thomashow et al. 2019). We have known since the 1928 discovery of penicillin by Alexander Fleming that microbes produce antibiotics in artificial culture, and screening for organisms that produce antibiotics *in vitro* (Figure 15.9) has driven decades of work in biological prospecting for natural products for medicine and agriculture (Aly et al. 2011, Maghembe et al. 2020), as well as in searches for bacteria and fungi that could be used in the biological control of plant pathogens (Section 17.6 and Raymaekers et al. 2020). Extracellular **lytic enzymes** including chitinases, β-1,3-glucanases, and cellulases produced by a wide diversity of microbes can degrade cell walls of fungi and oomycetes and impede their growth. Low-molecular-weight compounds produced by fungi, oomycetes, or bacteria that have detrimental effects

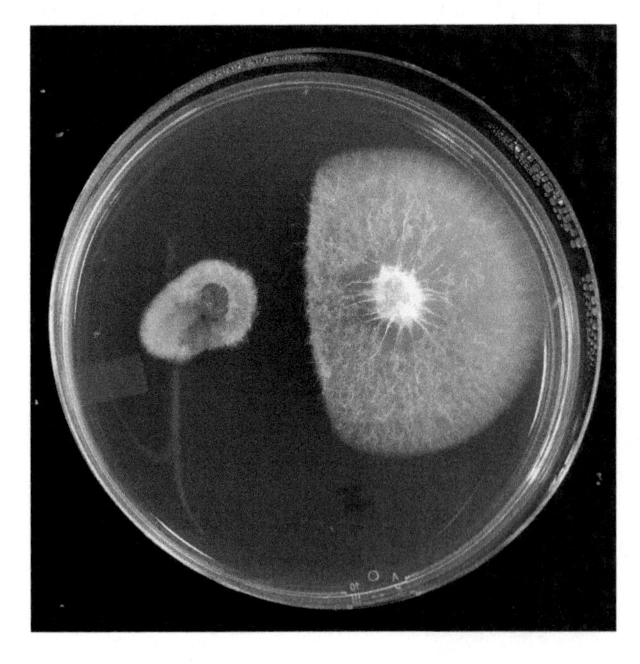

Figure 15.9 Antibiotics produced by the wood endophyte fungus *Pithomyces chartarum* (left) inhibit mycelial growth of pathogenic *Fusarium euwallaceae* (right).
Photo by Shannon Lynch.

on the metabolism, growth, or survival of other microbes are called **antibiotics** or **antimicrobials**. We know of many antibiotics because they have been harnessed for use in human and veterinary medicine: penicillin and cephalosporin from fungi; tetracycline, streptomycin, rifamycin, and chloramphenicol from bacteria.

Because antibiotics are so therapeutically important in protection against microbial infections, antibiotics are usually thought of as something that microbes must have evolved as weapons to use against other microbes. But the tremendous chemical diversity and specificity of impacts seen among microbially produced antibiotics suggest a more complex evolutionary history and ecological role (Fajardo and Martínez 2008). For example, some chemicals show concentration-dependent effects, where chemicals are stressful or lethal agents at high concentration but act as signaling compounds that modulate gene expression at low concentration. In some cases, signaling compounds involved in quorum sensing coordinate activities in populations of bacteria and allow colonies of single-celled organisms to emulate some aspects of the behavior of multicellular organisms (Section 5.4). The diversity of functions for the organisms that produce them spills over to a diversity of uses for people. For example, cyclosporin, a natural fungal product, has antimicrobial activity but derives its tremendous medicinal importance mostly from its efficacy as an immunosuppressant, used in preventing rejection of organ transplants.

Because microbes produce so many chemicals with so many potential functions, it is not trivial to prove that chemical antibiosis is a driving factor in shaping the plant microbiome—in the rhizosphere or phyllosphere—*in situ*. Early experiments showed a relationship between the abundance of particular rhizosphere bacteria known to produce antibiotics and negative impacts on other rhizosphere microbes. This suggested a functional role, but it was not clear whether microbes could produce antibiotics in quantities sufficient to cause changes in real, living microbiomes. Of course, those bacteria were more than just a bag of antibiotics, and other activities—like preemptive competition or siderophore production or induction of host defenses—might be more important than antibiosis. Later studies used targeted genetic manipulations, like directed knockout mutations in antibiotic production pathways, to show that the ability of a bacterium to produce an antibiotic was directly linked to its effects on the rhizosphere community. However, such studies still did not show that antibiosis *per se* (and not, say, an indirect effect through an induced plant response) was responsible for shaping the microbiome. Technological advances in transcriptomics and proteomics have finally brought together evidence for the direct role of antibiosis in shaping the composition and functionality of the plant microbiome (Thomashow et al. 2019). Understanding the scope and limits of such effects remains at the forefront of plant microbiome research.

15.11 Microbes working together

Not all microbe–microbe interactions in the plant microbiome are antagonistic, and there is a growing appreciation for how many plant-associated microbes thrive in ecological collaboration with other microbes. Complexes of microbes can produce qualitatively different outcomes for plant hosts than simple interactions with a single microbe.

Cross-feeding, or **microbial syntrophy**, is one way for microbes with different nutritional requirements to benefit from each other. For instance, a buildup of metabolic waste products from a microbe that degrades one substrate may be toxic or otherwise interfere with further activity of that microbe; other microbes that are able to metabolize those waste products do so for their own benefit, but in the process detoxify the substrate for the first microbe (Morris et al. 2013a). Similarly, early colonizers of complex substrates can cause changes that foster the growth of a different set of microbes with different nutritional palettes. For example, peptidoglycans produced in the rhizosphere by *Bacillus* foster the growth of rhizosphere bacteria in the *Cytophaga–Flavobacterium–Bacteroides* group (Peterson et al. 2006). In the heartwood of living trees, white-rot Basidiomycetes can degrade lignin at the same time they break down cellulose and hemicellulose. Lignin makes the cellulose and hemicellulose in heartwood inaccessible to most microbes, but the action of the white-rot Basidiomycetes releases simple sugars that allow for the growth of many bacteria, and delignification makes the wood accessible to a suite of Ascomycetes, accelerating heartwood decomposition. In some cases, the bacteria can even improve the growth of the white-rot fungi.

In addition to the epifoliar bacteria that grow adjacent to fungal hyphae (Section 15.7), some bacteria grow intracellularly as **endohyphal symbionts** (Shaffer et al. 2018). Such bacteria influence the behavior of their fungal hosts in many of the same ways as do mycoviruses, changing their growth patterns, and, in the cases of pathogenic fungi, either increasing or decreasing their virulence. Endohyphal bacteria are phylogenetically diverse and are known from a number of plant-associated fungi, from mycorrhizae to foliar endophytes (Hoffman and Arnold 2010). Our understanding of their impacts on plant–fungal interactions is still in its infancy.

Plant-associated microbes may also coexist in multi-species **biofilms** composed of species with complementary nutritional needs (Danhorn and Fuqua 2007). Bacteria produce **autoinducer compounds** that act as signals in **quorum sensing**, which is critical in bacterial biofilm production, sporulation, and virulence (Section 5.4 and Bogino et al. 2013). Most autoinducers serve for intraspecific communications, although some autoinducer compounds are more general and facilitate interactions among bacterial species, such as in the coaggregation of multispecies biofilms. The recent discovery of quorum sensing in fungi includes cross-kingdom fungal–bacterial biofilm formation (Deveau et al. 2018). Multispecies biofilms may play important roles in the growth, survival, syntrophy, and impacts of plant-associated microbial communities from PGPR to phylloplane bacteria, including virulence of pathogenic microbes. Much work on the role of biofilms in microbiome function has been pioneered in medical studies; the physiology and ecology of biofilms in plant microbiomes is a burgeoning area of research.

Finally, plants are always interacting with multiple microbes at the same time. We have looked at several ways in which interactions with one set of microbes (e.g., PGPR, mycorrhizae, hyperparasites) can mitigate the negative effects of another set of microbes (i.e., pathogens) on the host plant, through direct and indirect mechanisms. In other cases, the simultaneous interactions may have **synergistic** effects on the plant, where qualitatively different disease symptoms develop, or the impacts on the host are more severe than predicted by the impacts of each pathogen independently. Such synergistic interactions can form **disease complexes** (Lamichhane and Venturi 2015). In some cases, damage from one pathogen can open infection courts that allow opportunistic pathogens to infect hosts that might otherwise not be susceptible to them. In other cases, effects of infection by one pathogen may alter host immunity, allowing for additional growth or impacts of otherwise benign microbes. In some cases, infection by individual pathogens causes mild or no host symptoms, but coinfections by those same pathogens produce devastating disease. The etiology of diseases caused by complexes of pathogens is difficult to untangle using the usual approach to Koch's postulates, and the diversity and importance of such complexes is largely unknown.

It is important to remember that microorganisms involved in a disease complex are not collaborating *in order to* cause disease on a host plant. The actions of each microbe can directly affect the host as well as other cohabiting microbes, changing the relationships, behaviors, and responses of all the players. There is also the intriguing possibility of coevolution between cohabiting microbes to develop synergistic reliance on each other for growth and reproduction, mediated through disease impacts on a host plant. Distinguishing between casual synergistic effects among cohabiting organisms and deeper coevolutionary relationships requires a careful and creative scientific approach.

Further reading

Asiegbu, F.O. and A. Kovalchuk (editors). 2021. *Forest Microbiology. Tree Microbiome: Phyllosphere, Endosphere, and Rhizosphere*, Volume 1. Academic Press, New York.

Lindow, S.E., V.J. Elliott, and E.I. Hecht-Poinar (editors). 2002. *Phyllosphere Microbiology*. APS Press, St. Paul, MN.

Cardon, Z.G. and J.L. Whitbeck (editors) 2007. *The Rhizosphere. An Ecological Perspective.* Academic Press, New York.

References

Aasfar, A., A. Bargaz, K. Yaakoubi, A. Hilali, I. Bennis, Y. Zeroual, and I. Meftah Kadmiri. 2021. Nitrogen fixing *Azotobacter* species as potential soil biological enhancers for crop nutrition and yield stability. *Frontiers in Microbiology* **12**:354.

Abdollahzadeh, J., J. Groenewald, M. Coetzee, M. Wingfield, and P. Crous. 2020. Evolution of lifestyles in Capnodiales. *Studies in Mycology* **95**:381–414.

Aly, A. H., A. Debbab, and P. Proksch. 2011. Fifty years of drug discovery from fungi. *Fungal Diversity* **50**:3–19.

Arnold, A. E. and F. Lutzoni. 2007. Diversity and host range of foliar fungal endophytes: Are tropical leaves biodiversity hotspots? *Ecology* **88**:541–549.

Arnold, A. E., Z. Maynard, and G. S. Gilbert. 2001. Fungal endophytes in dicotyledonous neotropical trees: Patterns of abundance and diversity. *Mycological Research* **105**:1502–1507.

Arnold, A. E., Z. Maynard, G. S. Gilbert, P. D. Coley, and T. A. Kursar. 2000. Are tropical fungal endophytes hyperdiverse? *Ecology Letters* **3**:267–274.

Arny, D., S. Lindow, and C. Upper. 1976. Frost sensitivity of *Zea mays* increased by application of *Pseudomonas syringae*. *Nature* **262**:282–284.

Barge, E. G., D. R. Leopold, K. G. Peay, G. Newcombe, and P. E. Busby. 2019. Differentiating spatial from environmental effects on foliar fungal communities of *Populus trichocarpa*. *Journal of Biogeography* **46**:2001–2011.

Beattie, G. A. 2011. Water relations in the interaction of foliar bacterial pathogens with plants. *Annual Review of Phytopathology* **49**:533–555.

Beneduzi, A., A. Ambrosini, and L. M. Passaglia. 2012. Plant growth-promoting rhizobacteria (PGPR): their potential as antagonists and biocontrol agents. *Genetics and Molecular Biology* **35**:1044–1051.

Bennett, A. E. and K. Groten. 2022. The costs and benefits of plant–arbuscular mycorrhizal fungal interactions. *Annual Review of Plant Biology* **73**:649–672.

Berry, L., E. Jones, and J. Deacon. 1993. Interaction of the mycoparasite *Pythium oligandrum* with other *Pythium* species. *Biocontrol Science and Technology* **3**:247–260.

Blackwell, M. 2011. The Fungi: 1, 2, 3... 5.1 million species? *American Journal of Botany* **98**:426–438.

Bogino, P. C., M. D. l. M. Oliva, F. G. Sorroche, and W. Giordano. 2013. The role of bacterial biofilms and surface components in plant–bacterial associations. *International Journal of Molecular Sciences* **14**: 15838–15859.

Bradley, D. J., G. S. Gilbert, and I. M. Parker. 2003. Susceptibility of clover species to fungal infection: the interaction of leaf surface traits and environment. *American Journal of Botany* **90**:857–864.

Clay, K. 1990. Fungal endophytes of grasses. *Annual Review of Ecology and Systematics* **21**:275–297.

Clay, K. and C. Schardl. 2002. Evolutionary origins and ecological consequences of endophyte symbiosis with grasses. *American Naturalist* **160**:S99–S127.

Cope, K. R., A. Bascaules, T. B. Irving, M. Venkateshwaran, J. Maeda, K. Garcia, T. A. Rush, C. Ma, J. Labbé, and S. Jawdy. 2019. The ectomycorrhizal fungus *Laccaria bicolor* produces lipochitooligosaccharides and uses the common symbiosis pathway to colonize *Populus* roots. *Plant Cell* **31**:2386–2410.

Dalling, J. W., A. S. Davis, A. E. Arnold, C. Sarmiento, and P.-C. Zalamea. 2020. Extending plant defense theory to seeds. *Annual Review of Ecology, Evolution, and Systematics* **51**:doi: 10.1146/annurev-ecolsys-012120-115156.

Danhorn, T. and C. Fuqua. 2007. Biofilm formation by plant-associated bacteria. *Annual Review of Microbiology* **61**:401–422.

DeAngelis, K. M., S. E. Lindow, and M. K. Firestone. 2008. Bacterial quorum sensing and nitrogen cycling in rhizosphere soil. *FEMS Microbiology Ecology* **66**:197–207.

Decunta, F. A., L. I. Pérez, D. P. Malinowski, M. A. Molina-Montenegro, and P. E. Gundel. 2021. A systematic review on the effects of *Epichloë* fungal endophytes on drought tolerance in cool-Season grasses. *Frontiers in Plant Science* **12**:380.

Deveau, A., G. Bonito, J. Uehling, M. Paoletti, M. Becker, S. Bindschedler, S. Hacquard, V. Hervé, J. Labbé, and O. A. Lastovetsky. 2018. Bacterial–fungal interactions: ecology, mechanisms and challenges. *FEMS Microbiology Reviews* **42**:335–352.

Eskola, M., G. Kos, C. T. Elliott, J. Hajšlová, S. Mayar, and R. Krska. 2020. Worldwide contamination of food-crops with mycotoxins: Validity of the widely cited 'FAO estimate' of 25%. *Critical Reviews in Food Science and Nutrition* **60**:2773–2789.

Espinosa-Garcia, F. J. and J. H. Langenheim. 1990. The endophytic fungal community in leaves of a coastal redwood population: Diversity and spatial patterns. *New Phytologist* **116**:89–98.

Fajardo, A. and J. L. Martínez. 2008. Antibiotics as signals that trigger specific bacterial responses. *Current Opinion in Microbiology* **11**:161–167.

Farr, D. F., A. Y. Rossman, M. E. Palm, and E. B. McCray. 2003. Fungus–Host Distributions, Fungal Databases, Systematic Botany and Mycology Lab. ARS/USDA. http://nt.ars-grin.gov/fungaldatabases/.

Feau, N., C. Dutech, J. Brusini, D. Rigling, and C. Robin. 2014. Multiple introductions and recombination in Cryphonectria Hypovirus 1: Perspective for a sustainable biological control of chestnut blight. *Evolutionary Applications* **7**:580–596.

Flessa, F., J. Harjes, M. E. Cáceres, and G. Rambold. 2021. Comparative analyses of sooty mould communities from Brazil and Central Europe. *Mycological Progress* **20**:869–887.

Flinn, K. 2021. The idea the trees talk to communicate is misleading. *Scientific American* July 19. https://www.

scientificamerican.com/article/the-idea-that-trees-talk-to-cooperate-is-misleading/#.

Fonseca, A. and J. Inácio. 2006. Phylloplane yeasts. In: G. Péter, C. Rosa, editors. *Biodiversity and Ecophysiology of Yeasts*, pp. 263–301. Springer, Berlin.

Gabrielová, A., K. Mencl, M. Suchánek, R. Klimeš, V. Hubka, and M. Kolařík. 2018. The oomycete *Pythium oligandrum* can suppress and kill the causative agents of dermatophytoses. *Mycopathologia* **183**:751–764.

Genre, A. and G. Russo. 2016. Does a common pathway transduce symbiotic signals in plant–microbe interactions? *Frontiers in Plant Science* **7**:96.

Ghoul, M. and S. Mitri. 2016. The ecology and evolution of microbial competition. *Trends in Microbiology* **24**: 833–845.

Gilbert, G., D. Reynolds, and A. Bethancourt. 2007. The patchiness of epifoliar fungi in tropical forests: Host range, host abundance, and environment. *Ecology* **88**:575–581.

Gilbert, G. S., J. Handelsman, and J. L. Parke. 1994. Root camouflage and disease control. *Phytopathology* **84**: 222–225.

Gilbert, G. S., J. L. Parke, M. K. Clayton, and J. Handelsman. 1993. Effects of an introduced bacterium on bacterial communities on roots. *Ecology* **74**:840–854.

Gonçalves, B. L., C. F. S. C. Coppa, D. V. d. Neeff, C. H. Corassin, and C. A. F. Oliveira. 2019. Mycotoxins in fruits and fruit-based products: Occurrence and methods for decontamination. *Toxin Reviews* **38**:263–272.

Hartmann, A., M. Rothballer, and M. Schmid. 2008. Lorenz Hiltner, a pioneer in rhizosphere microbial ecology and soil bacteriology research. *Plant and Soil* **312**:7–14.

Hawksworth, D. 2001. The magnitude of fungal diversity: the 1.5 million species estimate revisited. *Mycological Research* **105**:1422–1432.

Henriksson, N., O. Franklin, L. Tarvainen, J. Marshall, J. Lundberg-Felten, L. Eilertsen, and T. Näsholm. 2021. The mycorrhizal tragedy of the commons. *Ecology Letters* 24:1215–1224.

Hillman, B. I., A. Annisa, and N. Suzuki. 2018. Viruses of plant-interacting fungi. *Advances in Virus Research* **100**:99–116.

Hoch, H. C., R. C. Staples, B. Whitehead, J. Comeau, and E. D. Wolf. 1987. Signaling for growth orientation and cell differentiation by surface topography in *Uromyces*. *Science* **235**:1659–1662.

Hoffman, M. T. and A. E. Arnold. 2010. Diverse bacteria inhabit living hyphae of phylogenetically diverse fungal endophytes. *Applied and Environmental Microbiology 76:4063–4075.*

Insausti, P., E. L. Ploschuk, M. M. Izaguirre, and M. Podworny. 2015. The effect of sunlight interception by sooty mold on chlorophyll content and photosynthesis in orange leaves (*Citrus sinensis* L.). *European Journal of Plant Pathology* **143**:559–565.

Jacott, C. N., J. D. Murray, and C. J. Ridout. 2017. Trade-offs in arbuscular mycorrhizal symbiosis: Disease resistance, growth responses and perspectives for crop breeding. *Agronomy* **7**:75.

Jiang, X., M. Xiang, and X. Liu. 2017. Nematode-trapping fungi. *Microbiology Spectrum* **5**:10.

Kamiya, T., K. O'Dwyer, S. Nakagawa, and R. Poulin. 2014. What determines species richness of parasitic organisms? A meta-analysis across animal, plant and fungal hosts. *Biological Reviews* **89**:123–134.

Kiers, E. T., M. Duhamel, Y. Beesetty, J. A. Mensah, O. Franken, E. Verbruggen, C. R. Fellbaum, G. A. Kowalchuk, M. M. Hart, and A. Bago. 2011. Reciprocal rewards stabilize cooperation in the mycorrhizal symbiosis. *Science* **333**:880–882.

Koskella, B. and T. B. Taylor. 2018. Multifaceted impacts of bacteriophages in the plant microbiome. *Annual Review of Phytopathology* **56**:361–380.

Lamichhane, J. R. and V. Venturi. 2015. Synergisms between microbial pathogens in plant disease complexes: a growing trend. *Frontiers in Plant Science* **6**:385.

Liu, K., J. Tian, M. Xiang, and X. Liu. 2012. How carnivorous fungi use three-celled constricting rings to trap nematodes. *Protein & Cell* **3**:325–328.

Lynch, J. and J. Whipps. 1990. Substrate flow in the rhizosphere. *Plant and Soil* **129**:1–10.

MacArthur, R. H. 1972. *Geographical Ecology. Patterns in the Distribution of Species*. Princeton University Press.

MacLean, A. M., A. Bravo, and M. J. Harrison. 2017. Plant signaling and metabolic pathways enabling arbuscular mycorrhizal symbiosis. *The Plant Cell* **29**:2319–2335.

Maghembe, R., D. Damian, A. Makaranga, S. S. Nyandoro, S. L. Lyantagaye, S. Kusari, and R. Hatti-Kaul. 2020. Omics for bioprospecting and drug discovery from bacteria and microalgae. *Antibiotics* **9**:229.

Martín, J. A., J. Witzell, K. Blumenstein, E. Rozpedowska, M. Helander, T. N. Sieber, and L. Gil. 2013. Resistance to Dutch elm disease reduces presence of xylem endophytic fungi in elms (*Ulmus* spp.). *PloS One* **8**: e56987.

Miller, Z. J. 2012. Fungal pathogen species richness: why do some plant species have more pathogens than others? *American Naturalist* **179**:282–292.

Morris, B. E., R. Henneberger, H. Huber, and C. Moissl-Eichinger. 2013a. Microbial syntrophy: interaction for the common good. *FEMS Microbiology Reviews* **37**: 384–406.

Morris, C. E., C. L. Monteil, and O. Berge. 2013b. The life history of *Pseudomonas syringae*: linking agriculture to earth system processes. *Annual Review of Phytopathology* **51**:85–104.

Müller, D. B., C. Vogel, Y. Bai, and J. A. Vorholt. 2016. The plant microbiota: systems-level insights and perspectives. *Annual Review of Genetics* **50**:211–234.

Neme, K. and A. Mohammed. 2017. Mycotoxin occurrence in grains and the role of postharvest management as a mitigation strategies. A review. *Food Control* **78**:412–425.

Parker, I. M. and G. S. Gilbert. 2007. When there is no escape: The effects of natural enemies on native, invasive, and noninvasive plants. *Ecology* **88**:1210–1224.

Peay, K. G., P. G. Kennedy, and J. M. Talbot. 2016. Dimensions of biodiversity in the Earth mycobiome. Nature Reviews Microbiology **14**:434–447.

Pérez, L. I., P. E. Gundel, I. Zabalgogeazcoa, and M. Omacini. 2020. An ecological framework for understanding the roles of *Epichloë* endophytes on plant defenses against fungal diseases. *Fungal Biology Reviews* **34**:115–25.

Peterson, S. B., A. K. Dunn, A. K. Klimowicz, and J. Handelsman. 2006. Peptidoglycan from *Bacillus cereus* mediates commensalism with rhizosphere bacteria from the *Cytophaga–Flavobacterium* group. *Applied and Environmental Microbiology* **72**:5421–5427.

Queiroz, J. M. and P. S. Oliveira. 2001. Tending ants protect honeydew-producing whiteflies (Homoptera: Aleyrodidae). *Environmental Entomology* **30**:295–297.

Quist, C. W., G. Smant, and J. Helder. 2015. Evolution of plant parasitism in the phylum Nematoda. *Annual Review of Phytopathology* **53**:289–310.

Rai, A., P. K. Rai, and S. Singh. 2017. Exploiting beneficial traits of plant-associated fluorescent pseudomonads for plant health. In: J. S. Singh, G. Seneviratne, editors. *Agro-Environmental Sustainability*, vol. 1, pp. 19–41. Springer, Cham.

Raymaekers, K., L. Ponet, D. Holtappels, B. Berckmans, and B. P. Cammue. 2020. Screening for novel biocontrol agents applicable in plant disease management—a review. *Biological Control* **144**:104240.

Rey, T., A. Chatterjee, M. Buttay, J. Toulotte, and S. Schornack. 2014. Medicago truncatula symbiosis mutants affected in the interaction with a biotrophic root pathogen. *New Phytologist* **206**:497–500.

Rodríguez, J., J. P. Elissetche, and S. Valenzuela. 2011. Tree endophytes and wood biodegradation. In: A. M. Pirttilä and A. C. I. Frank, editors. *Endophytes of Forest Trees*, pp. 81–93. Springer, Berlin.

Rollinger, J. L. and J. H. Langenheim. 1993. Geographic survey of fungal endophyte community composition in leaves of coastal redwood. *Mycologia* **85**:149–156.

Ryan, M. H. and J. H. Graham. 2002. Is there a role for arbuscular mycorrhizal fungi in production agriculture? *Plant and Soil* **244**:263–271.

Sang, M. K., J. D. Kim, B. S. Kim, and K. D. Kim. 2011. Root treatment with rhizobacteria antagonistic to Phytophthora blight affects anthracnose occurrence, ripening, and yield of pepper fruit in the plastic house and field. *Phytopathology* **101**:666–678.

Sepp, S. K., J. Davison, T. Jairus, M. Vasar, M. Moora, M. Zobel, and M. Öpik. 2019. Non-random association patterns in a plant–mycorrhizal fungal network reveal host–symbiont specificity. *Molecular Ecology* **28**: 365–378.

Shaffer, J. P., P.-C. Zalamea, C. Sarmiento, R. E. Gallery, J. W. Dalling, A. S. Davis, D. A. Baltrus, and A. E. Arnold. 2018. Context-dependent and variable effects of endohyphal bacteria on interactions between fungi and seeds. *Fungal Ecology* **36**:117–127.

Simard, S. 2021. Finding the Mother Tree: Uncovering the Wisdom and Intelligence of the Forest. Penguin, Harmondsworth, UK.

Song, Y. Y., M. Ye, C. Li, X. He, K. Zhu-Salzman, R. L. Wang, Y. J. Su, S. M. Luo, and R. S. Zeng. 2014. Hijacking common mycorrhizal networks for herbivore-induced defence signal transfer between tomato plants. *Scientific Reports* **4**:1–8.

Thakur, M. P. and S. Geisen. 2019. Trophic regulations of the soil microbiome. *Trends in Microbiology* **27**:771–780.

Thomashow, L. S., Y. S. Kwak, and D. M. Weller. 2019. Root-associated microbes in sustainable agriculture: models, metabolites and mechanisms. *Pest Management Science* **75**:2360–2367.

Tilman, D. 2007. Resource competition and plant traits: a response to Craine et al. 2005. *Journal of Ecology* **95**: 231–234.

Vacher, C., A. Hampe, A. J. Porté, U. Sauer, S. Compant, and C. E. Morris. 2016. The phyllosphere: microbial jungle at the plant–climate interface. *Annual Review of Ecology, Evolution, and Systematics* **47**:1–24.

Van't Padje, A., L. O. Galvez, M. Klein, M. A. Hink, M. Postma, T. Shimizu, and E. T. Kiers. 2021. Temporal tracking of quantum-dot apatite across in vitro mycorrhizal networks shows how host demand can influence fungal nutrient transfer strategies. *ISME Journal* **15**: 435–449.

Wilson, M. and S. E. Lindow. 1994. Ecological similarity and coexistence of epiphytic ice-nucleating (Ice+) *Pseudomonas syringae* strains and a non-ice-nucleating (Ice–) biological control agent. *Applied and Environmental Microbiology* **60**:3128–3137.

Wipf, D., F. Krajinski, D. van Tuinen, G. Recorbet, and P. E. Courty. 2019. Trading on the arbuscular mycorrhiza market: from arbuscules to common mycorrhizal networks. *New Phytologist* **223**:1127–1142.

Wohlleben, P. 2016. *The Hidden Life of Trees: What They Feel, How They Communicate—Discoveries From a Secret World*. Greystone Books, Vancouver.

Xin, X.-F., B. Kvitko, and S. Y. He. 2018. *Pseudomonas syringae*: what it takes to be a pathogen. *Nature Reviews Microbiology* **16**:316–328.

Zeng, X. Y., R. Jeewon, S. Hongsanan, K. D. Hyde, and T. C. Wen. 2020. Unravelling evolutionary relationships between epifoliar Meliolaceae and angiosperms. *Journal of Systematics and Evolution* **60**:23–42.

Zimmerman, N. B. and P. M. Vitousek. 2012. Fungal endophyte communities reflect environmental structuring across a Hawaiian landscape. *Proceedings of the National Academy of Sciences of the United States of America* **109**:13022–13027.

CHAPTER 16

Global change

Gregory S. Gilbert and Ingrid M. Parker

In the past century, ecological conditions on our planet have undergone rapid change, and much of this change is tied to human activity. Economic demands have driven extensive land use change, resource extraction, and increases in the environmental release of carbon, nitrogen, and sulfur compounds as well as other pollutants. Accelerating patterns of trade and travel influence the movement of species, including pathogens, that sometimes become devastating invasive pests. Global warming and other increasingly alarming manifestations of climate change have followed the rising concentration of CO_2 in our atmosphere. This chapter ties together many of the themes and concepts introduced in previous chapters to explore how different aspects of global change influence plant–pathogen interactions.

16.1 Land use change and fragmentation

Arguably the most dramatic anthropogenic change on Earth has been the widespread conversion of diverse ecosystems to agriculture and plantation forestry. Habitat conversion has created large areas with low species diversity and corresponding high density of single host species. Many pathogens respond positively to high host density, leading to greater levels of disease in these low-diversity systems (Chapters 10 and 14). Devastating disease outbreaks in agriculture and forestry are so iconic that early ecologists used them as foundational evidence to support the idea that simpler communities are less stable, and therefore diversity must promote stability (Elton 1958).

In addition to low species diversity, agriculture and forestry systems are also characterized by low genetic diversity, because only a few, highly selected lineages are planted. When host genetic diversity is low, pathogens evolve virulence to their host more quickly, because of strong selection for traits that promote infection of the one dominant genotype (Chapter 13). Taken together, low species diversity and low genetic diversity promote high disease intensity, including elevated incidence of major disease outbreaks. For example, beginning in the early 1900s, fortune-seeking "banana barons" cleared tropical rainforest across large areas in Costa Rica and other parts of Central America and planted monocultures of the Gros Michel banana cultivar (Ploetz 2015). Fusarium wilt (caused by the soil-borne *Fusarium oxysporum* f.sp. *cubense* and also known as Panama disease), spread throughout Gros Michel plantations, driving a wave of deforestation in Central America as the banana companies sought virgin, pathogen-free land where bananas could still be grown profitably (Section 17.2 and Soluri 2002). In the 1950s, Costa Rican plantations of the Gros Michel banana cultivar were also severely affected by bacterial wilt caused by *Ralstonia solanacearum*. This example is of particular interest to us, because *R. solanacearum* was acquired from reservoirs in the wild species *Heliconia* and *Eupatorium*, and it rapidly increased in virulence on the Gros Michel banana after sequential infections on that host (Sequeira and Averre III 1961). Eventually, disease pressure from *F. oxysporum* forced the global abandonment of cultivation of the Gros Michel banana in favor of the resistant Cavendish variety; unfortunately, a new race of the pathogen (tropical race 4, TR4) that

The Evolutionary Ecology of Plant Disease. Gregory S. Gilbert and Ingrid M. Parker, Oxford University Press. © Gregory S. Gilbert & Ingrid M. Parker (2023).
DOI: 10.1093/oso/9780198797876.003.0016

is pathogenic on Cavendish has emerged and led to similarly dramatic losses of commercial banana production in southeast Asia and Australia (Ploetz 2015). The story of the banana remains one of the best-known cautionary tales of the risks of agricultural monocultures.

Plantation forests and agricultural monocultures may in turn threaten vegetation in surrounding wildlands by building up a large pathogen inoculum reservoir that can infect natural stands of susceptible hosts through **pathogen spillover** (Section 14.5). For example, most of the crop-associated viruses that have been tested on wild plant hosts are pathogenic on those hosts and reduce survival, growth, or reproduction (Malmstrom and Alexander 2016). Recent genomic evidence from *Erwinia tracheiphila*, a pathogen of eastern North American origin that causes bacterial wilt of cucurbits, suggests that the introduction of cucumber (*Cucumis sativus*) by Spanish colonists in the early 1500s led to the rapid increase and expansion of *E. tracheiphila* in North America, with spillover and subsequent divergence on native North American *Cucurbita* species (gourds such as pumpkin and squash) (Shapiro et al. 2018).

With land conversion, remaining patches of wild ecosystems become more fragmented and isolated. The edges of fragmented forests are often warmer and drier, which may cause physiological stress in the host plant and enhance disease development (Chen et al. 1992). If, as described above, pathogens cross over into wild populations from a reservoir host in cultivation, then plants on the edges of wild fragments will be especially vulnerable to infection. Finally, as fragmentation increases and the remaining habitat patches become smaller, a highly virulent pathogen has a greater chance of causing extinction of a host population, simply because a small population is more vulnerable to extinction than a large one (Section 11.6). All of these mechanisms predict that fragmentation will increase the impact of plant disease in wild systems. However, the warmer and drier edges of fragmented forests may also reduce pathogen survival and host infection (Krishnadas and Comita 2018).

Ironically, fragmentation could mitigate rather than exacerbate extinction, when plant disease is involved. Fragmentation may reduce pathogen spread and catastrophic epidemics by isolating patches of individuals from each other. As described in Chapter 11, a study of extinction and colonization dynamics in naturally fragmented populations of *Plantago lanceolata* found that smaller and more isolated host populations were less likely to be infected by *Podosphaera plantaginis* (Laine and Hanski 2006). Here, isolation is an advantage for the host. Conservationists advocate for connecting small habitat fragments, arguing that corridors can bolster ecological diversity through a demographic rescue effect and help to maintain genetic diversity by promoting gene flow among populations (Damschen et al. 2019). Although these are important benefits, corridors also increase the risk that a pathogen (or plant pest) will spread from one fragment to another (Simberloff and Cox 1987, Hess 1994). To test this idea, Sullivan et al. (2011) took advantage of a large-scale study at the Savannah River National Environmental Research Park in South Carolina (USA), where forest openings were experimentally created with and without corridors (Figure 16.1). They quantified the prevalence of foliar fungal pathogens on three herbaceous plants that grow in forest clearings (*Lespediza stuevei*, *L. hirta*, and *Solidago odora*). For all three plant hosts they found differences (in varying directions) between the edge and interior of the patches, but no difference between isolated patches and patches connected by corridors. This lack of a corridor effect for the abiotically dispersed pathogens contrasted with a significant effect of corridors for insect galls. Sullivan and colleagues suggested that insect movement may be more directed along corridors than abiotically dispersed foliar pathogens. The generality of this result is unresolved; few studies have tested for the effects of habitat corridors on plant disease, and more research is needed.

16.2 Air pollution and nitrogen deposition

Human activities, especially the burning of fossil fuels in automobiles and industry, change atmospheric chemistry by releasing air pollutants such as sulfur oxides (SO_x) and nitrogen oxides (NO_x), both of which dissolve easily in water to form acid precipitation. Sulfur dioxide is strongly

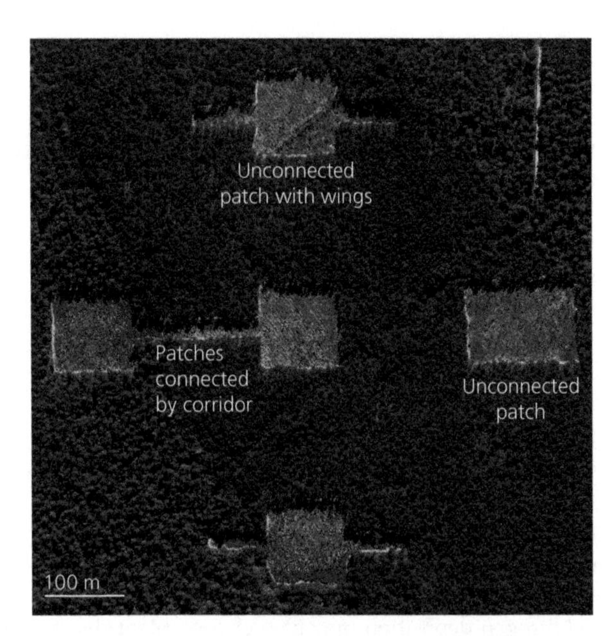

Figure 16.1 Fragmentation and corridor experimental study design at the Savannah River National Environmental Research Park, USA.
Figure by Gregory S. Gilbert, based on Google Earth image.

associated with the burning of sulfur-containing fossil fuels such as coal, and it affects photosynthesis, stomatal function, and other aspects of plant physiology. It also affects plant–pathogen interactions (Li et al. 2020). Different pathogens respond differently to sulfur dioxide pollution, with necrotrophs generally benefitting from the early senescence of leaf tissues, while biotrophs and hemibiotrophs are generally disadvantaged. Making use of a 160-year experiment on wheat cultivation in Rothampsted, UK, Bearchell et al. (2005) extracted fungal DNA from archived straw samples and quantified the relative abundance of two wheat pathogens, the necrotroph *Phaeosphaeria nodorum* and the hemibiotroph *Mycosphaerella graminicola*. They investigated variation in the ratio of *P. nodorum* to *M. graminicola* over the 160-year time series. Although they explored correlations with many agronomic (e.g., dominant cultivar, fungicide use) and climatic (e.g., rainfall, temperature) factors, they found that the relative importance of these two wheat diseases was most strongly explained by SO_2 concentration (Figure 16.2).

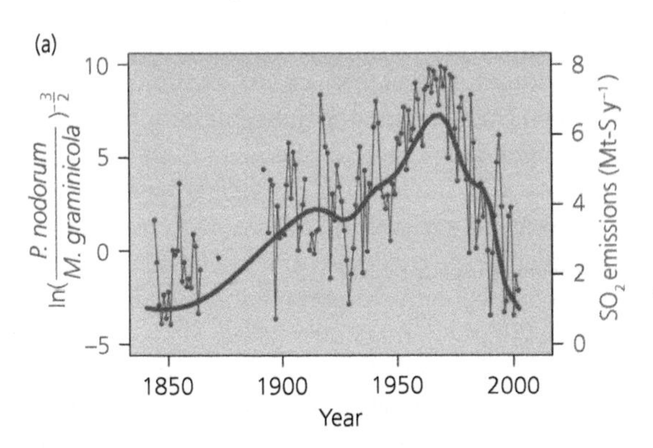

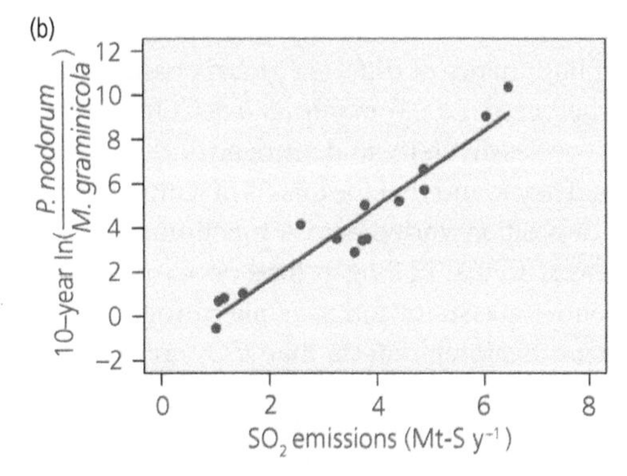

Figure 16.2 Increasing emissions of atmospheric SO_2 are associated with a shift from necrotrophs to hemibiotrophs. (a) During periods of high SO_2 emissions (in million tons of sulfur) from 1844 to 2003, disease on wheat caused by the necrotroph *Phaeosphaeria nodorum* increased, while disease caused by the hemibiotroph *Mycosphaerella graminicola* decreased, (b) leading to a higher ratio of the importance of *P. nodorum* to that of *M. graminicola*.
Figures by Gregory S. Gilbert based on data from Bearchell et al. (2005).

Nitrogen oxides are the main source of smog in areas with high vehicular traffic, and they can react with hydrocarbons in the presence of sunlight to generate ozone (O_3). Ozone affects the chemical and physical properties of leaf surfaces and may promote infection (Karnosky et al. 2002), among its other effects. Nitrogen oxides in the atmosphere react with OH to form nitric acid (HNO_3), which contributes to acid precipitation with the same effects described above for SO_2. Nitric acid may also be deposited in the soil and converted to nitrate (NO_3^-), which is actively taken up by plants. Because nitrogen is often a limiting nutrient for plants, nitrate from N-deposition can act as a fertilizer, increasing plant growth and overall productivity of the system, as well as changing community composition by favoring fast-growing, nutrient-loving species.

Nitrogen deposition may also change plant disease patterns. In both agricultural and wild ecosystems, nitrogen fertilization tends to increase disease severity (Huber and Watson 1974, Veresoglou et al. 2013), and several mechanisms may contribute to this relationship. An increased supply of nitrogen in host tissues can directly promote pathogen growth and reproduction. However, indirect mechanisms are also important; for example, larger, faster-growing plants may support larger pathogen populations (Whitaker et al. 2015), and rapid growth is often correlated with low investment in plant defense compounds (Coley et al. 1985). Conversely, nitrogen fertilization can increase host resistance or tolerance. In a classic mesocosm experiment designed to quantify the relative importance of different global change drivers, Mitchell et al. (2003) manipulated CO_2 concentration, species diversity, and nitrogen addition in field plots. They found that the effects of simulated nitrogen deposition varied across functional groups: it increased fungal pathogen load on C_4 grasses, but not on C_3 grasses or forbs. While nitrogen showed stronger treatment effects than CO_2 concentration, both effects were dwarfed by the importance of plant species diversity and species composition. This raises an important point: drivers like nitrogen deposition may have a stronger influence on plant disease pressure via indirect ecological effects such as plant community structure than via direct effects on individual pathogen–host interactions (Liu et al. 2017).

16.3 A framework for novel plant–pathogen interactions

Novel plant–pathogen interactions arise from one of four scenarios (Figure 16.3). Plant hosts and pathogens that have not previously interacted because they evolved in different geographic regions can be brought together either when the host species moves into the geographic range of a pathogen or when a pathogen moves into the geographic range of the host. Either of these can be accidental (e.g., weed seeds mixed in with imported animal feed or a pathogen that hitchhikes undetected on imported horticultural plants) or intentional (e.g., a new agronomic crop or a pathogen introduced for weed control that spills over onto resident species). Such novel plant–pathogen pairings are common; about a third of all surveyed plant and pathogen species in New Zealand had novel pairings with introduced partners (Bufford et al. 2016). Alternatively, both partners may be introduced into a novel region; within a new context of abiotic conditions and biotic interactions, a disease may emerge. A novel interaction can also develop among plant–microbe pairs in their overlapping native ranges when environmental conditions change. In these two latter cases, changing phenology or other factors may bring two symbionts together that were not previously interacting, or an existing interaction can go from benign to pathogenic because the environment has changed. Because both introduced plants and introduced pathogens have so many impacts on agriculture and resource management, predicting the outcomes of these novel interactions is fundamentally important for humans and the planet (Parker and Gilbert 2004).

The definition of terms in the field of biological invasions is important and somewhat complicated by politics, social context, and conflicting academic traditions. Here we use the term **introduced**

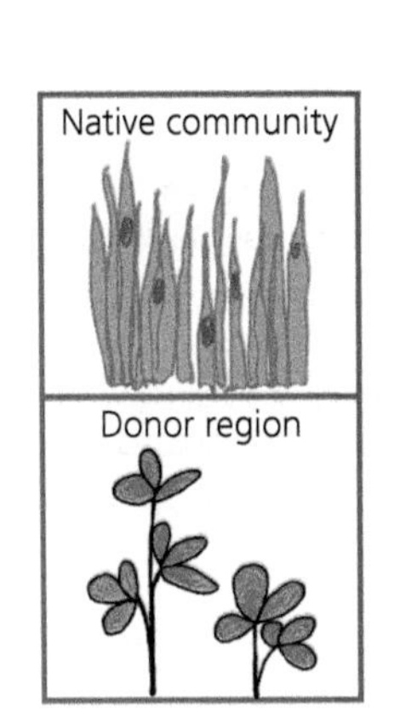

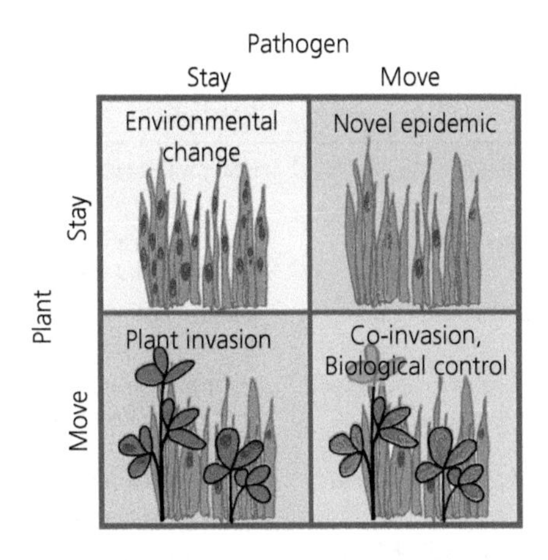

Figure 16.3 The emergence of novel plant–pathogen interactions through four scenarios. The native community includes both native plants and their native pathogens. First, a change in the environment may create novel conditions that promote an epidemic. Second, a novel pathogen moved from the donor region may be the cause of a new epidemic on a native host. Third, a plant moved from the donor region may invade and become a host for a native pathogen. Fourth, plant and pathogen may be moved independently or together from the donor region, or the pathogen may be subsequently introduced as a biological control agent.
Figure by Ingrid M. Parker, adapted from Parker and Gilbert (2004).

species to designate a species that has been moved between continents or regions by human actions, either intentional or unintentional, distinguishing the **native** range from the introduced range. Such species have been described as "non-native," "non-indigenous," "alien," or "exotic." We use the term **naturalized species** to refer to an introduced species that maintains self-sustaining populations, meaning that it does not require human propagation to persist. In some cases, a naturalized species may be limited to highly disturbed environments. The term **invasive** is sometimes used by biologists simply to mean that a species is spreading rapidly, especially away from these highly disturbed environments. However, **invasive species** are defined in the policy realm as introduced organisms that result in major ecological, social, or economic impacts (e.g., Executive Order No. 13112 1999). These invasive species have enormous ecological and economic costs worldwide (Hanley and Roberts 2019).

In this chapter, we consider the case of pathogens moving (Section 16.4) separately from the case of plants moving (Section 16.5), keeping in mind that many of the same biological mechanisms and questions underlie both of these topics.

16.4 Epidemics of novel pathogens

In the past century, increasingly rapid transport and global trade have led to the accelerated introduction of plant pathogens across and between continents. The success and spread of these novel pathogens have placed a major stress on agriculture and food systems. Many of our most famous examples of agricultural epidemics were caused by introduced pathogens, including many examples discussed in this book, such as Panama disease of banana (Sections 16.1, 17.2) and late blight of potato (Box 4.1).

In addition, there are a number of examples in which introduced pathogens have transformed wild ecosystems, decimating populations of dominant or charismatic host species, changing the physical structure of forests, increasing the threat and intensity of wildfires, and reducing important food or other resources for local wildlife (Table 16.1). While outbreaks can happen within pathogen native ranges, most transformative plant disease epidemics have occurred after a pathogen was introduced into a new geographic range with novel hosts, or with familiar hosts in a novel environmental context (Parker and Gilbert 2004).

Table 16.1 Examples of plant diseases caused by introduced pathogens that had dramatic impacts in wild ecosystems

Disease (pathogen and hosts)	Description
Chestnut blight (*Cryphonectria parasitica* on *Castanea dentata*)	Nearly wiped out chestnut populations that were the structurally dominant trees in eastern North American forests in the early 1900s (Anagnostakis 1987)
White pine blister rust (*Cronartium ribicola* on five-needle *Pinus* spp.)	Devastating high-elevation populations of several species of five-needle pines in western North America (Dudney et al. 2020)
Phytophthora cinnamomi in Australian forests and heathlands	Transformation of community structure and loss of half the plant species diversity in areas of Australian heathlands in just two decades (Wilson et al. 2020)
Sudden oak death (*Phytophthora ramorum*) on tanoak (*Notholithocarpus densiflorus*) and some oak species (*Quercus* spp.)	Selective decimation of tanoak and some oak species in coastal forests in California and Oregon (Meentemeyer et al. 2008)
Dogwood anthracnose (*Discula destructiva* on *Cornus florida*)	Rapid loss of a dominant species in the understory woody vegetation of hardwood forests in eastern North America (Jenkins and White 2002)
Pine wood nematode (*Bursaphelenchus xylophilus*) on pine species (*Pinus* spp.)	Major mortality of pines first throughout Asian and now European forests (Futai 2013)
Rapid 'Ōhi'a dieback (*Ceratocystis lukuohia and C. huliohia*) on *Metrosideros polymorpha*	Massive and rapid mortality of the most common and culturally important tree species in Hawai'i (Fortini et al. 2019)

These invasive pathogens are transformative precisely because they create a significant, novel disturbance in the recipient plant community. While many—even most—introduced pathogens may have minimal ecological effects, the challenge is to predict and effectively respond to the big ones. We now highlight the stories of three forest pathogens of big effect; each has had dramatic ecological, economic, and social impacts. Each one has also contributed in substantial ways to our understanding of plant–pathogen interactions, and each has prompted different kinds of management responses.

16.4.1 Chestnut blight

American chestnut (*Castanea dentata*, in the family Fagaceae) was once a dominant tree species through much of the Appalachian forest of the eastern United States from Mississippi to Maine. In many upland forests, the chestnut was a key structural element of the ecosystem, comprising nearly half of the individuals and much of the canopy. The massive tree, which could live to be over 600 years old and grew up to 4 m in diameter, was revered by indigenous tribes as "The Grandfather of the Forest" (Brewer 2018). The chestnut was used for food, shelter, medicine, and tools, and it plays a role in the stories and legends of many Native linguistic groups. It is likely that indigenous land management practices enhanced the populations of this useful tree. After European colonization, chestnut was arguably the most important east-coast timber wood, used for everything from telegraph poles to roof shingles; as observed by George Hepting (1974), "Not only was a baby's crib likely made of chestnut, but chances were, so was the old man's coffin." The bark provided tannin for the leather industry. The nuts were a food staple for bears, wild turkeys, squirrels, deer, as well as people. Yet this abundant, imposing, beloved tree was nearly eliminated within a 50-year timespan by a fungal pathogen.

In 1904, the chestnut blight fungus (*Cryphonectria parasitica*) was introduced to New York on logs from Europe, where it had already taken hold in populations of the European chestnut (*Castanea sativa*). The pathogen rapidly spread throughout the entire natural range of the American chestnut, killing about 90% of the trees (Anagnostakis 1987, Vandermast and Van Lear 2002). *Cryphonectria parasitica* has a fairly narrow host range, and it is most virulent on *Castanea* species. It kills via cankers, but the root systems of the chestnut remain alive and readily produce resprouts. When the resprouts reach the size of saplings, they are re-infected and die back.

East-coast forests are still full of the resprouting stumps of chestnut trees, essentially the ecological ghosts of a transformed forest. Other tree species such as oaks, maples, and hickories have replaced the chestnut across its former range.

There was not much plant pathologists could do to slow the spread of *Cryphonectria parasitica* or save infected chestnut trees. Decades of efforts to control chestnut blight using biological control with hypovirulent mycoviruses was largely unsuccessful because of difficulties getting the virus to spread through the fungal population (Section 15.10). Resistance was found in *Castanea* species from China where the pathogen is native, but backcross breeding (Section 17.3) is slow because of long generation times, and it proved difficult to create a hybrid that expressed the tall form of American chestnut with resistance from the small-statured Chinese species. An exciting new initiative is harnessing genetic engineering to create American chestnuts tolerant to chestnut blight (Section 17.3). Researchers have transformed American chestnut with a gene for oxalate oxidase (OxO), an enzyme found in many plant species that breaks down oxalic acid (oxalate) into CO_2 and hydrogen peroxide (Powell et al. 2019). *C. parasitica* produces oxalic acid at canker margins, which interferes with the ability of the host to mount a defense against the fungus. Critically, the production of OxO does not kill the fungus, but rather makes the tree able to tolerate infection, so that tree and fungus coexist. That means that there is no strong selection pressure on the pathogen to counteract the production of OxO; this is a good recipe for durable resistance (Sections 13.5 and 17.3). The American Chestnut Research and Restoration Project has already initiated the regulatory process to enable planting engineered chestnut seedlings in the United States (Newhouse and Powell 2021). Someday, future generations may experience the resurgence of the Grandfather of the Forest.

16.4.2 Sudden oak death

In coastal California and Oregon, the mixed evergreen forest is a patchwork of beautiful trees; alongside the famed coast redwoods (*Sequoia sempervirens*) are iconic coast live oaks (*Quercus agrifolia*) with their gnarled branches, California bay laurel (*Umbellularia californica*) with their heady fragrance, and tanoaks (*Notholithocarpus densiflorus*), which can grow to be 40 m tall and nearly 2 m in diameter, with fat acorns that were a preferred food of the first peoples of coastal California. In the mid 1990s, large numbers of tanoaks began to die in dramatic fashion; entire trees would suddenly turn golden-brown, followed by death. The disease was named sudden oak death (SOD), and its cause was discovered to be the oomycete pathogen *Phytophthora ramorum* (Rizzo and Garbelotto 2003). Public alarm only intensified when it was realized that *P. ramorum* has an exceptionally broad host range, and early predictions suggested that whole forest communities could be at risk.

The origin and epidemiology of *P. ramorum* as an emergent pathogen is still somewhat uncertain. It is most likely native to southeast Asia, but there are four separate lineages of the pathogen—two in North America and two in Europe—spreading independently and causing disease on different sets of hosts (Keriö et al. 2020). *P. ramorum* first appeared in a California nursery in Santa Cruz about the same time it was recognized in European nurseries, raising suspicion that it arrived in California on infected nursery stock, and pointing to plant commerce as a critical pathway of spread.

The interactions between *P. ramorum* and its many hosts actually encompass two types of host–pathogen interaction so distinct that they are now called different diseases. On tanoak and true oak hosts, the pathogen causes cankers on trunks, disrupting vascular tissue and killing the host. The pathogen reproduces on twigs and foliage when on tanoak. Curiously, true oaks are dead-end hosts on which the pathogen is unable to reproduce, even though they are extremely susceptible and often die rapidly from the infection. This is SOD.

In contrast, on most of the hundreds of forest and nursery plant species it can infect, *P. ramorum* causes Ramorum blight, limited to leaf necrosis and twig dieback, rarely killing its host. Some of the species susceptible to Ramorum blight are extremely competent hosts and are responsible for much of the pathogen reproduction in the forest. In particular, California bay is commonly found together with oaks and tanoaks and supports production of massive numbers of sporangia on infected leaves. This creates a strong spillover effect that propels the spread of sudden oak death (Section 14.5 and Kozanitas et al. 2022). The story of this complicated multi-host epidemiology has become an iconic example of the dual importance of **susceptibility** and **competence** (Section 14.8) and how disease dynamics across a plant community may be driven by one key host, even when that host shows only minor symptoms.

The broad host range of *P. ramorum*, impacts on ecologically and aesthetically important forest trees, rapid spread throughout coastal California, and association with horticultural nurseries have sparked many research efforts to find effective control measures for sudden oak death, but with limited success. The primary focus for management is on limiting its spread into new areas through phytosanitary exclusion. Actions include developing best practices in nurseries, requiring certification that nursery stock is free of *P. ramorum*, and imposing quarantine restrictions on all the counties in which the pathogen is established (Section 17.2).

16.4.3 Jarrah dieback/Phytophthora dieback

The southwest corner of Western Australia is a Gondwanaland refuge, full of plants with ancient origins going back to the time when Antarctica, Africa, India, and Australia were all joined in a single continent. The woodlands, heathlands, and jarrah-karri forests include a thick layer of flowering shrubs, with high species diversity and many endemic plants including species in the curious genera *Darwinia*, *Banksia*, and *Grevillea*.

Phytophthora cinnamomi is an invasive pathogen with an incredibly large host range. The root-infecting oomycete rapidly colonizes vascular tissue and causes plant wilting, dieback, and death. *P. cinnamomi* has an estimated host range of more than 5000 plant species (including asymptomatic hosts). *P. cinnamomi* joins *Cryphonectria parasitica* (Chestnut blight), *Ophiostoma ulmi* (Dutch elm disease), and *Banana bunchy top virus* as one of four plant pathogens included the International Union for Conservation of Nature list of the 100 worst invasive species (Lowe et al. 2000). Its center of origin is southeast Asia, probably Papua New Guinea, and it has been introduced to Europe, North America, South Africa, New Zealand, and Australia, where it causes significant damage to many perennial crops as well as wild plants (Keriö et al. 2020). *P. cinnamomi* was introduced into Western Australia in the 1920s, probably on infested nursery stock, and then spread throughout the region with soil on heavy machinery.

The transformation of Western Australia offers a dramatic illustration of the potential impacts of generalist pathogens. Foresters noticed rapid dieback of forests dominated by jarrah (*Eucalyptus marginata*, in the family Myrtaceae), and gave the disease the name Jarrah dieback. But the impacts of *P. cinnamomi* are remarkable and long-lasting not just because of its effect on jarrah trees, but because its broad host range has led to a total conversion of whole communities of native plants. In invaded areas, *P. cinnamomi* kills up to 50% of the understory vegetation, decimating particularly susceptible plant species in woody plant families like Myrtaceae and Proteaceae (Wilson et al. 2020). In the absence of competition for space and light, resistant understory species thrive and are joined by weedy, colonizing species from nearby ecosystems. A highly diverse community of trees and shrubs is converted to a species-poor community dominated by sedges and grasses. At Great Otway National Park, of particular concern is the decline of unique and ecologically important species such as the iconic Austral grass-tree (*Xanthorrhoea australis*) (Wilson et al. 2020).

Disease management for *Phytophthora* dieback includes quarantine, vehicle wash stations, and limiting use of heavy machinery to the dry season, which is less favorable to pathogen spread. However, *P. cinnamomi* produces copious zoospores to infect host roots in waterlogged soils, and it produces both oospores and chlamydospores that remain viable in soil or dead plant material for many years. Its long-term survival along with ready movement in water or soil make control of the pathogen extremely challenging. The case of *Phytophthora* dieback demonstrates how a single introduced pathogen with broad host range and an aggressive, necrotrophic life history can transform entire plant communities.

16.5 Biological invasions: Spread of invasive weeds

We have been discussing introduced pathogens, and now turn our attention to introduced plants. These plants include many of our worst agricultural weeds, and between 1970 and 2017 invasive plants caused an estimated global cost of US$8.9 billion (Diagne et al. 2021). Five centuries of accelerating human migration and global trade have resulted in widespread introduction of plants to new regions and continents. We move plants intentionally via the horticultural trade and when adopting novel crop and forestry species. We also move plants unintentionally via seeds and clonal propagules that end up in luggage, on shoes, and in shipping materials. Most introduced plants either fail to thrive altogether or are restricted to human-dominated landscapes. However, some are able to spread out of disturbed environments and establish self-sustaining populations in wild ecosystems. In a survey of 184 sites around the world, introduced plants make up from 1.3% to 64% (mean 16%) of plant species in local floras, with islands having a nearly three times greater proportion of introduced species than mainland sites (Lonsdale 1999). Invasive plants have significant impacts not just as agricultural weeds, but also as environmental weeds that cause harm to native plants, animals, and ecosystems (Parker et al. 1999, Vilà et al. 2011).

Plant pathogens can play different roles in the ecology of introduced plants. If plants arrive in a new locale that includes aggressive pathogens capable of infecting them, or where environmental conditions favor disease development, they may experience increased population regulation by pathogens. Similarly, if plants are strongly affected by pathogens in their native range and are introduced to an area where those pathogens are absent, then the plant population may be released from regulation by pathogens. Such changes in the prevalence and severity of diseases under different conditions have long been thought to shape the process of biological invasions in two hypothetical ways: (1) In wild habitats, native pathogens colonize and are highly virulent on naïve, introduced plant species and prevent plant population growth (biotic resistance); (2) Pathogens are important in regulating the plant population in its native range but not in its introduced range (escape from natural enemies, or "enemy release"). Let's look more closely at each of these mechanisms.

16.5.1 Biotic resistance

Exotic species are initially associated with disturbed environments and human-dominated landscapes such as agricultural fields, gardens, and industrial areas. Most introduced plants do not become invasive. Rather, they are **human commensals**, restricted to living under human care or in **anthropogenic habitats**, and they are unable to build self-sustaining populations in wild plant communities. One explanation for failed invasions is **biotic resistance**; native pests and pathogens may colonize introduced plants and prevent them from establishing viable populations (Elton 1958, Maron and Vilà 2001). Introduced plants may be more susceptible than local hosts because they have not evolved defenses against a local pathogen.

Which pathogens are likely to contribute to biotic resistance? Pathogens with a broader host range and a more generalist life history will be more likely than host-specialist pathogens to adopt an introduced plant species as a new host (Section 13.5). Of those local polyphagous pathogens that can infect an introduced host, species that are abundant (either because they infect one particularly common host or because they colonize a wide range of local species) will have the potential for ecologically significant spillover onto the introduced host (Sections 14.5 and 16.1). To cause biotic resistance, the pathogen must also have a high enough impact on the introduced host to reduce its fitness and its ability to compete in the local community.

Which communities are likely to show strong biotic resistance? Biotic resistance from pathogens likely depends on the diversity and composition of the plant community. Highly diverse plant communities should support more microbe diversity (with dampened disease outbreaks) and favor pathogens with more generalist life histories, because host population sizes are smaller on average and pathogens are more likely to disperse among host species rather than between individuals of the same species (Section 14.8). Therefore pathogen spillover may be one reason that hyper-diverse ecosystems like tropical rain forests seem to be more resistant to invasion by introduced plants (Fine 2002). In any community, pathogen spillover is most likely between closely related hosts (Figure 14.6), so we would expect that biotic resistance should be strongest in communities rich in close relatives to the introduced species, and invasion should be favored for introduced species with no close relatives. As predicted, when we placed randomized arrays of novel introduced hosts into a meadow community in Santa Cruz, California, disease pressure was lower on hosts with few close relatives (Parker et al. 2015). The idea that close relatives may help local communities repel invaders is called **Darwin's Naturalization Hypothesis** (because Darwin came up with it first), although natural enemies are only one possible mechanism, with competition between close relatives being a second (Mack 1996, Daehler 2001). Evidence for Darwin's Naturalization Hypothesis has been mixed, possibly because another important factor contributing to invasion is the fact that close relatives share ecologically important traits (Section 13.7) and are therefore likely to thrive in similar environments (Cadotte et al. 2018).

Understanding when and how pathogen-mediated biotic resistance limits the performance of introduced species is important beyond invasive weeds. Plants introduced for use in agriculture, horticulture, or forestry may also suffer from strong disease pressure from local pathogens, as described above for *Ralstonia solanacearum* on Gros Michel bananas (Section 16.1). Also consider the financial debacle of *Acacia mangium*. In Indonesia, massive investments were made into pulp-wood plantations of *A. mangium*, a tree native to Australia. Disease caused by two local pathogens, red root-rot disease (caused by *Ganoderma philippii*) and stem-wilt canker disease (caused by *Ceratocystis acaciivora*), led to unsustainable losses and the abandonment of *A. mangium* plantations (Nambiar et al. 2018).

16.5.2 Escape from natural enemies

Some plant species become much more abundant and more aggressive competitors after introduction to a new continent than they are in their native range. One of the central hypotheses to explain this phenomenon is that introduced species leave their natural enemies behind in their native range (Darwin 1859, Maron and Vilà 2001); this is called the **Natural Enemies Hypothesis** or **Enemy Release Hypothesis**. For example, Mitchell and Power (2003) used the USDA Fungus–Host Distributions database (Box 15.1) to show that plants were infected by 77% fewer rusts and smuts in their introduced range compared with their native range. They also found that plants that left a larger proportion of pathogens behind were more likely to be categorized as invasive or noxious. The Natural Enemies Hypothesis implies that herbivores, seed predators, and pathogens are a key regulating factor in plant populations in their native range. With a sterilization experiment, Reinhardt et al. (2003) found strong negative effects of soil pathogens on seedlings of *Prunus serotina* in its native range but not in its introduced range. They proposed that the much higher population densities they found in *P. serotina*'s introduced range compared to its native range are explained by release from pathogen control in the introduced range, and negative plant–soil feedbacks (Section 14.7) in the native range. A further prediction is that because of tradeoffs between allocation to plant defense and allocation to growth and reproduction, introduced populations will evolve low investment in defense in the absence of their enemies, giving them an advantage over native species. This is called the **EICA Hypothesis**, for Evolution of Increased Competitive Ability (Blossey and Nötzold 1995).

There are multiple ways to study "enemy release." You can study the number of pathogen species or disease pressure on a host in its new range compared to its native range, as Mitchell and Power and Reinhardt et al. did. You can also compare native to introduced plants, with the prediction that if enemy release explains something fundamental about biological invasions, then native species will experience greater overall disease pressure than invasive species. We studied a closely related suite of 10 native and 8 introduced species of clovers (*Trifolium* and *Medicago*) that all co-occur within the bounds of one ecological reserve in California (Parker and Gilbert 2007). We found that foliar pathogens were as diverse and as common on introduced hosts as on native ones (Table 15.1), that disease symptoms were not reduced on introduced hosts, and when we used fungicides to release plants from the effects of pathogens, introduced plants benefitted as much as native plants. Most importantly, although we found variability among species in their response to pathogens, escape from pathogens did not predict which of the introduced species were most successful at invading the local plant community. Why was there no "escape" for introduced clovers? The pathogens causing almost all of the foliar disease were necrotrophic generalists (fungi in the genera *Stemphylium*, *Alternaria*, *Cladosporium*, *Colletotrichum*, and *Phoma*). Pathogens were moving freely among the native and introduced hosts, with both pre-adaptation of the host to closely related pathogens in their native range and rapid adaptation of the pathogens to novel hosts (Gilbert and Parker 2010), likely playing a role in the story. Where generalist pathogens are ecologically dominant, it should be more difficult for introduced plants to escape disease.

Many studies have now tested the Enemy Release Hypothesis, mostly for herbivores but also for pathogens, both above and below ground. Heger and Jeschke (2014) reviewed 176 empirical studies of enemy release and found stronger support for certain subsets of the hypothesis than others. In particular, comparisons between the native and introduced ranges of the same species were more likely to find differences than comparisons between native and introduced species within a community, and studies on vertebrates and marine systems were more likely to show enemy release than plants or terrestrial systems. The big questions in this field now focus not on "Whether introduced species leave their enemies behind," but rather on understanding why the phenomenon appears to be more common or impactful in some situations than others.

Because invasive plants are by definition abundant, we should expect pathogens to find ways to take advantage of them as a resource, shifting onto a novel host gradually over time since introduction (Flory and Clay 2013). Phylogenetic rarity (not having close relatives among local species) can make some novel hosts more likely to escape pathogen attack and maintain that advantage for longer (Section 14.5). In addition, focusing on traits that make plants vulnerable to pathogen attack may help explain why the advantage of being novel persists longer for some plants than others (Fahey et al. 2022).

16.5.3 Biocontrol of invasive plants

If it is just a matter of time until local pathogens colonize invasive plants, one approach to managing invasive populations would be to try to promote that process by encouraging the evolution and spread of pathogens discovered on invaders in their new range. Another approach is to bring pathogens over from the native range; this is called **classical biological control**.

The Enemy Release Hypothesis provides the justification for classical biological control, in which natural enemies are brought from the native range to control weedy invasive plants (DeBach and Rosen 1991). For instance, several species of wild blackberries (*Rubus* spp.) are aggressive invaders and cause damage in wild ecosystems throughout the world. The obligate biotrophic rust *Phragmidium violaceum*, originally found in the native range of *Rubus ulmifolius*, shows great potential as a biological control agent against *R. ulmifolius* on Robinson Crusoe Island, Chile, where the introduced blackberry threatens native plant species in a biodiversity hot-spot (Vargas-Gaete et al. 2019). The rust is able to infect and reduce the performance of the invasive *R. ulmifolius*, but its narrow host range means that native species are not affected, including the closely related native species *R. constrictus*.

One risk of biological control introductions is the possibility of creating a new invasion problem with the introduced agent (Hinz et al. 2019). Therefore, careful testing of the host range of the pathogen is an essential part of any application of introduced plant pathogens for control of weeds. Non-target attacks by biological control agents are usually on close relatives of the target weed, so testing is usually focused on species within the same family. Increasingly thorough and thoughtful pre-release testing has reduced the occurrence of non-target attacks by weed biocontrol agents by 50% from the 1960s to the early 2000s, with only about 5% of biocontrol agent releases (of all kinds of agents) producing any non-target attack. Of the 29 plant pathogens that have been released for biological control, only one led to non-target attacks (Hinz et al. 2019).

To be successful, a biological control program must not only be safe, but also effective. Using pathogens for effective biological control of weeds requires understanding the environmental conditions favorable to disease development, the population dynamics of the pathogen, and how disease impacts host populations. Stage-structured population models are powerful tools that can be used to measure the impacts of pathogens and other natural enemies on population growth (Section 10.7, Box 10.2). These and other modeling approaches can help evaluate the potential for a biological control agent to control a target weed host (Shea and Kelly 1998, McEvoy and Coombs 1999, Parker 2000).

16.6 Climate change

We end this chapter with climate change, an anthropogenic driver that is likely to interact with all the other sources of global change we have considered so far. The climate of our planet is changing, accelerated by the burning of fossil fuels and the accumulation of greenhouse gases in the atmosphere. Across the globe, mean temperatures are on the rise: 2011–2020 was the warmest decade on record (World Meteorological Organization 2021). According to the United Nations' Intergovernmental Panel on Climate Change (IPCC), temperatures are predicted to increase more in northern latitudes than southern latitudes (Figure 16.4a), and minimum temperatures will increase more than maximum temperatures, leading to more frost-free days and longer growing seasons. Other aspects of climate are also changing and becoming more extreme; precipitation is increasing in some regions while in others it is decreasing (Figure 16.4b), and hurricanes, droughts, and other extreme weather events are becoming more common (IPPC 2014). How will these changes affect plant diseases in agriculture and in wild systems? As we have seen, the disease triangle (Figure 1.5) determines the outcome of plant–pathogen interactions, and as the environment changes, so will plant disease. But unfortunately, the complexity and variability of climate change across different regions means that impacts will vary, severely limiting our ability to make clear, generalizable predictions (Garrett et al. 2006, Ghini et al. 2011, Elad and Pertot 2014, Burdon and Zhan 2020)). Let's review some of the most commonly projected outcomes.

16.6.1 Shifting ranges

The most consistent climate change prediction for plant pathogens is that, as temperatures increase, the range of many pathogens will spread poleward with the increased probability of successful overwintering of primary inoculum (Bebber et al. 2013). The poleward extension of pathogen distributions may affect crops that currently experience relatively low disease pressure in the parts of their range with harsher winters (Boland et al. 2004). In addition, range expansions will bring pathogens into contact with new potential hosts. For example, oak wilt, a root rot caused by the introduced pathogen *Bretziella fagacearum*, has a large impact on oak species in the central region of the United States. Warmer, drier conditions allow the overwintering and development of root rots, and these diseases are expected to increase with climate change (Boland et al. 2004). Pedlar and colleagues (2020) used **species distribution models**, which translate current location data into information on the environmental tolerances of a species, to project how *B. fagacearum* and its two insect vectors will be able to spread north into Canada under future climate conditions. The resulting loss of oak timber products and removal of diseased trees in urban areas of Canada is expected to cost hundreds of millions of dollars, even without including any direct costs incurred by efforts to control the pathogen (Pedlar et al. 2020).

Wild plant hosts will also spread beyond the current limits of their ranges, possibly encountering new pathogens and escaping others. Whether hosts can escape some of their current pathogens depends on the niche requirements and dispersal ability of the pathogen relative to the host. For foliar pathogens, long-distance dispersal often exceeds that of their wild host plants, but soil-borne pathogens without adaptations for long-distance dispersal may lag behind their hosts' spreading populations (Section 11.1). Meanwhile, crop distributions will only expand into new areas that meet all the requirements for cultivation. As a result,

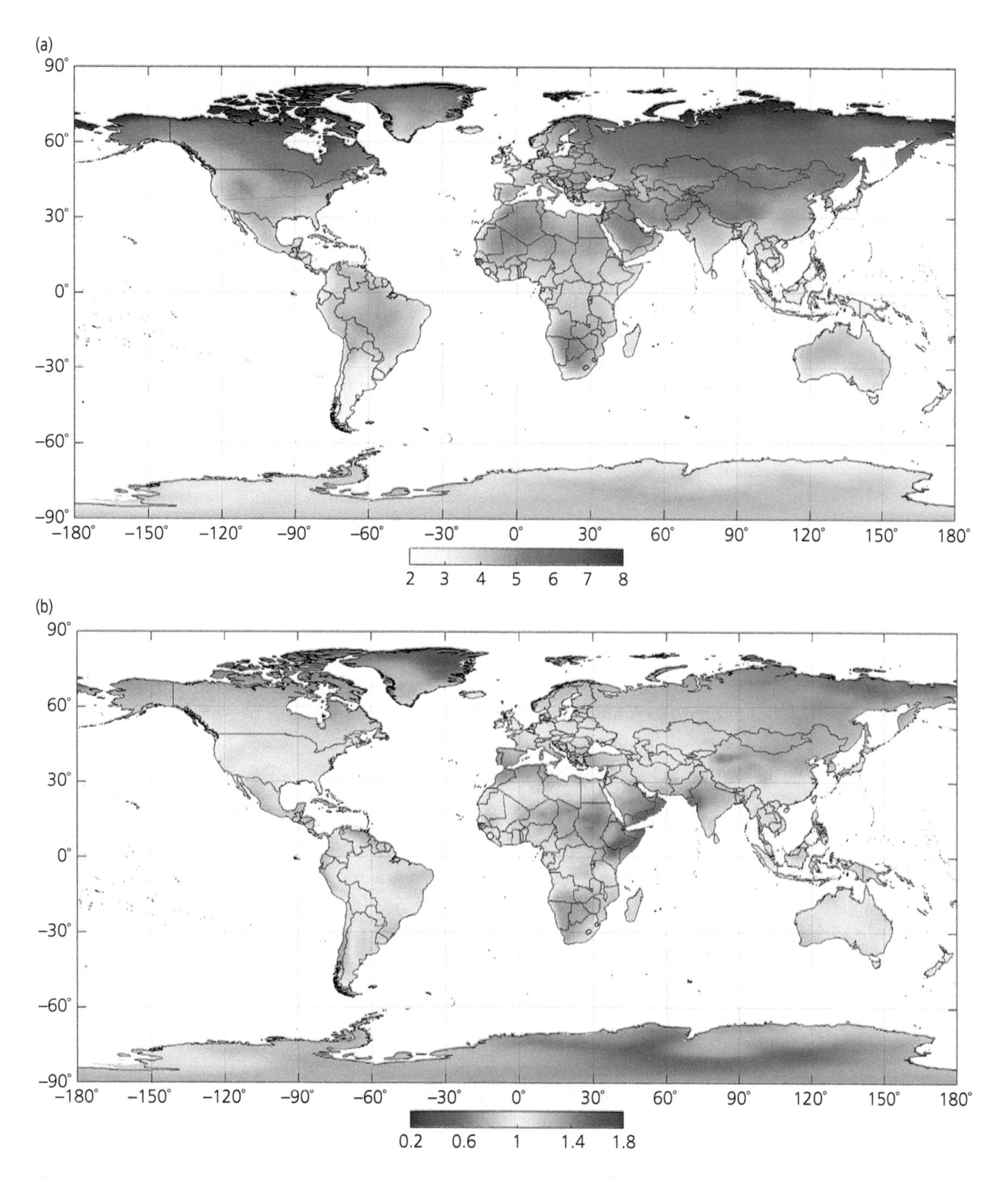

Figure 16.4 Projected changes in temperature and precipitation between 1980–2016 and 2071–2100 derived from climate model outputs. (a) mean air temperature change (°C) and (b) precipitation change. Means were calculated over all months and across models.
Reproduced with permission from Beck et al. (2018).

the area suitable for growing that crop may simply shrink. For example, Tibpromma et al. (2021) used models to predict that up to a third of the area currently suitable for growing tea (*Camellia sinensis*) may become unsuitable due to climate change, while three important fungal pathogens will have substantial overlap with tea in the remaining suitable regions.

16.6.2 Plant physiology

Changing climate will influence plant disease through the host, because weather strongly affects every aspect of plant physiology. This is so obvious, yet responses can also be idiosyncratic and therefore contribute to uncertainty about future scenarios. Increased frequency of high-temperature events and drought can result in greater plant susceptibility to infection, although abiotic stress can also induce generalized plant defense pathways that may increase resistance (Desprez-Loustau et al. 2006, Aung et al. 2018). Drought stress in the host will sometimes convert a commensal endophyte into a pathogen (Section 1.6). In comparison to predictions for temperature, predictions for future precipitation are highly variable both regionally and locally (Figure 16.2), making it hard to predict how infection and disease progress will respond. The number of wet days will increase in some places and decrease in others, while warmer temperatures may increase disease by exacerbating drought in dry regions or may suppress disease by reducing leaf wetness even in the face of more rainfall. An example of an unexpected outcome of climate change comes from the study of pitch canker (caused by *Fusarium circinatum*) on Monterey pine (*Pinus radiata*). The native range of Monterey pine is limited to a narrow coastal band in central California, where fog is an important source of moisture during the summer dry season. When pitch canker was first detected in native stands in the late 1980s, it was viewed as an imminent, significant threat to the endemic tree, endangering the three remaining native stands of Monterey pine in California. In recent years, however, the threat from pitch canker has diminished in what appears to be a response to the steady decline of summer fog (Gordon et al. 2020).

Beyond temperature and precipitation, other anthropogenic atmospheric changes will also affect plant physiology. The CO_2 concentration of the atmosphere was 316 parts per million (ppm) in 1960, 389 ppm in 2010, and at current rates of increase will exceed 550 ppm by 2060. This rise in global CO_2 concentration has a direct effect on plant physiology. Plants need CO_2 for photosynthesis, and to access CO_2 plants must open their stomata, allowing water molecules to escape; this is the underlying cause of drought stress in plants (Section 2.2). A higher CO_2 concentration allows the plant to achieve a higher rate of photosynthesis, or to achieve the same amount of photosynthesis while losing less water because its stomata are closed more of the time. This results in a noticeable positive effect on plant growth, called the "CO_2 fertilization effect." How CO_2 enrichment will affect a pathogen is tightly linked to the details of its biology—for example, a fungus that invades through stomatal openings may decrease in prevalence, while one that benefits from the moist microclimate associated with dense plant growth may increase. A global scientific effort began in the late 1980s to study the response of crops and whole ecosystems to an increase in atmospheric CO_2 (combined with manipulations of temperature and precipitation), using free-air CO_2 enrichment (FACE, Figure 16.5). The FACE experiments to date have frequently shown an increase in plant disease, even when a decrease was predicted (Ainsworth and Long 2021). For example, Kobayashi et al. (2006) found that rice (*Oryza sativa*) plants grown in elevated CO_2 were more susceptible to inoculation with the rice blast pathogen (*Magnaporthe oryzae*) and experienced higher natural incidence of rice sheath blight (caused by *Thanatephorus cucumeris*). The mechanisms behind the increased susceptibility seen in FACE experiments appear to be indirect, mediated by the physiology of the plant host, suggesting that more studies are needed before we can generalize across plant-pathosystems.

16.6.3 Pathogen populations

Climate change will affect pathogens in a range of ways, scaling up from cellular metabolism to population dynamics, influencing the spread and impact of plant disease. With increasing

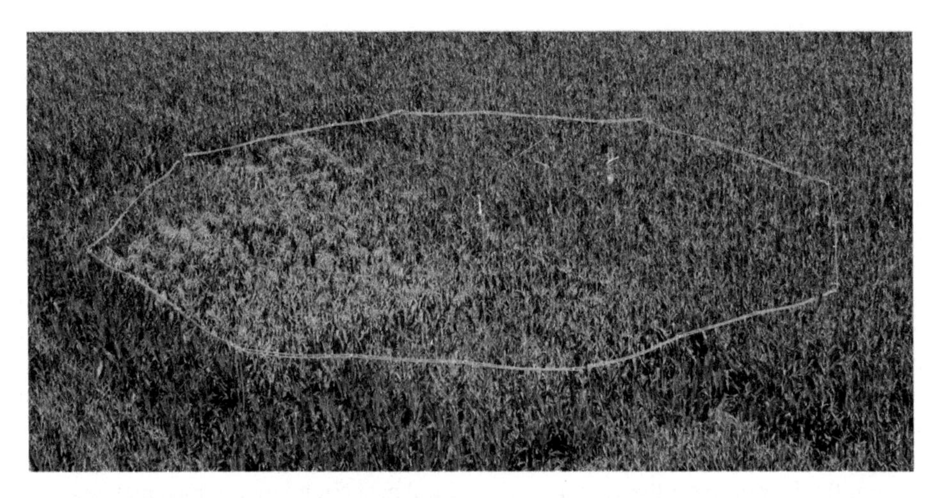

Figure 16.5 A free-air CO_2 enrichment (FACE) experiment at the University of Illinois measures the effect of increasing local CO_2 concentration on the growth of corn. Micropores in the bright green tubes in the photo deliver CO_2 gas at a rate moderated by wind direction and velocity. The whole system is controlled by a computer to maintain a constant level of enhanced CO_2.
Photo by James Baltz.

temperatures, all organisms experience higher cellular metabolic rates; this can also increase growth rates and reproductive rates (Figures 3.10 and 5.6). Growing seasons will lengthen, allowing additional generations in polycyclic pathogens. With faster demographic rates and additional generations, the population dynamics of pathogens will accelerate. Rates of adaptive evolution in pathogen populations may also increase. On the other hand, pathogens have upper limits to their temperature tolerance (Figure 3.10), and global warming may result in lower survival rates for some pathogen species. But wait ... adaptive evolution of thermal tolerance in the pathogen may allow the pathogen to catch up, again modifying the interaction between host and pathogen.

As increasing temperature accelerates the physiology and growth of pathogens, one outcome will be shorter **latent periods** (Figure 10.2). The shorter the latent period of a polycyclic pathogen, the more generations can fit into a single growing season. Latent period is often measured in **degree days**, where the number of days from infection to reproduction is multiplied by the temperature in degrees centigrade; a pathogen with a latent period of 120 degree days reproduces after 12 days at 10°C or after 8 days at 15°C. Pathogen life history also affects latent period, with necrotrophs generally having the shortest latent periods (<100 degree days), hemibiotrophs the longest (>200 degree days), and biotrophs having intermediate latent periods (Précigout et al. 2020); we may expect these groups of pathogens to respond differently to a warming climate.

16.6.4 Predicting responses to climate change

We use a range of analytical tools to predict how climate change will affect a particular plant–pathogen interaction in a particular place. We can model regions of overlap between hosts and pathogens using the **species distribution modeling** approach (also called "ecological niche modeling" or "habitat suitability modeling") described above for oak wilt and for tea and its pathogens. This approach extracts information about the environmental conditions in areas where the pathogen or host is found, drawing on global climate data interpolated between weather stations. Much of this work is now done with well-developed software packages such as CLIMEX and MaxEnt, but statistical models can also be tailored to a particular application. The simplest versions of these models predict current and future distributions based simply on "dots on a map" and large-scale climate factors such as monthly mean precipitation and maximum and minimum temperatures. However, for well-studied plant–pathogen interactions, the response variable can incorporate data on incidence and severity (not just occurrence), and explanatory variables can include detailed local and regional factors, like the number of days of frost, local topography, or the number of hours of morning fog.

While these quantitative approaches incorporate physiology only indirectly, we can also develop mechanistic ("process-based") models that rely on a more detailed understanding of how the pathogen and host respond to particular environmental variables. For example, in a study of the response to climate change of six forest disease systems in France, Desprez-Loustau et al. (2007) not only used field data on pathogen incidence but also incorporated information from the literature on the response of each pathogen to climate and seasonality. They then added their own laboratory experiments to quantify how fungal growth and survival respond to temperature. Using this synthetic approach, the authors were able to predict which forest diseases would be expected to expand and which should contract, and where disease pressure is likely to intensify or lessen. They found that warmer temperatures usually favor pathogens whereas changes in precipitation regimes have large but idiosyncratic effects, highlighting the critical importance of microbial hydrobiology in predicting responses to climate change.

For well-studied plant diseases, process-based models are powerful tools for making detailed projections, even at the global scale. For example, potato late blight, caused by the oomycete *Phytophthora infestans* (Box 4.1), is the most important disease in global potato production and has been well studied for its response to weather conditions (Harrison 1992). Models of potato late blight have been used for decades to predict disease risk as a function of temperature and hourly patterns of humidity, as well as to inform host resistance and disease management practices (Fry et al. 1983). Sparks et al. (2014) adapted these models to ask how climate change will affect potato yield under a range of predicted climate scenarios, especially in parts of the world where potato is a critical staple crop, like Himalayan Nepal and the Andean highlands. Their results suggest only moderate changes in potato production at the global scale. Some areas, such as Rwanda, will become more vulnerable, but others such as Nepal may see a reduction in the risk of late blight. In some areas, farmers will be able to plant potatoes earlier in the season, likely mitigating disease risk from late blight.

Wild systems may experience dramatic long-term impacts of climate change, particularly in forests, where host life-spans are long, and the abilities of the host to migrate and to evolve will both be slow (Zhu et al. 2012). Predicting the outcomes of climate change is difficult. Climate change in wild systems will affect plant–pathogen interactions not only because both pathogen and host distributions will track their physiological optima, but also via ecological changes like increases or decreases in local host abundance, which may respond to different climate factors or other species interactions such as plant competition or herbivory. For example, in the mountains of the Sierra Nevada of California, climate change from 1996 to 2016 shifted the climate optimum of *Cronartium ribicola*, the causal agent of white pine blister rust, upslope (Dudney et al. 2021). However, field surveys found a substantial overall decrease in mean disease prevalence; Dudney and colleagues suggest that this paradox could be explained by reduced infection risk driven by both increasing aridity and lower alternate host abundance.

While it can be overwhelming to contemplate all the possible responses of plant–pathogen interactions to climate change, and while it is impossible to predict them in aggregate, we do generally understand the fundamental biology behind these responses. Projecting the effects of climate change on particular disease systems draws on everything we have covered in this book, from basic plant physiology and pathogen life cycles, to epidemiology and the disease triangle, to rates of evolution and community ecology. Shaw and Osborne (2011) point out that normal weather patterns often have annual variation that exceeds changes described in the IPCC projections, and current agricultural practices already respond to factors that change even more quickly than projected rates of climate change. In this view, crop breeding and adoption of new agronomic methods should be able to keep up with climate change, at least for annual crops. Approaches to disease management in forests and wildlands that incorporate novel interactions over large geographic scales may open pathways to resilience. In other words, rather than an entirely new approach to disease management, human adaptation to climate change in agriculture will rely on ingenuity and people with broad training in plant pathology and disease ecology.

Further reading

Lockwood, J. L., M. F. Hoopes, and M. P. Marchetti. 2013. *Invasion Ecology*. John Wiley & Sons, Hoboken, NJ.

References

Ainsworth, E. A. and S. P. Long. 2021. 30 years of free-air carbon dioxide enrichment (FACE): What have we learned about future crop productivity and its potential for adaptation? *Global Change Biology* **27**:27–49.

Anagnostakis, S. L. 1987. Chestnut blight the classical problem of an introduced pathogen. *Mycologia* **79**: 23–37.

Aung, K., Y. Jiang, and S. Y. He. 2018. The role of water in plant–microbe interactions. *The Plant Journal* **93**: 771–780.

Bearchell, S. J., B. A. Fraaije, M. W. Shaw, and B. D. Fitt. 2005. Wheat archive links long-term fungal pathogen population dynamics to air pollution. *Proceedings of the National Academy of Sciences of the United States of America* **102**:5438–5442.

Bebber, D. P., M. A. Ramotowski, and S. J. Gurr. 2013. Crop pests and pathogens move polewards in a warming world. *Nature Climate Change* **3**:985–988.

Beck, H. E., N. E. Zimmermann, T. R. McVicar, N. Vergopolan, A. Berg, and E. F. Wood. 2018. Present and future Köppen–Geiger climate classification maps at 1-km resolution. *Scientific Data* **5**:1–12.

Blossey, B. and R. Nötzold. 1995. Evolution of increased competition ability in invasive nonindigenous plants: a hypothesis. *Journal of Ecology* **83**:887–889.

Boland, G., M. Melzer, A. Hopkin, V. Higgins, and A. Nassuth. 2004. Climate change and plant diseases in Ontario. *Canadian Journal of Plant Pathology* **26**:335–350.

Brewer, S. 2018. The life and death (and life) of the American chesnut. https://indiancountrytoday.com/archive/life-death-life-american-chestnut Indian Country Today.

Bufford, J. L., P. E. Hulme, B. A. Sikes, J. A. Cooper, P. R. Johnston, and R. P. Duncan. 2016. Taxonomic similarity, more than contact opportunity, explains novel plant–pathogen associations between native and alien taxa. *New Phytologist* **212**:657–667.

Burdon, J. J. and J. Zhan. 2020. Climate change and disease in plant communities. *PLoS Biology* **18**:e3000949.

Cadotte, M. W., S. E. Campbell, S.-p. Li, D. S. Sodhi, and N. E. Mandrak. 2018. Preadaptation and naturalization of nonnative species: Darwin's two fundamental insights into species invasion. *Annual Review of Plant Biology* **69**:661–684.

Chen, J., J. F. Franklin, and T. A. Spies. 1992. Vegetation responses to edge environments in old-growth Douglas-fir forests. *Ecological Applications* **2**:387–396.

Coley, P. D., J. P. Bryant, and F. S. Chapin, III. 1985. Resource availability and plant antiherbivore defense. Science **230**:895–899.

Daehler, C. C. 2001. Darwin's naturalization hypothesis revisited. *American Naturalist* **158**:324–330.

Damschen, E. I., L. A. Brudvig, M. A. Burt, R. J. Fletcher, N. M. Haddad, D. J. Levey, J. L. Orrock, J. Resasco, and J. J. Tewksbury. 2019. Ongoing accumulation of plant diversity through habitat connectivity in an 18-year experiment. *Science* **365**:1478–1480.

Darwin, C. 1859. *On the Origin of Species by Means of Natural Selection, or, The Preservation of Favoured Races in the Struggle for Life*. J. Murray, London.

DeBach, P. and D. Rosen. 1991. *Biological Control by Natural Enemies*. Cambridge University Press.

Desprez-Loustau, M.-L., C. Robin, G. Reynaud, M. Déqué, V. Badeau, D. Piou, C. Husson, and B. Marçais. 2007. Simulating the effects of a climate-change scenario on the geographical range and activity of forest-pathogenic fungi. *Canadian Journal of Plant Pathology* **29**: 101–120.

Desprez-Loustau, M. L., B. Marcais, L. M. Nageleisen, D. Piou, and A. Vannini. 2006. Interactive effects of drought and pathogens in forest trees. *Annals of Forest Science* **63**:597–612.

Diagne, C., B. Leroy, A.-C. Vaissière, R. E. Gozlan, D. Roiz, I. Jarić, J.-M. Salles, C. J. Bradshaw, and F. Courchamp. 2021. High and rising economic costs of biological invasions worldwide. *Nature* **592**:571–576.

Dudney, J., C. E. Willing, A. J. Das, A. M. Latimer, J. C. Nesmith, and J. J. Battles. 2021. Nonlinear shifts in infectious rust disease due to climate change. *Nature Communications* **12**:1–13.

Dudney, J. C., J. C. Nesmith, M. C. Cahill, J. E. Cribbs, D. M. Duriscoe, A. J. Das, N. L. Stephenson, and J. J. Battles. 2020. Compounding effects of white pine blister rust, mountain pine beetle, and fire threaten four white pine species. *Ecosphere* **11**:e03263.

Elad, Y. and I. Pertot. 2014. Climate change impacts on plant pathogens and plant diseases. *Journal of Crop Improvement* **28**:99–139.

Elton, C. S. 1958. *The Ecology of Invasions by Animals and Plants*. University of Chicago Press.

Executive Order No. 13112. 1999. Invasive Species. N. F. United States Federal Register Volume 64.

Fahey, C., A. Koyama, and P. M. Antunes. 2022. Vulnerability of non-native invasive plants to novel pathogen attack: do plant traits matter? *Biological Invasions* 24:3349–3379.

Fine, P. V. 2002. The invasibility of tropical forests by exotic plants. *Journal of Tropical Ecology* **18**:687–705.

Flory, S. L. and K. Clay. 2013. Pathogen accumulation and long-term dynamics of plant invasions. *Journal of Ecology* **101**:607–613.

Fortini, L. B., L. R. Kaiser, L. M. Keith, J. Price, R. F. Hughes, J. D. Jacobi, and J. Friday. 2019. The evolving threat of Rapid 'Ōhi'a Death (ROD) to Hawai'i's native ecosystems and rare plant species. *Forest Ecology and Management* **448**:376–385.

Fry, W., A. Apple, and J. Bruhn. 1983. Evaluation of potato late blight forecasts modified to incorporate host resistance and fungicide weathering. *Phytopathology* **73**:1054–1059.

Futai, K. 2013. Pine wood nematode, *Bursaphelenchus xylophilus*. *Annual Review of Phytopathology* **51**:61–83.

Garrett, K. A., S. P. Dendy, E. E. Frank, M. N. Rouse, and S. E. Travers. 2006. Climate change effects on plant disease: Genomes to ecosystems. *Annual Review of Phytopathology* **44**:489–509.

Ghini, R., W. Bettiol, and E. Hamada. 2011. Diseases in tropical and plantation crops as affected by climate changes: current knowledge and perspectives. *Plant Pathology* **60**:122–132.

Gilbert, G. S. and I. M. Parker. 2010. Rapid evolution in a plant–pathogen interaction and the consequences for introduced host species. *Evolutionary Applications* **3**: 144–156.

Gordon, T., G. Reynolds, S. Kirkpatrick, A. Storer, D. Wood, D. Fernandez, and B. McPherson. 2020. Monterey pine forest made a remarkable recovery from pitch canker. *California Agriculture* **74**: 169–173.

Hanley, N. and M. Roberts. 2019. The economic benefits of invasive species management. *People and Nature* **1**: 124–137.

Harrison, J. 1992. Effects of the aerial environment on late blight of potato foliage—a review. *Plant Pathology* **41**:384–416.

Heger, T. and J. M. Jeschke. 2014. The enemy release hypothesis as a hierarchy of hypotheses. *Oikos* **123**: 741–750.

Hepting, G. H. 1974. Death of the American chestnut. *Journal of Forest History* **18**:60–67.

Hess, G. R. 1994. Conservation corridors and contagious disease: a cautionary note. *Conservation Biology* **8**: 256–262.

Hinz, H. L., R. L. Winston, and M. Schwarzländer. 2019. How safe is weed biological control? A global review of direct nontarget attack. *Quarterly Review of Biology* **94**: 1–27.

Huber, D. and R. Watson. 1974. Nitrogen form and plant disease. *Annual Review of Phytopathology* **12**:139–165.

Jenkins, M. A. and P. S. White. 2002. *Cornus florida* L. mortality and understory composition changes in western Great Smoky Mountains National Park. *Journal of the Torrey Botanical Society* **129**:194–206.

Karnosky, D., K. E. Percy, B. Xiang, B. Callan, A. Noormets, B. Mankovska, A. Hopkin, J. Sober, W. Jones, and R. Dickson. 2002. Interacting elevated CO_2 and tropospheric O_3 predisposes aspen (*Populus tremuloides* Michx.) to infection by rust (*Melampsora medusae* f. sp. *tremuloidae*). *Global Change Biology* **8**:329–338.

Keriö, S., H. Daniels, M. Gómez-Gallego, J. Tabima, R. Lenz, K. Søndreli, N. Grünwald, N. Williams, R. Mcdougal, and J. LeBoldus. 2020. From genomes to forest management—tackling invasive *Phytophthora* species in the era of genomics. *Canadian Journal of Plant Pathology* **42**:1–29.

Kobayashi, T., K. Ishiguro, T. Nakajima, H. Kim, M. Okada, and K. Kobayashi. 2006. Effects of elevated atmospheric CO_2 concentration on the infection of rice blast and sheath blight. *Phytopathology* **96**:425–431.

Kozanitas, M., M. R. Metz, T. W. Osmundson, M. S. Serrano, and M. Garbelotto. 2022. The epidemiology of sudden oak death disease caused by *Phytophthora ramorum* in a mixed bay laurel-oak woodland provides important clues for disease management. *Pathogens* **11**:250.

Krishnadas, M. and L. S. Comita. 2018. Influence of soil pathogens on early regeneration success of tropical trees varies between forest edge and interior. *Oecologia* **186**:259–268.

Laine, A. L. and I. Hanski. 2006. Large-scale spatial dynamics of a specialist plant pathogen in a fragmented landscape. *Journal of Ecology* **94**:217–226.

Li, H.-R., H.-M. Xiang, J.-W. Zhong, X.-Q. Ren, H. Wei, J.-E. Zhang, Q.-Y. Xu, and B.-L. Zhao. 2020. Acid rain increases impact of rice blast on crop health via inhibition of resistance enzymes. *Plants* **9**:881.

Liu, X., S. Lyu, D. Sun, C. J. Bradshaw, and S. Zhou. 2017. Species decline under nitrogen fertilization increases community-level competence of fungal diseases. *Proceedings of the Royal Society of London, Series B: Biological Sciences* **284**:20162621.

Lonsdale, W. M. 1999. Global patterns of plant invasions and the concept of invasibility. *Ecology* **80**:1522–1536.

Lowe, S., M. Browne, S. Boudjelas, and M. De Poorter. 2000. 100 of the world's worst invasive alien species: a selection from the global invasive species database.

Invasive Species Specialist Group, Auckland. www.issg.org/booklet.pdf.

Mack, R. N. 1996. Biotic barriers to plant naturalization. In: V. C. Moran and J. H. Hoffman, editors. *Proceedings of the IX International Symposium on Biological Control of Weeds*, University of Cape Town, Cape Town, South Africa, pp. 39–46.

Malmstrom, C. M. and H. M. Alexander. 2016. Effects of crop viruses on wild plants. *Current Opinion in Virology* **19**:30–36.

Maron, J. L. and M. Vilà. 2001. When do herbivores affect plant invasion? Evidence for the natural enemies and biotic resistance hypotheses. *Oikos* **95**:361–373.

McEvoy, P. B. and E. M. Coombs. 1999. Biological control of plant invaders: regional patterns, field experiments, and structured population models. *Ecological Applications* **9**:387–401.

Meentemeyer, R. K., N. E. Rank, D. A. Shoemaker, C. B. Oneal, A. C. Wickland, K. M. Frangioso, and D. M. Rizzo. 2008. Impact of sudden oak death on tree mortality in the Big Sur ecoregion of California. *Biological Invasions* **10**:1243–1255.

Mitchell, C. E. and A. G. Power. 2003. Release of invasive plants from fungal and viral pathogens. *Nature* **421**: 625–627.

Mitchell, C. E., P. B. Reich, D. Tilman, and J. V. Groth. 2003. Effects of elevated CO_2, nitrogen deposition, and decreased species diversity on foliar fungal plant disease. *Global Change Biology* **9**:438–451.

Nambiar, E., C. Harwood, and D. Mendham. 2018. Paths to sustainable wood supply to the pulp and paper industry in Indonesia after diseases have forced a change of species from acacia to eucalypts. *Australian Forestry* **81**:148–161.

Newhouse, A. E. and W. A. Powell. 2021. Intentional introgression of a blight tolerance transgene to rescue the remnant population of American chestnut. *Conservation Science and Practice* **3**:e348.

Parker, I. M. 2000. Invasion dynamics of *Cytisus scoparius*; A matrix model approach. *Ecological Applications* **10**:726–743.

Parker, I. M. and G. S. Gilbert. 2004. The evolutionary ecology of novel plant–pathogen interactions. *Annual Review of Ecology, Evolution, and Systematics* **35**:675–700.

Parker, I. M. and G. S. Gilbert. 2007. When there is no escape: The effects of natural enemies on native, invasive, and noninvasive plants. *Ecology* **88**:1210–1224.

Parker, I. M., M. Saunders, M. Bontrager, A. P. Weitz, R. Hendricks, R. Magarey, K. Suiter, and G. S. Gilbert. 2015. Phylogenetic structure and host abundance drive disease pressure in communities. *Nature* **520**:542–544.

Parker, I. M., D. Simberloff, W. M. Lonsdale, K. Goodell, M. Wonham, P. M. Kareiva, M. Williamson, B. Von Holle, P. B. Moyle, J. E. Byers, and L. Goldwasser. 1999. Impact: towards a framework for understanding the ecological effects of invaders. *Biological Invasions* **1**:3–19.

Pedlar, J. H., D. W. McKenney, E. Hope, S. Reed, and J. Sweeney. 2020. Assessing the climate suitability and potential economic impacts of Oak wilt in Canada. *Scientific Reports* **10**:1–12.

Ploetz, R. C. 2015. Fusarium wilt of banana. *Phytopathology* **105**:1512–1521.

Powell, W. A., A. E. Newhouse, and V. Coffey. 2019. Developing blight-tolerant American chestnut trees. *Cold Spring Harbor Perspectives in Biology* **11**:a034587.

Précigout, P.-A., D. Claessen, D. Makowski, and C. Robert. 2020. Does the latent period of leaf fungal pathogens reflect their trophic type? A meta-analysis of biotrophs, hemibiotrophs, and necrotrophs. *Phytopathology* **110**:345–361.

Reinhart, K. O., A. Packer, W. H. Van der Putten, and K. Clay. 2003. Plant–soil biota interactions and spatial distribution of black cherry in its native and invasive ranges. *Ecology Letters* **6**:1046–1050.

Rizzo, D. M. and M. Garbelotto. 2003. Sudden oak death: endangering California and Oregon forest ecosystems. *Frontiers in Ecology and the Environment* **1**:197–204.

Sequeira, L. and C. Averre III. 1961. Distribution and pathogenicity of strains of *Pseudomonas solanacearum* from virgin soils in Costa Rica. *Plant Disease Reporter* **45**:435–440.

Shapiro, L. R., J. N. Paulson, B. J. Arnold, E. D. Scully, O. Zhaxybayeva, N. E. Pierce, J. Rocha, V. Klepac-Ceraj, K. Holton, and R. Kolter. 2018. An introduced crop plant is driving diversification of the virulent bacterial pathogen *Erwinia tracheiphila*. *Mbio* **9**:e01307-18.

Shaw, M. W. and T. M. Osborne. 2011. Geographic distribution of plant pathogens in response to climate change. *Plant Pathology* **60**:31–43.

Shea, K. and D. Kelly. 1998. Estimating biocontrol agent impact with matrix models: *Carduus nutans* in New Zealand. *Ecological Applications* **8**:824–832.

Simberloff, D. and J. Cox. 1987. Consequences and costs of conservation corridors. *Conservation Biology* **1**:63–71.

Soluri, J. 2002. Accounting for taste: Export bananas, mass markets, and Panama disease. *Environmental History* **7**:386–410.

Sparks, A. H., G. A. Forbes, R. J. Hijmans, and K. A. Garrett. 2014. Climate change may have limited effect on global risk of potato late blight. *Global Change Biology* **20**: 3621–3631.

Sullivan, L. L., B. L. Johnson, L. A. Brudvig, and N. M. Haddad. 2011. Can dispersal mode predict corridor effects on plant parasites? *Ecology* **92**:1559–1564.

Tibpromma, S., Y. Dong, S. Ranjitkar, D. A. Schaefer, S. C. Karunarathna, K. D. Hyde, R. S. Jayawardena, I. S.

Manawasinghe, D. P. Bebber, and I. Promputtha. 2021. Climate–fungal pathogen modeling predicts loss of up to one-third of tea growing areas. *Frontiers in Cellular and Infection Microbiology* **11**:610567.

Vandermast, D. and D. Van Lear. 2002. Riparian vegetation in the southern Appalachian mountains (USA) following chestnut blight. *Forest Ecology and Management* **155**:97–106.

Vargas-Gaete, R., H. Doussoulin, C. Smith-Ramírez, S. Bravo, C. Salas-Eljatib, N. Andrade, and B. Trávníček. 2019. Evaluation of rust pathogenicity (*Phragmidium violaceum*) as a biological control agent for the invasive plant *Rubus ulmifolius* on Robinson Crusoe Island, Chile. *Australasian Plant Pathology* **48**:201–208.

Veresoglou, S., E. Barto, G. Menexes, and M. Rillig. 2013. Fertilization affects severity of disease caused by fungal plant pathogens. *Plant Pathology* **62**:961–969.

Vilà, M., J. L. Espinar, M. Hejda, P. E. Hulme, V. Jarošík, J. L. Maron, J. Pergl, U. Schaffner, Y. Sun, and P. Pyšek. 2011. Ecological impacts of invasive alien plants: a meta-analysis of their effects on species, communities and ecosystems. *Ecology Letters* **14**:702–708.

Whitaker, B. K., M. A. Rua, and C. E. Mitchell. 2015. Viral pathogen production in a wild grass host driven by host growth and soil nitrogen. *New Phytologist* **207**: 760–768.

Wilson, B., K. Annett, W. Laidlaw, D. Cahill, M. Garkaklis, and L. Zhuang-Griffin. 2020. Long term impacts of *Phytophthora cinnamomi* infestation on heathy woodland in the Great Otway National Park in south-eastern Australia. *Australian Journal of Botany* **68**: 542–556.

World Meteorological Organization. 2021. State of the Global Climate 2020. WMO-No. 1264.

Zhu, K., C. W. Woodall, and J. S. Clark. 2012. Failure to migrate: lack of tree range expansion in response to climate change. *Global Change Biology* **18**:1042–1052.

CHAPTER 17

Disease management

Gregory S. Gilbert and Ingrid M. Parker

17.1 Introduction

Agricultural losses to plant disease are massive: for the five major crops that contribute 48% of global human caloric intake, Savary et al. (2019) calculated an estimated 19% annual crop loss to disease (wheat 19.8%; rice 21.0%; maize 18.8%; potato 16.0%; soybean 21.2%). Given the 2020 global production and market values, this represents an annual loss of $297 billion for just these five crops.[1] Forest ecosystems are similarly affected, and disease impacts are only growing; the UK alone faces roughly one outbreak of a novel forest pathogen every year (Roy et al. 2014). In a world that struggles with food security and environmental degradation, these challenges are consequential. Managing plant disease is not just economically important, it is also an ethical imperative.

Managing plant disease has similarities to how we manage infectious diseases of people, but also important differences. Society relies on public health measures to prevent and control the spread of human pathogens. We reduce our exposure: we inspect, wash, and cook our food, we avoid contact with contaminated bodily fluids, we build elaborate sewage systems, we use clothing and repellent to evade vectors like mosquitoes and ticks, we use masks to block aerial transmission, and we try to avoid or isolate disease outbreaks. We bolster our natural pathogen resistance by ensuring our bodies are well nourished and our microbiomes are healthy, and by activating our immune systems through vaccines. But when we do become ill, we count on medical interventions for a cure. This is where plant disease management diverges from its human counterpart. Although it is sometimes possible to cure an individual plant of infection through pruning or chemical application, intensive therapeutic approaches are seldom practical or effective means to control plant diseases. Therefore, plant disease management is practiced in a framework analogous to public health, one that stresses limiting contact with pathogens, working to prevent infection and to slow disease development and spread. We do this through a combination of cultural practices, breeding for resistance, and the use of agrochemicals and biological controls. Effective management practices derive from our understanding of the evolutionary ecology of plant disease, including the disease triangle (Chapter 1), pathogen life histories (Chapters 2–8), population growth and epidemiology (Chapters 10–11), plant immune systems and the evolution of virulence and susceptibility (Chapters 12–13), and the ecological context that moderates disease pressure (Chapter 14), including microbiomes that defend against pathogens (Chapter 15). Following the principles of evolutionary ecology can help create sustainable agricultural systems that favor the growth of desired plants while limiting the growth, spread, and impacts of plant pathogens.

We can think of plant disease management as being made up of two general *strategies*: avoidance and control; these are implemented through

[1] Wheat 775.9 Million Mg (Mg = megagram = metric ton) @ $221/Mg; Rice 505.4 M Mg @ $625/Mg; soybean 383 M Mg @ $407/Mg; maize 2792 M Mg @ $165/Mg; potato 359 M Mg @ $323/mt; Annual loss billion $US = $US per Mg/1000 * (1/(1 – % loss/100) * M Mg – M Mg). $297 billion is similar (in 2020 $US) to the often cited $220 billion attributed to the FAO but that was based on an estimate by Agrios (2004), p. 4.

The Evolutionary Ecology of Plant Disease. Gregory S. Gilbert and Ingrid M. Parker, Oxford University Press. © Gregory S. Gilbert & Ingrid M. Parker (2023).
DOI: 10.1093/oso/9780198797876.003.0017

Table 17.1 Strategies (bold) and tactics (plain font) for plant disease management

Avoidance	**Control**
Evasion	Resistance
Avoid geographic range of pathogen	Single-gene resistance
Avoid environmental conditions	Gene pyramiding
Soil testing for pathogen	Multilines
Adjust planting timing	Nutritional support
Exclusion	Diversity
Seed and stock certification	Intercropping
Phytosanitary regulations	Crop rotation
Best phytosanitary practices	Cultural practices
Pest-free zone certification	Plant spacing
Quarantine	Sanitation, tillage, and mulch
	Irrigation management
	Soil temperature and aeration
	Suppressive soils
	Preventative and curative applications
	Chemical control
	Biological control

diverse *tactics* that comprise smaller-scale, immediate actions (Table 17.1). The first strategy of managing plant disease is **avoidance**, keeping the plant and pathogen from interacting in a conducive environment. Because both pathogens and environmental conditions are geographically patchy, disease can be avoided by **evasion** (choosing where and when to plant a host to avoid pathogens) and by **exclusion** (keeping pathogens out of places where it is not present). When avoidance is not possible, the **control** strategy kicks in, taking measures to keep the pathogen population transmission rate $R < 1$ (Section 10.6). Such control can be accomplished by manipulating the host to make it unsuitable for pathogen infection or reproduction (e.g., through **genetic resistance**), by implementing **cultural practices** that interfere with survival, growth, reproduction, or spread of the pathogen (e.g., through sanitation, managing crop spacing and diversity, or soil and water management), or by protecting the host through **chemical pesticides** or **biological controls** that directly target pathogens or enhance host defenses. Here we explore how the evolutionary ecology of plant disease informs the contrasting strategies of avoidance and control, and their component tactics (Zhan et al. 2014). Underlying all decisions in plant disease management are tradeoffs that balance social, environmental, and financial considerations, agriculture's "triple bottom line" (He et al. 2021).

17.2 Biogeography and the disease triangle

Because most pathogen and plant species have limited biogeographic ranges, most disease interactions that *could* occur do not, simply because they do not have the opportunity. Even crops that are planted globally encounter different suites of pathogens in different regions (Figure 17.1). Patchiness in the distribution of pathogens occurs at much smaller scales as well. Such patchiness provides management opportunities to avoid disease; maintaining patchiness is itself an important goal. After all, if all plant pathogens were everywhere, it would truly be a disaster. The principle of maintaining patchiness can be applied both locally and globally.

17.2.1 Pathogen exclusion

This valuable patchiness is maintained by active exclusion of pathogens from places where they are not. At a local level, careful sanitation practices help prevent movement of infested soil or plant material into non-infested areas on vehicles, machinery,

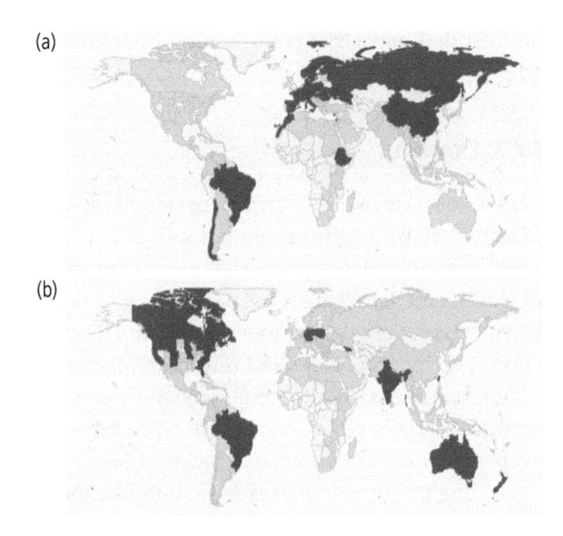

Figure 17.1 Reported biogeographic ranges (purple) of two fungal pathogens of garden pea (*Pisum sativum*): (a) *Neocosmospora pisi* and (b) *Uromyces pisi*. Light blue indicates countries or states with pea cultivation and a report of any fungal pathogen ($n = 1700$).
Maps by Gregory S. Gilbert based on data from the USDA ARS Fungal-Host Distributions Database.

tools, or boots (Section 17.5.2). Sanitation practices can limit pathogen introduction from known disease foci and contain localized outbreaks, but are less effective for polycyclic pathogens or those with long-distance aerial dispersal.

Many crop pathogens can be transmitted on seeds (or on the clonal propagules of vegetative crops), so planting pathogen-free seed is a key tool for avoiding disease. The system we have for controlling pathogens transmitted through seed and vegetative propagules is **certified seed production**. Certification generally includes confirming varietal traits and testing that seeds are clean of chemical contamination, pests, and pathogens. For instance, viruses and viroids of potatoes are frequently transmitted through seed potatoes, which are actually small potatoes (not actual seeds) used to generate potato plants. In the United States, seed potatoes are grown in 14 select regions where critical diseases are absent or readily controlled (Sun et al. 2004). The seed potatoes are then certified as pathogen-free through visual and molecular assessments, along with aggressive control measures such as **rogueing** of symptomatic plants, and distributed to potato-growing regions to produce potatoes for consumption. Rigorous seed certification can be highly effective. The *Potato spindle tuber viroid* (the first viroid recognized by plant pathologists, and once a globally important disease) is transmitted primarily through propagation of tubers and cuttings, and then subsequently by aphids. Effective potato seed certification programs have eradicated the disease from the United States (Sun et al. 2004).

Restoration of degraded natural areas often relies on seedlings of native plants that are propagated in nurseries and planted into restoration sites. Such practice has led to the inadvertent introduction of pathogens present on nursery stock into the vulnerable habitats undergoing restoration (Frankel et al. 2020). The introduced pathogens threaten not only the transplants themselves but also surrounding native vegetation. This has spurred the development of best practices for nurseries along with nursery certification networks focused on production of native plants for safe restoration efforts.

17.2.2 Phytosanitary measures

Global trade is steadily eroding the patchiness in the global biogeography of plant pathogens (Chapter 16). **Phytosanitary** regulations and practices are important tools to slow this erosion by excluding novel pathogens from areas in which they are not found and restricting the spread of emergent pathogens from areas with active disease. They are complicated by the big economic and political implications of restricting trade. Phytosanitary practices are governed by a hierarchical collection of state, national, regional, and global phytosanitary organizations, regulations, and agreements (Magarey et al. 2009).

Global phytosanitary governance can seem like a bowl of alphabet soup. The **International Plant Protection Convention** (**IPPC**) is an intergovernmental treaty that sets out the **International Standards for Phytosanitary Measures** (**ISPMs**), tools to protect global plant resources from the spread of pests and pathogens through trade (https://www.ippc.int). The IPPC is the source for phytosanitary guidelines recognized by the **World Trade Organization** (**WTO**), and it includes more than 180 countries as signatories. The IPPC is implemented by the **Commission on Phytosanitary Measures (CPM)**, which is hosted by the **Food and Agriculture Organization (FAO)** of the United Nations, based in Rome. It sets out measures, protocols, and phytosanitary treatments with the goals of protecting sustainable agriculture, food security, forests, biodiversity, and the environment, while facilitating economic and trade development. The IPPC serves as a coordinating clearing house for exchange of information among ten Regional Plant Protection Organizations (RPPO) such as the North American Plant Protection Organization (NAPPO), the European Plant Protection Organization (EPPO), and the mesoamerican *Organismo Internacional Regional de Sanidad Agropecuaria* (OIRSA), as well as the many National Plant Protection Organizations (NPPOs). The NPPO in the United States is the Animal and Plant Health Inspection Service (APHIS), which is part of the US Department of Agriculture; individual states also have governmental phytosanitary bodies that coordinate with APHIS.

The regional plant protection organizations play important roles in facilitating the safe global movement of agricultural goods (Magarey et al. 2009); they establish how to apply phytosanitary measures in international trade (ISPM 01), create frameworks for pest risk analysis (ISPM 02), generate guidelines for the release of biological control agents (ISPM 03), and coordinate a phytosanitary certification system (ISPM 07). They are responsible for developing lists of pests and pathogens that pose an economic or environmental danger, conducting surveillance of pest occurrence and absence in their regions, inspecting incoming goods, providing phytosanitary certificates for outgoing goods, and taking actions to eradicate emergent pests and pathogens. They can impose **quarantines** by defining infested areas from which potentially infected materials cannot be moved without appropriate treatment, or document **pest-free zones** where a pathogen is shown not to be present. They have limited powers, however. Signatories to the IPPC may only take actions necessary to protect plants from regulated pests and pathogens; the WTO, which operates global rules of trade and settles trade disputes, ensures that the NPPOs do not act in arbitrary or unjustified ways that could inhibit international trade.

17.2.3 Evasion

Evasion, as one of the tactics of an avoidance strategy, can be useful at both local and global scales. Spatial evasion of disease can be an effective management tool at the local scale for soil-borne pathogens, because their primary inoculum tends to be both patchy and long-lived (Figure 17.2). Growers can use bioassays to test soils for pathogen inoculum, which when combined with growers' site-specific knowledge, helps to reduce disease risk by avoiding planting in places with high likelihood of disease impacts (Muramoto et al. 2022). Those places with dense inoculum can then be planted with crops that are not susceptible.

Temporal evasion of disease is sometimes possible by planting crops early or delaying their planting to offset vulnerable growth stages from when pathogens are most active. For instance, in organic potato cultivation in the UK and the Netherlands, planting seed tubers early in the season allows the crop to produce harvestable potatoes before late blight (caused by the Oomycete *Phytophthora*

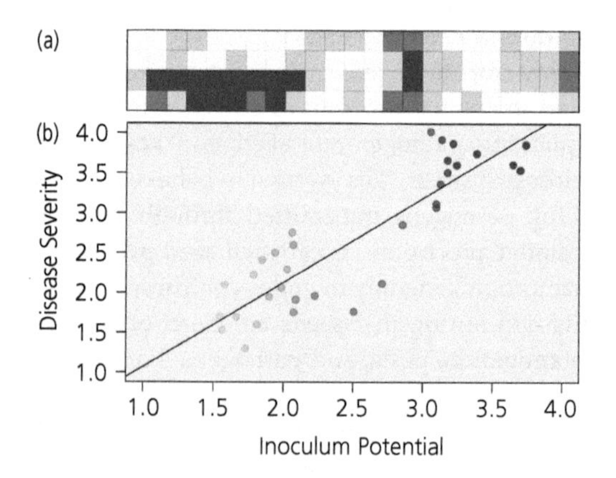

Figure 17.2 Variation in primary inoculum of soil-borne pathogens affects disease patterns. (a) Spatial variation in inoculum potential of the soilborne oomycete pathogen *Aphanomyces euteiches* across a ~1-ha field shows patchiness of primary inoculum. (b) Soil inoculum potential (intensity of purple) at a location in the field is a strong predictor of disease severity caused by *A. euteiches* on garden pea (*Pisum sativum*).

Figure by Gregory S. Gilbert based on data from Moussart et al. (2009).

infestans) substantially harms the plants (Hospers-Brands et al. 2008). In contrast, in Isfahan province, Iran, delaying planting of potatoes by a month significantly decreased severity of early blight (caused by the Ascomycete *Alternaria solani*) (Nasr-Esfahani 2022). While planting date can have a major impact on disease development, delayed planting can also come at a cost in yield because of a shortened growing season, risk of frost damage, etc. The vagaries of weather that affect the timing and severity of pathogen outbreaks make it difficult to get the temporal evasion of disease just right, especially when simultaneously managing multiple pathogens.

At a global scale, taking advantage of patchiness to avoid pathogens has been a force driving major historical changes in where commercial crops are grown. Banana cultivation in Latin America was devastated by the emergence of Fusarium wilt, caused by *Fusarium oxysporum* f.sp. *cubensis.*, a soil-borne fungus native to Indonesia (Section 16.1). In response to the emergence of the disease in Latin America, international banana companies abandoned at least 40,000 ha of infested plantations, and then purchased extensive new areas of rainforest and converted them to banana production in an attempt to evade the pathogen (Soluri 2002). Such evasion was a temporary relief, as the pathogen quickly followed into the new production areas. In the 1960s, commercial production shifted to a different, resistant clone called Cavendish, which is now threatened by a new race of the pathogen. Because bananas are clonal polyploids that do not make seed, breeding for resistance is extremely slow and difficult, so evasion and exclusion of the pathogen are essential tools in managing Fusarium wilt of banana.

Evasion of plant disease was also a major driver in a geographic shift in commercial production of natural rubber from its native Brazil to Southeast Asia. Rubber trees (*Hevea brasiliensis*) are native to the Amazon rain forest, where their thick latex was an important resource for indigenous peoples. After the late 19th century discovery of the industrial potential of the latex for tire production, there were various attempts in Brazil to grow *H. brasiliensis* in high-density plantations instead of the slow and labor-intensive harvesting from wild trees. Unfortunately, the high-density plantations suffered massive defoliating outbreaks of South American Leaf Blight (caused by the fungus *Pseudocercospora ulei*). This fungus contributed (together with insect pests, mismanagement, and labor unrest) to failure of the rubber plantations throughout Brazil. South American Leaf Blight continues to be the main impediment to commercial production of natural rubber in its native South America (Lieberei 2007). In contrast, natural rubber production thrives in Southeast Asia, where *P. ulei* is absent. Rubber production in Southeast Asia began with an act of **biopiracy** in 1876, when the British collector Henry Wickham smuggled *H. brasiliensis* seeds out of Brazil to Kew Gardens, propagated them, and then sent them on to the British colony Ceylon (now the independent nation of Sri Lanka). Production in Southeast Asia far outperformed that in Brazil and broke the Brazilian monopoly over the rubber industry. Today strict quarantine regulations help protect countries in Southeast Asia and Africa from importation of the pathogen.

17.3 Resistance

Just as wild plant populations reduce the impacts of disease primarily through the evolution of traits that confer resistance (Chapters 12 and 13), host resistance is the most important element in disease management in agricultural systems. All plants have many traits that help protect them from attack by most microorganisms, but traditional **selective breeding**, as well as much of genetic engineering, has centered on **major gene resistance**. Major genes refer to *R* genes with alleles that confer a large (also called "qualitative") effect on resistance, and which correspond to genes for pathogenicity (*Avr*) in the pathogen. Major gene resistance usually involves genes that code for pattern recognition receptors (PRR) or nucleotide-binding leucine-rich repeat (NLR) proteins that detect pathogen effectors and induce defense responses (Section 12.5). Major gene resistance is usually race-specific for particular pathogen strains and can be quickly overcome by evolution in the pathogen. Major *R* genes are most effective in breeding for resistance against biotrophic pathogens like rusts, smuts, downy mildews, and powdery mildews, because of the role of the hypersensitive response (Section 12.5).

Against necrotrophic pathogens, major gene resistance generally involves breaking susceptibility to host-specific toxins (Vleeshouwers and Oliver 2014).

17.3.1 Incorporating resistance genes into crops

All of agriculture is based on countless generations of artificial selection, in which humans select which plant genotypes are passed to the next generation by choosing which seeds to plant. Those seeds are usually from the plants with the best agronomic traits: rapid growth, tasty fruits, sturdy architecture, hardy and well suited to local conditions. Some traits are direct targets for artificial selection, while other traits are selected for by the agricultural process or just come along for the ride. This "unconscious" artificial selection can lead to either increased resistance to disease, such as in the gathering of seeds from more vigorous plants, or decreased resistance, such as in selecting against an *R*-gene variant that is negatively correlated with a color or flavor trait.

The development of permanent agricultural settlements, called the **neolithic revolution**, occurred around 10,000–12,000 years ago in the Near East, but artificial selection on edible wild plants and the beginnings of domestication probably predated the neolithic revolution by thousands of years and occurred independently in many parts of the world (Meyer et al. 2012). Indigenous and traditional agrarian communities managed and selected populations of cultivated species, giving rise to landraces. **Landraces** are locally adapted varieties, generally displaying particular traits suitable to local conditions within a diverse genetic background. Similarly, **heirloom varieties** are tended and handed down over generations of growers. However, the diversity of landraces and heirloom varieties has become threatened by globalization and the promotion of a small number of commercial elite lines, raising alarm and elevating the role of *ex situ* conservation of landrace germplasm (Ramirez-Villegas et al. 2022). Together with *ex situ* efforts, promoting the continued *in situ* conservation of landraces and heirloom varieties will be important to preserve crop diversity within its social and natural contexts (Raggi et al. 2022).

Throughout the history of agriculture, the desire to speed up evolution and to mold crop traits led to the development of a series of plant-breeding technologies. These include selective breeding, which is the mating of selected parents to create a particular combination of traits; transgenic technology, which uses genetic engineering to move genes from any source organism into a cultivar; and gene editing, which employs tools such as CRISPR-Cas9 to modify genes already found in the cultivar to express novel traits.

Backcross breeding

Often, a desired trait (e.g., resistance to a particular pathogen) is present in a crop genotype or wild relative species that otherwise has a poor profile of agronomic traits. Plant breeders create an F1 hybrid between a donor parent with the desired trait (such as an *R* gene) and an elite parent with desired agronomic attributes, and then use backcross breeding to get rid of undesirable traits from the donor parent. For each generation of breeding, the presence of the *R* genes in progeny is scored by eliciting the hypersensitive response or by the presence of linked molecular markers. In each backcross step, the proportion of elite alleles increases across the entire genome (Figure 17.3). The proportion of alleles from the elite parent in each generation can be calculated as $1 - 0.5^{n+1}$, where n = number of generations. Breeding programs often recommend six generations of backcrossing, at which point the genome will have 99.2% of elite-parent alleles, on average.

Genetic transformation

Backcross breeding has limitations. Hybridization is often only possible between close relatives, is not available for crops like bananas that do not reproduce through seed, and working through six generations can be slow, especially for perennial plants such as trees. Following the identification of useful resistance genes (Box 17.1), breeders have developed new ways to move specific genes from any source into elite cultivar genotypes using recombinant DNA technology, creating what is labeled a **transgenic** crop (Van Esse et al. 2020). When a resistance gene is identified in a potential donor organism, the first step is to isolate the DNA sequence and introduce many copies of it into the recipient plant.

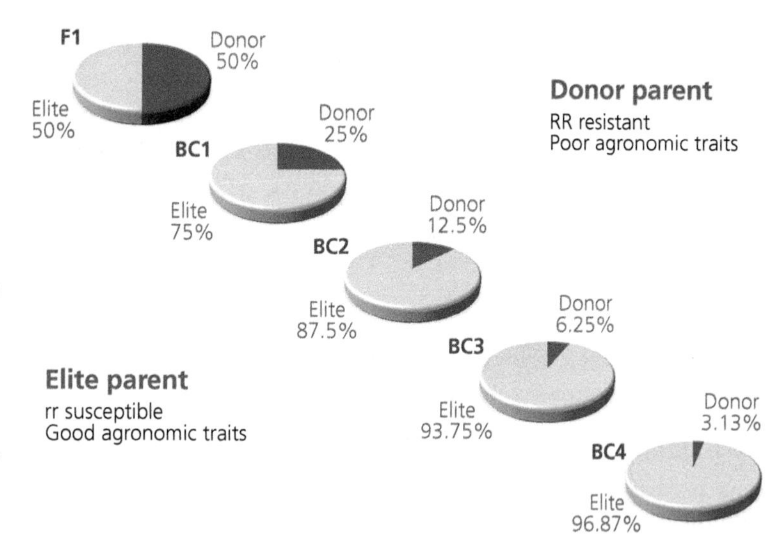

Figure 17.3 Back cross breeding to move a resistance gene from a donor (red) genotype that carries a desired resistance gene (*R*) into an elite cultivar line (light blue), which is susceptible (*rr*) but with desired agronomic traits. In each generation, the resistant phenotype is selected for continued breeding, and the other donor alleles decline in frequency by 50%.
Figure by Ingrid M. Parker.

Those many copies of the sequence were traditionally made by cloning into a bacterial plasmid, and then growing the bacterium in culture to make millions of copies. Now, the copies of the gene sequence are often made directly by amplification from the host genome through polymerase chain reaction (PCR; Box 9.3), skipping the bacterial cloning phase.

Moving that cloned gene into the plant usually takes one of two pathways: biolistics or *Agrobacterium* transformation. **Biolistics**, or particle bombardment, coats very small particles of gold or tungsten with copies of the DNA sequence for the resistance gene and uses a gene gun to shoot them, with great force, into undifferentiated embryonic plant callus grown on sterile agar medium with a precise ratio of auxin and cytokinin (Section 2.7). Some of the DNA-coated particles penetrate the plant nucleus, and some of those DNA sequences integrate into the nuclear DNA of the plant—often in multiple places in the genome. After biolistics, more cytokinins are added to prompt the callus to produce shoots, which are then transferred to medium with auxin to stimulate root formation, creating plantlets. The plantlets are screened for the transgene; the successful ones are transgenic plants, containing the resistance gene from the genome of another organism.

The other major pathway for creating transgenic plants is by taking advantage of the natural transformation abilities of the Ti plasmid in the pathogenic bacterium ***Agrobacterium tumefaciens*** (more correctly, but less commonly called *Agrobacterium radiobacter*, Figure 5.7). The cloned resistance gene can be inserted into the T-DNA region of the Ti plasmid, replacing the genes that would normally lead to disease development in an infected plant. The transformed *Agrobacterium* is then grown with embryonic plant callus in tissue culture, which lets the bacterium do its normal thing, transferring its T-DNA into the nuclei of plant cells. Because the T-DNA contains the cloned resistance gene, with luck that gene becomes integrated into the plant nuclear genome. The transformed plant callus is grown into plantlets in the same way as for biolistics.

Both biolistics and *Agrobacterium* transformation have been used to create transgenic plants that express a wide range of genes, from disease resistance to herbicide resistance to vitamin production. The tools are not without limitations, however. First, it is often not possible to transform an elite cultivar directly. Instead, a cultivar that performs well in the tissue-culture processes is sometimes used for the initial transformation, and then the gene has to be moved into the desired elite cultivar using the same, slow backcross breeding process described above. Second, both transformation methods lack tight controls over where, and how many times, a transgene is inserted into the host genome. This creates the possibility of undesirable nontarget effects of the transformation, requiring extensive screening and testing to assess the precise changes produced during the transformation. Finally, some people are concerned about the ecological risks and the ethics of releasing transgenic crops, also called **genetically modified organisms (GMOs)**. This has led to regulation, labeling, and sometimes banning of these crops (McHughen 2016).

Gene editing

Newer gene editing technologies promise to avoid some of the concerns involving GMOs by manipulating the plant genome without necessarily inserting genes from other organisms. The most prominent is that of **CRISPR-Cas nucleases** (along with zinc-finger nucleases (ZFNs) and transcription activator-like effector nucleases (TALENs)) (Lowder et al. 2016). The CRISPR-Cas systems derive from important antiviral systems in bacteria. They include two functional parts: the Cas endonuclease enzyme that cuts DNA, and the guide RNA (gRNA) that tells the endonuclease to cut where the DNA sequence complements its RNA sequence. The CRISPR-Cas system can then delete or replace parts of the host DNA at those cut sites—changing single base pairs or longer stretches of DNA. In this way, targeted changes in plant genes can be edited very specifically. Getting the CRISPR-Cas system into the plant nucleus can be done in a number of ways, including transforming the plant through particle bombardment or with *Agrobacterium* in which the T-DNA carries the Cas and gRNA sequences. (Since these integrate the bacterial Cas enzymes into the host genome, the resulting crop is technically transgenic.) Another approach is to package the Cas and gRNA as ssDNA or RNA into a plant virus and infect the plant cells with the virus. The virus then replicates and its genome is transcribed and translated within the plant cell, producing Cas endonuclease enzymes and gRNA, which then edit the host genome. The viral Cas and gRNA sequences themselves do not persist in the plant, leaving only the edited genes. Whether gene-edited crops should be regulated in the same ways as transgenic (GMO) crops, along with the overall role of gene editing in sustainable agriculture, are matters of continuing debate (Montenegro de Wit 2022).

RNA interference

Finally, RNA interference (**RNAi**; also called host **induced gene silencing**) is gaining traction as a promising approach to disease control because of its ability to precisely suppress the expression of particular genes in the host or an attacking pathogen (Koch and Wassenegger 2021). RNAi is a natural gene regulation mechanism found in all eukaryotes. Two kinds of small, single-stranded RNAs are produced: miRNAs (microRNA) are coded directly by the plant genome as gene regulators, and siRNAs (small interfering RNA) are produced as part of a cellular surveillance system. These small RNAs are incorporated as guide RNA (gRNA) into the RNA-induced silencing complex (RISC); the gRNAs guide the RISC to target complementary sequences of mRNA in the cell, which are bound and then cleaved or modified, preventing gene expression. This can be harnessed to protect plants against plant pathogens, especially RNA viruses, by engineering the host to produce miRNAs that serve as guides for RISC to degrade specific sequences in viral RNA. A papaya variety was engineered to use RNAi to recognize and silence viral coat protein RNA of the papaya ringspot virus. During a devastating outbreak of that disease in the late 1990s, the resistant transgenic variety saved the Hawaiian papaya crop (Gonsalves et al. 2004).

17.3.2 Durable resistance

In agriculture, host evolutionary dynamics are replaced by crop breeding programs and grower choices about what is planted where. Agronomic factors promote large-scale crop homogeneity, exerting strong selective pressure on pathogens that can lead to **selective sweeps** of *Avr* mutations that overcome resistance (Section 13.6). Because the protection offered by an *R* gene can be defeated by a single mutation in the corresponding *Avr* gene in the pathogen, resistance genes in crops often lack **durability**. Widespread deployment of single-gene resistance leads to **boom-and-bust cycles**, in which resistant genotypes that provide protection against aggressive pathogens generally fail after only a few years.

Box 17.1 How to find and keep track of genes associated with resistance or virulence

Genes associated with host resistance or microbe-associated molecular patterns and effectors can be identified and their chromosomal location mapped either through forward or reverse genetics (Boutrot and Zipfel 2017). Once localized, genes can be cloned, sequenced, or edited so they can then be used in breeding programs for the development of pathogen-resistant plant cultivars.

Forward genetics is the classical approach to matching genes to phenotypes; it begins with an observed phenotype, and then uses genetic tools to identify the associated gene. Finding the locus associated with a desired phenotype is conventionally done through gene mapping. A cross is done between two parents that differ both in the phenotype of interest (e.g., resistant and susceptible) as well as in numerous **genetic markers.** Originally, the genetic markers were readily observed phenotypic traits for which the locations of associated loci were already known; now the genetic markers are DNA sequence variants. A second generation of crossing produces an F2 generation in which traits and markers randomly recombine. The resulting offspring are examined for co-occurrence of the desired phenotype and the genetic markers—loci that segregate together are closer together on the chromosome. A large number of genetic markers is needed to hone in on the specific location of the target gene. Once mapped with high resolution, the gene can be cloned and sequenced.

The use of **single nucleotide polymorphisms** (SNPs) as genetic markers has revolutionized the speed and precision with which forward genetics can be done (Section 13.2). A SNP is a simple difference at a single base pair in the genome (e.g., ATAGC versus ATACC). Because genomes include millions of base pairs, there are always lots of SNPs that distinguish the two parents. Through genome sequencing, the locations of SNPs on chromosomes are known and are used to pinpoint the location of genes of interest. **Quantitative trait locus (QTL) mapping** is the application of the same approach to find multiple genes, sometimes on different chromosomes, that contribute to the expression of a complex (polygenic, also called quantitative) trait such as plant growth or fruit size. **Genome-wide association studies** are extending mapping possibilities without doing crosses by examining patterns across a large number of individuals. Statistical associations between a trait of interest (e.g., resistance) and molecular markers (generally SNP variants) suggest regions of DNA that might include genes involved in producing the trait. The researcher then uses bioinformatics to find candidate genes in those regions and designs further studies to explore the functions of those genes.

Reverse genetics differs from classical forward genetics approaches because it begins with an identified DNA sequence—perhaps one with sequence homology to a gene of known function in other organisms—and then seeks the function of that sequence in the organism. The low cost and speed of genome sequencing now makes it possible to search whole genomes of plants and pathogens for genes that are likely involved in resistance and virulence. To reveal its role, the sequence of interest is changed through a site-directed deletion or point mutation (often using CRISPR), or its expression is blocked (with RNAi gene silencing). The phenotype of the organism is then observed. Reverse genetics provides information about both gene location and gene function.

With reliable genetic markers or the genes themselves in hand, **marker-assisted selection** (Hasan et al. 2021) can readily facilitate breeding for resistance, especially for traits that are not expressed until plants mature, like adult plant resistance (APR). Large numbers of progeny can be screened for whether they will express the trait of interest by screening for the marker instead. This shortens the time for screening from weeks to days at a fraction of the cost. Multiple traits can be screened simultaneously, such as pathogen resistance and salt tolerance, or multiple resistance genes in a gene pyramiding breeding program.

Resistant cultivars with single *R* genes can initially provide effective protection against disease, but when deployed over large areas and in successive years, the pathogen will always evolve. The probability of a mutation to overcome a resistance gene is estimated at one in 100,000 to 10 million spores or cells; this sounds like a very low probability, but consider that a field moderately infested with powdery mildew can produce 10^{13} spores day^{-1} ha^{-1} (Stam and McDonald 2018). The mutation that circumvents the locally dominant *R* gene undergoes exponential increase across the genetically homogeneous population.

Therefore, *R* genes used in breeding for commercial agriculture generally have a rather short lifespan. Pathogen genotypes that evade a deployed resistance gene often appear within 2–3 years of release of the cultivar, and the spread of the virulence genotype throughout the pathogen population is so rapid that cultivars with the resistance gene are no longer useful after 5–10 years (Bockus et al. 2011). Because pathogens with polycyclic life cycles have more opportunities during each crop cycle for mutations to arise and then spread, major gene resistance against polycyclic pathogens is expected to be defeated more quickly than for monocyclic pathogens. Similarly, long-lived plants such as trees pose special challenges for deploying major-gene resistance. Their slow generation time makes breeding a logistical challenge (Pike et al. 2021), and many dozens of pathogen generations occur during the lifetime of a single tree—so they are particularly vulnerable to evolution in the pathogen to defeat whatever resistance they have.

It is possible to reduce disease pressure and extend the useful life of cultivars with particular *R* genes by releasing them sequentially. **Gene rotation** uses single-gene deployment until resistance is overcome, then retires that gene and deploys a cultivar with a different *R* gene; retired genes can be brought back into use when the frequency of the corresponding *Avr* gene in the pathogen population declines to low levels. This approach is possible but generally not very practical, because it requires intensive monitoring for gene frequencies in the pathogen population. It is also often ineffective, because for an earlier *R* gene to be reused, the *Avr* gene must be strongly selected against in the pathogen population and decline to low levels. But even when starting at low frequency, a virulent genotype will quickly spread to become the dominant genotype.

Breeding tactics to prolong the durability of resistance in crops tend to focus on the use of quantitative resistance traits, gene pyramiding, and multilines (Mundt 2014) (Figure 17.4).

17.3.3 Quantitative resistance

Increasingly, **quantitative resistance** traits are being used in breeding programs, with the benefit of greater durability. Quantitative resistance is **polygenic resistance** produced by the joint effects of several different genes that each contribute in a minor way (hence "quantitatively") to plant resistance (Niks et al. 2015). Such resistance is sometimes called **partial resistance** (phenotypically incomplete resistance) in contrast to the "yes/no" resistance of a gene-for-gene system, or it is sometimes simply referred to as **field resistance.** It may also be called **adult plant resistance** (APR), because it is often only apparent late in plant development. APR genes known as "slow-rusting genes" (such as *Sr2* or *Lr34*) provide partial resistance to rusts by prolonging the latent period, shortening the infectious period, or reducing pustule size and fitness of the pathogen. Quantitative resistance slows the spread of pathogens through the host population without inducing a hypersensitive response. Because quantitative resistance does not involve an "all-or-nothing" defense, there is less selection pressure on the pathogen to overcome resistance (Fonseca and Mysore 2019).

Mechanistically, quantitative resistance genes are often associated with signaling in defense responses, as well as traits associated with non-host resistance (Section 12.2). Though durable and effective, widespread use of quantitative resistance in commercial crops has been limited because the phenotype is difficult to detect in screening. The expression of resistance is not binary like a hypersensitive response, and it is often differently expressed in different genetic contexts: the resistant phenotype may be recessive, the phenotype may only be expressed in adult plants or in certain tissues, and quantitative resistance by definition requires the combination of several genes. However, advances in quantitative trait locus (QTL) mapping allow the identification of genes associated with partial resistance, and marker-assisted selection facilitates screening for desired genes in breeding programs (Box 17.1).

17.3.4 Gene pyramiding

Plant breeders can extend the longevity of *R* genes by making it more difficult for pathogens to overcome resistance. Traditional **gene pyramiding** (Mundt 2018) stacks multiple *R* genes together into a single cultivar (Figure 17.4); this can include genes that confer resistance to multiple genotypes of one pathogen as well as resistance against multiple different pathogens. Because any one *R* gene

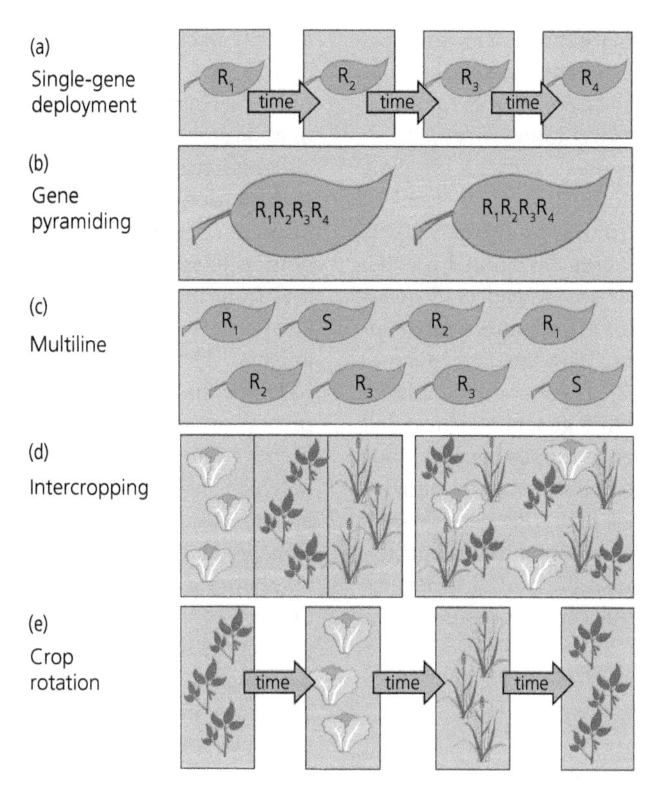

Figure 17.4 Managing the diversity of resistance genes and crop species in time and space to control plant disease. Resistance genes can be (a) deployed individually and replaced over time as effectiveness is lost (single-gene deployment), or (b) stacked within a crop genotype (gene pyramiding). (c) Different cultivars of the same crop that carry different resistance genes can be mixed together and grown as a multiline. (d) Different crop species with different ecological traits and pathogens can be grown together through intercropping of rows of each cultivar or with spatial mixtures of cultivars, creating diversity in space. (e) Diversity in time is created by crop rotation among different crop species in sequential growing seasons.
Figure by Gregory S. Gilbert and Ingrid M. Parker.

is enough to induce defenses against a pathogen with the corresponding *Avr* gene, for an asexually reproducing pathogen to overcome resistance to multiple *R* genes requires evolving mechanisms to evade all the *R* genes simultaneously. If the probability of each *Avr* gene mutating to evade its corresponding *R* gene is p, then the probability for two *Avr* genes is p^2, and for three *Avr* genes p^3; each additional gene in the pyramid greatly decreases the likelihood that an infectious pathogen will emerge. Although pathogens will sometimes overcome this multiplicative disadvantage (Mundt 2018), gene pyramiding greatly extends the useful lifetime of a resistant cultivar. Unfortunately, when an infectious genotype does emerge, it sweeps through the pathogen population as quickly as for single-gene resistance.

Gene pyramiding is widely used, and most commercial cultivars include resistance to multiple different pathogens. However, there are practical challenges to gene pyramiding. First, it is more difficult and time-consuming to breed cultivars with two or more novel *R* genes. New approaches employ cassettes of multiple resistance genes that are moved as a unit (acting as a single locus), which mitigates that constraint (Luo et al. 2021). Second, it is challenging to find novel *R* genes, and previously deployed *R* genes are often quickly overcome, as described above. Third, carrying extra *R* genes incurs a fitness cost when the host is not threatened by the corresponding pathogens (Section 13.5); growers can end up with a lower yield if they plant a cultivar with "extra" resistance genes. So, gene pyramiding is not a simple and universal solution.

So far, the most successful examples of gene pyramiding are largely limited to the control of wheat rust diseases (*Puccinia* spp.) (Mundt 2018), and these have an important deviation from the simple model in Figure 17.4b. Instead of *R* genes alone, they combine *R* genes in pyramids with quantitative resistance genes for "slow rusting," such as *Sr2* or *Lr34* (Section 17.2.2). These combinations of major *R* genes with minor APR genes have provided decades of durable resistance against several kinds of cereal rusts, but it is not completely clear why. It may be because *R* genes and APR genes confer resistance through fundamentally different mechanisms, or it may be because the partial resistance provided by APR reduces selection pressure on the pathogen.

17.3.5 Multilines

While the first step in the evolution of a pathogen to overcome resistance begins on an individual host plant, it is the subsequent spread of that novel genotype across fields and landscapes that defeats the usefulness of a resistance gene. Given that a 1-ha field of soybeans may have 680,000 plants, and with about 35 million ha planted to soybeans in the USA in 2021, that comes to 2.38×10^{13} soybean plants (that is 24 trillion plants!), mostly in the midwest. The high density and large-scale spatial genetic uniformity characteristic of much of commercial agriculture facilitates the rapid spread of pathogen genotypes that overcome resistance. To extend the durability of resistance, two types of tools enhance different aspects of diversity: diversity within and across the landscape and diversity across time.

Multilines, or **gene mixtures**, involve sowing mixtures of several different lines of cultivars, each with different sets of resistance genes (including susceptible lines) within a field (Figure 17.4c, Mundt 2002). For instance, Kristoffersen et al. (2022) found that multilines of wheat were effective in reducing severity of *Septoria tritici* blotch of wheat (caused by the Ascomycete *Zymoseptoria tritici*). Disease severity declined linearly with increasing numbers of cultivars in mixtures, with an average of 24% reduction in disease severity (and 3% yield increase) in mixtures of four cultivars, compared to those cultivars grown in monoculture.

Why does using multilines decrease disease pressure? There are various possible mechanisms, both ecological and evolutionary, and they follow from many of the ideas we have explored in Chapters 11–14. Borg et al. (2018) outline five mechanisms. [1] Multilines create a dilution effect by reducing the density of each host genotype. [2] Multilines can create a barrier effect, where resistant plants form a physical barrier to spore movement between susceptible individuals. [3] Multilines can foster induced resistance (also called **premunition**), in which non-compatible spores released from neighboring plants of susceptible cultivars induce defense responses in resistant plants, which then protect the host against further infection by spores of compatible pathogens. [4] The spatial mixture of genotypes in multilines causes disruptive selection in the pathogen, which slows adaptation in the presence of gene flow. [5] Finally, because plants compete with each other for limited resources, if more susceptible cultivars in the mixture have poor growth, compensatory growth of the resistant cultivars will lead to increased yield.

Why include susceptible varieties in the mixture at all? Including a susceptible variety in the mixture can slow evolution of the pathogen by providing it with a "refuge" of susceptible hosts. Evolutionary models that include such genetic refuges have been explored in other contexts, such as how to delay the emergence of insect pests resistant to anti-herbivore toxins in GMO crops (Bates et al. 2005). The **refuge effect** can delay the spread of a resistance allele through a sexually reproducing pest population by many generations. However, the potential efficacy of the refuge effect depends on many factors, for example the proportion of susceptible genotypes in the mixture, the spatial patterns of mixtures (homogeneous mixture, different sections within a field, or different genotypes in different fields), as well as overcoming logistical challenges to growers in planting and harvesting. The refuge effect requires sexual reproduction to work; e.g., toxin-sensitive individuals emerge in large numbers from susceptible plants and mate with any toxin-resistant individuals, swamping out the effects of selection by the toxin. This type of refuge effect will not be as effective for plant pathogens that mostly reproduce asexually. Susceptible lines may also be an important component of mixtures when they have a performance advantage over resistant lines—for example, when the cost of resistance leads to a reduction in yield.

17.4 Crop diversity

The effects on disease of genetic diversity through multilines and gene rotation has ecological parallels in crop species diversity deployed either spatially, through intercropping, or temporally, through crop rotation.

17.4.1 Intercropping

The species-level version of multilines is **intercropping**, a common agronomic practice most prominent in tropical regions, where combinations of crop species are grown together (Figure 17.4d). Intercropping takes advantage of complementary patterns of resource use (water, nutrients, light) to minimize competition, while facilitating the growth of one or more of the crops by enhancing resources (e.g., providing nitrogen) or reducing attack by natural enemies (Vandermeer 1992). For plant disease management, intercropping is an application of the dilution effect (Chapter 14). A review of over 200 studies comparing disease (primarily caused by foliar fungi) in monocrops and intercrops found that in 73% of the studies disease was reduced in the intercrop (Boudreau 2013). Phylogenetic signal in plant traits such as nutrient requirements, nitrogen fixation, and growth habit (Section 13.7) implies that the most complementary intercrop combinations should be distant relatives. Combinations used in traditional subsistence agriculture, such as maize–bean–squash, coffee–banana, sorghum–alfalfa, and cotton–groundnut, follow this rule of thumb. Because of the phylogenetic signal in pathogen host ranges, these distantly related intercrop combinations are also unlikely to share pathogens (Section 13.7 and Parker et al. 2015). Intercropping incorporates plant species of different sizes and growth habits, and a more complex crop structure can affect disease dynamics through changes in microclimate (Pérez-Hernández et al. 2021), plant physiology (Brooker et al. 2015), and barrier effects (Fernández-Aparicio et al. 2010). At the same time, such structural diversity is less amenable to mechanized crop production, limiting the scale at which it can be deployed.

For both multilines and intercropped species, there is a wide range of potential planting patterns across spatial scales, from a homogeneous mixture of different genotypes within a field or closely interplanted complementary crops (e.g., the classic three sisters maize–beans–squash system), to planting in alternating rows of crop species, to mosaics of fields of different crop species or varieties across the landscape (Figure 17.5). In general, increasing the local diversity of crop species and genotypes to break up the homogeneity of susceptible hosts will help reduce pathogen outbreaks and extend the lifetime of resistant varieties (Rimbaud et al. 2018).

17.4.2 Crop rotation

Crop rotation is a nearly universal agronomic practice where crops with complementary demands and contributions to soil nutrition are alternated to maintain soil health and productivity. Crop rotation also has significant effects on managing plant diseases. Just as intercropping uses host diversity across space to manage pests and pathogens and to improve crop yield, crop rotation aims for the same effects over time (Figure 17.4e). Whereas intercropping reduces the spread of secondary inoculum among plants in the same growing season, crop rotation reduces the amount of inoculum reaching crops planted in the same field in future seasons. Similar to intercropping, two major principles drive the development of rotation sequences: maximize differences in functional traits that shape how plants affect soil nutrients and minimize the chance that successive crops are susceptible to the same pathogens.

Repeated cropping of a single cultivar in the same field tends to create plant–soil feedbacks (Section 14.7) that generate long-lived primary inoculum of soil-borne pathogens. This increases disease prevalence for both monocyclic and polycyclic

Figure 17.5 Intercropping of a row of beans (Fabaceae), interplanted maize (Poaceae) and Amaranth (Amaranthaceae), and a row of squash (Cucurbitaceae) on the UC Santa Cruz certified organic farm.
Photo by Gregory S. Gilbert.

diseases. Crop rotation has four mechanisms that help break this cycle and reduce losses to plant disease. First, simply planting the same crop less frequently will delay the build-up of primary inoculum in the soil, provided that the intervening crops (or fallow) do not support the growth and reproduction of those pathogens. Second, the viability of long-lived soil-borne propagules (e.g., chlamydospores, sclerotia, cysts, oospores) declines over time through exposure to temperature extremes, desiccation, and predators and parasites; extending the time between susceptible hosts allows for the decay of inoculum in the soil. These two mechanisms work together to reduce the abundance of infectious propagules both by adding fewer and removing more.

Third, intervening crops may chemically suppress soil-borne inoculum. Crops in the mustard family (Brassicaceae) are widely planted in rotations or as **green manures** for such a purpose because of their effectiveness at reducing populations of soil-borne pathogens as well as weed seeds (Larkin and Griffin 2007). Brassicas naturally produce glucosinolates that provide the mustardy "bite" in cole crops (Section 12.2). When a brassica cover crop or crop residues are incorporated into the soil, bacterial enzymes act on the glucosinolates to release toxic isothiocyanates that actively kill soil pests and pathogens (Plaszkó et al. 2021). Brassica green manures can also change the microbial community of the soil in ways that reduce disease pressure, independently of the effect of glucosinolates (Mazzola et al. 2001).

A similar active effect is sometimes achieved through **trap crops**; plants that stimulate the germination of primary inoculum in the soil but do not support production of additional inoculum. For instance, potato cyst nematode (*Globodera pallida*) persists as viable but dormant cysts in the soil until eggs in the cysts are stimulated to hatch by root exudates of potatoes (*Solanum* spp.). Simple rotation to nonhost crops is not effective at reducing the pathogen population, because cysts can live dormant in soil for up to 15 years. However, root exudates of sticky nightshade (*Solanum sisymbriifolium*) stimulate egg hatching, but the nematodes are unable to parasitize the plant roots and perish without reproducing. Rotation planting of *S. sisymbriifolium* as a trap crop stimulates the dead-end hatching of *Globodera* cysts, leading to a rapid decline in potato cyst nematode populations and allowing the replanting of susceptible potato crops (Dandurand and Knudsen 2016).

Fourth and finally, crop rotation supports overall plant health through improved soil health and nutrition. Legumes are prominent elements of many rotation systems specifically because they are able to replenish soil nitrogen through symbiotic nitrogen fixation (Section 15.3). Richer soil nutrients foster robust plant growth, innate immunity systems, and diverse soil microbial communities that can be involved in disease suppression. The role of microbial communities is explored further in Section 17.5.5.

How long an interval of nonhost crops is needed to reduce disease pressure? That depends strongly on the crop and the life history of the pathogen. In some cases, a single year of planting a nonhost crop can significantly reduce pathogen populations and disease pressure (Larkin and Halloran 2014). On the other hand, many soil-borne pathogens produce robust propagules that can remain viable in the soil for years to decades. Rotation to non-hosts is not an effective way to reduce such inoculum (Lamour and Hausbeck 2003). Other methods of reducing primary inoculum, such as rotation with mustards or other cultural, biological, or chemical methods of control are needed to safely return to growing the susceptible host.

17.5 Cultural practices

Cultural control measures are agronomic practices that help manage disease (Katan 2000). They are the oldest of methods for reducing crop loss to disease, and they are essential complements to targeted disease-control tactics such as the choice of which crops to plant and the application of chemical or biological controls. Many cultural methods are motivated by agronomic needs other than disease management (e.g., irrigation) but can be fine-tuned to reduce disease. Other cultural methods are specifically intended to reduce disease pressure (e.g., sanitation). Cultural measures often involve tradeoffs in how they impact other agronomic priorities, like soil nutrient and water management, short-term yield, weed control, or the cost of labor. Understanding the life histories of local pathogens is key to evaluate how different practices either favor or limit disease.

17.5.1 Spacing

Because spacing between plants affects plant growth form, competition, crop microclimate, pathogen spread, and disease development, growers face tradeoffs in choosing a planting density (Section 11.4). At extremely low density, adding plants increases yield. But as density increases, intraspecific competition for water, nutrients, and light reduces plant performance and offsets increases in yield. For instance, as planting density increases from 0.4 to 2.8 plants m^{-2}, eggplant yield *per plant* decreases steadily from 16 kg to only 1.1 kg. Therefore, the total yield of eggplants per hectare is highest at an intermediate density (Ndereyimana et al. 2013).

Such simple tradeoffs between yield at the individual plant and field scales are complicated by density-responsive disease intensity (Section 10.8); most polycyclic plant pathogens cause more disease when their hosts are growing at greater density, both because shorter distance between plants increases pathogen transmission, and because crowding can affect crop microclimate in ways that favor disease development. For instance, the severity of chocolate spot of fava bean (caused by the Ascomycete fungus *Botrytis fabae*) increased linearly with planting densities from 6.25 plants m^{-2} to the recommended density of 25 plants m^{-2} (Bulson et al. 1997). But even with losses to severe disease, the greatest bean yield was at the recommended planting density.

Depending on the life cycle of the pathogen and the architecture of the host, increasing host density can sometimes reduce disease intensity. For example, the Basidiomycete fungus *Moniliophthora roreri* causes frosty pod of cacao, the plant that gives us life-affirming chocolate. Cacao is grown in the shady understory of the forest, and it is cauliflorous, which means it bears its fruits directly on the main trunk of the tree. Mitospores of *M. roreri* are produced on the pod surfaces and spread to other pods through wind and rain. Branches and foliage that surround the pods at high density provide a barrier that prevents the spread of propagules (Ngo Bieng et al. 2017). Thus, somewhat counterintuitively, disease intensity decreases with greater density of cacao trees, and planting cacao at greater densities could provide an effective cultural control measure against frosty pod. However, growers face a tradeoff because the productivity of each tree also declines with greater planting density.

Plant spacing affects microclimate under the crop canopy in ways that shape disease development (Cappaert and Powelson 1990). Closer plant spacing shortens the time until crops reach canopy closure (defined as when crop foliage covers >95% of the soil when viewed from above), and canopy closure affects light, temperature, and moisture beneath the canopy. For instance, canopy closure of carrots creates soil and subcanopy microclimate conditions that favor apothecia and ascospore production by the Ascomycete fungus *Sclerotinia sclerotiorum* which leads to Sclerotinia rot (Kora et al. 2005a, Foster et al. 2011). Simply trimming back the carrot canopies by 20% on each side between rows of carrots to create space between the plants increased air and soil temperatures by 9.2 and 3.1°C, and decreased daytime relative humidity by 30%; these changes in microclimate led to a 75% reduction in apothecia of *Sclerotinia* compared with a closed canopy, and essentially eliminated disease development (Kora et al. 2005b).

While optimal planting density for a particular crop is largely determined by host physiology and agronomic constraints, careful consideration of pathogen dispersal along with environmental requirements for disease development can help anticipate how changes in planting density may affect disease intensity.

17.5.2 Sanitation, tillage, and mulch

Field sanitation is about interrupting the life cycle of the pathogen. Various sanitation methods aim to remove or destroy infected plant material, and to prevent the dispersal of inoculum that can generate new disease foci (Figure 11.3). The practice of **rogueing** (removing individual symptomatic plants and their neighbors during the growing season) can help reduce the generation of new infections for polycyclic pathogens (Abrahamian et al. 2021). This requires significant investment in surveillance to detect diseased plants prior to entering the infectious period (Figure 10.2), which may not be feasible in large-scale systems or when plants become infectious before clear symptom expression.

Sanitation also involves preventing the movement of pathogen inoculum through disinfestation of tools, machinery, or people. Mechanically transmitted viruses and bacteria can be transferred from infected to susceptible plants on farm tools (Welde-Michael et al. 2008); regular **disinfestation** of tools can help reduce pathogen spread (Mackie et al. 2015). There is a wide variety of disinfectants available with different active ingredients: most are quaternary ammonium compounds, halogens (hypochlorites, like bleach), alcohols, or aldehydes (Mackie et al. 2015, Salacinas et al. 2022). Soil-borne pathogens can spread on muddy boots or vehicles from infested areas into areas free of the pathogen; washing stations use high-pressure water sprays or scrub brushes, sometimes followed by treatment with a disinfectant, when leaving infested areas and entering clean areas, to reduce the movement of pathogen propagules. Washing stations are often part of a rapid response to control emergent pathogen outbreaks (Goheen et al. 2012).

The need for soil conservation has led to widespread adoption of no-till farming methods, in which crops are seeded directly into the residue of previous crops. This is in contrast to conventional **tillage**, where plowing helps bury infected crop residue deep in the soil where growing roots are less likely to encounter inoculum, or shallow disking which helps break up, mix, and expose infested material to rapid decomposition or destruction of inoculum. For all the benefits of no-till methods, they present special challenges for disease management because of the buildup of primary inoculum of soil-borne pathogens, particularly those that produce long-lived resting structures like sclerotia, chlamydospores, and oospores, or those that can persist as saprotrophs (Paulitz 2006).

Mulching—a layer of shredded bark, compost, or plastic on top of the soil—is commonly used to control weeds and conserve moisture (Kader et al. 2017). Many pathogens survive in the soil as resting propagules or grow as saprotrophs, that then infect the above-ground part of plants when propagules are released into the air or splash up onto the stems, foliage, or fruits. Mulch creates a physical barrier between the soil and above-ground plant parts, and can help reduce splash dispersal and disease pressure. For example, a layer of rice-husk mulch is more effective than chemical treatments in controlling web blight of beans caused by *Thanatephorus cucumeris* (a basidiomycete known more commonly as its anamorph *Rhizoctonia solani*) (Galindo et al. 1983), by blocking dispersal from soil-borne sclerotia to susceptible tissues.

17.5.3 Irrigation

Irrigation is essential for crop production when natural precipitation does not meet the demands of plants to grow and reproduce. Globally, some 20% of croplands and 40% of food production is irrigated, either through surface irrigation (gravity-fed water provided to crops through canals, furrows, basins, or periodic flooding), overhead irrigation (pressurized sprinklers), or through subsurface or micro-irrigation (using pipes and hoses to apply water directly to plants at or below the soil level). Because moisture is such a critical component of the disease triangle, and because water can be such an important conduit for pathogen movement, irrigation has a large influence on disease development. Irrigation management offers numerous avenues for cultural control of disease.

Overhead irrigation (Figure 17.6a) exacerbates foliar disease intensity in three ways. First, free water on leaf surfaces is a critical requirement for infection by many fungi and bacteria that cause foliar diseases (Section 15.6), and overhead irrigation extends the time that plant leaves remain wet, facilitating infection. Abundant leaf moisture also promotes reproduction of foliar pathogens, generating secondary inoculum (Avenot et al. 2022). Finally, many foliar pathogens are splash-dispersed (Section 11.1), and overhead irrigation provides high-impact water droplets that dislodge pathogen propagules from leaf surfaces and transport them to neighbors. For many foliar pathogens, these effects lead to greater disease intensity with overhead irrigation (Café-Filho et al. 2019). Curiously, powdery mildew diseases are often less severe under overhead irrigation. Conidia of powdery mildews tend to be damaged or displaced by the impact of water drops, and somewhat unusually for fungi, germination and infection both decrease in free water.

Surface irrigation (also called open irrigation) is the most widely practiced form of irrigation globally. Controlled release of water through furrows between row crops, or short-term flooding of fields or orchards, provides water to crops when it is needed. Because it is gravity-fed, water moves from one part of a field to another, carrying soil particles along with it. This creates ideal circumstances for dispersal of numerous soil-borne plant pathogens, especially motile pathogens such as oomycete zoospores and some bacteria (Chapters 4 and 5). Dozens of plant pathogenic fungi, oomycetes, and bacteria have been detected in irrigation water, suggesting its potential importance in causing disease, but there is surprisingly little evidence that open irrigation systems drive pathogen epidemiology (Zappia et al. 2014, Lamichhane and Bartoli 2015). Unlike soil-borne pathogens, foliar pathogens are less likely to be enhanced by surface irrigation than overhead irrigation. For instance, epifoliar populations of the bacterium *Xanthomonas axonopodis* pv. *phaseoli* (cause of common bacterial blight of bean), were three orders of magnitude lower on leaves of beans grown with furrow irrigation than with sprinkler irrigation, and disease severity was similarly reduced (Akhavan et al. 2013).

(a)

(b)

Figure 17.6 Irrigation practices affect disease risk. (a) Overhead irrigation keeps plant surfaces wet for long periods, increasing opportunity for pathogen infection, compared to (b) drip irrigation, which places water at the plant root system, where it is absorbed.
Photos by Gregory S. Gilbert.

Micro-irrigation and subsurface irrigation, including buried irrigation systems and drip irrigation (Figure 17.6b), is the most water-conserving approach, placing water directly into the soil near the plant where and when it is needed. It reduces flow of irrigation water from one place to another (reducing the likelihood of pathogen spread) and avoids splashing and wetting of foliage (reducing spread and infection of foliar pathogens). For instance, the fungus *Colletotrichum acutatum* causes anthracnose of strawberry; conidia are splash-dispersed and infection is enhanced on wet leaves. A switch to drip irrigation eliminated splashing and kept leaves dry, reducing plant mortality from anthracnose by up to 85% compared to plants grown with overhead irrigation (Daugovish et al. 2012).

Recycling irrigation water, particularly in plant nurseries, is an increasingly common practice that helps reduce waste of both limited water resources and nutrients through leaching; unfortunately, recycled irrigation water can sometimes be laden with pathogens from infested plants and lead to spread of pathogens throughout a nursery. Ultrafiltration of the irrigation water, or sand filtration combined with chlorination, can be safe and effective ways to eradicate pathogens from irrigation water (Machado et al. 2013).

17.5.4 Temperature and oxygen

Most plant pathogens thrive under a limited temperature range and are aerobic, requiring adequate oxygen to grow (Figures 3.8 and 5.6). These pathogen requirements open opportunities for powerful cultural methods to manage plant disease.

The most direct way to use temperature is by **burning** material that contains the pathogen. Burning crop residues has long been used in both subsistence and commercial agriculture to prepare fields for planting. Burning crop residue can destroy pathogen reservoirs, and the heat generated can reduce pathogen populations in the first few centimeters of the soil (Dart et al. 2012). Unfortunately, burning also releases particulate matter and atmospheric carbon, resulting in substantial human health impacts and environmental impacts that limit the suitability of fire as a disease management tool in agriculture (McCarty et al. 2009, Kumar et al. 2020).

Fire in forest ecosystems has mixed effects on disease. Root- and butt-rot diseases may be enhanced because burned roots are more susceptible to fungal infection. In contrast, fire can suppress populations of dwarf mistletoes because they are killed directly by crown fires; branches also catch fire more readily when they are infected (Parker et al. 2006). How do prescribed burns that are used to reduce the threats of catastrophic wildfire in forest ecosystems interact with tree diseases? This is an open question and an active topic of research (Simler-Williamson et al. 2021).

Many soil-borne pathogens survive the time between successive assaults on their crop hosts as saprotrophs or dormant resting structures. If, in the time between crops, soil conditions can be managed to achieve unsuitable conditions for the pathogen, primary inoculum can be reduced for the next crop. One widely used approach is **solarization**, where plastic tarps cover the soil during times of hot, sunny weather. The tarps trap heat from the sun, raising soil temperatures above 50°C, high enough to kill many bacteria, fungi, oomycetes, nematodes, and parasitic plants (Katan et al. 1976). Unfortunately, solarization is a blunt tool that also kills beneficial microbes and creates dramatic shifts in the composition of the soil microbiome (Schreiner et al. 2001, Kanaan et al. 2018). So, crop health is not always improved by solarization. It is also most suited to regions that experience hot conditions outside the growing season. Solarization does not consistently generate temperatures high enough to kill pathogens in cooler climates.

A cousin of solarization that augments effects of high temperature while starving pathogens of oxygen is **anaerobic soil disinfestation** (**ASD**) (Mazzola et al. 2018). Like solarization, ASD involves tarping the soil between crop cycles. But ASD amends the soil with an easily degraded carbon source such as rice bran, and then saturates the soil with water beneath the tarps. With water-filled pore spaces and a tarp to reduce evaporation and gas exchange, rapid microbial respiration during decomposition of the rice bran turns the soil anaerobic. This creates conditions that are intolerable to aerobic members of the soil microbial community, including most plant pathogens. ASD is becoming an increasingly important alternative to chemical fumigation (Section 17.6.1) for management of soil-borne pathogens.

17.5.5 Suppressive soils

The phenomenon of **suppressive soils** highlights the role of soil microbial communities in modulating plant disease development. Suppressive soils contain microbial communities that inhibit the development of disease (Schlatter et al. 2017). Suppressiveness falls along a gradient of general to specific. **General suppressiveness** is provided by a robust, active, diverse microbial community, often associated with soils that are rich in organic matter, such as those found on well-tended organic farms. General suppressiveness is thought to reflect competition among microbes for key limiting resources, including the root exudates that stimulate and attract pathogens (Section 2.4). A robust microbial community rapidly consumes those exudates, making it difficult for pathogens to find and infect their hosts. Obviously, abiotic conditions in soils may also make them more or less conducive to pathogen activity. Therefore, to show that general suppressiveness is microbial in nature, the key experiment is to reduce or eliminate the microbial community through heating or gamma irradiation and test for a reduction of the suppressive effect. General suppressiveness can reduce disease caused by a broad range of pathogens.

In contrast, **specific soil suppressiveness** is generated by populations of particular soil microbes and is often effective against only a narrow range of pathogens. Unlike general suppressiveness, specific suppressiveness is easily transferable to other soils. A small inoculum of suppressive soil mixed in with other soil (e.g., 1% to 10% suppressive soil) can convert it to suppressive (Schlatter et al. 2017). Specific suppressive soils can develop in various ways, but curiously, one way is through the usually problematic practice of repeatedly planting the same crop in the same field. For example, Take-all (caused by the ascomycete fungus *Gaeumannomyces graminis*) is an important root disease of cereals, with mycelium surviving in dead roots and weedy grasses between crops. Successive monoculture crops of wheat or barley build up inoculum of *G. graminis* with increasing disease intensity in each crop. However, after 4–6 years, disease intensity then begins to decline, even with continued cultivation of the susceptible host. Disease suppression is associated with high abundance of microbial species that are antagonistic to *G. graminis*, most notably fluorescent *Pseudomonas* spp. These *Pseudomonas* produce the antibiotic 2,4-diacetylphloroglucinol (DAPG), to which *G. graminis* is highly sensitive.

In northern Europe, soils suppressive to *Thanatephorus cucumeris* (anamorph *Rhizoctonia*) on sugar beet have bacterial communities very different from those in soils that are conducive (i.e., not suppressive). The ectorhizosphere of sugar beets grown in the suppressive soil is enriched with certain bacteria, particularly in the *Pseudomonadaceae* (γ-Proteobacteria) and *Actinobacteria*, that prevent infection by *Thanatephorus* (Mendes et al. 2011). In addition, the endorhizosphere of sugar beets grown in suppressive soil is enriched with bacteria in the *Chitinophagaceae* and *Flavobacteriaceae* (Bacteroidetes) (Carrión et al. 2019). Suppressiveness is lost if bacteria are eliminated, and it can be transferred to conducive soils through inoculation (Mendes et al. 2011). When *Thanatephorus* manages to get past the ectorhizosphere bacteria and infects the root, the endorhizosphere bacteria activate expression of several biosynthetic gene clusters that produce peptides, polyketides, and enzymes such as chitinase that degrade fungal cell walls.

Changing soil nutrient and chemical conditions may create soils with general or specific suppressiveness by favoring fungi and bacteria that inhibit disease development. In addition to green manures and crop rotation (Section 17.4), soil amendments such as biochar, manure, and crustacean meal have shown promise in inducing soil suppressiveness to a variety of plant diseases (de Medeiros et al. 2021).

17.5.6 Landscape

Some disease management tactics are implemented at larger spatial scales. These tactics often focus on controlling alternate or alternative host species in weedy or wild plant populations, or managing environmental conditions to reduce pathogen spread.

Many important rust fungi are heteroecious and macrocyclic, and thus have an obligate **alternate host** (Box 3.2). The alternate host is often a wild species. This creates the opportunity to control the population of the "less important" host and break the pathogen life cycle. Indeed, efforts to eradicate barberry (*Berberis* spp.), the obligate aecial host of *Puccinia graminis* (stem rust) from around cereal fields, date back centuries. How interesting it is that farmers understood the importance of alternate hosts before microbes were even discovered! The importance of alternate host abundance for epidemiology and management varies widely, depending on the pathogen and the context. One of the most expansive (but ultimately unsuccessful) attempts at landscape-level management of a disease is of white pine blister rust (*Cronartium ribicola*), which has two obligate, alternate hosts: five-needle (white) pines (*Pinus* sect. *quinquefoliae*) and *Ribes* spp. (currants and gooseberries; Grossulariaceae). Beginning in 1912, an aggressive, US government-sponsored eradication of wild and cultivated species of *Ribes* (the telial host) attempted to remove the alternate host from within 300 m of pines, and in some regions commercial production of gooseberries was restricted (Maloy 1997). (The culinary affinity for gooseberries and currants in Canada and Europe meant that these measures were not adopted outside the US; they chose jam over lumber.) However, after decades of intense effort, the US eradication program was abandoned as ineffective, finding that to reduce blister rust impacts required nearly complete local extirpation of *Ribes*. This was particularly difficult in the western USA, where some 20 species of wild *Ribes* are found throughout the geographic range of pines. Disease control efforts for blister rust have since shifted toward breeding resistant pines and gooseberries.

Because most plant pathogens are moderately polyphagous (Section 13.5), it is common for a pathogen of a crop to also cause disease on some of the wild plants or even other crops in the area. This presents a management challenge when these plants serve as reservoirs for pathogen populations, leading to spillover onto the crops (Section 14.5 and Wisler and Norris 2005). For instance, the fastidious bacterium *Xylella fastidiosa*, transmitted by leafhoppers, causes serious disease in grapes, almonds, and olives, but infects, often asymptomatically, a wide variety of other plants (Kyrkou et al. 2018). Nearby riparian forests, weedy areas, fields of alfalfa, and citrus orchards are all reservoirs for the bacterium and vectors, and the landscape configuration of susceptible crops and reservoir species shapes disease risk. In this type of system, vegetation management and crop selection at the landscape scale are important tools for disease management.

Landscape-scale vegetation management can also be used to moderate the environmental conditions that favor disease. For instance, trees are grown as shelterbelt windbreaks around agricultural fields in many parts of the world. In addition to their benefits for soil conservation, crop production, and wildlife, shelterbelt windbreaks can contribute to reducing pathogen spread in crops. In citrus groves, arboreal windbreaks reduce the speed of strong, gusty winds, which cause branch wounds, allowing infection from the citrus canker bacterium *Xanthomonas citri* subsp. *citri* (Gochez et al. 2020). Planting trees for windbreaks is recommended as one of the primary measures for managing *Xanthomonas citri* in Florida.

17.6 Chemical and biological controls

Of the control strategies (Table 17.1), resistance and cultural practices are the core tactics for managing plant disease and serve to protect most crops from most pathogens. But as Dr. Malcolm says in *Jurassic Park*, "life finds a way." Disease outbreaks happen. We respond with preventative and curative measures.

Unlike human disease management, with plants we can almost never cure an individual that has already developed symptoms. That means our control of outbreaks must focus on protecting plants from infection. In addition to chemical control, which is the primary tactic used by growers to respond to an outbreak, there is the promise of biological control, which relies on a sophisticated ecological understanding of plant–microbe interactions. Here we take a brief look at the general principles of chemical and biological controls in the management of plant diseases.

17.6.1 Chemical control

For growers and the agricultural industry, the need to have predictable harvests from the crops they plant has led to widespread use of chemical treatments to prevent or manage plant diseases. Fungicide, bactericide, and nematicide treatments can be applied to control diseases at various points of contact between host and pathogen, including by suppressing primary inoculum in soil. Here we provide a brief overview of these approaches while we also consider issues of specificity, durability and the pesticide treadmill, the use of predictive systems, and protecting nontarget organisms and human health.

Fumigation of soil has long been practiced to rid fields of primary inoculum, including nematode cysts, oospores, fungal sclerotia, plant seeds, and saprotrophic mycelium. Fumigation is often used to allow continuous production of high-value monoculture crops like strawberries. The soil is covered with a tarp, and then broad-spectrum biocidal fumigants are injected into the soil beneath the tarp. One of the most widely used fumigants was methyl bromide; this highly effective fumigant did its intended job well, but it was also dangerous to handle, causing central nervous system and respiratory failure, and, alarmingly, it is also highly destructive to the atmospheric ozone layer that protects us from excessive UV radiation. In 2005, the Montreal Protocol mandated the phaseout of methyl bromide, but its limited application continues under some critical-use exemptions. Anaerobic soil disinfestation (Section 17.5.4) is a promising, environmentally friendly alternative to methyl bromide fumigation.

Damping-off is a significant challenge to growers planting seeds in cool, wet soils in early spring. The threat has a short time-window, generally a couple weeks from the moment seeds are planted and begin to absorb water until seedlings emerge and produce their first true leaves (Section 8.2). **Seed protectants** are anti-fungal or anti-oomycete chemicals with which seeds are coated before planting. The chemicals are present and active right where they are needed—at the germinating seed—and provide protection during the seedlings' vulnerable germination window.

Most fungicides and bactericides are sprayed onto plants to act as **surface protectants**, or **contact fungicides**. They prevent infectious propagules that land on a treated plant from growing and infecting the host, and they can act against pathogens that are reproducing on a plant surface before they disseminate to neighboring plants.

Systemic fungicides are taken up into the plant, where they act as protection both during and just after infection. Some systemic fungicides move locally within the plant (from one side of a leaf to the other by diffusion). Others are transported via the plant vascular system (Section 2.2); some, like the anti-oomycete metalaxyl, move unidirectionally from roots to leaves or fruits through xylem, and one (fosetyl-aluminum) moves bidirectionally through the phloem. Systemic fungicides can protect parts of plants distant to where the fungicide was applied, and in some cases, can cure plants after infection. Curative activity is generally limited to just after infection; by the time symptoms develop, it is too late.

The specificity of such fungicides varies widely. Some fungicides have very broad toxicity because they act on fundamental physiological processes common to most microbes. For instance, copper sulfate has been used as a surface protectant to control plant diseases since the 1700s, and it is one of the only pesticides certified for use in organic production systems. It is toxic to most kinds of bacteria and fungi, binding to cells, disrupting membrane function, and creating lethal leakage of cell contents. Other fungicides are more specific because they have mechanisms of action that target a specific trait shared by a restricted number of closely related pathogens. For instance, the systemic fungicide metalaxyl inhibits RNA polymerase I in Oomycetes, making it selectively effective against *Pythium*, *Phytophthora*, and downy mildews. (Metalaxyl is thus an oomyceticide and not a fungicide, but nobody calls it that). In general, it is difficult for microbes to evolve resistance against very broad-spectrum fungicides like copper sulfate, because it is hard to change, for instance, the fundamental way cell membranes work. But it is relatively easy to evolve resistance to fungicides with specific modes of action like metalaxyl; in fact, resistance to metalaxyl is widespread.

There is thus a trade-off in how to design and use chemical controls. Broad-spectrum fungicides are likely to have unwanted non-target effects but are persistently effective, while highly specific fungicides avoid non-target effects but lose their effectiveness quickly. Specific fungicides readily fall into a **pesticide treadmill**, where pathogen populations evolve resistance to the fungicide, and so a new fungicide with a different mode of action is needed. The pesticide treadmill is like the boom-and-bust cycles associated with pathogens overcoming host resistance. Like gene pyramiding, the use of combinations of multiple pesticides is sometimes recommended in order to delay evolution of the pathogen.

To help growers reduce pesticide use, there has been extensive research to develop **predictive systems** (also known as **forecast models** or **expert systems**), using pathogen or vector detection, and/or environmental data—temperature, precipitation, leaf wetness—to help make decisions about when to apply pesticides. The goal is to reduce both environmental impacts and grower costs by cutting the number of applications. In place of prophylactic or "calendar spray" programs (e.g., scheduled weekly sprays), the grower sprays only when the data suggest that disease pressure is high. Combining weather data with models of how moisture and temperature affect pathogen growth, dispersal, and infection has allowed the development of disease predictive systems for many crops. To meet grower needs, predictive systems must incorporate sophisticated details such as cost/benefit asymmetries, i.e., when the cost of making an unnecessary pesticide application is smaller than the cost in yield of missing an application during an imminent pathogen outbreak (Gent et al. 2013). While predictive models are useful tools to reduce chemical inputs in agriculture, there is a tension between model "success" from an academic perspective and the management of risk for growers protecting their livelihoods.

Fungicides, bactericides, and nematicides control pathogens by killing them; *cide* is Latin for kill. These pesticides would ideally kill only the target pathogens, but this is rarely the case. Their mechanisms of action have physiological targets that are shared with other related organisms, such as membrane integrity, respiratory activity, and formation of sterols or chitin in cell walls. Therefore, a fungicide or bactericide applied to control a pathogen may also kill many benign or beneficial symbionts in the plant microbiome, in the surrounding environment, or in the microbiomes of humans and other animals. In the United States, pesticides are regulated by the **Environmental Protection Agency** (**EPA**) under the **Federal Insecticide, Fungicide, and Rodenticide Act** (**FIFRA**), and the **Food and Drug Administration** (**FDA**) is responsible for ensuring that foods and animal feeds do not contain more residual pesticides than limits set by the EPA. While extensive testing and regulations help minimize direct risk to human health, the human and environmental impacts of long-term, widespread use of agrochemicals remains a significant concern. With a few limited exceptions, chemical controls are not allowed as part of certified organic agriculture.

17.6.2 Biological control

Biological control (or biocontrol) of plant diseases involves the intentional management of the plant microbiome to inhibit disease development. We have already considered how specific soil suppression can develop (and be encouraged) in the plant rhizosphere (Section 17.5.5). Beyond suppressive soils, biological control involves applying specific **biological control agents**—bacteria, fungi, oomycetes, or viruses—to soil or plants to establish living microbe populations that suppress pathogens. Biological control is an important tool in organic production systems, and the need for more effective nonchemical tactics for controlling plant pathogens is a key sustainability goal.

Seed and seedling diseases have been a major focus of research on biological control of plant pathogens. There are several reasons: not only does the moist soil environment support widespread seed and seedling diseases, but the rhizosphere of growing seedlings is also ideal for rapid growth of microbial biocontrol agents. In addition, protection is needed for only a short period of time until the plant outgrows its susceptible phase. Finally, coating seeds with biocontrol agents prior to planting is fairly easily integrated into farming practices. The range of agents shown to have strong potential for biological control of damping-off is impressive; bacterial agents include species in the Actinobacteria (*Streptomyces*), Firmicutes (*Bacillus*), Gammaproteobacteria (*Pseudomonas*), and Betaproteobacteria (*Berkholderia*), Fungal agents include the Ascomycetes *Trichoderma*, *Gliocladium*, and *Beauveria*, and even an Oomycete, *Pythium oligandrum*.

The phylloplane bacterial community is another potential target for biocontrol of foliar diseases. Suspensions of bacterial control agents can be sprayed onto plant leaves to establish epifoliar populations that protect plants. Sometimes the goal is to introduce one specific agent with key traits, and sometimes to manipulate the entire community. At the specific end, inoculating leaves with a single-gene deletion mutant strain of *Pseudomonas syringae* can prevent leaf colonization by wild-type *P. syringae* that nucleate ice formation, protecting plants against frost damage (Section 15.7). At the other extreme, **constructed communities** (also called **synthetic communities**, or SynComs) of up to 12 strains of bacteria sprayed onto tomato leaves reduced populations of pathogenic *P. syringae* pv. *tomato* on the phylloplane and protected the plants against bacterial speck disease (Berg and Koskella 2018). In this study, the diversity of bacterial strains in the constructed community was less important than the particular composition of the communities, i.e., "Who, not how many." There are intriguing efforts to design SynComs of microbes that can be grown in culture and applied to plants as community-level biological control agents.

Successful biological control involves a range of possible mechanisms. First, all of the microbe–microbe interactions explored in detail in Chapter 15 are potentially at play. Microbial biocontrol agents may directly challenge the pathogen through competition for resources, predation, hyperparasitism, or the production of antimicrobials (Section 15.10). Second, they may operate indirectly through **priming** or **induced systemic resistance**. Here the biological control agent stimulates host defenses by triggering the equivalent of the systemic acquired resistance (SAR) response, resulting in upregulated defenses throughout the plant (Section 12.7). For instance, the adaptable biological control bacterium *Pseudomonas simiae* WCS417, in addition to direct interference with fungal pathogens in the rhizosphere, can colonize the endorhizosphere and elicit induced systemic resistance that protects above-ground plant parts against a variety of pathogens (Pieterse et al. 2021). One goal of designing SynComs as mentioned above is to build plant microbiomes that provide both direct control of pathogens and induced systemic resistance (Vannier et al. 2019).

The research literature on potential biological control agents is voluminous and has been the source of great advances in our understanding of seed, rhizosphere, and phyllosphere microbial communities. Unfortunately, high variation in performance in the field underlies a disappointing paucity of commercial applications. Biocontrol agents are themselves living symbionts, and their activities are shaped by their environment and interactions with their host plant (last chance to cite the disease triangle! Figure 1.5), as well as interactions with other microbes. The effectiveness of biocontrol agents in terms of both establishment and control of their pathogen antagonists is often highly context dependent. While there are a handful of biological control agents available commercially, variable performance has been a major limitation to their widespread adoption (O'Brien 2017, Raymaekers et al. 2020). A fascinating and vital application of the evolutionary ecology of plant disease in the future will be to discover new approaches to effective biocontrol.

Further reading

American Phytopathological Society. Diseases and Pests Compendium Series (numerous crop-specific compendia). https://apsjournals.apsnet.org/series/compendia

Fry, W. E. 1982. *Principles of Plant Disease Management*. Academic Press, New York.

Gliessman, S. R. 2015. *Agroecology. The Ecology of Sustainable Food Systems*, 3rd edition. CRC Press, Boca Raton, FL.

Sheaffer, C. C. and K. M. Moncada. 2012. *Introduction to Agronomy. Food, Crops, and Environment*, 2nd edition. Delmar, Albany, NY.

References

Abrahamian, P., A. Sharma, J. B. Jones, and G. E. Vallad. 2021. Dynamics and spread of bacterial spot epidemics in tomato transplants grown for field production. *Plant Disease* **105**:566–575.

Agrios, G. N. 2004. Plant Pathology, 5th edition. Elsevier, Amsterdam.

Akhavan, A., M. Bahar, H. Askarian, M. R. Lak, A. Nazemi, and Z. Zamani. 2013. Bean common bacterial blight: pathogen epiphytic life and effect of irrigation practices. *SpringerPlus* **2**:41.

Avenot, H. F., A. Baudoin, and C. Hong. 2022. Conidial production and viability of *Calonectria pseudonaviculata* on infected boxwood leaves as affected by temperature, wetness, and dryness periods. *Plant Pathology* **71**: 696–701.

Bates, S. L., J.-Z. Zhao, R. T. Roush, and A. M. Shelton. 2005. Insect resistance management in GM crops: past, present and future. *Nature Biotechnology* **23**:57–62.

Berg, M., and B. Koskella. 2018. Nutrient- and dose-dependent microbiome-mediated protection against a plant pathogen. *Current Biology* **28**:2487–2492.e3.

Bockus, W. W., E. D. De Wolf, B. S. Gill, D. J. Jardine, J. P. Stack, R. L. Bowden, A. K. Fritz, and T. J. Martin. 2011. Historical durability of resistance to wheat diseases in Kansas. *Plant Health Progress* **12**:25.

Borg, J., L. P. Kiær, C. Lecarpentier, I. Goldringer, A. Gauffreteau, S. Saint-Jean, S. Barot, and J. Enjalbert. 2018. Unfolding the potential of wheat cultivar mixtures: A meta-analysis perspective and identification of knowledge gaps. *Field Crops Research* **221**:298–313.

Boudreau, M. A. 2013. Diseases in intercropping systems. *Annual Review of Phytopathology* **51**:499–519.

Boutrot, F. and C. Zipfel. 2017. Function, discovery, and exploitation of plant pattern recognition receptors for broad-spectrum disease resistance. *Annual Review of Phytopathology* **55**:257–286.

Brooker, R. W., A. E. Bennett, W. F. Cong, T. J. Daniell, T. S. George, P. D. Hallett, C. Hawes, P. P. Iannetta, H. G. Jones, and A. J. Karley. 2015. Improving intercropping: a synthesis of research in agronomy, plant physiology and ecology. *New Phytologist* **206**:107–117.

Bulson, H., R. Snaydon, and C. Stopes. 1997. Effects of plant density on intercropped wheat and field beans in an organic farming system. *Journal of Agricultural Science* **128**:59–71.

Café-Filho, A. C., C. A. Lopes, and M. Rossato. 2019. Management of plant disease epidemics with irrigation practices. In: G. Ondrasek, editor. Irrigation in Agroecosystems. IntechOpen. DOI: 10.5772/intechopen.78253

Cappaert, M. R. and M. L. Powelson. 1990. Canopy density and microclimate effects on the development of aerial stem rot of potatoes. *Phytopathology* **80**:350–356.

Carrión, V. J., J. Perez-Jaramillo, V. Cordovez, V. Tracanna, M. De Hollander, D. Ruiz-Buck, L. W. Mendes, W. F. van Ijcken, R. Gomez-Exposito, and S. S. Elsayed. 2019. Pathogen-induced activation of disease-suppressive functions in the endophytic root microbiome. *Science* **366**:606–612.

Dandurand, L. M. and G. Knudsen. 2016. Effect of the trap crop *Solanum sisymbriifolium* and two biocontrol fungi on reproduction of the potato cyst nematode, *Globodera pallida*. *Annals of Applied Biology* **169**:180–189.

Dart, N. L., S. M. Arrington, and S. M. Weeda. 2012. Flaming to reduce inocula of the boxwood blight pathogen, *Cylindrocladium pseudonaviculatum*, in field soil. *Plant Health Progress* **13**:33.

Daugovish, O., M. Bolda, S. Kaur, M. J. Mochizuki, D. Marcum, and L. Epstein. 2012. Drip irrigation in California strawberry nurseries to reduce the incidence of *Colletotrichum acutatum* in fruit production. *HortScience* **47**:368–373.

de Medeiros, E. V., N. T. Lima, J. R. de Sousa Lima, K. M. S. Pinto, D. P. da Costa, C. L. Franco Junior, R. M. S. Souza, and C. Hammecker. 2021. Biochar as a strategy to manage plant diseases caused by pathogens inhabiting the soil: a critical review. *Phytoparasitica* **49**: 713–726.

Fernández-Aparicio, M., M. Amri, M. Kharrat, and D. Rubiales. 2010. Intercropping reduces *Mycosphaerella pinodes* severity and delays upward progress on the pea plant. *Crop Protection* **29**:744–750.

Fonseca, J. P. and K. S. Mysore. 2019. Genes involved in nonhost disease resistance as a key to engineer durable resistance in crops. *Plant Science* **279**:108–116.

Foster, A. J., C. Kora, M. R. McDonald, and G. J. Boland. 2011. Development and validation of a disease forecast model for Sclerotinia rot of carrot. *Canadian Journal of Plant Pathology* **33**:187–201.

Frankel, S. J., C. Conforti, J. Hillman, M. Ingolia, A. Shor, D. Benner, J. M. Alexander, E. Bernhardt, and T. J. Swiecki. 2020. *Phytophthora* introductions in restoration areas: Responding to protect California native flora from human-assisted pathogen spread. *Forests* **11**:1291.

Galindo, J., G. Abawi, H. Thurston, and G. Galvez. 1983. Effect of mulching on web blight of beans in Costa Rica. *Phytopathology* **73**:610–615.

Gent, D. H., W. F. Mahaffee, N. McRoberts, and W. F. Pfender. 2013. The use and role of predictive systems in disease management. *Annual Review of Phytopathology* 51:267–289.

Gochez, A. M., F. Behlau, R. Singh, K. Ong, L. Whilby, and J. B. Jones. 2020. Panorama of citrus canker in the United States. *Tropical Plant Pathology* **45**:192–199.

Goheen, D. J., K. Mallams, F. Betlejewski, and E. Hansen. 2012. Effectiveness of vehicle washing and roadside sanitation in decreasing spread potential of port-orford-cedar root disease. *Western Journal of Applied Forestry* **27**:170–175.

Gonsalves, D., C. Gonsalves, S. Ferreira, K. Pitz, M. Fitch, R. Manshardt, and J. Slightom. 2004. Transgenic virus resistant papaya: From hope to reality for controlling papaya ringspot virus in Hawaii. APSnet DOI:10.1094/APSnetFeature-2004-0704.

Hasan, N., S. Choudhary, N. Naaz, N. Sharma, and R. A. Laskar. 2021. Recent advancements in molecular marker-assisted selection and applications in plant breeding programmes. *Journal of Genetic Engineering and Biotechnology* **19**:1–26.

He, D.-c., J. J. Burdon, L.-h. Xie, and Z. Jiasui. 2021. Triple bottom-line consideration of sustainable plant disease management: From economic, sociological and ecological perspectives. *Journal of Integrative Agriculture* **20**:2581–2591.

Hospers-Brands, A., R. Ghorbani, E. Bremer, R. Bain, A. Litterick, F. Halder, C. Leifert, and S. Wilcockson. 2008. Effects of presprouting, planting date, plant population and configuration on late blight and yield of organic potato crops grown with different cultivars. *Potato Research* **51**:131–150.

Kader, M., M. Senge, M. Mojid, and K. Ito. 2017. Recent advances in mulching materials and methods for modifying soil environment. *Soil and Tillage Research* **168**: 155–166.

Kanaan, H., S. Frenk, M. Raviv, S. Medina, and D. Minz. 2018. Long and short term effects of solarization on soil microbiome and agricultural production. *Applied Soil Ecology* **124**:54–61.

Katan, J. 2000. Physical and cultural methods for the management of soil-borne pathogens. *Crop Protection* **19**:725–731.

Katan, J., A. Greenberger, H. Alon, and A. Grinstein. 1976. Solar heating by polyethylene mulching for the control of diseases caused by soil-borne pathogens. *Phytopathology* **66**:683–688.

Koch, A. and M. Wassenegger. 2021. Host-induced gene silencing–mechanisms and applications. *New Phytologist* **231**:54–59.

Kora, C., M. McDonald, and G. Boland. 2005a. Epidemiology of Sclerotinia rot of carrot caused by *Sclerotinia sclerotiorum*. *Canadian Journal of Plant Pathology* **27**: 245–258.

Kora, C., M. R. McDonald, and G. J. Boland. 2005b. Lateral clipping of canopy influences the microclimate and development of apothecia of *Sclerotinia sclerotiorum* in carrots. *Plant Disease* **89**:549–557.

Kristoffersen, R., L. B. Eriksen, G. C. Nielsen, J. R. Jørgensen, and L. N. Jørgensen. 2022. Management of *Septoria tritici* blotch using cultivar mixtures. *Plant Disease* **106**:1341–1349.

Kumar, I., V. Bandaru, S. Yampracha, L. Sun, and B. Fungtammasan. 2020. Limiting rice and sugarcane residue burning in Thailand: Current status, challenges and strategies. *Journal of Environmental Management* **276**:111228.

Kyrkou, I., T. Pusa, L. Ellegaard-Jensen, M.-F. Sagot, and L. H. Hansen. 2018. Pierce's disease of grapevines: a review of control strategies and an outline of an epidemiological model. *Frontiers in Microbiology* **9**:2141.

Lamichhane, J. R. and C. Bartoli. 2015. Plant pathogenic bacteria in open irrigation systems: what risk for crop health? *Plant Pathology* **64**:757–766.

Lamour, K. and M. K. Hausbeck. 2003. Effect of crop rotation on the survival of *Phytophthora capsici* in Michigan. *Plant Disease* **87**:841–845.

Larkin, R. P. and T. S. Griffin. 2007. Control of soilborne potato diseases using *Brassica* green manures. *Crop Protection* **26**:1067–1077.

Larkin, R. P. and J. M. Halloran. 2014. Management effects of disease-suppressive rotation crops on potato yield and soilborne disease and their economic implications in potato production. *American Journal of Potato Research* **91**:429–439.

Lieberei, R. 2007. South American leaf blight of the rubber tree (*Hevea* spp.): new steps in plant domestication using physiological features and molecular markers. *Annals of Botany* **100**:1125–1142.

Lowder, L., A. Malzahn, and Y. Qi. 2016. Rapid evolution of manifold CRISPR systems for plant genome editing. *Frontiers in Plant Science* **7**:1683.

Luo, M., L. Xie, S. Chakraborty, A. Wang, O. Matny, M. Jugovich, J. A. Kolmer, T. Richardson, D. Bhatt, and M. Hoque. 2021. A five-transgene cassette confers broad-spectrum resistance to a fungal rust pathogen in wheat. *Nature Biotechnology* **39**:561–566.

Machado, P. d. S., A. C. Alfenas, M. M. Coutinho, C. M. Silva, A. H. Mounteer, L. A. Maffia, R. G. de Freitas, and C. d. S. Freitas. 2013. Eradication of plant pathogens in forest nursery irrigation water. *Plant Disease* **97**:780–788.

Mackie, A., B. Coutts, M. Barbetti, B. Rodoni, S. McKirdy, and R. Jones. 2015. Potato spindle tuber viroid: stability on common surfaces and inactivation with disinfectants. *Plant Disease* **99**:770–775.

Magarey, R. D., M. Colunga-Garcia, and D. A. Fieselmann. 2009. Plant biosecurity in the United States: Roles, responsibilities, and information needs. *Bioscience* **59**:875–884.

Maloy, O. C. 1997. White pine blister rust control in North America: a case history. *Annual Review of Phytopathology* **35**:87–109.

Mazzola, M., D. M. Granatstein, D. C. Elfving, and K. Mullinix. 2001. Suppression of specific apple root pathogens by *Brassica napus* seed meal amendment regardless of glucosinolate content. *Phytopathology* **91**:673–679.

Mazzola, M., J. Muramoto, and C. Shennan. 2018. Anaerobic disinfestation induced changes to the soil microbiome, disease incidence and strawberry fruit yields in California field trials. *Applied Soil Ecology* **127**:74–86.

McCarty, J. L., S. Korontzi, C. O. Justice, and T. Loboda. 2009. The spatial and temporal distribution of crop residue burning in the contiguous United States. *Science of the Total Environment* **407**:5701–5712.

McHughen, A. 2016. A critical assessment of regulatory triggers for products of biotechnology: Product vs. process. *GM Crops & Food* **7**:125–158.

Mendes, R., M. Kruijt, I. De Bruijn, E. Dekkers, M. Van Der Voort, J. H. Schneider, Y. M. Piceno, T. Z. DeSantis, G. L. Andersen, and P. A. Bakker. 2011. Deciphering the rhizosphere microbiome for disease-suppressive bacteria. *Science* **332**:1097–1100.

Meyer, R. S., A. E. DuVal, and H. R. Jensen. 2012. Patterns and processes in crop domestication: an historical review and quantitative analysis of 203 global food crops. *New Phytologist* **196**:29–48.

Montenegro de Wit, M. 2022. Can agroecology and CRISPR mix? The politics of complementarity and moving toward technology sovereignty. *Agriculture and Human Values* **39**:733–755.

Moussart, A., E. Wicker, B. Le Delliou, J.-M. Abelard, R. Esnault, E. Lemarchand, F. Rouault, F. Le Guennou, M.-L. Pilet-Nayel, and A. Baranger. 2009. Spatial distribution of *Aphanomyces euteiches* inoculum in a naturally infested pea field. *European Journal of Plant Pathology* **123**:153–158.

Mundt, C. C. 2002. Use of multiline cultivars and cultivar mixtures for disease management. *Annual Review of Phytopathology* **40**:381–410.

Mundt, C. C. 2014. Durable resistance: a key to sustainable management of pathogens and pests. *Infection, Genetics and Evolution* **27**:446–455.

Mundt, C. C. 2018. Pyramiding for resistance durability: theory and practice. *Phytopathology* **108**:792–802.

Muramoto, J., D. M. Parr, J. Perez, and D. G. Wong. 2022. Integrated soil health management for plant health and one health: Lessons from histories of soil-borne disease management in California strawberries and arthropod pest management. *Frontiers in Sustainable Food Systems*:doi.org/10.3389/fsufs.2022.839648.

Nasr-Esfahani, M. 2022. An IPM plan for early blight disease of potato *Alternaria solani* Sorauer and *A. alternata* (Fries.) Keissler. *Archives of Phytopathology and Plant Protection* **55**:785–796.

Ndereyimana, A., S. Praneetha, L. Pugalendhi, B. Pandian, and P. Rukundo. 2013. Earliness and yield parameters of eggplant (*Solanum melongena* L.) grafts under different spacing and fertigation levels. *African Journal of Plant Science* **7**:543–547.

Ngo Bieng, M. A., L. Alem, C. Curtet, and P. Tixier. 2017. Tree spacing impacts the individual incidence of *Moniliophthora roreri* disease in cacao agroforests. *Pest Management Science* **73**:2386–2392.

Niks, R. E., X. Qi, and T. C. Marcel. 2015. Quantitative resistance to biotrophic filamentous plant pathogens: concepts, misconceptions, and mechanisms. *Annual Review of Phytopathology* **53**:10.1146.

O'Brien, P. A. 2017. Biological control of plant diseases. *Australasian Plant Pathology* **46**:293–304.

Parker, I. M., M. Saunders, M. Bontrager, A. P. Weitz, R. Hendricks, R. Magarey, K. Suiter, and G. S. Gilbert. 2015. Phylogenetic structure and host abundance drive disease pressure in communities. *Nature* **520**:542–544.

Parker, T. J., K. M. Clancy, and R. L. Mathiasen. 2006. Interactions among fire, insects and pathogens in coniferous forests of the interior western United States and Canada. *Agricultural and Forest Entomology* **8**:167–189.

Paulitz, T. C. 2006. Low input no-till cereal production in the Pacific Northwest of the US: the challenges of root diseases. *European Journal of Plant Pathology* **115**: 271–281.

Pérez-Hernández, R. G., M. J. Cach-Pérez, R. Aparicio-Fabre, H. V. d. Wal, and U. Rodríguez-Robles. 2021. Physiological and microclimatic effects of different agricultural management practices with maize. *Botanical Sciences* **99**:132–148.

Pieterse, C. M., R. L. Berendsen, R. de Jonge, I. A. Stringlis, A. J. Van Dijken, J. A. Van Pelt, S. Van Wees, K. Yu, C. Zamioudis, and P. A. Bakker. 2021. *Pseudomonas simiae* WCS417: star track of a model beneficial rhizobacterium. *Plant and Soil* **461**:245–263.

Pike, C. C., J. Koch, and C. D. Nelson. 2021. Breeding for resistance to tree pests: successes, challenges, and a guide to the future. *Journal of Forestry* **119**:96–105.

Plaszkó, T., Z. Szűcs, G. Vasas, and S. Gonda. 2021. Effects of glucosinolate-derived isothiocyanates on fungi: A comprehensive review on direct effects, mechanisms, structure–activity relationship data and possible agricultural applications. *Journal of Fungi* **7**:539.

Raggi, L., L. C. Pacicco, L. Caproni, C. Álvarez-Muñiz, K. Annamaa, A. M. Barata, D. Batir-Rusu, M. J. Díez, M. Heinonen, and V. Holubec. 2022. Analysis of landrace cultivation in Europe: A means to support in situ conservation of crop diversity. *Biological Conservation* **267**:109460.

Ramirez-Villegas, J., C. K. Khoury, H. A. Achicanoy, M. V. Diaz, A. C. Mendez, C. C. Sosa, Z. Kehel, L. Guarino,

M. Abberton, and J. Aunario. 2022. State of ex situ conservation of landrace groups of 25 major crops. *Nature Plants* **8**:491–499.

Raymaekers, K., L. Ponet, D. Holtappels, B. Berckmans, and B. P. Cammue. 2020. Screening for novel biocontrol agents applicable in plant disease management—a review. *Biological Control* **144**:104240.

Rimbaud, L., J. Papaïx, J.-F. Rey, L. G. Barrett, and P. H. Thrall. 2018. Assessing the durability and efficiency of landscape-based strategies to deploy plant resistance to pathogens. *PLoS Computational Biology* **14**:e1006067.

Roy, B. A., H. M. Alexander, J. Davidson, F. T. Campbell, J. J. Burdon, R. Sniezko, and C. Brasier. 2014. Increasing forest loss worldwide from invasive pests requires new trade regulations. *Frontiers in Ecology and the Environment* **12**:457–465.

Salacinas, M., H. J. Meijer, S. H. Mamora, B. Corcolon, A. Mirzadi Gohari, B. Ghimire, and G. H. Kema. 2022. Efficacy of disinfectants against Tropical Race 4 causing Fusarium wilt in Cavendish bananas. *Plant Disease* **106**:966–974.

Savary, S., L. Willocquet, S. J. Pethybridge, P. Esker, N. McRoberts, and A. Nelson. 2019. The global burden of pathogens and pests on major food crops. *Nature Ecology & Evolution* **3**:430–439.

Schlatter, D., L. Kinkel, L. Thomashow, D. Weller, and T. Paulitz. 2017. Disease suppressive soils: new insights from the soil microbiome. *Phytopathology* **107**: 1284–1297.

Schreiner, P. R., K. L. Ivors, and J. N. Pinkerton. 2001. Soil solarization reduces arbuscular mycorrhizal fungi as a consequence of weed suppression. *Mycorrhiza* **11**: 273–277.

Simler-Williamson, A. B., M. R. Metz, K. M. Frangioso, and D. M. Rizzo. 2021. Wildfire alters the disturbance impacts of an emerging forest disease via changes to host occurrence and demographic structure. *Journal of Ecology* **109**:676–691.

Soluri, J. 2002. Accounting for taste: Export bananas, mass markets, and Panama disease. *Environmental History* **7**:386–410.

Stam, R. and B. A. McDonald. 2018. When resistance gene pyramids are not durable—the role of pathogen diversity. *Molecular Plant Pathology* **19**:521.

Sun, M., S. Siemsen, W. Campbell, P. Guzman, R. Davidson, J. L. Whitworth, T. Bourgoin, J. Axford, W. Schrage, and G. Leever. 2004. Survey of potato spindle tuber viroid in seed potato growing areas of the United States. *American Journal of Potato Research* **81**:227–231.

Van Esse, H. P., T. L. Reuber, and D. van der Does. 2020. Genetic modification to improve disease resistance in crops. *New Phytologist* **225**:70–86.

Vandermeer, J. H. 1992. The Ecology of Intercropping. Cambridge University Press.

Vannier, N., M. Agler, and S. Hacquard. 2019. Microbiota-mediated disease resistance in plants. *PLoS Pathogens* **15**:e1007740.

Vleeshouwers, V. G. and R. P. Oliver. 2014. Effectors as tools in disease resistance breeding against biotrophic, hemibiotrophic, and necrotrophic plant pathogens. *Molecular Plant–Microbe Interactions* **27**:196–206.

Welde-Michael, G., K. Bobosha, T. Addis, G. Blomme, S. Mekonnen, and T. Mengesha. 2008. Mechanical transmission and survival of bacterial wilt on enset. *African Crop Science Journal* **16**(1). DOI: 10.4314/acsj.v16i1.54349.

Wisler, G. C. and R. F. Norris. 2005. Interactions between weeds and cultivated plants as related to management of plant pathogens. *Weed Science* **53**:914–917.

Zappia, R. E., D. Hüberli, G. S. J. Hardy, and K. L. Bayliss. 2014. Fungi and oomycetes in open irrigation systems: knowledge gaps and biosecurity implications. *Plant Pathology* **63**:961–972.

Zhan, J., P. H. Thrall, and J. J. Burdon. 2014. Achieving sustainable plant disease management through evolutionary principles. *Trends in Plant Science* **19**:570–575.

Epilogue

OK, so here we are at the end of the book. Plant diseases are pretty cool, and now you see them everywhere, right?

What we love about evolutionary ecology is that it embraces a search for generalizations in the living world, providing a predictive framework to help us think about novel situations. Mysterious forest diebacks, introduced pathogens spreading across landscapes, crop diseases intensifying under climate change: plant diseases provide us with many novel challenges. Traditionally, plant pathologists have dug deeply into individual plant–pathogen interactions, developing rich, multi-faceted biological understanding in the service of subduing those pathogens. Our attention to plants of economic and social importance fueled our powers of observation, from Micheli's original discovery of biogenesis on a slice of melon through nearly all plant pathology today. Evolutionary ecology, with its search for generalizations in the living world, helps us apply knowledge from those case studies more broadly. We see evolutionary ecology as a sort of theory-rich meta-analysis, bringing together data from many independent sources to help make sense of the big picture. It provides opportunities for creative approaches to disease management in support of sustainable food systems, helps us prioritize responses to emerging pathogens that pose threats to human wellbeing, and allows us to recognize that not all diseases need to be managed.

Over a hundred years of plant pathology research has yielded a huge body of knowledge about particular diseases caused by particular pathogens on particular host plants—how, where, and why those diseases develop, and how to stop them. The explosion of content available on the Internet has made this information more accessible than ever. Which is a great thing, because the pure quantity of information is humbling; we could not possibly cover it adequately in this book. We hope our readers will forgive our omission of many important case studies and many technical details. We hope you will accept our invitation to use this book as a foundation for exploring the principles of the evolutionary ecology of plant disease in the much broader universe of phytopathology. We have tried to shine a light on the overarching questions, the approaches, the intellectual mysteries and debates, and to provide a useful framework to think about any of the countless plant–pathogen interactions awaiting further study.

What is the future of the evolutionary ecology of plant disease as a field? Trying to predict the future can be hazardous and arrogant, and we are reluctant to do so . . . yet it is tempting to peer into the crystal ball. Here are a few areas we think are particularly compelling.

1. Evolution of molecular plant–microbe interactions

The tools of molecular biology have rapidly advanced our understanding of the mechanisms of plant immune responses and signaling pathways, and the pathogen MAMPs and effectors that trigger them. Pieces of the puzzle that were completely missing two years ago are now falling into place; "facts" we thought we knew are being replaced with new knowledge. We are now able to study the evolution of plant and pathogen traits at the molecular level. Embedding those new facts about plant–microbe communications into a phylogenetic framework will help us understand the tempo and

The Evolutionary Ecology of Plant Disease. Gregory S. Gilbert and Ingrid M. Parker, Oxford University Press. © Gregory S. Gilbert & Ingrid M. Parker (2023).
DOI: 10.1093/oso/9780198797876.003.0018

mode of molecular trait evolution. Designing elegant gene-knockout experiments with pathogens and their hosts will illuminate the mechanisms of coevolution in plant–microbe symbiosis. And integrating gene-editing tools such as CRISPR with a deep understanding of both molecular physiology and evolutionary theory will support the design of safe and durable approaches to disease resistance in agriculture and forestry.

2. Plant and soil microbiomes

High-throughput sequencing technologies have transformed our knowledge of plant microbiomes. They have also helped us think more clearly about how plant–pathogen interactions are embedded in a broader framework of symbioses. Metagenomics analysis is shining a light into the black box of plant and soil microbiomes, linking plant–microbe interactions more directly to observed ecological phenomena. At the same time, a deepening appreciation of how entire microbiomes shape disease development, not just in plants but also in humans, propels a growing interest in microbiomes. The ability to study plant and soil microbiomes in wild ecosystems provides exciting new opportunities to test hypotheses about pathogens as drivers of plant population dynamics and spatial patterns, competitive interactions, community succession, and the maintenance of diversity. Our growing understanding of what shapes the structure of microbial communities also opens up opportunities in agroecosystems.

Traditional knowledge contains great wisdom about sustainable agricultural practices. New technologies to study microbiomes may help uncover the mechanistic basis for effective management strategies long used to manage diseases in traditional agriculture.

3. Beyond model systems

Until recently, plant pathology has relied heavily on (and benefited greatly from) a concerted focus on a few model systems for both plant hosts and pathogens, and model systems still play an important role in scientific discovery. However, one could say that we have entered a new era in which novel technological and analytical approaches are taking us beyond this dependence on model systems. On the evolution side, the powerful and increasingly accessible tools of genomics have opened the door to understanding evolutionary change in all sorts of plants, not just major crops and *Arabidopsis*. At the time we write this, whole-genome sequences are already available for over 800 species of plants and over 800 species of fungi. The plant genomes allow for comparative studies of, for example, *R* gene diversity or plant–microbe signaling across many types of hosts. Similarly, the pathogen genomes facilitate comparative studies of evolutionary trajectories between parasitism and mutualism in plant–microbe symbiosis. On the ecology side, growing interest in plant disease ecology outside of agroecosystems has fueled an explosion of studies in wild ecosystems—tropical to polar and everything in between; forests and grasslands and marshes and abandoned lots and alpine meadows. To synthesize this growing collection of empirical studies, analytical tools like meta-analysis will be essential to generate new insights into broad ecological patterns.

4. Trait-based ecology and phylogenetic ecology

Trait-based ecology focuses on the traits of individuals, rather than species, as the driving force in ecological patterns and processes. Trait-based ecology has a long tradition in microbial ecology, where traits like the ability to use different kinds of organic compounds or the production of particular antibiotics were recognized as more important than species identity; after all, most microbes are more readily characterized by what they do than by what they are named. More recently, a focus on functional traits has surged in the broader ecological community, linking organisms and their traits to ecosystem properties in a way that is general and predictive. Because traits are shaped by evolutionary trajectories and relationships, phylogenetic ecology similarly provides novel tools to understand and predict how species interact with their environment and with each other. The availability of robust phylogenetic supertrees for Plants and Fungi, together with new analytical tools, opens fresh lines of inquiry in the study of plant disease.

As we better understand how traits govern plant–pathogen interactions, we gain confidence in how to make appropriate inferences about what to expect from novel pathogens, how climate change will affect familiar plant–pathogen systems, and how to manage ecosystems to reduce damaging impacts of disease.

5. New ways to meld theory and data

Plant pathology has a rich quantitative history in epidemiology and statistics, and this field continues to grow. Sophisticated new disease prediction models in agricultural science combine meteorological data with detailed information about crop development and pathogen physiology and dispersal. High-resolution sensing technologies offer pathways to parameterize and test such detailed models, and deep-learning approaches such as artificial neural networks can derive patterns from immensely complex data. The lessons learned from these intensive case studies will provide entries to understand processes in aerobiology and spatial ecology that have been elusive. Ecological theory also has a lot to offer plant disease ecology. We could do a much better job of leveraging stage-structured population models to evaluate pathogens for classical biocontrol or even simply to understand the impacts of pathogens in wild populations. Niche modeling can project the potential impacts of climate change on plant disease or predict the range expansion of introduced pathogens, and it is being integrated into many current studies. The melding of plant–soil feedback theory with traditional competition theory offers a quantitative framework for testing whether soil microbiomes drive plant species coexistence. Finally, network theory is an exciting tool for studying the complexity of interacting communities like plants and their pathogens. Advances in DNA sequencing, bioinformatics, phylogenetic ecology, and network theory are all converging to lay open the fascinating, diverse universe of microbes interacting with plants.

Writing this book has been quite a journey. Along the way, we have learned so much, especially about how many important and exciting questions still need to be answered! It makes us wish we were at the beginning of our careers, instead of in sight of the end. But we wrote this book to help launch the next generation of creative inquiry and plant disease problem-solving, and we can't wait to see what you do.

Thank you for being a part of this journey.

Index